Rope Rescue

Principles and Practice

FIFTH EDITION

Loui McCurley
Chief Executive Officer
Pigeon Mountain Industries, Inc.
Denver, Colorado

Tom Vines
Civilian Deputy for Search and Rescue (ret.)
Carbon County Sheriff's Search and Rescue
Red Lodge, Montana

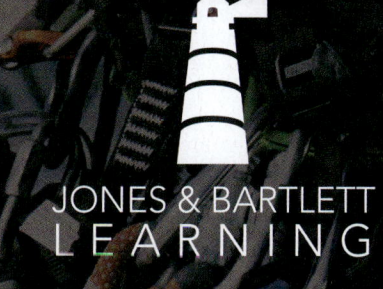

JONES & BARTLETT
LEARNING

World Headquarters
Jones & Bartlett Learning
25 Mall Road
Burlington, MA 01803
978-443-5000
info@jblearning.com
www.jblearning.com
www.psglearning.com

Jones & Bartlett Learning books and products are available through most bookstores and online booksellers. To contact the Jones & Bartlett Learning Public Safety Group directly, call 800-832-0034, fax 978-443-8000, or visit our website, www.psglearning.com.

Substantial discounts on bulk quantities of Jones & Bartlett Learning publications are available to corporations, professional associations, and other qualified organizations. For details and specific discount information, contact the special sales department at Jones & Bartlett Learning via the above contact information or send an email to specialsales@jblearning.com.

Copyright © 2023 by Jones & Bartlett Learning, LLC, an Ascend Learning Company

All rights reserved. No part of the material protected by this copyright may be reproduced or utilized in any form, electronic or mechanical, including photocopying, recording, or by any information storage and retrieval system, without written permission from the copyright owner.

The content, statements, views, and opinions herein are the sole expression of the respective authors and not that of Jones & Bartlett Learning, LLC. Reference herein to any specific commercial product, process, or service by trade name, trademark, manufacturer, or otherwise does not constitute or imply its endorsement or recommendation by Jones & Bartlett Learning, LLC and such reference shall not be used for advertising or product endorsement purposes. All trademarks displayed are the trademarks of the parties noted herein. *Rope Rescue: Principles and Practice, Fifth Edition* is an independent publication and has not been authorized, sponsored, or otherwise approved by the owners of the trademarks or service marks referenced in this product.

There may be images in this book that feature models; these models do not necessarily endorse, represent, or participate in the activities represented in the images. Any screenshots in this product are for educational and instructive purposes only. Any individuals and scenarios featured in the case studies throughout this product may be real or fictitious but are used for instructional purposes only.

The publisher has made every effort to ensure that contributors to *Rope Rescue: Principles and Practice, Fifth Edition* materials are knowledgeable authorities in their fields. Readers are nevertheless advised that the statements and opinions are provided as guidelines and should not be construed as official policy. The recommendations in this publication or the accompanying resource manual do not indicate an exclusive course of action. Variations taking into account the individual circumstances, nature of medical oversight, and local protocols may be appropriate. The publisher disclaims any liability or responsibility for the consequences of any action taken in reliance on these statements or opinions.

23011-6

Production Credits
Vice President, Product Management: Marisa R. Urbano
Vice President, Content Strategy and Implementation: Christine Emerton
Director, Product Management: Laura Carney
Director, Content Management: Donna Gridley
Manager, Content Strategy: Kim Crowley
Content Strategist: Jennifer Deforge-Kling
Director, Project Management and Content Services: Karen Scott
Manager, Project Management: Kristen Rogers
Project Manager: Dan Stone
Senior Digital Project Specialist: Angela Dooley
Director, Marketing: Brian Rooney
Vice President, International Sales, Public Safety Group: Matthew Maniscalco

Director, Sales, Public Safety Group: Brian Hendrickson
Content Services Manager: Colleen Lamy
Vice President, Manufacturing and Inventory Control: Therese Connell
Composition: S4Carlisle Publishing Services
Cover Design: Scott Moden
Text Design: Scott Moden
Media Development Editor: Faith Brosnan
Rights & Permissions Manager: John Rusk
Rights Specialist: Liz Kincaid
Cover Image (Title Page, Chapter Opener): © Jones & Bartlett Learning.
  Courtesy of Loui McCurley.
Printing and Binding: LSC Communications

Library of Congress Cataloging-in-Publication Data
Names: McCurley, Loui, 1965- author. | Vines, Tom, author.
Title: Rope rescue principles and practice / Loui McCurley, Tom Vines.
Description: Fifth edition. | Burlington, Massachusetts : Jones & Bartlett
  Learning, [2023] | Includes bibliographical references and index.
Identifiers: LCCN 2021057393 | ISBN 9781284195101 (paperback)
Subjects: LCSH: Rappelling. | Rope. | Knots and splices. |
  Mountaineering--Search and rescue operations--United States. | Rescue
  work--United States.
Classification: LCC GV200.19.R34 V56 2023 | DDC 363.14--dc23
LC record available at https://lccn.loc.gov/2021057393

6048

Printed in the United States of America
26 25 24 23 22   10 9 8 7 6 5 4 3 2 1

Brief Contents

Acknowledgments	xviii
From the Author	xx
Dedication	xxi

SECTION 1
Awareness

CHAPTER 1 Becoming a Rope Rescuer	3
CHAPTER 2 Size-Up	21
CHAPTER 3 Hazards Associated with Rope Rescue	33
CHAPTER 4 Initiating a Response	49
CHAPTER 5 Supporting the Operations- or Technician-Level Rescue Incident	65

SECTION 2
Operations

CHAPTER 6 Developing the Incident Action Plan	83
CHAPTER 7 Hazard-Specific Personal Protective Equipment	97
CHAPTER 8 Rescue Equipment	113
CHAPTER 9 Ropes, Knots, Bends, and Hitches	147
CHAPTER 10 Principles of Rigging	183
CHAPTER 11 Anchorages	197
CHAPTER 12 Fall Protection and Belay Operations	219
CHAPTER 13 Patient Evacuation	243
CHAPTER 14 Lowering Systems	267
CHAPTER 15 Mechanical Advantage Systems	299
CHAPTER 16 Working in Suspension	321

SECTION 3
Technician

CHAPTER 17 Horizontal Systems	341
CHAPTER 18 Personal Vertical Skills	369
CHAPTER 19 Pickoff and Litter Management	415
CHAPTER 20 Special Rescue Disciplines for Additional Training	443
APPENDIX A: Correlation Grid	451
GLOSSARY	456
INDEX	464

Contents

Acknowledgments	xviii
From the Author	xx
Dedication	xxi

SECTION 1
Awareness

CHAPTER 1
Becoming a Rope Rescuer — 3

Introduction	4
The Authority Having Jurisdiction	4
NFPA Resources	5
Levels of Rescue	6
Using Job Performance Requirements (JPRs)	7
What Is Rope Rescue?	8
Fire Service Rescue	8
Where Do Rope Rescues Occur?	9
Recreational Incidents	9
Rock Climbing	9
Rappelling	9
Mountaineering	9
Caving	9
Ice Climbing	9
Workplace Incidents	10
Work at Height and Rope Access	10
Industrial Sites	11
Wind Turbine Rescue	11
Confined Space Rescue	11
Tower Rescue	12
Water Rescue	12
Becoming a Rope Rescuer	12
Aptitude for the High-Angle Environment	13
Characteristics of an Effective Rescuer	13
Safety	13
Personal Red Flags	14
Physical Safety Concepts	14
Systems Thinking	14
Use of Low-Risk Methods First	15
Preparation for Self-Rescue	16
Backup of Other Rescuers	16
Care of Equipment	16
Attention to Detail	16
Team Concepts	17
Warning Call	17
Training Expectations	17

CHAPTER 2
Size-Up — 21

Introduction	22
Determine the Scope of the Rescue	22
Identify the Number of Subjects	23
Establish the Last Known Location of Subjects	23
Identify and Interview Witnesses and Reporting Parties	24
Assessing Resource Needs	25
Preplanning	25
Identifying Appropriate Resources	26
LAST	27
Obtain Information Required to Develop an Incident Action Plan	27
Using Reference Materials	27

Identify Search Parameters	28
Lost Person Behavior	28
Time of Day	28
Initial Goals	29
Containment	29
Attraction	29
Hasty Search	29
Detailed Search Methods	29
Relay Information	30
Who–What–When–Where–Why–How	30

CHAPTER 3
Hazards Associated with Rope Rescue   33

Introduction	34
Hazard Identification and Risk Assessment	34
Preparing for Hazards	35
Hazards Associated with Rope Rescue	37
Rescue Incident Safety Control Chart	37
Specific Hazards	37
Protection Against Falls and Other Hazards	38
Additional Hazards	39
Situational Awareness	39
Risk Management	39
Communicating for Risk Management	40
Verbal Communications	41
Hand Signals	41
Clear Communication	41
Creating a Shared Mental Model	41
Team Briefings	42
Explicit Communication	42
Incident Safety Officer	43
Stress Management	44

CHAPTER 4
Initiating a Response   49

Introduction	50
Preplanning	50
Initial Response	51
Recognizing the Need	51

Activate the Emergency Response System	51
Command Structure	52
Incident Command System	52
Multiagency Operations	54
Additional Command Staff	54
Site Control and Scene Management	54
Site Control	54
Scene Management	55
Evacuation Signals	55
Recognize and Mitigate Hazards	55
Personal Protective Equipment	56
Rope Rescue Resources	56
Awareness Level	56
Operations Level	56
Technician Level	57
Incident Action Plan	57
Incident Planning Forms	59
LAST	59
Locate	59
Access	59
Stabilize	60
Transport	60
After Action Review	60
Hot Debrief	60
Structured Debrief	61
Threat and Error Management System	61
Five Why's	61

CHAPTER 5
Supporting the Operations- or Technician-Level Rescue Incident   65

Introduction	66
Incident Location Concerns	66
Guidance on Providing Support	66
Personal Protective Equipment Selection	67
National Fire Protection Association Standards	67
Common Protective Equipment at Rope Rescues	68
Harnesses	68
Gloves	70
Helmet	72

Hauling Systems Overview … 73
   How Hauling Systems Work … 73
   The Haul Team … 74
   Communications for Hauling Operations … 74
Supporting the Incident Action Plan … 74
Incident Facilities and Stations … 76
Accountability … 76
   Briefings … 76
Reporting Progress … 77
Personnel Rehabilitation … 77

SECTION 2
Operations

CHAPTER 6
Developing the Incident Action Plan — 83

Introduction … 84
Incident Action Plan … 84
   Incident Objectives Form … 85
   Planning P … 85
      Phase 1: Understanding the Situation … 85
      Phase 2: Establishing Incident Objectives … 87
      Phase 3: Developing the Plan … 88
      Phase 4: Preparing and Disseminating the Plan … 88
      Phase 5: Executing, Evaluating, and Revising the Plan … 89
Implementing the IAP … 89
   Span of Control … 90
   Chain of Command … 90
   Multijurisdictional/Multiagency Operations … 90
Communications … 90
   Communicating for Safety … 92
   IC Communications … 92
   Incident Name … 92
Demobilization … 92
   Manage Rescuer Risk and Site Safety … 93
   Scene Security and Custody Transfer … 93
   Recordkeeping and Documentation … 94
      Data Collection and Management Systems … 94
   Conduct Postincident Analysis … 94

CHAPTER 7
Hazard-Specific Personal Protective Equipment — 97

Introduction … 98
Rope Rescue Personal Protective Equipment … 98
   Protective Garments … 99
   Footwear … 101
   Gloves … 102
   Helmet … 102
Harnesses for Rescue … 103
   Victim Extrication Device … 105
   Emergency Harnesses … 105
      Diaper Emergency Harness … 105
      Swiss-Seat Emergency Harness … 106
   Pickoff Seat (Rescue Triangle) … 106
   Escape Belts … 107
      A Note on Suspension Trauma … 107
Other Considerations … 107
   Preventing Dropped Objects … 107
   Light Sources … 107
   Knives … 108
Inspection, Maintenance, and Recordkeeping … 108
   Signs of Damage … 109
   Maintenance … 109

CHAPTER 8
Rescue Equipment — 113

Introduction … 114
Rope Rescue Equipment Program … 115
Equipment Selection Considerations … 115
Equipment Strength and Safety Factors … 117
   Load Ratios … 117
Carabiners … 119
   Material … 119
   Carabiner Components … 119
   Basic Carabiner Shapes … 120
   Strength … 120
      Testing … 121
      Strength Rating for Carabiners … 122

Gate Openings	122
Locking Mechanisms	122
Use Considerations	124
Inspection and Maintenance	125
Screw Links	126
Braking Devices and Descenders	127
Autolocking and Manual Devices	128
Additional Considerations	128
Friction and Heat	129
Guidelines for Selecting a Braking Device	129
Types of Braking Devices	129
Braking Devices for Rescue	129
Autolocking	129
Climbing Technology Sparrow	130
CMC MPD	130
Harken Clutch	130
ISC D4/D5	130
Petzl ID	130
Petzl Maestro	130
Skylotec Lory	131
Skylotec Sirius	131
SMC Spider	131
Precautionary Notes	131
Nonautolocking	131
Brake Bar Racks	131
Scarab	133
Alpine Brake Tube	134
Figure-8 Descender	134
Escape Devices	135
Escape Devices: Autolocking	135
Petzl EXO	135
Sterling F4	135
Escape Devices: Nonautolocking	135
Escape 8	135
PMW Hook	136
Belay Devices	136
Rope Grabs and Ascenders	137
Fall Arrest	137
Rigging	138
Personal Handled Ascender	138

Anchorage Connectors	139
Anchor Slings	139
Beam Clamps	139
Bolt Anchors	140
Portable Anchors	140
Pickets	140
Pulleys	140
Specialized Pulleys	141
Other Hardware	141
Rigging Plates	141
Swivels	142
Maintenance and Inspection	142
Care	142
Inspection and Recordkeeping	142
How to Inspect	143
Damaged Equipment	144

CHAPTER 9
Ropes, Knots, Bends, and Hitches — 147

Introduction	148
Fibers Used to Make Rope	148
Natural Fibers	148
Synthetic Fiber Ropes	149
Polyolefins (Polypropylene and Polyethylene)	149
Polyester	149
Nylon	149
UHMPE (Extended Chain, High-Modulus Polyethylene)	149
Aramids	150
Rope Construction	150
Laid	150
Plaited	150
Braided	150
Double Braid	151
Kernmantle	151
Static and Low-Stretch Kernmantle	151
Performance Characteristics	152
Rope Selection	152
Diameter	152
Large-Diameter Ropes	154

Breaking Strength	154
Elongation	155
Abrasion Resistance	156
Hand	157
Color	157
Length	157
Anatomy of a Knot	**157**
Manufacturer-Supplied Eye Termination	158
Knots	**159**
Qualities of a Good Knot	159
Removing Knots	159
Completing the Knot	159
Specific Knots for the Rope Rescue Environment	**160**
Overhand Knot	160
Figure 8 Family of Knots	161
Figure 8 Stopper Knot	161
Figure 8 On a Bight	161
Figure 8 Follow-Through	162
Double Figure 8 Loop	162
Optional Approach: Bowline Knots	163
Inline Figure 8	166
Butterfly Knot	167
Figure 8 Bend Knot	168
Optional Approach: Grapevine Bend	168
Double Sheet Bend	168
Ring Bend (Water Knot)	168
Hitches	169
Prusik Hitch	169
Clove Hitch	170
Additional Hitches	172
Care of Ropes	**172**
Keeping a Rope History Log	173
Identification and Marking of Ropes	173
Storing Ropes	174
Bagging Ropes	174
Coiling	175
How Ropes Are Damaged	**175**
Harmful Substances	175
Overloading a Rope	175
Damage from Falling Objects	175
Abrasion	176

Thermal Damage	176
Rope Damage Through "Flash" Rappels	177
Rotation of Ropes	177
Strength Loss Through Knots	177
Effects of Bending a Rope	177
The 4:1 Rule	177
Inspecting a Rope	**177**
Establishing Responsibility for Life Safety Ropes	**178**
Retiring a Rope	178
Washing Rope	**178**
Rope-Washing Devices	178
Cleaning Ropes with a Washing Machine	179
Fabric Softeners	180
Special Cleaning Problems	180
Dressing Rope Ends	**180**

CHAPTER 10
Principles of Rigging 183

Introduction	184
Principles	185
Force	185
Vector	186
Resultant	186
Angles in Rescue Systems	186
Tension	187
Friction	187
Edge Protection	188
Fall Line	191
Putting It Together	191
Safety Factors	191
Rope Systems	192
Dual-Tensioned Rope Systems (DTRS)	192
Optimizing Force	193

CHAPTER 11
Anchorages 197

Introduction	198
Anchor Terminology	198
Principles of Anchoring	199
Anchorage Strength	199
Direction of Pull	199

Positioning of Anchorage Systems	200
Redundancy	200
Connecting to an Anchorage	201
Single Point Anchorages	202
Tensionless Hitch	202
Knots	203
Soft Slings as Anchorage Connectors	203
Presewn Slings	203
Anchor Straps	203
Portable Anchorages	205
Rigging Plates	206
Multi-Point Anchor Systems	206
Directionals	212
Location of Directionals	213
Back-Tie	213
Pretensioned Back-Ties	213
Structural Anchorages	214
Less-Obvious Anchorages	214
Elevator and Machine Housings	214
Scuppers (Roof Drain Holes)	214
Wall Sections Between Windows and/or Doors	215
Stairwell Beams	215
Installed Anchorages	215
Adapting to a Lack of Anchors	215
Extending Anchor Points	215
Using Vehicles for Anchorages	216
Placement of Anchor Systems	216
Evaluating an Anchor System	216

CHAPTER 12
Fall Protection and Belay Operations — 219

Introduction	220
Situations Requiring a Belay	220
Decisions on When to Belay	220
Belays as Safety Backups	221
Self-Belay	221
Active Belay	223
The Belay System	223

Light Load Versus Rescue Load Belays	224
Active Belay Systems: Light Loads	224
The Münter Hitch	225
Using a Free-Running Belay Device	225
Assisted-Braking Belay Devices	225
Practicing a Light Load Belay System	226
Communications	228
Belaying a Rescue Load	229
Manually Operated Assisted Belay for Rescue	229
Rescue Load Belay Practice	230
Tandem Prusik Belay System	231
Load-Releasing Hitches	231
Constructing the Tandem Prusik Belay System	232
Operating the Tandem Prusik Belay	234
Releasing the Radium-Release Hitch	236
Prusik Minding Pulley	236
Additional Words of Caution for Belayers	238
Arranging the Belay Direction	238
Maintaining Proper Slack in the Belay Rope	238
Securing the Belayer	238
Bottom Belay (for Rappelling Only)	238
Body Belays	239
Belay Failure: The Human Element	239
Dual-Tension Rope Systems as Backup Systems	239
Two-Tensioned Rescue System (TTRS)	239
Operating the TTRS	240

CHAPTER 13
Patient Evacuation — 243

Introduction	244
Litter Function	244
Types of Litters	244
Rigid Litters	244
Metal Basket Litters	245
Enclosed Basket Litters	245
Basket Litter Strength and Performance	246
Flexible Litters	246
Sked Litter	246
Choosing a Litter for Rescue Operations	247

Medical Care	247
Medical Technologies	247
Benefit to Subjects	248
Technical Rescue Medical Kit	249
Assessment and Interventions	249
Triage	249
Severe Bleeding	250
Airway and Breathing	250
Head and Spine Considerations	250
Packaging the Subject in a Litter	251
Spinal Motion Restriction and Medical Considerations	251
Airway Management	251
Immobilization (or Spinal Motion Restriction)	252
Splinting	253
Protecting the Subject in the Litter	254
Securing the Subject in the Litter	255
Improvised Litter Restraint	256
Manufactured Litter Restraint	259
Carrying the Litter	259
Subject Care	259
Litter Team Roles	259
Lifting and Carrying the Litter	260
Rest Breaks	260
Litter Team Transitions	261
Lift-Assist Sling Carry	261
Litter Wheels	262
Litter Handles	263
Negotiating Obstacles with a Litter	263
Traveling Control Lines	263
Hand Pass	263
Transferring the Subject to EMS	264
Patient Transfer Report	264
Verbal Communication	264
Written Communication	264
Delivering the Subject to EMS	265

CHAPTER 14
Lowering Systems   267

Introduction	268
Low-Angle Rope Rescue Overview	269
Elements of a Low-Angle Evacuation System	270
Roles and Responsibilities in a Low-Angle Lowering Operation	270
Operations Leader	270
Rope Rescuer	270
Rigger	271
Brakeman	271
Safety Officer	271
Optional Additional Roles	271
Ambulatory Subjects	272
Ambulatory Subject Lower Method	273
Litters and Nonambulatory Subjects	274
Low-Angle Rope Rescue Systems	274
Low-Angle System Braking Devices	274
Autolocking Devices	274
Non-Autolocking	274
Adjustable Friction Devices	275
Friction Wrap	275
Rope	275
Litter	275
Tying the Mainline Rope Directly to the Head of the Litter	275
Tying a Closed Loop Bridle to the Head of the Litter	276
Low-Angle Litter Management	277
Low-Angle: No Tie-In	277
Low-Angle: Tie-In to Rope	278
Low-Angle: Tie-In to Litter	278
Low-Angle Evacuation with a Litter Wheel	279
Practicing Low-Angle Litter Evacuation	279
Communications	280
A Typical Low-Angle Lowering Operation	280
Preparation	280
Evacuation of More Than One Rope Length	281
Safety Considerations in Low-Angle Rope Rescues	283
Slope Safety Lines	283
The Belay Question	284
Summarizing Low-Angle Evacuations	284
High-Angle Rope Rescue Overview	285
High-Angle Rope Rescue System	285
Primary Anchor System	286
Ropes for Lowering	286

Braking Systems for High-Angle Lowering	286
Belays for High-Angle Lowering	286
Two-Tensioned Rescue Systems	287
Rigging a System for Lowering	288
Practicing a High-Angle Litter Lower	290
Litter Orientation	290
Securing the Subject in the Litter	291
Attaching the Litter to the Main Rescue Line	291
Rigging Without a Litter Attendant	292
Practicing High-Angle Litter Lower Techniques	293
Aerial Ladder Slide Lowers	295

CHAPTER 15
Mechanical Advantage Systems — 299

Introduction	300
Principles of Mechanical Advantage	300
Direction Change	301
Counterbalance Haul System	301
Progress Capture	302
Positioning the PCD	302
Load Release	303
Using a Winch for Mechanical Advantage	304
Mechanical Advantage Systems	304
Simple Mechanical Advantage Systems	305
2:1 Simple Systems and Beyond	305
Rigging a 2:1 Mechanical Advantage with Direction Change	306
Converting a 2:1 Mechanical Advantage to 4:1	307
3:1 Simple In-Line Systems and Beyond	307
In-Line Haul System with Braking Device PCD	309
Operating a Mechanical Advantage System	310
Piggyback Rope Systems	311
Compound Systems	312
Two-Tensioned Rescue Systems	315
Safety Check	316
Operating a Haul System	317
Communications in Raising Operations	317

CHAPTER 16
Working in Suspension — 321

Introduction	322
Suspension-Induced Injury	322
Third Man Operations	323
Communications	324
Connecting to the Rope Rescue System	325
Negotiating Edges	326
Lowering Over an Edge	326
Raising Over an Edge	327
Landing	327
Moving a Litter While Suspended	327
Horizontal Litter, Single Attachment Point	328
Using the Litter Spider	328
Commencing a Lower	330
Body Management with Litter Movement	331
Vertical Litter Configuration	332
Pike and Pivot Litter Bridle	332
Pike and Pivot Lower	333
Pike and Pivot Raise	334
Suspended Work on A Highline	335
Rigging to a Horizontal System	335
Riding with a Litter	336

SECTION 3
Technician

CHAPTER 17
Horizontal Systems — 341

Introduction	342
Track-Line Systems	343
Rigging a Track-Line System	344
Spanning the Gap	344
Ropes for Highlines	344
Track-Line(s)	344
Tag-Line(s)	345
Anchorages	345
Far-Side Track-Line Anchorage	347
Near-Side Track-Line Anchor/Tensioning System	347

Carriage System	348
Tag-Line Anchor Systems	349
High Directionals	349
Parts of a High Directional	350
Head	350
Legs	350
Feet	351
Hobbles	351
Additional Rigging Points	351
Configurations	351
Tripod	351
Quad-Pod	352
Bi-Pod	352
Monopod	352
Horizontal Span	352
Using Multi-Pods	352
Assembly	352
Raising and Securing the Portable High Anchorage	352
Managing Forces	353
Forces in a Highline System	354
The Anchorages	354
The Load	354
The Sag	354
Analyzing the System for Force	354
Pre-Tension Guidelines	355
10 Percent Sag Rule (Pre-Loading)	355
10 Percent Loaded Line Sag Rule	355
Loaded Line Guidelines	357
Attaching a Load to a Highline	357
Rigging a Litter to a Highline	358
Guiding-Lines	362
Dynamic Systems	363
English Reeve	363
Skate Block	363
Opposing Systems	365

CHAPTER 18
Personal Vertical Skills — 369

Introduction	370
Single Rope Technique (SRT)	371
Descending	371
Importance of Control	372
How Descending Works	372
Descending Techniques	373
Setting Up the System	373
Throwing the Rope	374
Descending with a Bagged Rope	375
Initiating Descent	375
Practicing Descending	376
Rappelling with a Brake Bar Rack	376
Locking Off the Midline	376
Ascending Basics	384
Choosing an Ascender	385
Ascending with Prusiks	385
Choosing a Cord for Prusiks	385
Creating a Prusik Ascending System	386
Ascending with a Prusik	387
Ascending with Mechanical Ascenders	388
Parts of a Mechanical Ascender	388
Ascender Slings	388
Footloops	389
Creating an Ascending System	389
Customizing the Slings	390
Basic Ascending System	391
Other Ascending Systems	391
Ascending over an Edge	391
Edge with a Gradual Rollover	391
Undercut Edge	393
Tying Off Short	393
Developing Vertical Skills	393
Effect of the Rope Angle	393
Equipment Management	394
Stance	394
Edge Transitions	394
Undercut Edges	394
Gaining Extra Friction When on Rappel	396
Gaining Extra Control with Rope Angle	397
Gaining Extra Control with a Spare Carabiner	397
Gaining Extra Control with the Body	397

Contents xiii

Changeovers	397
Changeover from Ascending to Descending	398
Changeover from Descending to Ascending	400
Passing Obstacles	400
Rope-to-Rope Transfers	403
Passing a Knot	404
Passing a Deviation	404
Passing a Deviation on Ascent	405
Passing an Intermediate Anchorage Point	405
Getting Off the Rope	406
Troubleshooting	406
Self-Belay Techniques	406
Using a Prusik Safety	406
Secondary Backup Systems	406
Self-Rescue	407
Emergency Descent Systems	407
Arm Rappel	407
Body Rappel (Dulfersitz)	410
Carabiner Wrap	412
Münter Hitch Rappel	412

CHAPTER 19
Pickoff and Litter Management 415

Introduction	416
The Suspended Rescuer	416
Selecting the Rescue Approach	417
Accessing the Subject	417
Rescue of a Subject Clinging to a Surface	418
Applying the Evacuation Seat to a Clinging Subject	418
Rescue Chest Harness	420
Pickoff Rescue	422
Medical Considerations	423
To Belay or Not	423
Teamwork and Communication	423
Protecting the Rescuer	423
Pickoff by Lowering	424
Pickoff by Rappel	426
Pickoff by Traveling Brake	428
Releasing a Suspended Subject	429
Managing a Loaded Litter	430
A Note on Litter Bridles (aka Spiders)	430
Tending a Litter Lower	430
Traveling Brake	432
Scaffold-Type Litter Lowers	433
Litter Bridles for Scaffold-Type Lowers	434
Passing Knots for Litters	434
Subject Care	439
Managing a Litter Across a Highline	439
Managing a Distraught Subject	439
Communicating with the Subject	440

CHAPTER 20
Special Rescue Disciplines for Additional Training 443

Introduction	444
Rope Rescue for Structural Collapse	444
Rope Rescue for Confined Spaces	444
Rope Rescue for Swiftwater Environments	445
Rope Rescue for the Wilderness Rescue Organization	446
Rope Rescue for Trench and Excavation	446
Rope Rescue for Cave Rescue	447
Rope Rescue for Mine and Tunnel Incidents	447
Rope Rescue for Tower Incidents	447
Rope Rescue for Animal Rescue Incidents	448

APPENDIX A: Correlation Grid 451

Glossary	456
Index	464

Skill Drills

CHAPTER **7**

SKILL DRILL 7-1 106
Creating a Diaper-Style Emergency Harness

SKILL DRILL 7-2 106
Creating a Swiss Seat Emergency Harness

SKILL DRILL 7-3 109
Inspecting Soft Goods

CHAPTER **9**

SKILL DRILL 9-1 160
Tying a Simple Overhand Knot

SKILL DRILL 9-2 161
Tying a Figure 8 Stopper Knot

SKILL DRILL 9-3 162
Tying a Figure 8 on a Bight

SKILL DRILL 9-4 163
Tying a Figure 8 Follow-Through Knot

SKILL DRILL 9-5 164
Tying a Double Figure 8 Loop

SKILL DRILL 9-6 165
Tying a High-Strength Bowline

SKILL DRILL 9-7 165
Tying an Interlocking Long-Tail Bowline

SKILL DRILL 9-8 166
Tying an Inline Figure 8

SKILL DRILL 9-9 167
Tying a Butterfly Knot

SKILL DRILL 9-10 168
Tying a Figure 8 Bend Knot

SKILL DRILL 9-11 169
Tying a Grapevine Bend

SKILL DRILL 9-12 170
Tying a Ring Bend

SKILL DRILL 9-13 171
Tying a Prusik Hitch

SKILL DRILL 9-14 172
Tying a Clove Hitch

SKILL DRILL 9-15 175
Bagging a Rope

CHAPTER **11**

SKILL DRILL 11-1 202
Tying a Tensionless Hitch

SKILL DRILL 11-2 208
Creating a Fixed and Focused Multi-Point Anchor System

SKILL DRILL 11-3 210
Creating a Simple Self-Adjusting Anchor System

SKILL DRILL 11-4 210
Creating a Multi-Point Anchor System

CHAPTER **12**

SKILL DRILL 12-1 226
Practicing Using a Light Load Belay Practice System

SKILL DRILL 12-2 230
Practicing Rescue Load Belays

SKILL DRILL 12-3 232
Constructing a Radium-Release Hitch

Skill Drills

SKILL DRILL 12-4 234 Constructing a Tandem Prusik Belay System	**SKILL DRILL 14-5** 278 Constructing an Adjustable Tie-In
SKILL DRILL 12-5 235 Operating (Tending) a Tandem Prusik Belay	**SKILL DRILL 14-6** 282 Performing a Knot Pass
SKILL DRILL 12-6 236 Releasing the Radium-Release Hitch	**SKILL DRILL 14-7** 289 Practicing a High-Angle Lowering of a Rescuer with a Dual-Tension Lowering System
SKILL DRILL 12-7 237 Rigging the Prusik Minding Pulley	**SKILL DRILL 14-8** 289 Practicing a Lower with a Mainline/Belay System
SKILL DRILL 12-8 237 Monitoring the Prusik Minding Pulley	**SKILL DRILL 14-9** 293 Connecting a Tagline to a Litter

CHAPTER 13

SKILL DRILL 13-1   254
Removing a Helmet

SKILL DRILL 13-2   256
Performing a Burrito Warp

SKILL DRILL 13-3   257
Applying Upper 30 Tubular Restraint

SKILL DRILL 13-4   258
Applying Lower 30 Tubular Restraint

SKILL DRILL 13-5   260
Offseting the Litter

SKILL DRILL 13-6   261
Performing the Tap Litter Attendant Rotation

SKILL DRILL 14-10   293
Rigging a Litter for a Single Point Suspension High-Angle Lower

SKILL DRILL 14-11   293
Preparing a Two-Tensioned Rescue System

SKILL DRILL 14-12   294
Practicing a Lower with a Two-Tensioned Rescue System

SKILL DRILL 14-13   294
Practicing a Litter Lower with a Mainline and Belay

SKILL DRILL 14-14   295
Performing an Aerial Ladder Slide

CHAPTER 15

SKILL DRILL 15-1   306
Rigging a 2:1 Mechanical Advantage with Direction Change

CHAPTER 14

SKILL DRILL 14-1   272
Tying an Improvised Full Body Harness

SKILL DRILL 14-2   274
Performing an Ambulatory Subject Lower

SKILL DRILL 14-3   276
Tying a Mainline Rope to the Head of a Litter

SKILL DRILL 14-4   277
Tying a Closed Loop Bridle to the Head of a Litter

SKILL DRILL 15-2   307
Converting a 2:1 Mechanical Advantage to 4:1

SKILL DRILL 15-3   309
Rigging a 3:1 (Z-Rig) Hauling System

SKILL DRILL 15-4   310
Rigging an In-Line 3:1 System with an Integrated Brake

SKILL DRILL 15-5   310
Operating a Mechanical Advantage System

SKILL DRILL 15-6 313
Rigging the 4:1 System

SKILL DRILL 15-7 315
Rigging a Compound 4:1 Piggybacked to a Mainline

SKILL DRILL 15-8 316
Rigging a Two-Tensioned Inline 3:1 Raising System

CHAPTER 16

SKILL DRILL 16-1 328
Constructing the Litter Spider

SKILL DRILL 16-2 329
Attaching a Rescue Line to a Litter

SKILL DRILL 16-3 330
Crafting the Attendant Rig

SKILL DRILL 16-4 332
Managing a Litter while Suspended

SKILL DRILL 16-5 333
Rigging a Pike and Pivot Litter Bridle

SKILL DRILL 16-6 333
Negotiating the Edge Using a Pike and Pivot

SKILL DRILL 16-7 335
Performing a Pike and Pivot Raise

CHAPTER 17

SKILL DRILL 17-1 358
Rigging a Highline

SKILL DRILL 17-2 360
Moving a Load Across a Highline

SKILL DRILL 17-3 362
Rigging a Guiding-Line System

CHAPTER 18

SKILL DRILL 18-1 374
Throwing a Rope

SKILL DRILL 18-2 377
Practice Rappelling

SKILL DRILL 18-3 378
Threading a Brake Bar Rack

SKILL DRILL 18-4 380
Descending with a Brake Bar Rack

SKILL DRILL 18-5 382
Locking off the Descent

SKILL DRILL 18-6 383
Releasing a Locked-Off Descender

SKILL DRILL 18-7 383
Locking Off a Brake Bar Rack

SKILL DRILL 18-8 384
Releasing a Locked-Off Brake Bar Rack

SKILL DRILL 18-9 386
Attaching a Prusik Loop to a Host Rope

SKILL DRILL 18-10 388
Ascending Using a Prusik Ascending System

SKILL DRILL 18-11 392
Rigging a Basic Ascending System

SKILL DRILL 18-12 395
Performing a Standing Edge Negotiation

SKILL DRILL 18-13 396
Performing a Kneeling Transition Method

SKILL DRILL 18-14 398
Performing a Change Over (Ascending to Descending)

SKILL DRILL 18-15 401
Performing a Change Over (Descending to Ascending)

SKILL DRILL 18-16 403
Performing a Rope to Rope Transfer

SKILL DRILL 18-17 404
Passing a Knot

SKILL DRILL 18-18 404
Passing a Knot in an Access Line

SKILL DRILL 18-19 405
Passing a Deviation on a Descent

SKILL DRILL 18-20 405
Passing a Deviation on an Ascent

SKILL DRILL 18-21 406
Passing an Intermediate Access Point

SKILL DRILL 18-22   408
Extricating an Obstruction from a Jammed Rappel Device

SKILL DRILL 18-23   411
Wrapping the Rope for a Body Rappel

CHAPTER 19

SKILL DRILL 19-1   419
Applying a Diaper Style Harness

SKILL DRILL 19-2   419
Tying a Hasty Seat Harness

SKILL DRILL 19-3   421
Tying a Rescue Chest Harness

SKILL DRILL 19-4   424
Pickoff by Lowering

SKILL DRILL 19-5   426
Rappelling by Pickoff

SKILL DRILL 19-6   428
Performing a Pickoff Rescue

SKILL DRILL 19-7   429
Practicing a Travelling Brake Pickoff

SKILL DRILL 19-8   431
Preparing the Litter for Lower

SKILL DRILL 19-9   432
Performing a Dual Tension Litter Lower

SKILL DRILL 19-10   433
Practicing Lowering on a Travelling Brake System

SKILL DRILL 19-11   435
Performing a Scaffold-Type Litter Lower

SKILL DRILL 19-12   437
Performing a Knot Pass with a Litter

Acknowledgments

Authors

Loui McCurley
Chief Executive Officer
Pigeon Mountain Industries, Inc.
Denver, Colorado

Tom Vines
Civilian Deputy for Search and Rescue (ret.)
Carbon County Sheriff's Search and Rescue
Red Lodge, Montana

Reviewers

Jason T. Allen
Camano Island Fire and Rescue
Stanwood, Washington

Hudson Babler
TEEX US&R Rescue Specialist Certificate 2018-01
Dallas Fire Rescue
Dallas, Texas

Toby Ballard
Captain
Missoula Rural Fire District
Stevensville, Montana

Erik Baynard
Firefighter, Paramedic, Rope Rescue Technician, SPRAT II
Indianapolis Fire Department
Elevated Safety LLC
Indianapolis, Indiana

Christoph A. Berndt, BA, CEM, EMT-P, Fire Chief
Western Taney County Fire District
Branson, Missouri

Roger Brafford II
Fire and Rescue Instructor
West Virginia Public Service Training
Old Fields, West Virginia

David Frank Briggs, BAS
Wausau Fire Department
Mid-State Technical College
Wisconsin Rapids, Wisconsin

Richard K. Caudill, AAS, Industrial Management
IFSAC Level III Instructor
NCDOI OSFM Retired Instructor
Alleghany County Rescue Squad
Sparta, North Carolina

Jason Caughey
Fire Chief
Laramie County Fire District #2
Cheyenne, Wyoming

Kent Courtney
Paramedic, Firefighter, Educator
Essential Safety Training and Consulting
Rimrock, Arizona

Steve Disick
ITRA Instructor Rope Level 3, #15350
Capital Technical Rescue and Safety Consultants, LLC
Albany, New York

Timothy J. Dorsey
Lake Ozark Fire Protection District
Missouri US&R Task Force 1
Lake Ozark, Missouri

Rob Gaylor
Deputy Chief
City of Westfield Fire Department
Westfield, Indiana

Tye Herron, AAS
Technician
Casper, Wyoming

Wayne Allen Hodge
KCTCS Kentucky Fire Commission
State Fire Rescue
Louisville, Kentucky

Dustin Housewright, NRP
Captain
Eastman Chemical Fire Department
Kingsport, Tennessee

Scot Hughes
Captain, Fire Training Officer
Williamson EMA /FIRE
Franklin, Tennessee

Addis Kendall
Chief Special Operations
Nashville Fire Department
Nashville, Tennessee

James Lannan, MS, EFO, NRO
Assistant Chief
Western Taney County Fire Protection District
Branson, Missouri

Joshua W. Livermore
Battalion Chief
Bullhead City Fire Department
Bullhead City, Arizona

Captain Phil Lopez
Missouri Fire Service Instructor II
Rope Rescue Instructor
Missouri State Evaluator
Greater St. Louis County Fire Academy
Clayton Fire Department
Wentzville, Missouri

Jason L.P. McMillan, EMT-P, FO1
Springfield Fire Dept
Illinois Fire Service Institute
Champaign, Illinois

Assistant Fire Chief Daniel Manning, PhD
Pohakuloa Fire Emergency Services
Hilo, Hawaii

Allen Michel
Nebraska State Fire Marshal
Chappell, Nebraska

John Newlin
Assistant Chief
Gainesboro Fire and Rescue Department
Stephenson, North Carolina

Jerry L. Mills
Fire Chief/Training Officer
Coahoma County Fire Department
Clarksdale, Mississippi

Jerry A. Nulliner
Division Chief (Retired)
Fishers Fire Department
Fishers, Indiana

Eldon G. Offutt
IFSAC Certified Instructor
New Mexico Firefighters Training Academy
Bingham, New Mexico

John O'Neal
Noblesville Fire Department
Noblesville, Indiana

Battalion Chief Derrick Phillips
Chief of Training & Homeland Security
St. Louis Fire Department
St. Louis, Missouri

Lieutenant Richard E. Presley
French Broad Volunteer Fire/Rescue
Broad River Volunteer Fire/Rescue
Asheville, North Carolina

Stephen V. Rinehart, AAS, BS
Firefighter II
Fire Officer II
Maryland Heights Fire Protection District
Maryland Heights, Missouri

Louie Robinson, MS FP-C
HealthNet Aeromedical
Cross Lanes, West Virginia

Bradley J. Smith
Battalion Chief
Oklahoma City Fire Department
Task Force Leader – OK-TF1
Edmond, Oklahoma

Marquis R. Solomon, NREMT, MA, SPHR
West Columbia Fire Department
West Columbia, South Carolina

Raleigh Sprouse, BPS, MPA
Oxford, Mississippi

Captain Dennis Thurman
City of Rogers Fire Department
Special Operations
Rogers, Arkansas

Captain Darin Virag
Charleston Fire Department
Culloden, West Virginia

Richard Vober, BA, EFO
Akron Fire Department
Akron, Ohio

Peter C. Webb
Battalion Chief
Dothan Fire Department
Dothan, Alabama

Don Werhonig
Lead Rope Rescue Instructor
Lynden Training Center
Fairbanks, Alaska

Chris Zawierzeniec
Training Officer, First Class Firefighter
Greater Sudbury Fire Services
Sudbury, Ontario, Canada

Josh Zent, BS
Deputy Chief of Operations/Special Operations
Northwest Fire District
Tucson, Arizona

From the Author

Neither this, nor any other of my writing efforts, would ever have been possible without the relentless encouragement and enduring support of my incredible husband, Bob McCurley. He is in every way the one who makes me complete.

Bob is my Rock. He's also my number one physics/engineering consultant, technical critic, proofreader, cheerleader, and friend, and he's the spiritual leader without whom I would simply not have capacity to grasp, appreciate nor even begin to attempt to apply the gifts and talents that God has graciously bestowed on me.

Thank you to our cover model, Krista Stippelmans. Krista is a certified rope access bridge inspector and engineer with Michael Baker International. Her career is a perfect intersection of her passion for bridges, climbing and the outdoors, and she can often be found hanging on ropes from rock faces and bridge tops.

A special thanks to the technical reviewers who helped to critique and form the content of this work, including Mike Railsback, Ben Waller, Frank Brennan, Cliff Freer, Bob McCurley, and the team at PMI (especially Lurii Chizh) who helped to provide content for graphics and images.

Loui McCurley
February 3, 2022

Dedication

This book is dedicated to the Technical Rope Rescue Trailblazers who, knowingly or otherwise, have helped to shape, impact, and influence my perspectives on rope rescue over the years—most notably my dear friend and mentor, Steve Hudson.

This book is also dedicated to other vanguards of our craft who have influenced me deeply over the last four decades, including Tom Vines, John Dill, Hamish MacInnes, Rod Willard, Steve Attaway, Arnor Larson, Bill Clem, Hal Murray, Bill Cuddington, Mike Roop, Mike Brown, John McKently, Tom Patterson, Tom Fiore, Drew Davis, and yes, you too Reed Thorne!

Perhaps you recognize some of these names, perhaps you don't. There are so many technical rescue experts in our industry today, movers and shakers whose names may be more recognizable to you than some of those listed above—but these are some of the technical rescue mavericks who I respect and admire most simply because our paths converged in space and time in the formative years of rescue when ideas and practice were young and flexible, and so was I. The work of these groundbreaking rescuers set the stage for how we think and do rescue today; by sharing their names here, I hope to encourage you—whoever you are—to pour into the young rope technicians around you who are likewise hungry for knowledge. You can and do make a difference.

I feel blessed to have been encouraged from my earliest recollections in rescue to understand the *concepts* behind the systems, rather than just the systems themselves.

I am thankful to those who taught me to *question* whether the newfangled equipment offered by manufacturers was for the benefit of the rescue operation, or for the benefit of the seller.

I am thankful to those who taught me to *question* whether the newfangled techniques offered by trainers were for the benefit of the rescue operation, or for the benefit of the trainer.

I am thankful to those who taught me to tuck the useful bits of information I find along the way into my *toolbox*, and let the rest blow away in the wind.

I am thankful to those who taught me to continually *learn* from mistakes—those of others and of myself—quietly and with dignity, without the need to criticize or self-aggrandize.

I am thankful to those who taught me that there are *many* right ways of doing things, and that the only "wrong" way is one that is unsafe.

If you are holding this book in your hands, I hope that you will use it as an opportunity to affirm the knowledge you already have, and to explore some new ideas; I hope you will find some different equipment and techniques to try and to consider; and I hope that you will tuck the bits you find useful in your toolbox, and let the rest blow away in the wind.

Let me encourage you to help inspire and equip those around you who are hungry for knowledge to do the same.

Climb High, Stay Safe
Loui McCurley

SECTION 1

Awareness

CHAPTER **1** **Becoming a Rope Rescuer**

CHAPTER **2** **Size-Up**

CHAPTER **3** **Hazards Associated with Rope Rescue**

CHAPTER **4** **Initiating a Response**

CHAPTER **5** **Supporting the Operations- or Technician-Level Rescue Incident**

CHAPTER 1

Awareness

Becoming a Rope Rescuer

KNOWLEDGE OBJECTIVES

After studying this chapter, you should be able to:

- Define the term authority having jurisdiction. (**NFPA 1006: 5.1.5**, p. 4)
- Identify the National Association of Fire Protection (NFPA) standards that apply to rope rescue. (p. 5)
- Identify the three levels of rope rescuer: awareness, operations, and technician. (**NFPA 1006: 5.1.2**, p. 6)
- Explain how the job performance requirements in **NFPA 1006**, *Standard for Technical Rescue*, support training and skills maintenance. (**NFPA 1006: 5.1**, p. 7–8)
- Describe the common types of rope rescue incidents. (pp. 9–13)
- Identify the mindset characteristics of an effective rope rescuer. (pp. 13–17)
- Explain how to alert others to a falling object. (p. 17)

SKILLS OBJECTIVES

There are no skills in this chapter.

You Are the Rescuer

Your organization is responding to a technical rescue incident at a local natural amphitheater. Several teenagers climbed up the rocks surrounding the amphitheater for a drinking party, and when security staff tried to disperse them, a fight ensued. Dispatch tells you that a dozen people are now in various forms of peril, including being stranded on a rock ledge, injured from fighting, and passed out at the top of the cliffs. Your agency needs mutual aid to assist with this many potential patients.

1. How do you first request mutual aid?
2. How do you determine whether the responders from a neighboring agency possess adequate skills to address this incident?

 *Access Navigate for more practice activities.*

Introduction

Training an individual, and ultimately an organization, to effectively achieve the required level of competency in rope rescue, while at the same time maintaining compliance with industry standards and best practices, can be a daunting task (**FIGURE 1-1**). What is the minimum level of training required, and is it enough? This text is designed help you and your organization achieve competency as well as compliance. It has been written and reviewed by highly experienced, field-oriented rescuers and drafted in accordance with requirements for **technical rescue** established by the National Fire Protection Association (NFPA). A technical rescue is the application of special knowledge, skills, and equipment to resolve unique and/or complex rescue situations.

The NFPA is a nonprofit organization that generates codes, standards, recommended practices, and guides through consultation and consensus with experienced fire professionals in order to prevent injury, death, and property loss due to fire and additional hazards. When you see the term "NFPA approved," it should be understood as meaning that something is *approved to meet an NFPA standard*, not *approved by the NFPA* itself. There is no real consistency as by whom or why this designation is conferred. Sometimes this is self-designated; other times, there may be a third-party approval process in place. Most NFPA **equipment standards** call for third-party testing and approval of products, but this does not stop some manufacturers from stating that their equipment *meets the requirements of* a certain standard, even if it has not been tested, listed, or approved by a third-party testing organization.

It is up to the authority having jurisdiction (AHJ) to determine the acceptability of installations, procedures, equipment, or materials based on compliance with NFPA or other appropriate standards. In the absence of such standards, the AHJ may require evidence of proper installation, procedure, or use.

The Authority Having Jurisdiction

The term **authority having jurisdiction (AHJ)** is used frequently and liberally throughout NFPA documents to refer to whichever entity (whether an organization or an individual) is responsible for approving and/or enforcing guidelines with respect to the actions of a response agency.

The AHJ may be a federal, state, local, or other regional agency; it may be a group such as a board of directors or operating committee; it may be an individual such as a fire chief, commanding officer, or executive-in-charge; or it may be any other entity that has statutory authority. Because jurisdictions and

FIGURE 1-1 Training is achieved at both an individual and a team level.
Courtesy of Steve Hudson.

approval agencies vary, as do their responsibilities, it is left to the responding agency to determine to whom they answer.

Typically, the AHJ has authority over the actions of a response agency as well as a modicum of responsibility for the agency. NFPA standards require that the AHJ provide necessary personal protective clothing and equipment to responders congruent with their assignments. In addition, NFPA 1006 calls for the AHJ to establish certain criteria as a prerequisite for technical rescue training, from age requirements to medical requirements.

NFPA Resources

Although NFPA standards on technical rescue do not carry the weight of law, they are used by agencies all over the world as a baseline for achieving industry best practice. NFPA standards are developed through a consensus process that brings together volunteers representing varied viewpoints and interests. This helps to ensure a balanced perspective and the viability of the information represented as truly being best practice.

The NFPA standards referenced in this text include the following:

- NFPA 1006, *Standard for Technical Rescue Personnel Professional Qualifications (2021 Edition)*
- NFPA 2500, *Standards for Operations and Training for Technical Search and Rescue Incidents and Life Safety Rope and Equipment for Emergency Services*. NOTE: NFPA 2500 now includes content from the following standards:
  - NFPA 1670: *Standard on Operations and Training for Technical Search and Rescue Incidents*
  - NFPA 1983, *Standard on Life Safety Rope and Equipment for Emergency Services*
  - NFPA 1858, *Standard on Selection, Care & Maintenance of Life Safety Rope and Equipment for Emergency Services*

In an effort to improve its usefulness as a training adjunct, this text presents training requirements for rope rescue in the context of NFPA 1006. These requirements are harmonized with the organizational requirements historically found in NFPA 1670, so references to that document may also be found.

Equipment used by rescuers has historically been required by the AHJ to meet the requirements of NFPA 1983, so there are many references to those requirements in this text as well.

Finally, NFPA 1858 is considered the user guide for NFPA 1983, so this document is also referenced.

Effective 2022, the contents of NFPA 1670, NFPA 1983, and NFPA 1858 are collected together in a single volume, NFPA 2500, so in future years rescuers will need to become accustomed to referencing that standard. In the meantime, NFPA 2500 also includes parenthetical references to the respective source documents, so the same will be true throughout this text. However, in time, NFPA 2500 will become the ultimate reference for all things related to technical rescue.

While this text will focus primarily on Chapter 5 of NFPA 1006, it is important to understand that in NFPA terminology, technical rescue refers to more than just ropes. In fact, NFPA 1006 and NFPA 2500 (1670) identify 20 different, unique, and/or complex search and rescue situations as technical rescue specialty disciplines:

- Tower rescue
- Rope rescue
- Structural collapse rescue
- Confined space rescue
- Common passenger vehicle rescue
- Heavy vehicle rescue
- Animal technical rescue
- Wilderness search and rescue
- Trench rescue
- Machinery rescue
- Cave rescue
- Mine and tunnel rescue
- Helicopter rescue
- Surface water rescue
- Swiftwater rescue
- Dive rescue
- Ice rescue
- Surf rescue
- Watercraft rescue
- Floodwater rescue

Rope rescue is but one of these, and as a discipline, it rarely exists in isolation. Typically, rope rescue occurs in the context of some other environment, whether urban, wilderness, tower, water, ice, or confined space. In context of NFPA 1006 and NFPA 2500 (1670), rope rescue as a discipline is inclusive of both low-angle and high-angle rescue concepts. It should be understood that these various technical rescue disciplines are not mutually exclusive, and two or more may be intertwined in any given incident.

Because of the nature and specific knowledge and skills required during a technical rescue incident, it is not realistic for any one person, or even a single organization, to maintain expertise in all disciplines. Instead, it is recommended that the AHJ establish written guidelines for each discipline and identify the level of operational response capability desired based on the probability of incidence, frequency, and risk within their response area.

Levels of Rescue

NFPA 2500 (1670) provides guidelines to assist the AHJ in assessing **technical search and rescue** hazards within their response area, identifying the level of operational capability for the organization/team, and establishing operational criteria for each discipline. The levels of organizational response capability are defined as follows:

- **Awareness level.** This level represents the minimum capability of organizations that provide response to technical search and rescue incidents. Responders at this level are expected to know just enough about all disciplines to prevent further injury or harm, perform initial size-up, and activate appropriate response resources (**FIGURE 1-2**).
- **Operations level.** At this level, organizations may respond to technical search and rescue incidents. At the incident, they may identify hazards, support technician-level rescuers with specific techniques, and assist in search and rescue. As it applies to rope rescue, rescuers at this level must have the same basic capabilities as awareness-level rescuers as well as the knowledge and skills to perform rescues where they do not need to personally get on a rope rescue system in a vertical environment (**FIGURE 1-3**).
- **Technician level.** This level represents the capability of organizations to respond to technical search and rescue incidents and to identify hazards, use equipment, and apply advanced techniques as necessary to coordinate, perform, and supervise technical search and rescue incidents. As it applies to rope rescue, responders at the technician level are expected to be capable of everything an operations-level responder does, plus perform duties associated with putting themselves in the vertical plane—whether suspended from a litter, ascending/descending rope, or climbing a structure (**FIGURE 1-4**).

All agencies should maintain capability at the awareness level for all disciplines so that incident

FIGURE 1-2 Awareness-level rescuers are trained to activate appropriate response resources at an emergency.
© sirtravelalot/Shutterstock.

FIGURE 1-3 Operations-level rescuers must be able to perform rescues where they do not need to personally get on a rope rescue system in a vertical environment.
© Ken Hawkins/Alamy Stock Photo.

CHAPTER 1 Becoming a Rope Rescuer

FIGURE 1-4 Technician-level rescuers must be able to identify hazards, use equipment, and apply advanced techniques as necessary to coordinate, perform, and supervise technical search and rescue incidents.
Courtesy of Skedco, Inc.

types are recognized, rescuers avoid risk of injury, and appropriate resources are summoned. Maintaining an operations- or technician-level capability as an organization is more difficult. For any given discipline, maintaining an operations- or technician-level capability requires that the organization be capable of mounting an operations- or technician-level response whenever such an incident might occur.

This is where NFPA 1006 dovetails with NFPA 2500 (1670). While NFPA 2500 (1670) defines organizational capabilities, the number of NFPA 1006 technicians required to achieve the desired response capability at the organizational level under 2500 (1670) will vary depending on circumstances. For this reason, NFPA does not specify how many responders are required to be trained at a given level to maintain operational capability. It is left up to the AHJ to consider and determine how many responders are required to mount a rescue at their intended level, based on their jurisdictional needs.

Emergency services organizations may have to invest considerable resources to provide the equipment and training needed to perform technical rescues safely and efficiently at each level. This is not to say that organizations with limited resources cannot provide technical rescue services, only that the individuals charged with performing technical rescues should be qualified at the operations- or technician-level according to this standard.

Responding organizations should focus their resources on the types of incidents and levels of response that offer the greatest benefit for their response area. Where a need for operations- or technician-level response is identified, a written plan should be developed.

One way to achieve greater capability when resources are limited is for the organization to establish a system for cooperation with other organizations, local experts, specialized resources, and/or mutual aid. Where the likelihood of incident or anticipated number of responses in a given discipline does not support the time and expense of maintaining the necessary level of proficiency, building cooperative relationships with another responding organization can be the best choice. This is permitted—and even encouraged—within the context of NFPA standards.

Using Job Performance Requirements (JPRs)

The NFPA uses **job performance requirements (JPRs)** to describe the personnel knowledge and skills relative to each identified requirement. NFPA 1006 establishes tasks for each level and discipline that represent the minimum capabilities that personnel must be able to perform for a specific specialty area; these are grouped according to the duties of the job. This text is focused specifically on rescue in the high-angle environment, and is based on the complete list of JPRs for rope rescue found in Chapter 5 of NFPA 1006. It covers what an individual must be capable of to achieve each level of response capability (awareness, operations, or technician) relative to rope rescue. Additional requirements may be set by the AHJ depending on local need.

Within the standard, three critical components are provided for each JPR:

- The task to be performed: This is a concise statement of what the person is required to do.
- Tools, equipment, or materials: These are items that the evaluating entity must provide to ensure that all individuals completing the task are given the same (or equivalent) tools, equipment, or materials when they are being evaluated.
- Evaluation parameters and performance outcomes: This portion of the JPR promotes consistency in evaluation by reducing the variables used to measure performance.

Individually, maintaining an operations- or technician-level of capability requires more than just a one-time course or passing a test. Sending a responder to a technician-level course is not enough in and of itself to qualify that individual as a technician. A technician is one who knows the subject matter thoroughly (or knows where to get the answer quickly) and who has the skills to execute tasks consistent with the need. They must be able to think through problems that have never (or rarely) been seen before, and be comfortable in the high-angle environment.

Developing such depth requires hands-on experience and repetition. Only through a combination of ongoing study, training, skill, and frequency of operations in that discipline can an individual maintain an advanced level of capability.

The standards do not specify frequency of training; they state only that competency be demonstrated on an annual basis. This demonstration of competency should be documented. Demonstration of competency is an exercise that should be taken to show existing capability, and it should occur separately from the training itself. Learning a new skill and then mimicking it is not the same as demonstrating competence. Maintaining competence involves problem solving and necessitates frequent training and practice. Rescuers who may be called upon to perform rope rescue should be provided the opportunity to practice skills monthly, or at least quarterly, to help maintain sufficient skill to perform the annual demonstration of competency, as well as respond to any incidents likely to occur within their response area.

What Is Rope Rescue?

Depending on the environment and the rescuers involved, rope rescue may also be called vertical rescue or high-angle rescue. Although some people may refer to rope rescue using the term *technical rescue*, this is something of a misnomer, as we have seen that this term encompasses several other disciplines besides rope rescue. The defining factor in rope rescue is that a rope and other associated gear are necessary to move the subject from the hazard, provide safety, and/or protect the rescuer(s) and subject(s) from falling.

Historically, the term *victim* has been commonly used to refer to the subject of a rescue. However, in recent years industry professionals have gravitated away from this terminology, given the idea that the victim of an unfortunate circumstance should cease to be a victim as soon as rescuers arrive. At this point, the term *casualty* might well apply, particularly if they are badly maimed or deceased, and if they are injured, they might be better referred to as a *patient*. However,

FIGURE 1-5 In this text, we will use the term subject to refer to the person being rescued.
© Napa Valley Register/ZUMA Press, Inc./Alamy Stock Photo.

the term *patient* connotes medical intervention, which may or may not actually be required. For these reasons, in this book we will use the more contemporary term **subject** to refer to the individual who is being rescued, with the understanding that technical evacuation methods, not medical interventions, are the focus of this text (**FIGURE 1-5**).

Originally, most of the equipment and techniques used in rope rescue were the same as those used in mountaineering. However, in recent decades, a major shift has occurred away from recreational types of equipment and techniques toward methods and gear that are more specialized for rescue.

Fire Service Rescue

The use of life safety rope and associated equipment in the fire service has undergone tremendous change during recent decades. This is partly because of well-publicized tragedies, the best known of which was arguably one that occurred in June 1980 in New York City. During this incident, memorialized in a white paper called *Line to Safety* published by the International Association of Fire Fighters (IAFF), a rope was severed during a rescue attempt and two fire fighters fell to their deaths. Further impetus for improved equipment and training has come from fire service organizations such as NFPA, the IAFF, and the International Society of Fire Service Instructors (ISFSI).

Perhaps the most significant bit of modernization of rope rescue equipment has been the migration away from natural fiber laid ropes and toward the use of synthetic fiber kernmantle ropes. Experts now recognize that the use of natural fiber rope for life safety is a dangerous and irresponsible practice, a realization that resulted in a massive changeover by fire departments as well as workers in fall protection worldwide.

Where Do Rope Rescues Occur?

A need for rope rescue may occur whenever someone becomes injured at, stranded at, or falls from a point above or below grade. These kinds of incidents can occur as a result of recreational pursuits or in a workplace environment.

Recreational Incidents

Rock Climbing

Rock climbing is a common form of recreation throughout the world. It may take place in outdoor settings where there are well-developed recreational climbing areas, as well as more remote opportunities for climbing. Climbing accidents can also occur on manmade climbing walls in gyms, parks, recreational facilities, or even in homes (**FIGURE 1-6**). There is relatively little consistency among rock climbers in terms of what equipment they are using or what skills they might have, so rescuers must be prepared for almost anything when responding to these kinds of calls. Recreational equipment is generally lighter weight, not as robustly constructed, and more prone to failure under heavy loads than is rope rescue equipment. This makes recreational ropes and gear very suspect after they have sustained a fall or other damage. Recreational climbers are often found in remote, backcountry locations, so even when the rope rescue part of a response is completed, there may still be a lot of work to do before the incident is finished. This can include medical care; preventing or treating heat/cold stress; feeding and hydration; and long, rugged evacuations.

Rappelling

Rappelling is another backcountry activity, but is a sport all its own, more common in some parts of the country than others. These days, even urban responders are likely to respond to rappelling incidents, whether in climbing gyms or urban interface parks and recreation areas. Rappellers come in all shapes and sizes, and with a wide range of experience. Rescuing a stranded or injured rappeller in a remote environment can be quite the undertaking and may require extensive trained resources.

Mountaineering

Mountaineering is not the same as recreational climbing or rappelling, even though at times it may involve both of these activities. Typically, mountaineers get themselves much further into the backcountry, often in complicated natural terrain, and with exposure to hazards that are either uncommon or more severe than similar hazards experienced in an urban setting. It is a fine line, but for rescues involving backcountry response, specialized and experienced backcountry rescue teams should be called upon. Mountaineering can involve extremely high altitudes, the need for multiday expeditions, major logistics support, and specialized rescue and medical procedures that are beyond the capability of those not trained and experienced at operating in those environments.

Caving

Caves are more than just an underground wilderness; they are unique environments where moving an incapacitated subject can be complicated by long distances, tight passageways, vertical drops, water, mud, and total darkness. Although some confined space rescuers or mine rescue professionals might think of caves as simply an extension of the environment(s) in which they are accustomed to working, caves are unique (**FIGURE 1-7**). Personnel with specialized training and equipment for cave rescue should always be utilized for cave rescue operations.

Ice Climbing

Recreational ice climbing is a growing sport in some areas—so much so that frozen waterfalls in areas

FIGURE 1-6 Rock climbing can vary significantly in altitude and difficulty.
© Roberto Caucino/Shutterstock.

FIGURE 1-7 Caves are unique environments with unique hazards.
© Sementer/Shutterstock.

FIGURE 1-8 A worker using a fall arrest system.
© M2020/Shutterstock.

FIGURE 1-9 Rope access is a specialized approach to safety for work at height.
© noomcpk/Shutterstock.

such as quarries, road cuts, and parks are increasingly used for sport climbing. Whether easily accessible or remote, rescue of ice (or steep snow) climbers poses some very unique hazards and requires special training for responders. While the rules of physics still apply in these situations, specialized equipment for personal safety, anchoring, and mitigating potential forces are essential.

Workplace Incidents

Work at Height and Rope Access

Emergency responders should become familiar with commercial uses of **fall protection** and **rope access** within their jurisdiction. The form and function of this equipment is different from that used by rescuers, and certainly differs from equipment used for recreational pursuits. In some cases, special knowledge of fall protection equipment is required for rescue.

The term *fall protection* is used in workplace vernacular to refer to methods of eliminating or controlling hazards so that workers are prevented from an accidental fall from elevation and/or are protected from the potential consequences of such a fall (**FIGURE 1-8**). In general, the term can be applied to everything from guardrails to catch nets to harness-based systems, but for the purposes of rope rescue, our focus is typically those workers who are using harness-based fall protection systems (also sometimes referred to as active fall protection). Fall arrest, one common type of fall protection, utilizes ropes and lanyards to mitigate the effects of a fall.

Rope access, a very specialized form of work at height, now enables workers to access hard-to-reach work sites more quickly and easily than using traditional structural installations such as scaffolding and platforms.

Rope access has a wide range of uses, including window cleaning, bridge painting, and engineering inspection of building exteriors and ocean oil platforms (**FIGURE 1-9**). In most rope access techniques used on a work site, the worker has two ropes, or lines: a working line and a backup safety line, with a harness as the primary means of attachment.

Although self-rescue and coworker-assisted rescue are integral parts of all rope access training, rope rescue teams increasingly may be called upon to extract these workers from accidents that onsite workplace teams cannot handle.

The Society of Professional Rope Access Technicians (SPRAT) develops consensus safety standards for vertical rope access work. SPRAT has developed the *Safe Practices for Rope Access Work* and *Certification Requirements for Rope Access Work*, both of which apply to the types of work and worker positions in rope access. The American Society of Safety Professionals (ASSP) Fall Protection Code also has published guidance for rope access, in the form of *ANSI Z459 Safety Requirements for Rope Access Work*. The equipment and methods set out in these standards differ from those used by professional rescuers. By making an effort to familiarize themselves with these systems and equipment, responders can better prepare themselves to effectively and safely rescue workers at heights.

Industrial Sites

Industrial rescue incidents can occur in high-angle and confined areas—such as refineries, chemical plants, and open-pit and underground mining sites—that pose the risk of falls, entrapment, medical emergencies, and exposure to dangerous materials. The principles of high-angle rescue in these industrial environments are similar to those in natural environments, but environmental circumstances may either help or hinder industrial rescuers. In some industrial rescue situations, abundant structural anchors often are available, and the response time could be shorter. Rescue teams responding to industrial incidents may be municipal agencies responding from a local department, or they may be onsite, either in the form of an industrial fire department within a facility or simply predesignated emergency responders pulled from other roles within the organization. Industrial workers are often exposed to additional hazards such as energized electrical equipment, machinery, razor-sharp metal elements, very hot surfaces, and pressurized hydraulic and pneumatic lines and equipment. There can also be chemical, radiation, and biologic hazards present in industrial environments.

The Occupational Safety and Health Administration (OSHA) requires that employers have a rescue plan for accidents. Emergency responders to an industrial accident site may find the rescue already in progress. Professional responders to industrial rescue need to preplan how they will integrate with other possible rescuers (coworkers) already underway.

Wind Turbine Rescue

With the development of alternative forms of energy, wind turbines are found in increasing numbers. Wind turbines offer their own challenges for access, hazards, equipment, and rescue techniques. At heights of 200 feet (61 m) and greater and only a ladder inside the tubular tower for access, simply getting to a patient can be a challenge. Once in the nacelle, the cramped quarters, moving machinery, dust, and fumes make patient care, packaging, and extrication particularly difficult. If the subject is inside or outside the blades or on the outer parts of the nacelle or turbine, access and evacuation are further complicated by heights; wind; extremes of temperature; and slick surfaces covered with snow and ice, leaking hydraulic oil, or rain.

Confined Space Rescue

Confined spaces, simply defined as a spaces with limited access and not designed for continuous human occupancy, may be found on both industrial and municipal sites (**FIGURE 1-10**). These require special rescue methods and unique safety considerations. Confined spaces often contain hazardous materials, contaminated air, and dangerous machinery. Most significant

> ### TIP
>
> **OSHA Regulations for Confined Spaces**
>
> As a result of the high death rate among those attempting rescues in confined spaces, OSHA has established regulations specific to rescuers in confined spaces. These can be found in 29 CFR 1910.146, *Permit-Required Confined Spaces*. Additionally, NFPA 350, *Guide for Safe Confined Space Entry and Work*, is a tremendous resource for personnel entering confined spaces, as well as for municipal and industrial rescue teams. Specific to rescue, NFPA 1006, *Standard for Technical Rescuer Professional Qualifications*, and NFPA 2500 (1670), *Standard on Operations and Training for Technical Search and Rescue Incidents and Life Safety Rope and Equipment for Emergency Services*, include sections on confined spaces. Rescuers who may respond to confined spaces should be aware of these standards and regulations.

FIGURE 1-10 Confined spaces require specialized rescue techniques.
© Napa Valley Register/ZUMA Press, Inc./Alamy Stock Photo.

are the hazardous atmospheres that require rescuers to safeguard themselves with lockout/tagout procedures (LOTO), atmospheric monitoring and ventilation techniques, chemical protective clothing, and a positive-pressure self-contained breathing apparatus (SCBA) or supplied air breathing apparatus (SABA). In addition, the presence of hazardous materials can require rescuers to use specialized equipment such as explosion-resistant or intrinsically safe light sources.

Tower Rescue

Other industrial situations may require rope access procedures to allow workers to do their jobs. An example would be a worker climbing up a structural tower. Tower hazards—and the methods used for rescue from them—will vary widely depending on the type and use of the tower. Rescuers should consult, and preferably work closely with, experts who have experience in the type of tower where the incident is taking place. In some areas, water towers comprise an extraordinarily high proportion of tower rescue incidents. Power distribution and transmission towers differ from television transmission towers, which in turn differ from communications cell towers. The National Association for Tower Erectors has additional training guidelines for tower rescue.

Water Rescue

Water rescue in oceans, lakes, or slow-moving rivers has often been performed from boats, but in swiftwater, the safest and most efficient method of conducting a rescue is often shore based. Shore-based techniques include deploying throwbags with 50 to 75 feet (15 to 23 m) of positively buoyant water-rescue rope to retrieve the subject, along with constructing highlines, tension diagonal lines, or boat control rope systems over the water (**FIGURE 1-11**).

FIGURE 1-11 Ropes are a vital piece of equipment in water rescue.

High-tension rope systems such as highlines, tension diagonal lines, and Telfer systems should use ropes rated for high-angle rescue. Swiftwater rescue is usually an emergency, but potential rescuers must have the appropriate personal protective equipment and rescue gear before attempting a rescue. Ice rescue involves a whole separate range of hazards for rescuers and requires specialized rescuer protection and equipment. In most cases, rope for in-water rescue should float so that it is visible to rescuers and less likely to snag on hazards under the water. American Society of Testing and Materials (ASTM) F1739, *Standard Guide for Performance of a Water Rescuer—Level I*, and ASTM F 1824, *Standard Guide for Performance of a Water Rescuer—Level II,* address water rescue requirements in greater detail.

Becoming a Rope Rescuer

Operations- and technician-level rope rescuers are trained in and have experience with the skills needed in the rope rescue environment. They use specialized equipment to conquer gravity and to move about with ease in any direction needed. They also may use equipment specifically designed for rope rescue to carefully move subjects away from dangerous situations and into safe areas, where any injuries can be treated. However, it should be noted that rope rescue in unique environments—including confined spaces, towers, wilderness, cave, mines, etc.—requires additional training relative to that environment. This is because of the unique hazards posed by different environments, more difficult anchorage selection criteria, and unique skill sets and methods required to prevent further harm.

This text is not hazard specific, but focuses on foundational principles that awareness-level, operations-level, and technician-level rope rescuers can follow to be both safe and successful in their work. This should not be assumed to replace instruction by a competent rescuer, as rope skills cannot be learned simply by reading a book. This text is designed as a training manual to be used under the guidance of a competent instructor experienced in the field of rope rescue techniques. Such instructors can be found in established training organizations, such as government organizations and fire training centers, or independent schools with a long history of training in high-angle techniques. An instructor or training school should have experience in the environment in which the candidate will be working, and the instructor should be knowledgeable about the relevant standards and regulations. For example, if a candidate expects to be working in an area that could involve confined spaces, the instructor must understand the specific hazards and the relevant regulations that apply to confined-space entry and confined-space hazards.

Instruction in rope rescue techniques is only the first step toward becoming a rope rescuer. Rope rescue skills deteriorate quickly without frequent practice and regular training. A training schedule is an indispensable part of every rope rescuer's routine. The difference between a competent rope rescuer and one who can only talk about doing it is that the competent rescuer really has the skills for the job. It is essential not only that a person understand the skills, but also that they use them instinctively. In sudden emergencies, people react with actions that are instinctive. The only way to make rope rescue skills instinctive is practice.

Aptitude for the High-Angle Environment

A fear of heights is natural to human beings, and a degree of it is necessary for survival. Individuals without any fear or respect for the hazards involved in high-angle work are a danger to themselves and to others. However, until a person feels at ease operating in the high-angle environment, their discomfort could prevent them from being effective in these activities. As with any unaccustomed environment, the approach to movement, using equipment, and working alongside others is completely new. Responders who expect to be fielded in such a situation must be able to work at the highest structure or cliff face in their response area. The only way to become accustomed to working hundreds of feet in the air and to become effective doing so is to spend time hundreds of feet in the air—in other words, practice (**FIGURE 1-12**).

Characteristics of an Effective Rescuer

Although it is difficult to define exactly all the characteristics of an effective rescuer, certain traits stand out. One of the most significant is the rescuer's concern for the subject, the person being rescued. The rescuers must realize that the rescue subject is the reason for everyone's involvement. This is a human being in distress, a person in physical and perhaps emotional pain. A rescuer may contribute to that distress by regarding the rescue subject more as an object than as a person. Rescuers must continually communicate with the rescue subject in their care, even when the subject is unconscious. Hearing is the last sense to go in an unconscious person, and many rescue subjects have

FIGURE 1-12 Practice is key to becoming comfortable in a high-angle environment.
Courtesy of Tom Wood.

subsequently reported having heard conversations that took place around them while they were noted to have been unconscious.

Safety

Beyond a doubt, the primary objective in any rope rescue activity is to perform it safely. This primary objective is achieved through mental and physical excellence and through team organization. Safety is also achieved by comprehensive scene size-up, evaluating options, and choosing the best option. Failure to do this can have devastating consequences. An example is a 1989 incident in Georgia in which a rock climber had fallen and broken his leg. He was lying at the bottom of a cliff approximately 110 feet (34 m) high, but had not fallen that entire distance. Paramedics rappelled to him, packaged him for evacuation, and were tending him as he was rescued by rope. One of the paramedics was using a non–life-safety rope for access, and did not have a redundant belay. As the subject neared the top, rope slack was not minimized.

The paramedic slipped and fell over the edge, the rope broke, and the paramedic fell the entire distance to the bottom of the cliff. The paramedic was critically injured as a result of the fall and later died at a regional trauma center. After additional rescuers reached him, he was evacuated by four-wheel-drive vehicle to a helicopter landing zone. This option had been available for the original rescue as well.

Personal Red Flags

An individual should be aware of a number of red flags, or danger signs, when working in the rope rescue environment, including the following:

- Tunnel vision (an obsession with a detail or unexpected problem that causes the rescuer to lose situational awareness)
- Physical fatigue
- Mental or physical impairment caused by environmental factors such as heat or cold
- Impairment caused by alcohol or drugs
- Failure to speak out when seeing a dangerous situation
- Overconfidence, leading to a lack of attention to the task at hand
- Inability to concentrate because of outside concerns, such as family or personal problems
- Overexcitement (adrenaline rush): Feel your pulse; if it is racing well beyond your normal rate, sit down and cool off.
- Poor physical condition
- Medical conditions that impair the ability to maintain energy levels or situational awareness
- Failure to ask for help or assistance: People sometimes have a dangerous tendency not to admit they are in over their heads, because doing so would make them look inexperienced. If you feel you are getting in over your head, ask for help (a belay, if appropriate) or back off. If you are at all uncertain about how to tie a knot or use a piece of equipment, ask for advice or get some feedback.

Physical Safety Concepts

The following are some physical safety rules that should be observed at all times:

- Establish safety lines; everyone near the edge must be attached to a safety system.
- When appropriate, use redundant systems (e.g., more than one anchor).

FIGURE 1-13 All rescuers should wear protective gear to reduce the chance of injury.
© Nancy Greifenhagen/Nancy G Fire Photography/Alamy Images.

- Wear a helmet and situationally appropriate personal protective equipment (this applies to all personnel on site) (**FIGURE 1-13**).
- Check equipment frequently to make sure it is in safe condition and is being used properly.

Systems Thinking

This text examines the necessary elements of the rope rescue system and the ways in which their strengths and weaknesses determine the effectiveness of the system and, ultimately, the outcome of the operation.

One suggestion for anyone working in the rope rescue environment is to think in terms of systems. Activities in the rope rescue environment are rarely completed with only one rope, only one piece of hardware, or only one person. The rope rescue environment system consists of many elements, such as rope, hardware, anchors, and other elements, that cannot be viewed in isolation. For the elements to operate safely and properly, they must be viewed together as a system.

Just as a chain is only as strong as its weakest link, the rope rescue system is only as robust as its weakest element. For example, if your rescue system uses a rope that has a tensile test strength of 9000 lbf (40 kN) but it is attached to an anchor that pulls out at 500 lbf (2.2 kN), the effective strength of the entire system is only 500 lbf (2.2 kN).

Through careful study of this text and training under a competent instructor, candidates should learn to assess the strengths of the component pieces of equipment and the systems being used. As candidates come to understand the various rigging systems, they will learn that each system component can be rigged in different ways. Some ways will result in maximum strength, while rigging in other ways might weaken the overall system. Candidates should learn to engineer the systems used in an agency and build them in a way that yields optimum strength as a total system.

Use of Low-Risk Methods First

There are a great many ways to accomplish any given rescue. A competent rope rescuer must have sufficient knowledge and depth to be capable of sifting through the various options rapidly and in context of the circumstance to select an appropriate approach. The optimum approach to any rescue subject is one that effectively accomplishes the rescue task while minimizing danger both to the subject(s) and rescuer(s). To this end:

- Always evaluate the situation before approaching it. Depending on the specific environment, dangers to rescuers could include hazardous atmospheres, energized wires, falling rocks, bad weather, or hostile individuals.
- Do not rush. Rushing causes mistakes that endanger both the rescuers and the subject. Move carefully and meticulously. Do not crowd other rescuers.
- Choose the least dangerous route to the subject.
- Use the simplest rescue system that works effectively and safely. In rope rescue, the simplest way of doing the job is usually the most effective way. The more complicated the rescue system, the greater the chance of something going wrong.
- Instead of setting up operations directly above the subject, where rocks or hardware may fall on the person (and other rescuers), move slightly off to one side. Keep all nonessential personnel away from the area because they may dislodge rocks or other objects.
- Ensure that the team leader is well-qualified and experienced in the type of rescue operation at hand. If the incident commander is not experienced at the specific type of rescue taking place, they should appoint a rescue group supervisor who will oversee the rigging and rope rescue tactics and tasks.
- Appoint a team safety officer to oversee equipment use, anchoring, rigging belays, personnel rappelling and ascending, and harness tie-ins. The safety officer should be among the most experienced of the team members, must be able to oversee all aspects of

FIGURE 1-14 Safety lines reduce the risk of accidental detachment.
Courtesy of Ken Phillips.

the operation, and must have the authority to interrupt any dangerous activity.
- Ensure that all rescuers wear protective gear, such as helmets, to prevent head injury from falling objects and to reduce injury if the rescuer falls. All personnel qualified for high-angle operations should wear their harnesses so that they are ready for any immediate needs.
- Before rescue operations begin, set up safety lines (**FIGURE 1-14**). Anyone near an edge must maintain a positive connection to one of these at all times during the rescue and incident termination. If continued access up and down a

steep slope is needed, a securely anchored safety line should be set from top to bottom off to the side, slightly away from the rescue site.

Preparation for Self-Rescue

Whatever the activity, a person who works in the rope rescue environment must always keep in mind that anything can go wrong, and someday it probably will. Therefore, rescuers should be ready for the worst-case scenario and be ready to extricate themselves. Rescuers should prepare and train for an emergency, and they should always carry sufficient gear to perform self-rescue, including a small assortment of carabiners, a couple of slings made from webbing or rope, and either a set of Prusik loops or other ascenders. (See Chapter 8, *Rescue Equipment* for an explanation of the specific purpose of this gear.)

Backup of Other Rescuers

In addition to being ready for self-rescue, each rescuer should be ready at any time to extricate a fellow rescuer who gets into difficulty. Whenever any team member is not at work performing a task, all attention should be focused on the activity at hand. Everyone, no matter how intelligent and experienced, has an occasional lapse. All team members should be alert to the development of any unsafe condition and be ready to make appropriate corrections.

Care of Equipment

One sign of good mechanics or carpenters is their meticulous care of the tools of their trade. The same is true of a rope rescuer, who must care for rope and hardware. Proper care of rope rescue equipment is especially important because lives depend on these tools. Considerations for avoiding loss and damage to rescue gear include the following:

- Do not lay unsecured equipment at the edge of a drop. It can easily be kicked or knocked over the edge and be damaged or lost. Even worse, the object could injure those working below.
- When using equipment on a vertical face, keep all gear secured. It should be either attached to the system or secured to the equipment sling that many harnesses have.
- Do not lay equipment on the ground or floor. Hardware is easily lost in debris or dirt, and grit can damage it. Prepare for this when you first arrive on the site; either lay out an equipment tarp or hang a sling on a tree or other structure.

All hardware not in use should be placed on the tarp or clipped into the sling.

- Inspect all gear after each operation in addition to periodic, regularly scheduled inspections. Document all inspections in rope logs and gear logs, every time. The time to discover that gear is defective is during inspection, not during a rescue.
- Ropes need special care. A traditional rule is never step on a rope. Many experienced vertical ropeworkers are fanatical about never stepping on a rope—although in truth, modern life safety ropes are not nearly as vulnerable as the three-strand laid constructions and natural fibers once used. NOTE: *While a good rope rescuer will take rope care seriously, stepping on a rope is not an automatic retirement sentence. If the rope passes robust inspection even after it has been stepped on, it may be able to be placed back into service.*
- Ropes that have been exposed to petroleum hydrocarbons, hazardous materials, excessive wear, or visible damage should be retired if there is any doubt about future safe use.
- Take all defective gear out of service immediately.
- Belay all lowering and raising of rescue subjects. Any rescuers who request them must be provided with belays. Double rope systems (main line and belay line or shared tension systems) should be used for all rescues where access above the subject is practical.

Attention to Detail

Attention to detail in the rope rescue environment is necessary for rescuers to work effectively and to prevent injury to themselves and others. This environment is unforgiving, and a lapse that may go unnoticed on level ground could result in severe injury or death in the rope rescue environment.

There is, of course, a need for balance between big-picture awareness and observation of detail. Rescuers must be capable of maintaining enough big-picture perspective so that they are able to complete a task in context of the overall objective. However, only a focused mind can examine a system quickly and immediately determine if all the necessary details are in place. This comes only with practice.

Although attention to detail is an essential trait for the rope rescuer, the nature of rope rescue operations requires that a person have the ability to improvise. Every situation is different, with varying circumstances in terms of weather and terrain. Therefore, a well-trained individual with good judgment who is able to adapt to any situation will be a more capable and valuable resource than one who is trained to perform in only one way.

Team Concepts

Many accidents or near misses occur because of organizational and management failures. Rescuers can help avoid this by training together often so that individual skills are combined into one working unit. An important position of any operation is the safety officer. The safety officer is an individual who makes sure that safe procedures are followed during training and real rescues. The safety officer should be someone other than the team leader/rescue group supervisor, because the leader often is too busy to check all safety systems. (Chapter 4, *Initiating a Response* presents more information about the role of the safety officer.)

Warning Call

One of the most common dangers in the rope rescue environment is falling objects, either hardware dropped by others or dislodged rocks. Whenever a hard object begins to fall, even if no one is thought to be below, the universal warning is to yell very loudly, "Rock! Rock! Rock!" Repeating the warning three times helps to ensure that the message gets through. Do not say, "Look out," "Heads up," or anything other verbiage, as this could result in the rescuer looking up and/or responding verbally rather than protecting themselves. Likewise, rescuers should be well trained to respond immediately to the warning "Rock! Rock! Rock!" by covering and protecting themselves and their subject to the best of their ability.

Training Expectations

A rope rescuer is one who is trained in the necessary skills for rope rescue, has shown that they competent in those skills, and continually trains to maintain them. An awareness-level rope rescuer should be able to do the following:

- Recognize a rope rescue incident and activate the necessary number and level of resources

An operations-level rope rescuer should be able to do the following:

- Demonstrate the proper use and care of rope
- Demonstrate the proper use and care of other equipment needed in the rope rescue environment
- Demonstrate the ability to tie correctly and without hesitation the 10 necessary knots, plus the Münter hitch and the Prusik hitch (see Chapter 9, *Ropes, Knots, Bends, and Hitches* for the 10 required knots)
- Demonstrate the ability to rig safe and secure anchors
- Demonstrate the ability to calculate system safety ratios on various systems with common technical rescue gear, ropes, and rigging
- Demonstrate the ability to belay another person safely and confidently

A technician-level rope rescuer should be able to do the following:

- Demonstrate the ability to rappel safely, confidently, and under control; the ability to tie off a rappel device to operate safely with hands free of the rope and then return to a safe and controlled rappel; and the ability to operate on the rope with the body in any position, including being inverted
- Demonstrate the ability to ascend safely; the ability to tie a friction hitch correctly and without hesitation, as well as how to use it; the uses and limitations of mechanical ascenders; and the ability to ascend a fixed rope safely using both friction hitches and mechanical ascenders
- Demonstrate the ability while on rope to change over from rappelling to ascending and from ascending to rappelling confidently and safely, and the ability to self-extricate from a jammed rappel device (or similar problem) without using a knife

In addition, there are a range of other areas in which the competent rope rescuer should train and maintain skills. Among these are the following:

- Emergency medical skills. Team members should be trained at least to a level of Department of Transportation (DOT) emergency medical responder (EMR).
- Team skills. Some examples are litter-handling techniques and lowering and hauling systems.
- Communication skills. These include not only the ability to communicate electronically, but also the standard voice communications required in specialized team operations such as rescuer lowering and hauling systems.
- Incident operations. All rescuers should have a well-rounded understanding of logistics, management, and operations such as may be found in the National Incident Management System (NIMS), at least to the level at which they are expected to operate.

After-Action REVIEW

IN SUMMARY

- Training is essential to a good rope rescue program, both at an individual level and at the team level.
- NFPA 1006 addresses individual skills for all three levels of rope rescue, and should be used in conjunction with NFPA 2500 (1670) for team capabilities and requirements by the authority having jurisdiction (AHJ).
- Rope rescue skills are foundational to many other specific disciplines of technical rescue, but should be augmented with specific discipline/environment training for the best outcome.
- The three levels of rescue are awareness, operations, and technician.
  - Awareness-level rescuers are expected to know just enough about all disciplines to prevent further injury or harm, perform initial size-up, and activate appropriate response resources.
  - Operations-level rescuers must have the same basic capabilities as awareness-level rescuers as well as the knowledge and skills to perform rescues where they do not need to personally get on a rope rescue system in a vertical environment.
  - Technician-level rescuers are expected to be capable of everything an operations-level responder does, plus perform duties associated with putting themselves in the vertical plane—whether suspended from a litter, ascending/descending rope, or climbing a structure.
- Depending on the environment and the rescuers involved, rope rescue may also be called vertical rescue or high-angle rescue.
- The defining factor in rope rescue is that a rope and other associated gear are necessary to move the subject from the hazard, provide safety, and/or protect the rescuer(s) and subject(s) from falling.
- In rope rescue, subject is the preferred term for a victim requiring rescue.
- Rope rescues occur in a variety of environments, from recreational incidents such as rock and ice climbing to workplace incidents such as industrial towers to water rescue to confined space incidents.
- In addition to specific required skills and knowledge, rope rescuers must have an aptitude for work at height.
- Rope rescuers should be safety-oriented team players who are inclined to hone their skills for their own safety and that of others.

KEY TERMS

Authority having jurisdiction (AHJ) An organization, office, or individual responsible for enforcing the requirements of a code or standard, or for approving equipment, materials, an installation, or a procedure.

Awareness level A functional level of technical rescue capability that represents the minimum capability of an organization or an individual to provide response to technical search and rescue incidents.

Equipment standards NFPA equipment standards are test methods developed by consensus agreement among a balanced committee of persons representing user, regulatory, manufacturer, and other interests.

Fall protection Any equipment, device, or system that prevents an accidental fall from elevation or that mitigates the effect of such a fall.

Job performance requirements (JPRs) Written statements that describe a specific job task, list the items necessary to complete the task, and define measurable or observable outcomes and evaluation areas for the specific task.

Operations level A functional level of technical rescue capability that represents the capability of an organization or an individual to respond to technical search and rescue incidents and to identify hazards, use equipment, and apply limited techniques specified in this standard to support and participate in technical search and rescue incidents.

Rope access A set of techniques where hardware is used as the primary means of providing access and support to workers. Generally, a two-rope system is employed: the working (main) rope supports the worker, and the safety (belay) rope provides backup fall protection. The two ropes may be switched back and forth between being the main and belay, but the worker must always be attached to two ropes. Rope access workers must be trained in self- and partner rescue; outside rescue sources may also be needed.

Subject The individual who is being rescued; the preferred term for a victim in rope rescue.

Technical rescue The application of special knowledge, skills, and equipment to resolve unique and/or complex rescue situations.

Technician level A functional level of technical rescue capability that represents the capability of an organization or an individual to respond to technical search and rescue incidents and to identify hazards, use equipment, and apply advanced techniques specified in this standard necessary to coordinate, perform, and supervise technical search and rescue incidents.

Technical search and rescue The application of special knowledge, skills, and equipment to resolve unique and/or complex search and rescue situations.

On Scene

1. What's the difference between being trained in a given skill versus being competent in that skill?

2. What rope rescue disciplines are covered by NFPA 1006 and 2500 (1670)?

3. Which of these technical rescue disciplines exist(s) in your response area?

4. Which of these technical rescue disciplines might involve rope rescue?

5. You respond to a car-over-the-edge incident, where a rope is draped down the slope as a handline for rescuers to hold onto while scrambling down the bank. Does this constitute rope rescue?

6. What are some red flags in a person's behavior or attitude that might suggest to you that they may not be an ideal candidate to become a rope rescuer?

Section and Chapter Opener: © Jones & Bartlett Learning. Courtesy of Loui McCurley; On Scene siren: © Bildgigant/Shutterstock.

CHAPTER 2

Awareness

Size-Up

KNOWLEDGE OBJECTIVES

After studying this chapter, you should be able to:
- Describe the process of performing a scene size-up at a rope rescue. (**NFPA 1006: 5.1.2, 5.1.3**, pp. 22–24)
- Explain the process of determining the scope of a rope rescue. (**NFPA 1006: 5.1.2**, pp. 22–23)
- Explain the considerations for gathering information about subjects at a rope rescue. (**NFPA 1006: 5.1.2**, pp. 23–24)
- Explain the considerations for determining what additional resources may be needed at an incident. (**NFPA 1006: 5.1.2**, pp. 25–27)
- Explain how scene size-up informs the creation of an incident action plan. (**NFPA 1006: 5.1.2**, p. 27)
- Describe the purpose and components of an incident action plan. (**NFPA 1006: 5.1.2**, p. 27)
- Identify rope rescue reference and resource materials. (**NFPA 1006: 5.1.2**, pp. 27–28)
- Describe how search parameters for a rope rescue are identified. (**NFPA 1006: 5.1.2**, pp. 28–29)
- Describe size-up information that must be communicated to command. (**NFPA 1006: 5.1.2**, p. 30)
- Analyze reference and resource materials. (**NFPA 1006: 5.1.2**, pp. 27–28)
- Communicate information concisely and accurately during a rope rescue incident. (**NFPA 1006: 5.1.2**, p. 30)

Additional Standards
- **NFPA 472**, *Standard for Competence of Responders to Hazardous Materials/Weapons of Mass Destruction Incidents, 2013 edition*
- **NFPA 1006**, *Standard for Technical Rescuer Professional Qualifications, 2013 edition*
- **NFPA 1500**, *Standard on Fire Department Occupational Safety and Health Program, 2013 edition*
- **NFPA 1521**, *Standard for Fire Department Safety Officer, 2008 edition*
- **NFPA 1561**, *Standard on Emergency Services Incident Management System and Command Safety, 2014 edition*
- **NFPA 1561**, *Standard on Emergency Services Incident Management System and Command Safety, 2014 edition*
- **NFPA 2500**, *Standards for Operations and Training for Technical Search and Rescue Incidents and Life Safety Rope and Equipment for Emergency Services, 2022 edition*
  - **NFPA 1983**, *Standard on Life Safety Rope and Equipment for Emergency Services*
  - **NFPA 1858**, *Standard on Selection, Care, and Maintenance of Life Safety Rope and Equipment for Emergency Services*
  - **NFPA 1670**, *Standard on Operations and Training for Technical Search and Rescue Incidents*

SKILLS OBJECTIVES

After studying this chapter, you should be able to:
- Conduct a rope rescue incident scene size-up. (**NFPA 1006: 5.1.2**, pp. 22–24)
- Describe how to interview witnesses. (**NFPA 1006: 5.1.2**, pp. 24–25)

You Are the Rescuer

You are first on scene at a motor vehicle crash where a 12-passenger van has gone over an embankment. The van is on its top, about 60 yards (55 m) down a 30-degree slope, wheels still turning. You immediately notice two people lying midway down the slope, who appear to have been thrown from the vehicle as it rolled. Another person is wandering around on the shoulder of the road looking dazed and confused, cradling their left arm with their right. You see blood on their face.

1. How would you call this into dispatch?
2. How would you describe the incident, the number of subjects, and their condition?
3. What hazards would you report, and kinds of resources would you request to respond?

 Access Navigate for more practice activities.

Introduction

Size-up is a term used by rescuers to refer to the process of mentally assessing a situation and evaluating factors at an incident prior to determining a course of action or committing resources. The goal of size-up is to take both a long view and a short view to identify and consider anything that could influence an impending rescue. This could include environmental conditions, scene safety, location and condition of the subject(s), and evaluation of whether or not the subject(s) might be able to assist in their own rescue. This initial report should be taken with something of a grain of salt. Responders to a call for a cat stranded on a windowsill might actually find the 10-year-old owner of the cat is also in danger as she tries to reach it. Rescue parameters can change quickly as the incident progresses—for example, if the cat jumps from windowsill to fire escape, and/or other family members become involved.

Responders should be trained to perform a thorough size-up quickly and efficiently, and to continue to reevaluate these same thought processes on an ongoing basis during the course of the rescue. A good size-up will include at least the following information:

1. Scope, magnitude, and nature of the incident
2. Location, number, and condition of subjects
3. Risk versus benefit analysis (body recovery vs. rescue)
4. Access to the scene
5. Environmental factors
6. Available/necessary resources
7. Ability to make contact with the subject without endangering either responders or subjects

This is not an exhaustive list. Other factors may need to be considered.

The purpose of the size-up is to perform a quick and efficient analysis of relevant information to provide incoming resources and leadership with information on which to base a plan for operational decisions and actions. A range of alternatives may be considered in any given operation, so having more information from which to derive decisions is better.

Size-up is more than a one-time event. The first responder arriving on scene should perform an initial size-up so that they can effectively communicate and request resource needs, and then continue with ongoing observations and analysis of size-up topics. Likewise, subsequent rescuers arriving on scene should also continue to perform size-up analysis, especially those who will serve in management roles.

Beginning with this chapter, we will focus on what rescuers should do after the call goes out from dispatch. The actions taken during the initial phase of an incident response can contribute to overall success of the mission as it progresses—or not.

Determine the Scope of the Rescue

Prior to resources being selected and deployed, a determination needs to be made as to what, exactly, the intended outcome of the operation will be. Determining what needs to be achieved, and what work needs to be done to reach the desired outcome, is known as the **scope** of the operation.

Decisions regarding resources, including team size and capability, equipment requirements, anticipated time to completion, and need for additional resources, all depend on having a clearly defined and communicated scope. Ultimately, the personnel deployed, the equipment and techniques used, and the procedures followed will be driven by scope.

At a very high level, the scope of most rescue operations is grounded in the following:

1. Removing the subject(s) from harm, while maintaining safe practices, policies, and procedures
2. Minimizing loss of life, personal injury, and property damage
3. Executing the rescue in a timely and efficient manner

What this might look like for any given incident will vary depending on the incident itself. For example, in the case of a window cleaner suspended from a collapsed scaffold 20 stories up a 30-story building, the scope of the rescue might be to remove the window at the level nearest and above the subject's location, and to utilize a rope rescue system to retrieve and raise the subject to that level within 20 minutes without dropping any debris on lower balcony levels (**FIGURE 2-1**).

Within the scope of the entire operation, each team will have a narrower scope. For example, one field team might be responsible for setting an anchor, another might be responsible for building a raise/lower system, while yet another might be preparing for medical intervention. Likewise, during a multiagency response, the scope of any one agency will be narrower than the scope of the entire incident. It is the function of the rescue leader to coordinate the segmented efforts and draw them together to a successful conclusion.

Identify the Number of Subjects

The number of subjects involved in an incident may initially appear to be obvious, but do not fall into the trap of making assumptions. It is not uncommon, for example, for persons involved in a motor vehicle incident to be thrown a great distance from the accident site. In a car-over-the-edge scenario, subjects may have been ejected at multiple points along the path of travel. In a rock-climbing incident, there will almost always be at least one additional climber involved, if not more. Likewise, in the case of workplace incidents, there may be any number of coworkers involved in ways that are not immediately obvious.

Although there may or may not be more than one person stranded or injured at height and in need of rope rescue, there may also be other impacted persons who are not stranded, injured, or in obvious distress, but are still in need of care. Determining how many subjects will need to be rescued or otherwise cared for during an incident is one of the first and most important items that the rescue leader must consider so that an appropriate rescue plan can be developed. An incident where available personnel or rescue resources are outnumbered by the number and severity of subjects is known as a **mass-casualty incident (MCI)**.

In any rescue incident, safety is paramount. Regardless of the number of casualties or subjects, rescuer safety follows standard protocol:

1. Your own safety is FIRST priority.
2. Safety of fellow rescuers is your SECOND concern.
3. Safety of the subject is your THIRD priority.
4. Safety of bystanders is FOURTH in this list.

Establish the Last Known Location of Subjects

Sometimes a rescue scenario begins with an assumption of a subject whose actual location is unknown. In this case, the rescue may begin with a search. Searching for a subject who is likely to need rope rescue can be difficult, as steep or high-angle surfaces are often

A

B

FIGURE 2-1 The scope of the operation depends upon the emergency.
(A) © vivalapenler/iStock Editorial/Getty Images Plus/Getty Images; (B) © Peter Seyfferth/imageBROKER/Alamy Stock Photo.

not conducive to close examination. While backcountry or wilderness environments may initially come to mind as the most likely environments where search might be necessary, any rescue can begin with a search. This might include rural, suburban, or urban environments, and may involve difficult terrain, confined spaces, caves, water, buildings, bridges, towers, or other structures. Regardless of search environment, having a good, solid point of reference where the subject was last seen or known to be can expedite the process.

The point where a person was last definitively known to be, with positive identification, is often called the **last known point (LKP)**. Usually this will be where someone actually physically saw the person, and for this reason some agencies prefer to use the term **point last seen (PLS)** instead. However, in this age of advanced technology, this information may also be derived from other available data. In wide-area search, cell phone forensic data has become a useful tool for helping to determine the LKP of a subject. To be anywhere close to accurate, cell phone forensics require a combination of several adequate cell towers in the area and rapid, effective cooperation of authorities. Even then, it may not be completely accurate. Subscription-based satellite technology, such as Iridium, is another potential resource that is rapidly becoming an effective method for triggering SOS with integrated location information via a global satellite network. However, unless the person was actually physically seen by a reliable source, a derived LKP must be taken with a certain amount of uncertainty.

Ideally, the subject will still be located at their LKP. This makes impending rescue much more efficient. If they are not at their LKP, finding them is the first task and forms the basis for the initial scope of the incident.

Identify and Interview Witnesses and Reporting Parties

The person who reported the need for rescue is known as the **reporting party (RP)** (**FIGURE 2-2**). This could be the subject of the rescue themself or another person. When the RP is another person, the RP may or may not even be at the incident site. Regardless of the specifics, ensuring that the RP is appropriately isolated and interviewed is a top priority. Then, others who may have first-hand information or relevant knowledge of the subject or the incident should also be interviewed.

Isolation of the RP and other relevant interviewees helps to keep them readily available for interviewers, while at the same time keeping them out of reach of media and separated from incident operations. Where

FIGURE 2-2 Interviewing a reporting party.
© ZUMA Press Inc/Alamy Stock Photo.

there are multiple RPs or potential interviewees, also keeping them separate from one another is best practice. Interviewing people separately and discouraging them from comparing information can help preserve the integrity of what they know. Of course, in most cases an interviewee will be free to come and go as they please, but a cooperative RP will often be happy to stay and help.

TIP

Good candidates for interview might include the following:
- Reporting party
- Incident eyewitnesses
- Family members
- Coworkers/schoolmates
- Healthcare or welfare workers
- Friends of the subject

Interviewing an RP is unlike any other type of interview that you will ever perform. A person reporting an incident may be especially stressed and more than a little emotionally distraught, particularly if the subject is a close family member or friend or if they have witnessed a disturbing incident. When interviewing an RP under these circumstances, information can be disjointed and facts may not hang together in a normal way. Although we do not like to think in these terms, sometimes an RP may have ulterior motives, or may even intentionally try to distort information. Good interview techniques can make all the difference to a successful outcome.

RPs should be encouraged to share whatever information they can think of relative to the situation, even things that they may not consider important. A savvy interviewer might pick up on important details, and sometimes one bit of information can trigger other important memories as it is recounted. Cognitive interview methods that have been shown to help improve information recall and analysis include the following:

- Alternative narrative order: This method, long used by law enforcement personnel, relies on the recency effect, which suggests that the human mind recalls recent events more clearly than older ones. Describing events from end to beginning (rather than beginning to end) can improve accuracy and recall of specific details.
- Contextual restatement: The interviewer using this method tries to help stimulate accurate recall by asking questions that re-create context for the RP, including what they were doing at the time of the incident; how they felt; and what sights, sounds, smells, and even emotions they remember experiencing at the time. Answering questions in this context can help trigger recall of important details.
- Perspective: In this interview style, the RP is asked to describe what happened from different perspectives. By thinking about and explaining how they think that others, or even the subject themselves, might have seen the incident, they may be able to offer insights that might not otherwise occur to them.

Using cognitive interview methods such as these places information outside the RP's normal frame of reference or biased interpretations, which is believed to be a key factor in improving recall.

Forms may be used to document interview information and results. Forms not only help to retain information for later reference, they can also help to keep the process more structured and thorough. General questions about the subject, their habits and characteristics, and general information are a good starting point. Then, the conversation can move into questions that might help the interviewee to recall specific details that may have led to the incident at hand.

Some of the kinds of information that may be useful to gather include the following:

- Name(s) of the subject(s)
- Age of the subject(s)
- Clothing worn by the subject(s)
- Skills or familiarity of the subject(s) relative to local area and activity
- Time, place, and circumstances under which last seen
- General state of physical and mental health
- Specific information regarding known injuries or health concerns
- Conditions or situation under which the incident occurred
- Relevant conditions between the time of occurrence and the present
- Response activities already performed or currently underway

As information is gathered, a subject profile will begin to emerge. Having a greater understanding of the subject helps rescuers to have a better idea of what the subject might do or how they might react to the situation they are in. If the location of the subject is known, this profile will help rescuers to make educated guesses about how cooperative, incapacitated, or threatening the subject might be. If a search is required, the profile will help in developing a search plan.

Assessing Resource Needs

As information is gathered about the scope, subject, and situation, the incident commander and members of the command staff can begin the ongoing process of determining resource needs. Resources can include any combination of personnel and equipment, the needs for which should be expected to evolve throughout the course of the incident. Having a prearranged system for understanding and even classifying the capabilities and limitations of available resources can help facilitate this constantly changing aspect of rescue.

Preplanning

Understanding scope, frequency, and magnitude of potential incidents will help the authority having jurisdiction (AHJ) better determine what response capabilities to request when sizing up an incident. Where applicable, capabilities should be in alignment with other relevant portions of NFPA 1006 and NFPA 2500.

The technical rescue disciplines recognized and identified by NFPA 1006, *Standard for Technical Rescue Personnel Professional Qualifications*, and NFPA 2500 (1670), *Standards for Operations and Training for Technical Search and Rescue Incidents and Life Safety Rope and Equipment for Emergency Services*, and the relevant chapters where they are addressed in each standard, are as follows (as of 2022, NFPA 1670 will be rolled into the new NFPA 2500 standard, so the chapter numbers noted above are as referenced in the 2022 edition of NFPA 2500):

- Tower rescue (NFPA 1006 Chapter 4; NFPA 2500 (1670) Chapter 23)
- Rope rescue (NFPA 1006 Chapter 5; NFPA 2500 (1670) Chapter 5)

- Structural collapse rescue (NFPA 1006 Chapter 6; NFPA 2500 (1670) Chapter 6)
- Confined space rescue (NFPA 1006 Chapter 7; NFPA 2500 (1670) Chapter 7)
- Vehicle rescue (NFPA 1006 Chapter 8; NFPA 2500 (1670) Chapter 8)
- Heavy vehicle rescue (NFPA 1006 Chapter 9; NFPA 2500 (1670) included in Chapter 8)
- Animal rescue (NFPA 1006 Chapter 10; NFPA 2500 (1670) Chapter 9)
- Wilderness rescue (NFPA 1006 Chapter 11; NFPA 2500 (1670) Chapter 10)
- Trench rescue (NFPA 1006 Chapter 12; NFPA 2500 (1670) Chapter 11)
- Machinery rescue (NFPA 1006 Chapter 13; NFPA 2500 (1670) Chapter 12)
- Cave rescue (NFPA 1006 Chapter 14; NFPA 2500 (1670) Chapter 13)
- Mine and tunnel rescue (NFPA 1006 Chapter 15; NFPA 2500 (1670) Chapter 14)
- Helicopter rescue (NFPA 1006 Chapter 16; NFPA 2500 (1670) Chapter 15)
- Surface water rescue (NFPA 1006 Chapter 17; NFPA 2500 (1670) Chapter 16)
- Swiftwater rescue (NFPA 1006 Chapter 18; NFPA 2500 (1670) Chapter 17)
- Dive rescue (NFPA 1006 Chapter 19; NFPA 2500 (1670) Chapter 18)
- Ice rescue (NFPA 1006 Chapter 20; NFPA 2500 (1670) Chapter 19)
- Surf rescue (NFPA 1006 Chapter 21; NFPA 2500 (1670) Chapter 20)
- Watercraft rescue (NFPA 1006 Chapter 22; NFPA 2500 (1670) Chapter 21)
- Flood rescue (NFPA 1006 Chapter 23; NFPA 2500 (1670) Chapter 22)

Although the focus of this text is specific to the discipline of rope rescue, NFPA 1006 and 2500 (1670) clearly suggest it is very likely for some of the disciplines overlap with one another. For example, the hazard identification may identify that there are cell tower sites within the response area where a rope rescue response may be required (**FIGURE 2-3**). In this case, the responders tasked to such an incident should possess the necessary levels of training outlined in the chapters related to both disciplines: rope rescue and tower rescue.

For maximum benefit in preplanning, the AHJ should refer to NFPA 2500 (1670) for training requirements at the organizational level, and to NFPA 1006 for individual job performance requirements. Together, these two documents comprise a well-rounded approach for the AHJ to take in developing rescue preparedness. NFPA 2500 (1670) helps the AHJ to identify what organizational capabilities are necessary—and to what degree—while NFPA 1006 describes individual qualifications required to perform at each identified level.

A

B

FIGURE 2-3 Identified hazards may include cell phone towers or water towers.
(A) © Aleks Viking/Shutterstock; (B) © Emily on Time/Shutterstock.

Identifying Appropriate Resources

A key part of scene size-up is recognizing which type(s) of responders are needed for a given incident, so the appropriate resources can be requested. This requires a baseline understanding of different types of incidents, as well as capabilities and limitations of resources within the response area, which is why pre-planning by the AHJ is important.

Even when a need for rope rescue appears obvious, considerations such as environment and other apparent risks may necessitate activation of multiple types of responders. When requesting additional resources for rope rescue, it is important to also note what other environmental conditions are present so that an appropriate response may be mustered. Environments often coupled with rope rescue incidents include the following:

- Buildings and structures
- Structural collapse incidents
- Wilderness/mountain
- Communications or utility towers
- Confined spaces
- Trenches or excavations
- Caves
- Mines and tunnels
- Lakes, rivers, and dams
- Flood areas

LAST

Especially where a response encompasses several different disciplines, organizing resources can be made more efficient by segmenting the response effort into phases. One common approach to segmentation is known by the mnemonic LAST, where each letter represents one phase of the response:

- L = Locate
- A = Access
- S = Stabilize
- T = Transport

Resources might differ for each phase of the operation, or the same personnel and equipment may be utilized across all phases. The amount of time, number of personnel, and equipment will vary depending on the situation. A well-staffed and thoroughly equipped organization may have an advantage in being able to assign one group of rescuers to prepare an upcoming phase of the rescue while the previous phase is still being accomplished by another team. Thinking of the operation in this phased approach can also be useful when considering utilization of external resources.

Obtain Information Required to Develop an Incident Action Plan

Every rescue operation should have an **incident action plan (IAP)**. The purpose of the IAP is to document and track the strategy and management of an operation. This plan should encompass search (if necessary) as well as rescue and should be considered a living document, evolving as the incident progresses.

The IAP should include the following:

- Operational goals and objectives
- Record of the agencies and resources involved
- Leadership roles and responsibilities
- Objectives by **operational period** and assigned resources
- Strategic and operational plans
- Transfer of authority (for multiple operational periods or jurisdictions)
- Demobilization plans

Information for the development of the IAP should be taken from whatever resources are available, starting with the incident size-up and encompassing information obtained from reporting parties, reference to any existing preplans, the **incident command system (ICS)** of the AHJ, standard operating procedures, experienced personnel, and leadership staff. The experience of other available personnel should not be overlooked or underestimated.

IAPs and the working within the ICS are important for rescuers, because most operations will be managed under some form of predefined and established ICS under the direction of the AHJ. An ICS defines roles and responsibilities of personnel and standard operating procedures to be used in the management and direction of emergency operations. Development and use of an IAP, and associated information about information management systems, are more thoroughly covered in Chapter 4 of this text, *Initiating a Response*.

Using Reference Materials

Rescue planning cannot be effectively performed in a vacuum. Existing knowledge, information, and resources should be studied and considered, both prior to and during an incident. Materials generally will fall into one of two categories:

1. Those that can be studied and researched in advance, as in preplanning
2. Those that are designed for easy reference during an incident

Item one includes resources such as textbooks (like the one you are reading now), regulatory documents and standards, and industry manuals.

Item 2 includes field guides. Field guides can be a bit difficult to use in the case of rope rescue operations, because full attention and all appendages are generally required when rigging. If anything, they are best used in checklist format or a visual reference to confirm safety, rather than in an instructional format during rigging. If any such reference tool is used, care should be taken to ensure that it is not dropped from height during the operation.

Identify Search Parameters

If a search will be required, the incident commander will need to determine where to search and what methods will be used. Generally speaking, the sooner a search can be initiated, the greater the chance of a quick and successful outcome. The sense of urgency with which a search should be approached should be relative to the risk. Factors that might increase the urgency of a search include the following:

- Age: A very young, or very old, subject warrants a higher urgency.
- Medical condition: Search for subjects who have known illnesses, especially those requiring medication, is more urgent.
- Weather: If hazardous weather is predicted, search should be approached with greater urgency.
- Experience/skill: Search for a subject who is inexperienced or suspected to be incapable of caring for themselves in the environment is more urgent.
- Equipment: An ill-equipped subject warrants a higher sense of urgency.

Defining the search area is an ongoing process. The area is likely to expand and shrink during the course of the operation. In the initial phases, the search area will consist primarily of the highest probability areas, then expand to a wider area, and subsequently narrow down to smaller subareas as clues are found and pursued. A good understanding of the search area as it develops will help search managers better assign responsibilities and monitor progress.

Land search, whether urban or wilderness, is a specialty area all its own, with many great resources available for further study. The information in this chapter offers only a small taste of the kinds of information available.

Lost Person Behavior

Lost person behavior is a study unto itself, and rescuers would do well to utilize available data in planning operations. In his book, *Lost Person Behavior*, Robert Koester identifies 41 different types of persons who might become lost, along with corresponding approaches to mounting a search—all the while acknowledging that 41 is probably not the final number.

Some key data points to consider in developing a search plan include the following:

- Mental status. Generally people who are depressed or suicidal do not travel far, and are likely to be found at the interface between two types of terrain.
- Age. Lost adults tend to stick to a trail, but may seek out a high point to get a view of their surroundings. They rarely backtrack.
- Children (15 years and younger): Young people behave quite differently from adults when lost. They are a poor judge of time, distance, and direction, but are inclined to look for something familiar (**FIGURE 2-4**).
- Activity. Hunters are often stranded by darkness or lost while tracking or field dressing game; most are content and prepared to build a shelter for the night and self-extricate in the morning. Berry pickers and mushroom hunters are also often lost while looking at the ground, but are less likely to be prepared to spend the night, and are more likely to keep wandering and become further lost.

All of this is to say that search is a unique field of study and a person who is interviewing an RP can have tremendous impact on the outcome of a search if they understand some of the basics of lost person behavior and tailor their questions accordingly.

Time of Day

The earlier a search is initiated, the more likely a positive outcome. However, the safety of rescuers must be considered when mounting a response. Many

FIGURE 2-4 Young people are inclined to look for familiar surroundings when lost.
© palidachan/Shutterstock.

backcountry lost-person reports come at or near the end of a day, which begs the question: Is it appropriate to search at night?

There is no single right answer to this question. It depends completely on circumstances, including risk to the subject, the training and capabilities of responders, risk to rescuers, and probability of success. At least 50 percent of searches can be resolved within 3 hours, most within 2 miles of the LKP. Postponing a search to the following morning derails this opportunity. That said, even teams who are trained and capable of night operations must weigh the risk of injury or disorientation of rescuers and trampling clues against the potential benefits of early containment or detection.

Initial Goals

The initial goals of any search operation are the following:

- Containment
- Attraction
- Hasty search

Containment

Containment efforts are intended to confine the movement of a lost subject in as close a search area as possible to the LKP. Methods are based on the assumption that a subject is unlikely to cross over a trail, road, river, or similar linear land feature. By establishing patrols along such routes, setting track traps, placing lookouts at high points with good visibility, and establishing perimeter controls, the likelihood of the subject wandering further is reduced. One method that is sometimes used is that of creating **string lines**, which are, quite literally, lines of string placed approximately waist high in areas where a subject might walk, with arrows pointing toward a road or trail. Taking early steps to prevent the search area from becoming enlarged will reap dividends later.

Attraction

In the time shortly after a subject goes missing, using attraction methods such as noise and lights from the LKP, or in other high-probability areas, can be especially useful for getting the attention of the subject. Common methods include vehicle horns, sirens, whistles, or even voice. At night, lights including emergency lights, vehicle lights, flashlights, and headlamps can be seen from long distances (**FIGURE 2-5**). In this way, a disoriented subject can be alerted to the

FIGURE 2-5 Lights and sirens may help to attract lost persons.
© Photo Spirit/Shutterstock.

whereabouts of help, and a mobile subject may even be able to walk out. When using attraction, be sure to pause and listen from time to time in case the subject is making noise in return.

Hasty Search

In the initial phases of a search, the places with the highest **probability of area (POA)** should be a priority. Simply put, POA is an estimation of the likelihood that a subject is in a given area. While there are complex computer models available for determining POA, in the initial phases of a search it is acceptable to base POA on "gut feel" of one or more officers leading the search. Some examples of areas that might commonly have high POA include the actual route(s) the subject is expected to have taken; points of interest such as buildings, vehicles, scenic overlooks, bathrooms, bodies of water, and other natural stopping points; and terrain features like trails, drainages, and terrain channels. Hasty search is generally performed with speed as a priority. While the prime goal of search is always to find the subject, searchers should also be on the lookout for relevant clues such as scent items, footprints, and discarded items. Care should be taken during hasty search to avoid damaging potential clues.

Detailed Search Methods

If initial search efforts are unsuccessful, more detailed search methods may need to be employed. As an area search expands, it is typically segmented into subareas and searched using refined methods that improve the **probability of detection (POD)**. POD refers to how likely it is that, if the subject were in the search area, they would have been found. POD may refer to either the subject themselves, or to clues.

Relay Information

Obtaining an accurate initial report from dispatch is just the beginning. Using a standardized format to collect appropriate information helps to ensure continuity of information, and will also help when the initial responder is providing size-up information to dispatch or other resources (**FIGURE 2-6**). No communication process is complete without also allowing opportunity for clarification and questions from the information recipient. Always provide a means for dispatch to reach back to you for additional details if needed.

When reporting size-up information to dispatch, at a minimum the following information should be relayed:

- Incident location and access route
- Number of subjects
- Additional people involved
- Suspected injuries
- Current rescue efforts
- Distances involved
- Anticipated hazards
- Reporting party and callback number for additional information

FIGURE 2-6 Use a standard format when relaying information.
© racheldonahue/E+/Getty Images.

Who–What–When–Where–Why–How

It is important to communicate key facts discovered during size-up to the command staff, including the typical Who–What–When–Where–Why–How group of questions, and should include the following:

- Who
  - Who is impacted?
  - Who is the subject and any details (number, age, gender, relevant information)
  - Who else is on scene with the subject?
- What
  - What happened?
  - What is the mechanism of injury/nature of illness?
  - What is the present situation?
- When
  - When did this occur?
  - When was the call for help initiated?
  - When did changes in subject condition occur?
- Where
  - Where is the incident (location)?
  - Where will rescuers access the subject (and distance)?
  - Where is the evacuation route?
- Why
  - Why is the subject incapacitated?
  - Why is help required?
  - Why is the situation dangerous?
- How
  - How will subject need to be moved?
  - How can responders obtain more information about the subject?

Using a consistent format and practicing verbalization of this information in training scenarios will help information to flow more accurately and succinctly in real situations.

After-Action REVIEW

IN SUMMARY

- Initial size-up sets the stage for a successful rescue.
- Ongoing size-up is a critical tool for ongoing success.
- The first unit arriving should take immediate responsibility for completing a preliminary size-up and assuming command of the incident.
- Determining the scope of the rescue is key in defining what the intended outcome of the operation will be and what resources will be required to complete the operation.

- Identifying the number of subjects will help determine to scope of the incident.
- Establishing the last known location of subjects may require interviewing the reporting party.
- Assessing resources needs begins with preplanning and includes identifying appropriate resources based on the needs of the community,
- Every response has the potential of needing to address all four phases of a rescue: Locate, Access, Stabilize, and Transport.
- The incident action plan should include:
  - Operational goals and objectives
  - Record of the agencies and resources involved
  - Leadership roles and responsibilities
  - Objectives by operational period and assigned resources
  - Strategic and operational plans
  - Transfer of authority (for multiple operational periods or jurisdictions)
  - Demobilization plans
- Once the last known location is determined, the search perimeter may be determined.
- The initial goals of a search operation are:
  - Containment
  - Attracting
  - Hasty search
- It is important to communicate key facts discovered during size-up to the command staff, including the typical Who–What–When–Where–Why–How group of questions.

KEY TERMS

Hazard identification Often referred to in context of risk assessment, this is an evaluation and analysis of environmental and physical factors influencing the scope, frequency, and magnitude of technical rescue incidents and the impact and influence they can have on the ability of the authority having jurisdiction to respond to and safely operate at these incidents.

Incident action plan (IAP) A written preplan that outlines key aspects of preparing for, responding to, and managing a given type of incident.

Incident command system (ICS) A standardized on-scene emergency management construct promulgated by the Federal Emergency Management Agency and specifically designed to provide for the adoption of an integrated organizational structure that reflects the complexity and demands of single or multiple incidents, without being hindered by jurisdictional boundaries.

Land search The search of terrain by Earth-bound personnel.

Last known point (LKP) This is the point where the person was last known to have been, with a positive identification. It might be a trailhead, hunting camp, boat dock, parking lot, etc. Some agencies use the term *point last seen* instead.

Mass-casualty incident (MCI) An emergency situation where the number of patients exceeds the capacity of available resources. Generally defined as at least three patients.

Operational period A period of time scheduled for execution of a given set of operational actions that are specifically specified in the incident action plan (IAP). Operational periods can vary in lengths, although do not normally exceed 24 hours.

Point last seen (PLS) This is the point on the map where the person was last spotted by a witness with a positive identification. It might be a trailhead, hunting camp, boat dock, parking lot, etc. Some agencies use the term *last known point* instead.

Probability of area (POA) The likelihood that a subject is in a given area or search segment. POA changes for each segment after any portion of the total search area is searched

Probability of detection (POD) The likelihood of a subject (or a clue) being found if it is in a given search area.

Reporting party (RP) The person(s) who report an incident.

Scope As pertains to search and rescue, the extent of the area or subject matter that is relevant to the incident.

Size-up The ongoing observation and evaluation of factors that are used to develop strategic goals and tactical objectives.

String line A search area containment method that utilizes lines of string to bound an area.

On Scene

1. You've been called to a park where a 7-year-old girl was reported to have gone for a walk with her grandmother 1 hour ago. The girl is now about one-third of the way from the top of a 40-foot rocky outcrop, clinging to a shrub, shrieking. As you assess the scene, what key points are you going to be considering? What questions will you ask?

2. How would you expect interviewing the reporting party for a rescue incident to differ from interviewing for a search incident?

3. A group of kids who often congregate at a nearby railroad bridge report that one of their friends was despondent and threatened to jump from the bridge 90 minutes ago. He has not been seen or heard from since. Whom would you interview, and what questions would you ask, to initiate a search effort?

4. Discuss what the mnemonic LAST means, and describe how you would apply it to at least two different incident types in your response area.

5. How specific does an incident action plan (preplan) need to be in order to be useful?

CHAPTER 3

Awareness

Hazards Associated with Rope Rescue

KNOWLEDGE OBJECTIVES

After studying this chapter, you should be able to:
- Explain the purpose of a hazard identification and risk assessment. (**NFPA 1006: 5.1.3**, pp. 34–35)
- Explain the purpose of a rescue preplan. (pp. 35–37)
- Identify the hazards associated with rope rescue incidents. (**NFPA 1006: 5.1.3**, pp. 35–39)
- Describe how to maintain situational awareness. (p. 39)
- Describe the considerations for maintaining an accurate perception of the rescue environment. (pp. 39–40)
- Identify the methods of managing the risks associated with rope rescue. (**NFPA 1006: 5.1.3**, pp. 37–40)
- Identify the methods of communication that may be used during a rope rescue incident. (pp. 40–41)
- Explain ways to ensure rescue teams create a shared mental model. (pp. 41–42)
- State the role of the incident safety officer at rope rescue incidents. (**NFPA 1006: 5.1.3**, pp. 43–44)
- Explain the importance of critical incident stress management. (pp. 44–46)

SKILL OBJECTIVES

There are no skill objectives for this chapter.

You Are the Rescuer

You are tasked with creating a rescue preplan for a cement materials processing/grinding plant in your jurisdiction. The plant has a cement mill, three silos for the materials, and high walkways.

1. What kinds of questions will you ask the company safety manager as you prepare to develop your preplan?
2. With whom will you share your preplan?

 Access Navigate for more practice activities.

Introduction

Rescue generally follows a precipitating event; someone has been injured, fallen ill, or gotten themselves into a dangerous predicament. Another term for a situation that poses a threat to life or property is an emergency. In the United States, Canada, parts of the Caribbean, and now Mexico, the public is programmed to respond to an emergency by dialing 9-1-1. In the member states of the European Union, it is 1-1-2. Other numbers are used in various countries throughout the world, but they all have one thing in common: People call these numbers with the expectation that a responding agency will send personnel who possess the training, tools, and skills to resolve the emergency.

While most people of sound mind are naturally inclined to distance themselves from a dangerous situation, an emergency responder is trained to enter the scene of an emergency for the specific purpose of providing assistance to others. Those who run toward, rather than away from, emergencies rely upon their training and experience to maintain their own safety, and that of others, during a response. Rescuers often feel a heightened sense of responsibility, fueled by expectations of the person(s) involved in the emergency, or even bystanders, to prioritize speed and action. Sometimes called Go-Fever, the term was popularized by the National Aeronautics and Space Administration (NASA) in relation to various incidents, including the Apollo 1 fire in 1967, the Space Shuttle Challenger disaster in 1986, and the Space Shuttle Columbia disaster in 2003. In short, **Go-Fever** speaks of a tendency toward hurrying to execute a plan or take action without giving full attention to the potential risks or errors because the drive to act overshadows clear thinking. Such commitment to a predefined course of action is often the result of not wanting to be perceived as uncaring or incapable of addressing a given situation, and may even be perceived as noble. Go-Fever has no place among emergency responders. As rescuers, maintaining your own safety is paramount, as an injured responder cannot offer assistance.

There are two elements to maintaining safety during an incident response: first, understanding the types of incidents to which the department may be called, and second, understanding the specific risks associated with each type of incident.

Hazard Identification and Risk Assessment

Aside from the obvious duties of fire fighters responding to fires, NFPA 1006, *Professional Qualifications for Technical Rescue,* identifies 20 different specialty rescue disciplines. For each of these, the authority having jurisdiction (AHJ) should conduct a **hazard identification and risk assessment** exercise within their designated response area to identify potential risks and develop corresponding response plans. The specialty rescue disciplines are discussed in Chapters 1, *Becoming a Rope Rescuer* and 2, *Size-Up*.

The phrase *hazard identification and risk assessment* is not redundant. Each of the elements carries a unique and specific meaning, and neither should be ignored. **Hazard identification** refers to an examination of *what* things cause danger, while **risk assessment** refers to making a determination of *how great* a danger is posed by the hazard.

A hazard identification and risk assessment for a community begins with an analysis of the surrounding area, its terrain and other features, and the people in it. Based on the hazards identified, and the level of risk associated with each, the AHJ should determine the scope, frequency, and magnitude of technical rescue incidents likely to occur in their area and develop the necessary resources to provide safe and effective response.

In short, this exercise seeks to answer the following questions:

1. **Scope:** What would an incident look like, how likely is it to occur, how big or far reaching would it likely be, and what rescue capabilities might be required? The scope of an incident is really just a big-picture view. Determining scope should take into consideration various locations where different types of incidents might be most apt to occur, how many people could be reasonably expected to be involved, what specifically might happen, how long it might take to resolve, and what response interventions will likely be needed. Methods of predicting the scope of an incident might include evaluation of the physical characteristics of the response area, demographic information, types of businesses and social venues in the area, and a review of historical response data.

2. **Frequency:** How often would such an incident be likely to occur (**FIGURE 3-1**)? Historical response data, geographic details, and demographic information may help to predict the frequency of incidence. For incidents that are likely to occur frequently, at different times of day or night, or even simultaneously, a higher number of resources may need to be trained and prepared for that type of response. Where a particularly high—or particularly low—frequency of incidence may be estimated, an **interagency agreement (IAA)**, **memorandum of understanding (MOU)**, or other working agreement with other agencies should be considered to augment response capabilities.

3. **Magnitude:** What would be the likely impact of such of an incident? Every jurisdiction has inherent risks, but the impact of a potential incident will vary with consideration to the potential for critical injury, loss of life or property, interruption to infrastructure, and the ability of the response agency to continue to provide services during a response. Response agencies should have a plan to ensure continued coverage within their jurisdiction even during a critical response.

Preparing for Hazards

All responders should possess at least awareness-level training in each discipline identified as a potential risk within their response area. That is, every responder should be able to recognize the need for technical rescue resources at an incident, identify what type of incident it is, initiate the appropriate response system for that jurisdiction, and secure the scene until additional resources arrive.

A **rescue preplan** should be developed by the AHJ for every hazard where a significant likelihood of incidence is identified. The preplan need not be an extensive or exhaustive detailed analysis, but should address common or unique factors that might aid in the development of technical rescue capabilities and reduce risks related to the types and scope of incidents most likely to occur. A preplan may include such things as a location map, access and egress information, details about hazards specific to that location, guidance on personal safety, anchorage identification, and suggested rescue procedures.

There is no magic formula to a preplan, nor is there a required format that must be used, but it should be

A

B

FIGURE 3-1 The agency's location will help determine the types of emergencies that occur. A mountainous area will have a greater number of recreational emergencies, while a flat, industrial area may have more workplace emergencies.
(A) © f11photo/Shutterstock; **(B)** © Allen Allnoch/iStock/Getty Images Plus/Getty Images.

documented in writing and then deliberately and objectively revisited at least annually. Where the need for rope rescue is identified, the AHJ should identify availability and training levels of personnel, as well as type, quantity, and location of equipment. Either internal or external resources may be utilized.

For incident types where a high likelihood is identified, a preplan will aid in preparation and in responding to the incident. A preplan can be generic in nature, as in generalized standard operating procedures for certain types of evacuation methods, or it can be site specific. A range of capabilities should be developed so that they may be applied to different circumstances as needed. In cases where multiple hazards overlap (i.e., rope rescue and confined space, rope rescue and tower rescue) or where there is an increased hazard or potential for incidence, more specific preplanning may be warranted (**FIGURE 3-2**). For example, if your response area includes a grain processing facility where rope rescue involving entrapment, confined space, and evacuation from heights is likely, advance collaboration with the proprietors can build mutual understanding and better preparedness on the parts of both the employer and responders.

A preplan should address interoperability with trained workplace personnel who may already be on-site, potential need for specialized equipment, and provision of medical care. These capabilities may come from within the response agency or through interagency cooperation. Interagency cooperation might involve various municipal agencies, or a combination of municipal and private entities, working together.

In the case of employers whose personnel are exposed to risks such as falls from heights, the Occupational Safety and Health Administration (OSHA) requires that the employer make provision for prompt rescue. The definition of "prompt" is grounded in the concept of being expedient enough to prevent further harm. For this reason, most employers will rely on some combination of internal resources and municipal responders. Municipal agencies and responders should recognize that the employer is striving to be in compliance with workplace safety regulations and consider how best to integrate the employer's preplan with their own.

While it is normal and prudent for an employer to activate local municipal response agencies in the event of an incident, typically an employer's rescue plan will also include provision for the fallen worker to attempt self-rescue, as well as some sort of assisted rescue attempt by assigned coworkers, to be undertaken while awaiting the arrival of local responders. In the case of hazardous workplaces, it is to the benefit of both the employer and the response agency to preplan for integration of resources when they arrive.

Better preparedness of responders equals better understanding of equipment used, access methods available, and specific risks and inherent hazards—all of which improve safety and efficiency. Preplanning also clarifies expectations in terms of anticipated response times, capabilities, and methods of interfacing once on scene. One example commonly found in nearly every jurisdiction is that of communications towers. Although most towers are not climbed often, they do require periodic access for maintenance; however, tower maintenance workers (**FIGURE 3-3**) rarely know

FIGURE 3-2 Responding agencies should train periodically in accordance with preplans, especially for the low-probability/high-risk incidents that could occur.
© AlenaPaulus/E+/Getty Images.

FIGURE 3-3 Unique environments such as towers pose special hazards and warrant training events with local employers.
© Kevin Steele/Aurora Photos/Cavan Images/Alamy Stock Photo.

in advance which towers they will access, or when, exactly, the climb will occur. The fact that most tower and antennae maintenance companies are trained for coworker-assisted rescue is a blessing to response agencies, as numerous unique hazards can be present on these sites. That said, response agencies who coordinate in advance and preplan with tower site owners in their jurisdiction will have greater familiarity with hazards, anchorages, and rescue methods appropriate to each unique site before an incident occurs. The same principle holds true for other workplaces where fall hazards might exist, such as factories, construction sites, grain and material processing sites, shipyards, etc.

Hazards Associated with Rope Rescue

Rescue Incident Safety Control Chart

While a community hazard identification and risk assessment exercise takes place in advance of an emergency to try to identify what kinds of responses may become necessary, there is another type of hazard identification and risk assessment that should take place at the onset of any actual rescue operation, be revisited throughout the incident, and utilized in planning and operations. The **rescue incident safety control (RISC) chart** is a simple, visual tool to help identify specific hazards at a rescue site, acknowledge specific associated risks, and visualize strategies for contending with them (**FIGURE 3-4**). For every identified risk, it should be identified who, specifically, is potentially at risk; what would trigger the risk; what efforts can be made to mitigate the risk; and what defense is to be provided for protecting against risks that slip through the mitigation attempts.

Some of the more obvious hazards associated with rope rescue are those related to gravity. Rope rescue operations typically take place in the vertical realm, where steep-to-vertical terrain presents difficulty in accessing a patient, risk of falling, and potential for dropped objects. With this in mind, one entry in the RISC chart might be as shown in (**FIGURE 3-5**).

Specific Hazards

Access to a subject who may require rope rescue can be made more difficult by poor footing, lack of entry points to the location of the subject, and natural or manmade obstacles. Rescuers should avoid entering terrain or environments where they are not sufficiently prepared to assess, mitigate, and defend against specific hazards. Where possible, access to the subject should be made from an adjacent location, rather than above or below, to minimize risk of dropping or knocking things onto the subject and of the rescuer being struck by dropped objects. When access must be made from above or below, rescuers should make every effort to stay out of the fall line (directly above or below the subject or other rescuers) as much as possible.

Limited entry points often result in at least some delay in getting a sufficient number of rescuers to the subject. In such cases, rescuers should avoid the

	Risk	Persons	Trigger	Mitigation	Defense
Risk #1					
Risk #2					
Risk #3					
Risk #4					

FIGURE 3-4 A sample rescue incident safety control chart.

	Risk	Persons	Trigger	Mitigation	Defense
Risk #1	Dropped objects	Rescuers in HOT zone Subject(s) Bystanders who get too close	1. Debris dislodged by persons above 2. Debris dislodged by moving rope	1. Establish no-entry zone above 2. Assign security to keep bystanders away	• Rescuers and subjects all wear helmets, eye protection at all times • Assign spotter to make verbal warning as needed

FIGURE 3-5 A sample completed RISC chart.

temptation to create alternate routes or take greater risks in accessing the patient, as this could increase the potential for dropped objects and fall hazards. Where access is limited, evacuation routes are, most probably, likewise limited. Begin scouting evacuation routes from the outset when accessing a subject in a particularly difficult location.

Protection Against Falls and Other Hazards

Awareness-level rescuers should be provided with appropriate personal protection against falls and other hazards. The level of protection provided should be at least consistent with what would be expected for a worker to make access or perform similar work in that terrain and environment. To be clear, this is to say only that the rescuer must be adequately protected—not that they must use exactly the same methods used in industry. Fall protection equipment used in construction or general industry may not be appropriate for use by emergency responders. For example, workers often use vertical lifelines with rope grabs, self-retracting lanyard devices, or even catch-nets for protection against falls, whereas manual belays are a much more common and appropriate means of fall protection for technician-level rope rescuers. Although methods of protection may vary for emergency responders, the degree of protection should be equivalent to that which a worker in the same environment might expect.

For example, industrial fall protection regulations may sometimes stipulate that falls on a front attachment point be limited to 2 feet or less, but more important than distance is impact force. When a rescuer is entering a hazard, these same principles should apply. Whatever equipment or device is used, the potential impact force should be limited to 1800 lbf **maximum arrest force** (MAF), and 900 lbf **average arrest force**.

When working in steep or vertical terrain, rescuers must remain vigilant at all times for items falling from above (**TABLE 3-1**). Prevention and avoidance provide

TABLE 3-1 Common Hazards in Technical Rescue Tactical "Watch Outs"

Hazard	Corrective Action
Dropped objects	Wear a helmet; secure loose objects.
Falls from height	Always maintain adequate fall protection.
Lack of awareness of hazards	Perform operational briefing, to include hazard awareness.
Complacency	Team members are encouraged to speak up when they observe a safety deficiency.
Ineffective communications	Give briefings in person and aggressively maintain incident communications with all personnel.
Failure to protect against sharp edges	Aggressively deploy edge protection.
Misuse and inattention to equipment; improperly tied rigging	Repeatedly examine rigging, both visually and tactilely, during rescue operation.
Cross-gate forces on carabiners	Rig correctly and recheck rigging frequently.
Carabiner gates unlocked or held open by webbing or rocks	Visually check rigging frequently.
Loose Prusiks	Inspect Prusiks visually and tactilely.
Rescuer fatigue, thirst, boredom, and distraction	Change out the belayer, rehydrate often, and promote disciplined behavior.
Environmental injuries	Wear appropriate hand, eye, and body protection.

the first line of defense against dropped objects. Helmets provide some degree of protection, but they should not be considered an adequate substitute for prevention and avoidance. Loose equipment should be secured using tethers or, for heavy objects, a separate rope system.

Additional Hazards

In addition to height-specific hazards, it is quite likely that additional hazards may be present. These might include environmental conditions such as climate (temperature, humidity, etc.), exposure to hazardous substances (air quality, nearby toxins, etc.), or physical hazards (machinery, natural disasters, structural instability, etc.). Such hazards can be difficult to assess, and they have a tendency to change rapidly. Rescuers can help protect themselves against changing hazards by maintaining good situational awareness.

Situational Awareness

Situational awareness is most simply defined as the ability to know what is going on around us. For situational awareness to be of use to emergency responders, this awareness must be coupled with the ability to quickly and accurately assess and utilize the information to make good decisions and take appropriate actions.

Research abounds on situational awareness in the field of human factors engineering, and the military community, in particular, has dedicated much attention to the topic. Although municipal emergency response situations may not always be as high stress as military operations, in both there is a shared sense of urgency and a propensity for numerous distractions. Situational awareness goes far beyond simply being aware of numerous bits of information, and it is a skill that can be honed and developed at different levels to support operational objectives.

There are three different levels at which situational awareness can be developed:

- **Level 1.** This is the ability to accurately recognize the moving parts in the current situation.
- **Level 2.** This is the ability to understand the moving parts in the current situation.
- **Level 3.** This is the ability to understand how the moving parts will shift during the situation and anticipate how to address the changes.

Human factors are perhaps the most difficult part of technical rope rescue to train, yet arguably the most important. Equipment failure in rope rescue is extremely rare; most operational accidents are attributed to human error. All levels of rope rescuers must learn to maintain an accurate perception of the rescue environment, and to react appropriately to any problems that might arise.

Risk Management

Risk is a probability that can be assessed, calculated, and even managed. Managing risk is a matter of managing the points at which hazards intersect with potential vulnerabilities. Two areas where rescuers can make the most difference include susceptibility (the proximity or likelihood of being exposed to the hazard) and resilience (the mitigation of the effects, or ability to maintain safety within the hazard).

As a simple example, rope rescuers are inevitably exposed to the hazard of rope burn when working with moving ropes. A two-fold approach to managing the risk associated with rope-burn might be as follows:

1. Reduce susceptibility by managing friction to ensure ropes move slowly
2. Increase resilience by wearing gloves with double or triple layers of protection in the palm, where heat buildup is most likely

This same concept of risk management—focusing on reducing susceptibility and increasing resilience—may be applied to all hazards, large or small.

Risk management became a focus of the U.S. Coast Guard (USCG) when, between 1991 and 1993, the organization experienced four major Search and Rescue (SAR)–related tragedies in the marine environment. As a result, in 1996 the USCG executed a systematic process to assess and manage risks continuously, known as **Operational Risk Management (ORM)**. In ensuing years, the U.S. Navy evolved the concept to a system now known as **Time Critical Risk Management (TCRM)**. This system is designed with the understanding that risk management is especially challenged when time and resources are limited.

In the context of TCRM, risks are identified and controlled using an established set of key operational policies and procedures throughout the evolution of an incident. The principles underpinning TCRM mimic those identified in the original ORM models:

1. **Accept risk when benefits outweigh the cost.** The ability to weigh risks against opportunities and benefits helps to maximize unit capability. Even high-risk endeavors may be undertaken when decision makers acknowledge that the sum of the benefits exceeds the sum of the costs.

2. **Accept no unnecessary risks.** While all SAR operations inevitably entail some level of risk, "unnecessary risk" is that which inappropriately increases the mission's risk level. In other words, rescuers are directed to expose personnel and resources to the lowest possible risk while accomplishing the mission.
3. **Anticipate and manage risk by planning.** Risks associated with rescue response can change rapidly during the mission. Preplanning provides a frame of reference and can prepare responders in advance with an ongoing strategy relevant to changing conditions.
4. **Make risk decisions at the right level.** Decision-making authority is allocated at varying degrees to personnel in accordance with levels of responsibility. Every responder must remain aware of their own limitations and know when to refer a decision to a higher level.

Research has shown that a breakdown often occurs with operational risk management in emergency situations due to the unique stresses that responders face. It is impossible to accurately foresee or predict all possible variations in a mission, yet meanwhile there is often life-and-death pressure to effect a positive outcome. To help responders better analyze and apply risk management principles during the course of an operation, the U.S. Navy developed a simple ABCD mnemonic **TABLE 3-2**.

In recent years, forest service, fire service, military, and others have integrated these and other principles into their incident risk management processes.

TIP

Remember your priorities for operational safety:
- You are number ONE!
- Your fellow rescuers and the public are your SECOND concern.
- The subject is your THIRD priority.
- Bystanders are a distant FOURTH, and are best directed to remove themselves from the area.

Safety is of paramount importance at all times. If you see any action that is unsafe, your responsibility is to speak up! Remember that no one is infallible, including you. The worst-case scenario is having a rescuer injured, resulting in two victims. Plan and execute a safe response so that you do not create an incident within an incident.

Modified from Klein, G. (1999). *Sources of power: How people make decisions* (2nd ed.). MIT Press.

TABLE 3-2 ABCD Mnemonic for TCRM

Assess the Situation	- What's the mission? - What are the hazards/risks/controls? - What will happen next? - Situational awareness!
Balance Resources and Options	- What personnel are available, and at what level of training? - What equipment resources do I have? - How does time need to play into my plan? - What's the best use of my resources?
Communicate Intentions	- Update your team on the plan—up and down the chain. - Foster open communication. - Consider barriers to good communication. - Who can help? Who can provide backup?
Do it	- Execute the plan. - Manage change as it occurs. - Revisit ABCD throughout. - Revise as needed.
(. . .and then **D**ebrief, using ABCD)	- **A**ssess how things went. - How well were resources **B**alanced? - How was the **C**ommunication? - What would you **D**o differently next time?

Communicating for Risk Management

Emergency responders rely on one another for safety under duress in all sorts of situations. While technical skills are of prime importance, the importance of communications cannot be overemphasized. The two parts of communications that should be honed are communication methods and communication skills. Methods often used for communication in rope rescue include verbal communication, to include radio, and hand signals.

Verbal Communications

Verbal communications are generally used for briefings before and during an incident, dialogue regarding operational issues, discussion of concerns, and perhaps even commands during the rope rescue evolution. Limiting communications to that which are necessary, and keeping chatter to a minimum, is generally a good rule that contributes positively to safety. Unless it pertains to the mission, it probably does not need to be said, and even if it does pertain to the mission, make sure it needs to be said before you say it. Information should be stated clearly, firmly, and loudly enough to be heard, but not necessarily shouted.

Radio communications are often the most appropriate and necessary means of transmitting information. In this case, it is even more important to avoid unnecessary chit-chat. Radio communications are, at best, more difficult to understand, and by nature the broadcast is distributed to everyone in radio range (**FIGURE 3-6**). This is potentially distracting and can create confusion when out of context. Every transmission also draws against available battery power of every receiving unit. That said, when radio communications are necessary, speak clearly and concisely, using plain language as much as possible. Especially when communications are urgent, the risk of misunderstanding is increased when codes are used.

FIGURE 3-6 Radio communications should be concise.
Courtesy of Ken Phillips.

Hand Signals

Hand signals can be useful in loud environments where voice or radio communications are not effective, or where there is some reason for not wanting information broadcast to a wide audience. Hand signals must be preplanned and agreed upon in advance to be most effective.

Clear Communication

Of course, other communication methods may be used in the rescue environment; the important thing is to remember to keep communications clear and concise. While emphasis is placed here on avoiding unnecessary chit-chat, this does not mean that team leaders or members should be secretive or withhold information. Information pertaining to the mission should be shared readily, as rescuers can function more effectively and safely if they are thoroughly informed.

Rescue leaders, especially, should ensure that information is clearly understood. Communications regarding situation, strategy, tactics, and hazards are essential to safety as well as operational success, so do whatever it takes to ensure that information is provided and that it is understood. Likewise, the recipient of a communication bears some responsibility for ensuring that their understanding of the information is accurate. When you are given instructions and assignments, repeat the information to ensure that it is understood.

Creating a Shared Mental Model

Per a 2014 study by McComb and Simpson, the concept of shared mental models has been long used in health care to improve patient outcomes. By ensuring that all members of a team have a common understanding of the task and what needs to be done to achieve that task, the mission is clarified and collaborative efforts are improved. Translating this concept to technical rescue, teams can improve performance by ensuring the following of team members:

1. They are familiar with one another's roles and responsibilities.
2. They are able to anticipate the needs of other team members.
3. They are able to adapt to changing needs.

In healthcare collaboration, it has been found that using a shared mental model approach helps to facilitate teamwork by breaking down boundaries and reducing the perception of power differentials between regular responders and outside specialists. Verbal acknowledgments and updates on situational status and goals help to prevent individual team members going off on counterproductive tangents, such as a rescue leader starting an evacuation before the system or necessary personnel are ready.

Creating a foundation for a shared mental model begins with a good team briefing.

Team Briefings

Even before the rescue operation begins, responders should be briefed by the incident commander, the operations officer, or a strike team leader, depending on the size of the incident and number of rescuers. This thorough but concise briefing should include information relevant to the operation so that when the action starts, everyone is functioning from the same baseline. The team briefing should include at least the following topics:

- Subject situation
- Environmental conditions
- Safety concerns
- Goal of the operation
  - Individual roles and responsibilities
  - Anticipated plan
  - Identification of known tasks
- Review foreseeable "what-if" scenarios
- Allow team members to ask questions and gain clarification.

FIGURE 3-7 Communications should be direct within teams.
© AlenaPaulus/E+/Getty Images.

stating the obvious rather than risking misunderstanding or missed information. This can be more difficult than it sounds, as in tightly knit teams, members can often anticipate one another's actions and expectations, even without words (**FIGURE 3-7**). One example of this might be an experienced technician-level rope rescuer assuming that the person setting the anchor has oriented it with the anticipated evacuation route. There is a tendency, especially in high-stress situations, for experienced team members to gravitate toward thinking that others are thinking the same as they are. Delays and poor outcomes can be avoided with simple ongoing dialogue, using direct statements such as "John, while I am... I would like for you to be..."

> **TIP**
>
> **Briefing Format for Emergencies**
> 1. Here's what I think we face.
> 2. Here's what I think we should do (assignments, communications, and contingencies).
> 3. Here's why.
> 4. Here's what we should keep an eye on.
> 5. Now, talk to me.
>
> Modified from Klein, G. (1999). *Sources of power: How people make decisions* (2nd ed.). MIT Press.

> **TIP**
>
> **Six components of direct statements:**
> 1. Address the person to whom you are talking by name.
> 2. Begin with "I," as in "I would like" or "I believe."
> 3. State your message as clearly as possible.
> 4. Use the appropriate tone of voice for your message so that it is delivered as you intended.
> 5. Require a response.
> 6. Do not disengage with the other person until you have achieved an understanding.

Explicit Communication

Communications during emergency operations should be made clearly and overtly, leaving no room for ambiguous interpretation. Team members should err toward

One effective means of ensuring clear communication during an incident is to use direct **closed-loop communication**. In this method, the person requesting an action makes their request directly and explicitly, and then the person on the receiving end acknowledges the request and, if applicable, states when it has been completed.

Example:

Sarah: Jim, I would like for you to set an anchor for a low-angle litter evacuation in direct line with the subject, OK?

Jim: OK, I will set an anchor for a low-angle evac in direct line with the subject.

Jim: (a few minutes later) Sarah, the anchor is set and ready for low-angle evacuation of the subject in direct line with his present location.

Preceding an instruction with the name or identity of a recipient ensures that the request is not made to empty space or lost in the ether. When possible, making eye contact can also help ensure clarity and confirm understanding. It may sound elementary, but a direct statement gets the person's attention and forces the individual to deal with your concern rather than allowing them to ignore your message, and closed-loop communication ensures that requests have been heard and understood. This approach also provides opportunity for clarification if needed, removes ambiguity, and contributes toward developing a shared mental model for everyone involved.

Regardless of methods used, all rescuers, not just team leaders and managers, would do well to brush up on the interpersonal relations parts of communication skills. Two foundational concepts upon which good emergency communications rest are as follows:

1. Maintain good communications with leadership, team members, and others involved.
2. Provide clear instructions and ensure they are understood.

Incident Safety Officer

A discussion of hazards associated with rope rescue is not complete without at least some reference to the **incident safety officer (ISO)** (**FIGURE 3-8**). This position is an identified Command staff function within the Incident Command System used by many organizations, with the purpose of directly supporting the **incident commander (IC)** and contributing to the overall management of the incident.

The ISO is an operational necessity during a potentially hazardous activity. Rope rescue, by its nature, is hazardous. Therefore, every rope rescue operation should have an assigned ISO. The ISO immediately performs a reconnaissance of the incident for hazards and provides the incident commander with a risk assessment by evaluating the scene conditions, rescuer activities, and incident operations. The ISO can initiate actions to mitigate possible hazards—for example,

FIGURE 3-8 The incident safety officer functions at the level of the incident commander.
© Military PCF/Alamy Stock Photo.

through apparatus placement or use of personal protective equipment.

The ISO is responsible for making sure that all rescuers at a scene know the established safety zones and hazards, and the ISO has the authority to stop any operation they see as a threat to safety. The ISO must also be alert to transmission barriers that could result in missed, unclear, or incomplete communication. Selection of an ISO during rope rescue should be based on the individual's technical expertise in rope rescue procedures.

The ISO functions at the level of the IC. Optimally, this dedicated position means that the ISO does not get involved in other activities, such as operations, so that they can stand back and be objective about any threats to safety. An ISO should be present at every geographic location or sector where a potentially hazardous operation is taking place.

It is essential that the ISO be an individual experienced in rescue who thoroughly understands equipment, rigging, and rescue techniques. If possible, they should not be pulled into the operation itself. With a small team response, however, the ISO position may have to be a collateral duty for a team member.

The ISO monitors the following:

- Personal safety equipment, including helmets, gloves, and breathing apparatus
- Safe work practices for specialized equipment, such as helicopters and watercraft
- Rescue rigging, including belays, anchors, rope padding, and unlocked carabiners
- Personal rigging, including harnesses, locked carabiners, and correct tying of knots
- Identifiable hazards, including hazardous atmospheres and falling debris
- Fatigue, hydration levels, and environmental exposure

Effective ways to reduce the risk to rescue personnel include the following:

- Effective training that simulates conditions encountered in emergencies
- Rest and rehabilitation of emergency personnel
- Continuous evaluation of changing conditions
- Reliance on the accumulated experience of team members

These safety principles should also be applied to training drills and exercises that involve hazards similar to those encountered at actual emergencies.

TIP

Dirty Dozen of Human Errors

1. Lack of communication: A failure to exchange information
2. Complacency: Loss of awareness and the development of overconfidence
3. Lack of knowledge: Lack of experience or training in the task
4. Distraction: Anything that takes your mind off the job
5. Lack of teamwork: Without teamwork, we are only a group of individuals involved in a similar task.
6. Fatigue: Considered to be the number one contributor to human error
7. Lack of resources: Insufficient or not fully operational equipment and manpower to safely perform a task
8. Pressure: External and self-imposed psychological pressure
9. Lack of assertiveness: Failing to speak up when things do not seem right
10. Stress: Being overwhelmed by stress leads to human error.
11. Lack of awareness: A lack of alertness and vigilance in observing; failing to ask the "what if?" question
12. Norms: The "normal" accepted way things actually are done in an organization, regardless of whether their practices are valid and safe. Unaccepted practices occur without corrective action.

Reproduced from Dupont, G. (1997). The dirty dozen errors in maintenance. *11th symposium on human factors in aviation maintenance*, San Diego, CA.

TIP

The following guidelines can help ensure the safety of those participating in the rescue operation:

- Do not rush! Maintain a sense of "controlled urgency."
- Provide thorough briefings. Use standard communications terminology and techniques.
- Choose well-trained, experienced rescuers for the core of the team. The most highly skilled rescuers should be used in decision-making roles and where technical competence is absolutely essential. Less-experienced or less-skilled personnel can easily function as low-angle litter bearers, hauling team personnel, or equipment runners.
- Make sure rescuers are prepared for contingencies and have adequate personal gear to stay warm, dry, fed, and well hydrated.
- Establish a well-marked hazard (exclusion) zone at a cliff edge (**FIGURE 3-9**). Consider using flagging or a chalk line, as well as chemical light sticks at night, to identify this line.
- Make sure that rescuers are tied in when working within 6 feet (1.8 m) of an exposed edge. Establish a marked safety perimeter with flagging, chalk, or chemical light sticks.
- Minimize the number of personnel working near an exposed edge.
- Designate an ISO (this may be a collateral role for a rescuer).
- Perform a safety check before using a system; recheck equipment when in use.
- Engineer a redundant system; rescue systems have backups.
- Aggressively use appropriate personal protective equipment (e.g., gloves, footwear, helmet, harness, personal flotation device, hearing protection, Nomex clothing, safety goggles, sunscreen) for all incident hazards, environments, and tasks. Have spare equipment available. Wear helmets in the rock-fall zone.
- Use edge protection to protect lines from sharp edges.
- Establish safety lines where exposure places personnel at significant risk of injury.
- Make a point of not standing inside a bight under tension during a raising operation. Take steps to prevent personal injury in case a change of direction fails in a hauling system. Stand to the outside of the rope bight.
- Never get on a rope without adequate gear to go up the rope and down the rope. If you are rappelling, make sure you have ascenders with you.
- Inspect gear before stowing it away.
- Secure loose gear in a cache adjacent to the rescue operations area.
- Keep a Prusik and trauma scissors with you to handle a self-rescue situation.
- Make sure to avoid cross-gate forces and three-way forces on carabiners.
- Derig gear after the rescue, starting at a cliff edge and working away from it.

Modified from Phillips, K. (2005). *National park service technical review manual* (10th ed.). Author. http://kristinandjerry.name/cmru/rescue_info/National%20Park%20Service/Basic%20Technical%20Rescue%20-%20Phillips.pdf.

Stress Management

Over the years, it has been determined that some of the most damaging hazards are the long-term psychological impacts that extend beyond the actual rescue.

READY (green)	REACTING (yellow)	INJURED (Orange)	ILL (red)
Definition • Optimal functioning • Adaptive growth • Wellness **Features** • At one's best • Well trained and prepared • In control • Physically, mentally, and spiritually fit • Mission focused • Motivated • Calm and steady • Having fun • Behaving ethically	**Definition** • Mild and transient distress or impairment • Always goes away • Low risk **Causes** • Any stressor **Features** • Feeling irritable anxious, or down • Loss of motivation • Loss of focus • Difficulty sleeping • Muscle tension or other physical changes • Not having fun	**Definition** • More severe and persistent distress or impairment • Leaves a scar • Higher risk **Causes** • Life threat • Loss • Moral injury • Wear and tear **Features** • Loss of control • Panic, rage, or depression • No longer feeling like normal self • Excessive guilt, shame, or blame	**Definition** • Clinical mental disorder • Unhealed stress injury causing life impairment **Types** • PTSD • Depression • Anxiety • Substance abuse **Features** • Symptoms persist and worsen over time • Severe distress or social or occupational impairment
Leader Responsibility	Individual, Shipmate, Family Responsibility		Caregiver Responsibility

FIGURE 3-9 Stress continuum.
Modified from USMC. (2016, May 2). *Maritime combat and operational stress control (MCTP 3-30E)*. https://www.marines.mil/Portals/1/Publications/MCTP%203-30E%20Formerly%20MCRP%206-11C.pdf?ver=2017-09-28-081327-517

Critical incident stress management (CISM) describes the practice of assisting those who have been involved in a critical incident to share their experiences and perspectives, give release to their emotional response, and find ways to deal with the experience. The issue was first recognized in military applications, and subsequently the principles were applied to civilian emergency responders. A formalized approach to critical incident stress debriefing for paramedics, fire fighters, and law enforcement was developed by Jeffrey T. Mitchell, PhD in the mid-1970s, but more recent research suggests that simply debriefing does not always improve outcomes. It has also been learned that stress injuries go far beyond a specific incident, and are the result of collective influences over time and across all aspects of life.

The U.S. Marine Corps and the U.S. Navy have placed a great deal of emphasis in recent years on psychological stress injuries, and have developed a Maritime Combat and Operational Stress Control program to support personnel. This program has been emulated by other military and civilian programs, with varying degrees of success. A foundational element of this program is that psychological stress is seen not as an isolated event, but as a continuum. Every rescuer—and every person—is always at some point on this continuum, moving back and forth between the stages. These stages include the following:

- Ready: This is a stage of wellness, characterized as times when a person is functioning at their best.

- Reacting: At this stage, there is a low level of stress. It may be caused by any stressor, but is relatively mild in nature and always goes away.

- Injured: Initiated by ongoing wear and tear, loss, or a life threat, this is a mode of distress that generally calls for intervention.

- Ill: Resulting from persistent injury, in this stage a person's life is significantly disrupted and quality of life is impaired by the effects.

The goal is for every person to maintain status at the healthier end of this spectrum—that is, as close to Stage 1 as possible. Reality is that the stuff of life pulls us constantly toward the other end, and emergency workers must be particularly cognizant of this. Self-care is an important part of healthy stress management. At the third and fourth stages of the stress continuum, coworkers and family members play an important role in helping a person to recognize and seek appropriate care. In times of crisis and near crisis, psychological stress intervention is particularly critical.

The goals of psychological stress intervention are as follows:

1. Promote a sense of safety
2. Promote calming
3. Promote connectedness
4. Promote a sense of self and collective efficacy (competence)
5. Promote confidence and hope

Again, these are actions that rescuers and family members can take to support one another on an ongoing basis. At times of crisis, it will be necessary to focus more on the basic needs like safety and calming, while times of good health may be conducive to building competence and confidence.

After-Action REVIEW

IN SUMMARY

- The authority having jurisdiction (AHJ) should conduct a hazard identification and risk assessment to identify potential risks and develop corresponding response plans.
- The hazard identification and risk assessment should determine the scope, frequency, and magnitude of potential incidents.
- A rescue preplan should be developed for each potential incident identified during the hazard identification and risk assessment.
- A rescue incident safety control (RISC) chart is a visual tool to help identify specific hazards at a rescue site, acknowledge specific associated risks, and visualize strategies for contending with them.
- Specific hazards at rope rescue incidents include the potential to fall and limited entry points.
- Situational awareness is the ability to know what is going on around us and must be coupled with the ability to quickly and accurately assess and utilize the information to make good decisions and take appropriate actions. It is essential to maintain situational awareness to assess and mitigate hazards.
- Managing risk is a matter of managing the points at which hazards intersect with potential vulnerabilities.
- A mindset of ongoing risk–benefit analysis is an asset to any rescuer.
- Methods used for communication in rope rescue include verbal communication, radio communication, and hand signals.
- A pre-briefing ensures that all rescuers know:
  - Subject situation
  - Environmental conditions
  - Safety concerns
  - Goal of the operation
  - Review foreseeable "what-if" scenarios
  - Allow team members to ask questions and gain clarification.
- Clear communication is essential during an incident and depends upon using direct closed-loop communication.
- The incident safety officer (ISO) is a command level position and is responsible for making sure that all rescuers at a scene know the established safety zones and hazards. The ISO has the authority to stop any operation they see as a threat to safety.
- It has been determined that some of the most damaging hazards are the long-term psychological impacts that extend beyond the actual rescue. Take active measures to decrease stress, just as you would take active measures to don personal protective gear.

KEY TERMS

Average arrest force The average amount of force that a person using a fall arrest system will experience over the duration of the catch.

Closed-loop communication A direct communication style in which a person requesting an action makes their request directly and explicitly, and then the person on the receiving end acknowledges the request and, if applicable, states when it has been completed.

Critical incident stress management (CISM) A program designed to reduce acute and chronic effects of stress related to job function (NFPA 450).

Go-Fever The response to take action without thoroughly considering potential risks or errors, to the extent that the drive to act overshadows clear thinking.

Hazard identification An examination of *what* things might cause danger.

Hazard identification and risk assessment Intentional exercise to identify all potential hazards within a given context, and then to assess the likelihood and possible severity of injury or harm.

Incident commander (IC) The person responsible for managing an emergency response, including (but not limited to) developing incident objectives, assigning other roles and duties, and applying available resources to resolve the incident.

Incident safety officer (ISO) An individual assigned by the incident commander (IC) to help identify hazards and manage risk during an operation.

Interagency agreement (IAA) A legal instrument used between two response agencies to formalize the terms under which resources might be shared or exchanged.

Operational Risk Management A systematic process used to assess and manage risks continuously during an operation.

Maximum arrest force The largest amount of force that a person using a fall arrest system will experience as a result of being caught by the safety system.

Memorandum of understanding (MOU) A formal agreement between two or more parties that outlines conditions or terms of a mutual agreement. Typically not legally binding.

Rescue incident safety control (RISC) chart A visual tool to help incident managers and responders to identify specific hazards at a rescue site, and to visualize strategies for contending with them.

Rescue preplan An emergency response plan that is prepared prior to an incident occurring, based on a concept and generalized parameters.

Risk assessment A determination of *how great a danger* is posed by a given hazard.

Situational awareness The ability to know what is going on around us.

Time Critical Risk Management (TCRM) An Operational Risk Management system that takes into account the understanding that risk management is especially challenging when time and resources are limited.

REFERENCE

McComb, S, Simpson, V. The concept of shared mental models in healthcare collaboration. *Journal of Advanced Nursing* 70 (2014): 1479–88.

On Scene

1. What types of environments exist in your jurisdiction that might lend themselves to rope rescue? How frequently are these likely to occur?
2. What kinds of risks can you imagine that might affect rescuers who are performing rope rescue in the environments identified in #1?
3. What distractions are likely to create additional hazards for rescuers during rope rescues?
4. Discuss the five goals of psychological stress intervention, and some possible methods to achieve them.

Chapter Opener: © Jones & Bartlett Learning. Courtesy of Loui McCurley; On Scene siren: © Bildgigant/Shutterstock.

CHAPTER 4

Awareness

Initiating a Response

KNOWLEDGE OBJECTIVES

After studying this chapter, you should be able to:

- Describe how to apply a preplan to the incident action plan. (pp. 50–51)
- Explain the initial rope rescue response actions that can be taken by an awareness-level rope rescuer. (**NFPA 1006: 5.1.4**, p. 51)
- Identify the signs that indicate an incident requires rope rescuers. (**NFPA 1006: 5.1.4**, p. 51)
- Describe how the emergency response system is activated. (**NFPA 1006: 5.1.4**, pp. 51–52)
- Describe the purpose and components of the Incident Command System. (**NFPA 1006: 5.1.4**, pp. 52–54)
- Describe the methods of isolating hazards and hazardous areas. (**NFPA 1006: 5.1.3**, pp. 54–55)
- Describe the capabilities of operations- and technician-level rope rescuers. (**NFPA 1006: 5.1.4**, pp. 56–57)
- Define the Planning P. (**NFPA 1006: 5.1.4**, pp. 57–59)
- Identify the forms that can be utilized to build and implement an incident action plan. (**NFPA 1006: 5.1.4**, pp. 57–59)
- Describe the debriefing purpose and potential processes. (pp. 60–61)
- Compare a hot debrief to a structured debrief. (pp. 60–61)

SKILL OBJECTIVES

After studying this chapter, you should be able to:

- Describe how to utilize scene control barriers. (**NFPA 1006: 5.1.3**, pp. 54–55)
- Isolate an area to control a hazard. (**NFPA 1006: 5.1.3, 5.1.4**, pp. 54–55)

You Are the Rescuer

You are called to the scene of a stranded swimmer at a local natural park swimming hole. When you arrive, you find a 30-foot (9-m) waterfall with several swimmers in the pool below. Off to one side of the waterfall, about 20 feet (6 m) up, is a shivering, crying young girl, clinging to the wet rocks.

1. What steps will you take to control the scene?
2. What resources will you request from dispatch?
3. How can you best prepare while you are waiting for additional resources?

 Access Navigate for more practice activities.

Introduction

The success of a mission begins long before the call goes out. All of the preparatory processes, including the hazard analysis, training of rescuer, and preplanning specific incident types, are tasks that can be performed prior to an incident occurring. Prepared in advance and implemented in the initial moments of a response, these actions can contribute positively to the eventual outcome. The specifics of preplanning and other preincident activities are described in the previous chapters.

Preplanning

There are advantages to utilizing a written preplan to identify hazards within the local jurisdiction, specify required types of rescue capabilities, and establish operational procedures for conducting rescue tasks. The advantage of a formal document is that it provides an excellent reference for consistently training new personnel on how the organization operates and eliminates the shortcoming of assuming team members know how the task will be accomplished. A written preplan can also provide legal documentation to show compliance with regulations and assist with documentation in case of liability. Ultimately, a well-written plan provides a framework from which field personnel can adapt their operational response to meet the needs of a particular type of emergency. The preplan and the incident action plan (IAP) can also provide a common reference point among different agencies to communicate goals, intentions, roles, and responsibilities as they relate to certain types of operations (e.g., rescue vs. recovery, isolate vs. evacuate, fire vs. law enforcement). Topics covered in the preplan will be more specifically addressed in the IAP that is developed during the response (**FIGURE 4-1**).

FIGURE 4-1 Command staff prepares the IAP.
© Aled Llywelyn/Alamy Stock Photo.

A well-written preplan is a useful tool for guiding the development of an IAP. Preplans should provide useful insight and consistency, but should not be so specific as to be limiting in their application. Rescues seldom appear the same way every time, so too heavy a reliance on a standard response format can backfire when the incident does not fit a preestablished profile. The key concepts to be clearly articulated in a preplan are the mandatory safety practices of the organization. Tragically, deficiencies in safety and intentional deviations from established safety practices are common factors in fatal rescue accidents.

When initiating a response, key items to look for in the preplan document to help in preparation of the IAP include the following:

- Review of jurisdictional and operational responsibility for rescue
  - List of external resources with specialized capability, and how to reach them.
  - Radio frequency information for intra- and inter-agency cooperation
- Hazard documentation of likely incident types/locations

- Maps of known hazard areas
- Notes on local municipal and business operations where hazards exist
- Site-specific information on known hazards (power shutoffs, structural information, etc.)
■ Standardized rescue operating protocols
- Activating appropriate resources
- Operational guidelines for different types of rescues
■ Established air ambulance landing zones, including coordinates, and known aerial hazards
■ Incident review procedures

Initial Response

The incident-specific IAP begins to form with the first responders to arrive at an incident. These may not be the same responders who will ultimately perform the rope rescue, so analyzing the scene and specifying appropriate resources can be a challenge. Any responder who might be exposed to a situation requiring rope rescue must be trained at the awareness level to do the following:

1. Recognize the need for a rope rescue
2. Identify resources necessary to conduct rope rescue operations
3. Activate the emergency response system where rope rescue is required
4. Carry out site control and scene management
5. Recognize general hazards associated with rope rescue and the procedures necessary to mitigate these hazards
6. Identify and use personal protective equipment (PPE) appropriate to a rope rescue incident

Recognizing the Need

The initial report of an incident may not always identify the need for rope rescue. A call may come in simply as a person who has fallen, a report of a medical condition, or even as a missing person. Initial responders should be alert to cues that could indicate a potential need for rope rescue so that necessary actions and resources can be initiated sooner than later. Clues that may suggest that rope rescue might become necessary could include presence of a known or potential fall hazard, knowledge of previous fall incidents in that area, or activity in which the subject was reportedly engaged at the time of the incident. Rope rescue methods may be appropriate whenever a fall hazard exists either for the subject or for rescuers, or when the subject is positioned on terrain or a surface upon

FIGURE 4-2 Rope rescue methods may be needed in areas where rescuers need additional support to maneuver.
© Iopurice/iStock/Getty Images Plus/Getty Images.

which rescuers cannot comfortably walk unaided (**FIGURE 4-2**). Bystanders may or may not recognize when a need for rope rescue exists, so it is especially important that the first-arriving resources be capable of making this determination. Where either the subjects or rescuers are exposed to a potential fall, rope rescue protocols should be initiated.

In such situations, untrained personnel should not be tempted to enter the hazard area, and the initial responders should be prepared to discourage them from doing so. An injured or ill subject who is in need of rope rescue requires rescuers who have special skills, equipment, and experience to deal with the inherent hazards posed by the environment. The primary goals of awareness-level rescuers are to activate appropriate resources and to prevent further harm. Under no circumstance should the awareness-level rescuer enter a hazard for which they are not properly prepared.

Once need has been identified, awareness-level rescuers can strive to achieve the following goals:

1. Size up the incident and determine appropriate resources (see Chapter 2)
2. Gather information from the responsible party (see Chapter 2)
3. Activate the emergency response system and establish an appropriate command structure
4. Provide for rescuer and bystander safety

Activate the Emergency Response System

Information provided by the awareness-level rescuer in the initial phases of a response helps to ensure that resources deployed to a rope rescue incident are adequately prepared and adept in functioning in the environment where the incident occurred. Methods for activating the appropriate emergency response system may vary by jurisdiction, but generally involve reaching out to dispatch to request appropriate resources. These resources may come from within the agency, aid agreements with neighboring agencies, or from the private sector. The following are examples of situations where additional external resources should be considered:

- Nighttime car-over-the-edge incidents can benefit from additional lighting on scene.
- Utilities incidents, where power shut-off procedures should be activated
- Landslides or excavations requiring heavy equipment operators
- Collapse incidents requiring shoring supplies and specialists
- Tower incidents where owner input is integral to structural or other safety
- Extended time incidents where food service and/or sanitation is needed for rehabilitation
- Farm or machinery incidents where chemical or engulfment hazards exist

This is by no means a comprehensive list, but should be used as a primer to encourage the awareness-level rescuer to think through the what-if's and the types of unique hazards that might require technical specialists with unique skills.

Command Structure

In the United States, most responding organizations function under some version of the **National Incident Management System (NIMS)**, as formulated and disseminated by the Federal Emergency Management Association (FEMA). NIMS incorporates best practices applicable to all levels of government, the private sector, tribes, and nongovernmental organizations, and provides a national incident management capability framework that is scalable and applicable to nearly any incident.

The NIMS has three guiding principles:

- Flexibility
- Standardization
- Unity of effort

The idea behind this unified approach is to enable organizations with specific jurisdictional responsibilities to better support one another while maintaining their own distinct and respective authorities. NIMS is broken down into three distinct components:

- Resource management
- Command and coordination
- Communications and information management

Resource Management is the term used within the NIMS to refer to a systematic approach toward managing resources (personnel, equipment, etc.) both before and during incidents. As with all response management protocol, the authority having jurisdiction (AHJ) serves as the primary control point for resource management. This means that the AHJ is primarily responsible for determining the necessary qualifications and credentialing required for resources responding to certain types of incidents as long as they meet the basic criteria established by NIMS. In this case, the AHJ may be a municipal agency or a private sector organization.

Command and Coordination is the term used to describe leadership roles, processes, and recommended organizational structures for incident management at the operational and incident support levels and explains how these structures interact to manage incidents effectively and efficiently. Ensuring that local command and coordination are harmonized with the nationally recognized NIMS procedures is a big part of what makes the system scalable as an incident grows. Incident command systems (ICSs) and **multiagency coordination (MAC) groups** used at the local level fall under NIMS Command and Coordination.

Communications and Information Management refers to the ongoing resources, systems, and methods that help to ensure incident personnel and stakeholders have appropriate information to make and communicate decisions.

Incident Command System

As previously discussed, the ICS is the systematic approach most commonly used for on-scene management of incidents. The first-arriving emergency responder with jurisdictional authority sets the stage for the rest of the incident by establishing the ICS framework from which to manage operations. This person immediately becomes the initial **incident commander (IC)**. As additional personnel arrive, the initial IC may be relieved of this role by an equally or more-qualified responder, or may continue in it, depending on the needs created by the emergency and the qualifications of the individual. Ultimately, there should be no question of who is in command.

The initial IC should establish a position from which to work where they can hear all radio communications; maintain accountability of and access to available resources; and, if possible, observe rescue operations. If it is not feasible to locate the command center near the actual rescue site, other command staff may become the eyes and ears of the IC.

One problem that can develop quickly in the initial stages of an operation is one person having too many

CHAPTER 4 Initiating a Response

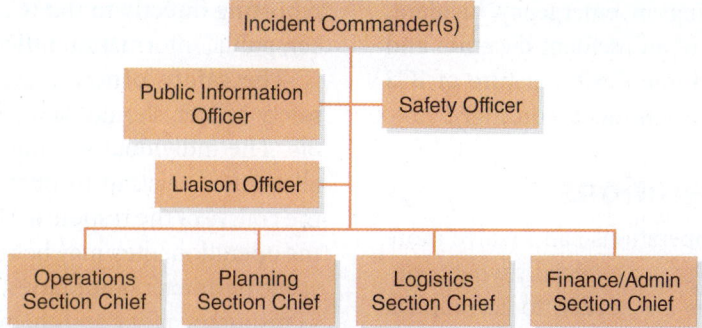

FIGURE 4-3 The five major functional areas within ICS are Command, Operations, Planning, Logistics, and Finance/Administration.

people to manage. Within the ICS, the optimal **span of control** (that is, the number of people that any one person manages directly) is recommended as being one supervisor to five subordinates; however, this is only a guideline, and actual distribution of subordinates to supervisors can vary significantly depending on circumstance and need. In any case, care should be taken to ensure that no one person is tasked beyond what is reasonable to ensure appropriate management and supervision. As additional resources arrive, roles and responsibilities should be adjusted, and resources divided into strike teams, resource teams, and other functional units.

There are five major functional areas defined within the ICS (**FIGURE 4-3**):

- Command
- Operations
- Planning
- Logistics
- Finance/Administration

Initially, these areas are the responsibility of the IC, but as an incident expands, the ability of that individual to effectively manage all aspects of the operation is diminished. As this occurs, the IC may assign a "chief" to one or more of these sections. These chiefs (overseen by the IC) collectively form what is known as the **general staff** for the incident. Some method of distinctive identification (e.g., helmet, vest, clip-on tag) can help to identify general staff and make their assignments known as additional emergency personnel arrive (**FIGURE 4-4**). Assignments should also be identified on the radio to enhance incident-wide understanding of who is in command.

Typically, the Planning section is responsible for gathering and analyzing operational information and sharing situational awareness, while the Operations section is responsible for executing tactical activities. In smaller incidents, the roles of Logistics and Finance/Administration may never come into play.

FIGURE 4-4 A piece of identification such as a helmet or vest will highlight the incident commander.
© zulkamalober/Shutterstock.

The IC is responsible for developing an IAP, which may be verbal or, for incidents of longer duration, put in writing. This plan establishes the overall strategic decisions and assigned tactical objectives for the incident, such as a decision to raise or lower a subject. This plan is a constantly evolving resource, changing as information is exchanged and circumstances develop. The individual who initially assumes command may not maintain command for very long, but the resources and actions they put into place form the beginnings of the plan—if well-considered and properly documented and communicated—can lend to a more rapid and successful outcome.

One of the most important advantages of the incident command function is its flexibility. The ICS framework can be expanded or contracted, depending on the size and nature of the incident. This is particularly important when the response to an incident grows, involving many more people. The organization adapts so that no one is overtasked in their span of control.

Using the ICS for rescue operations permits efficient use of resources, ensures personnel accountability, and promotes improved mutual aid responses. The modular nature of ICS allows it to grow with the complexity and scope of the incident. ICS permits multiple agencies to join forces in a seamless manner and to produce a coordinated effort during an emergency incident. As the operational needs of an incident decrease and resources are demobilized, the flexible nature of ICS permits the organization to contract accordingly.

Multiagency Operations

Even small-scale rescue operations can involve more than one jurisdiction or agency, especially where specialized skills or resources are needed to accomplish a mission. To prevent conflict, ICS management protocol includes clear processes for utilizing agency liaisons, unified command practices, and emergency operations centers (EOCs) to meld leadership and develop collective strategies and tactics for a coordinated mission response.

Where an IC requests assistance from another agency or organization that does not necessarily have jurisdictional responsibility, the assisting organization typically assigns an **agency liaison** to act as a point of contact for the AHJ. This representative, who may be the IC from the assisting agency, becomes the channel through which information is transferred to and from that agency. Typically, the agency liaison participates in planning meetings, monitors resource status, transmits agency-assigned objectives, and provides agency-specific demobilization information and requirements.

Unified command is an ICS term for a process in which leadership from multiple agencies works closely together to form an integrated, unified command team. Using this approach, each agency with responsibility for an incident assigns an IC to work together with ICs from other agencies with responsibilities related to that incident. Together, they work to form a common set of incident objectives and strategies, with each agency continuing to maintain authority, responsibility, and accountability relative to their respective role.

Particularly in larger operations where extensive unified command efforts are undertaken, agency representatives may work from an EOC. EOCs are the physical locations where staff from multiple agencies assemble to provide coordinated support to incident command, on-scene personnel, and/or other EOCs. There is a great deal of flexibility in how EOCs are composed and run. The IC can determine how to best staff the EOC to collect and share information toward a decision process that utilizes available resources and disseminates information appropriately.

Additional Command Staff

Aside from the five major functional areas of the ICS structure, there are three established positions that function as independent members of Command staff, reporting directly to the IC. These are the safety officer, public information officer, and liaison officer.

The **safety officer (SO)**, also called the incident safety officer, should be assigned as soon as practicable. The individual serving in this capacity monitors all matters relating to operational safety throughout the course of the response. During the initial phases of the operation, they may be responsible for establishing scene perimeter boundaries, with their role expanding as resources arrive and the rescue gets underway.

The **public information officer (PIO)** serves an equally vital role in supporting an incident by gathering, verifying, coordinating, and disseminating the right message to the right people at the right time. Target audiences may range from bystanders to agency representatives to media (**FIGURE 4-5**). The operation can benefit greatly from a PIO who also provides feedback to the ICS as appropriate.

The liaison officer (LO) is a Command staff position that is typically activated only when an operation scales to include other cooperating organizations or agencies. This position acts as a point of contact for agency representatives and maintains awareness of the present status of each, anticipating next steps and potential interorganizational challenges and needs.

Site Control and Scene Management

Site Control

One of the most useful things awareness-level rescuers can do is to establish and maintain control of the scene and perimeter areas. Identifying and marking a controlled hazard zone helps to preserve safety of the

FIGURE 4-5 The public information officer may update the media on an emergency.
Courtesy of Rob Schnepp.

subject, and can also keep well-intentioned bystanders from being exposed to hazards. Law enforcement resources, if available, can be a tremendous asset here. The area to be controlled should extend well beyond the subject and any identified hazards. It can be identified and blocked off with hazard tape, rope, cones, or other boundary markers; soliciting assistance from bystanders to help monitor these controlled hazard zones can help them to buy in to the concept of staying out of the controlled hazard zone themselves (**FIGURE 4-6**).

As a reminder, generally accepted criteria for controlled hazards zones are as follows:

- Hot zone: 100 feet for critical functions
- Warm zone :200 feet for support functions
- Cold zone: 300 feet for command and control

Actual distances may vary given site-specific hazards and other considerations.

Scene Management

Initial responders should also take note of related actions that can be taken to assist in scene management. Determination of what safety measures are appropriate is specific to the incident and situation, but should include all aspects of safety. For example, aside from rope- or height-specific criteria, it may be helpful to reduce noise or vibration from heavy equipment in the area, provide protection from the elements, source lighting, or to control nearby traffic.

The early phases of site control are also the right times to implement a personnel accountability system to ensure that all responders are accounted for as they arrive, are deployed into the incident, and leave the site. As part of this site control process, the need for evacuation protocols should be considered and, if applicable, implemented. Evacuation protocols are typically necessary only where additional hazards may pose a risk, such as a landslide, avalanche, chemical release, etc. Where such is identified, all responders at the scene must be briefed in how to react if an evacuation order is given.

Evacuation Signals

Guidance for training and operations provided by FEMA includes recommendations for the use of LCES (Lookouts, Communications, Escape routes, and Safe zones) during both training and emergency response evolutions. Using a preestablished group of communication signals established by FEMA and Urban Search and Rescue will help to ensure everyone's safety:

- Evacuate: Three short blasts (1 second each)
- Cease operations: One long blast (3 seconds long)
- Resume operations: One long and one short blast

Recognize and Mitigate Hazards

Controlling and limiting access to the rescue site and traffic around it is a big part of hazard mitigation, but it is by no means the end. For the awareness-level rescuer, the primary recommended method for hazard mitigation is to stay away from hazards and keep others from them as well. Rescuers with additional levels of training will use different methods of mitigating hazards, as exposure will differ in accordance with skills and methods used at that level. It is incumbent on the awareness-level rescuer to take this topic seriously, as effective management of hazards has a strong influence on the eventual outcome of an operation. This process of analyzing hazards never stops; any rope or technical rescue can change in a matter of moments, and only constant vigilance will keep rescuers adequately protected. Chapter 3 of this text is wholly dedicated to identifying and mitigating hazards in a broad, general sense.

FIGURE 4-6 Caution tape may be used to create a border around a scene.
Courtesy of Ken Phillips.

Personal Protective Equipment

Awareness-level responders should be familiar enough with equipment to know what does—and what does not—constitute appropriate PPE for rope rescue so that they are able to better protect themselves and to safely contribute support functions during an incident. Personal protective equipment and rigging gear specific to operations- and technician-level rope rescue operations is discussed in detail in Chapter 7 of this text.

Generally, protective equipment used for rope rescue operations will differ from that required in a fire-ground environment, unless the rope rescue is occurring in the fire-ground environment. In most cases, environment dictates what is and what is not appropriate PPE for rope rescue. Aside from rope systems gear, basic PPE requirements for awareness-level rescuers include the following:

1. Clothing appropriate to the task and environment: Clothing should provide adequate physical protection while not presenting undue thermal hazards. It should not be so tight-fitting as to be restrictive, but neither should it be so baggy as to potentially become caught in moving ropes and equipment.
2. Footwear appropriate to the task and environment: Boots should fit securely, protect against hazards, have sturdy soles, and be suited to the environment where the response is taking place. Rubber boots designed for fire-ground response are typically not the best choice for rope rescue on buildings and other structures, towers, water, or wilderness environments.
3. Helmet appropriate to vertical work: The rope rescue helmet should be robust enough to protect from impact and dropped objects, but not heavy. It should fit securely with a retention system that secures the helmet with straps at both sides and at the back of the head to keep the helmet secure. Fire helmets are generally unsuited to rope rescue operations.
4. Gloves with dexterity: Hand protection primarily protects against abrasion and rope burn when handling a moving rope (**FIGURE 4-7**). Multiple layers across the rope path in the palm are ideal, while lighter coverage in the fingers facilitates knot tying and fine motor actions. **Body substance isolation** (BSI) should be used when in close proximity to the subject.

FIGURE 4-7 NFPA 1951-compliant tech rescue utility gloves.
Courtesy of Pigeon Mountain Industries.

5. Eye protection: Rope rescue incidents can occur in any environment, so the eye protection chosen should be appropriate to the environment where the work is being done, and should also provide adequate BSI as needed.

Rope Rescue Resources

When considering what level of rope rescue resources to request/activate, the awareness-level rescuer should have a clear understanding of NFPA's guidelines for capabilities for responders operating at rope rescue incidents.

Awareness Level

As alluded to already, the goals of a rescuer trained at the awareness level are to prevent further harm and to initiate appropriate resources. Awareness-level rescuers may be incorporated into incident response, but only at the level to which they are adequately trained and prepared. Typically, this will enable them to perform support roles as needed. Awareness-level rescuers may not be adequately prepared to go into the field unless specific training provided by the AHJ has prepared them. Follow all local standard operating procedures.

Operations Level

Generally, the operations level reflects the ability to perform rigging and related operations for the movement of rescuers or the subject from one stable location to another. While the description of a stable environment may vary based on the context, in general it is intended to reflect conditions where the rescuer would not be expected to perform patient care or related tasks unaided while supported by a rope system.

While it may be necessary to employ rope systems for the movement of operations-level rescuers to access the subject, in many cases it is quite possible to resolve rope rescue incidents without the need for rescuers to enter the high-angle environment at all, instead limiting the use of ropes to the movement of a nonambulatory subject. In these circumstances, rescuers can access the subject using conventional means, such as stairways, ladders, or trails, then upon reaching the subject, employ rope rescue techniques to accomplish the removal (**FIGURE 4-8**).

Operations-level rescuers are able to establish anchorages and perform lowering and raising operations. These functions may be performed from above the subject or below. However, the skill set at the operations level is intended to exclude conditions that require the rescuer to work independently while suspended or protected by a rope rescue system. Examples would include performing a pickoff of a subject in a vertical environment, functioning as a litter attendant in a vertical setting, or making access to a subject using climbing techniques. Skills and knowledge relevant to operations-level rescue are covered in Chapters 6 through 16 of this text.

Technician Level

A technician-level response is required where rescuers need to work independently to perform a rescue-related task while suspended or protected by a rope system. This would include activities like performing rescues where a subject is clinging to a natural or manmade feature or suspended by rope, as well as circumstances where the rescuer must provide their own protection to access the subject, such as using lead climbing or work positioning techniques (**FIGURE 4-9**). In contrast to the operations level, where the rope system is used to transport the rescuer to an environment where they can work relatively free of the risk of falling and have the support of other team members, the skill set at the technician level requires the rescuer to operate in the vertical environment performing tasks while suspended, often independently with little immediate support. Additionally, technician-level rescuers must be capable of managing multiple concurrent hazards, such as communicating with subjects in emotional or psychological crisis, as well as the combination of horizontal movement of a suspended load through the use of offsets such as tracking lines and highlines. Rescuers performing at this level must be adept in all of the skills and knowledge relevant to operations, as well as additional topics covered in Chapters 17 through 20 of this text, and should also possess extensive field experience.

FIGURE 4-8 Operations level rescuers may access subjects directly by a ladder, stairway, or trail.
© CandyRetriever/Shutterstock.

FIGURE 4-9 Technician-level rope rescuers need to be able to work in suspension.
Courtesy of Steve Hudson.

Incident Action Plan

While the organizational preplan is a useful tool for advance preparation of resources and response capabilities, each incident requires its own specific IAP. The IAP is the collective plan generated by the IC. The IC may choose to obtain input from section chiefs as applicable. For small operations, the IAP need not be a written document, but for larger incidents—especially where multiple agencies are involved—a written plan helps to ensure continuity of information and contributes to a shared mental model. A written IAP is absolutely essential when an incident extends to multiple operational periods.

In the IAP, incident objectives for a given operational period are stated and a plan is formed to support these objectives using available resources. A well-executed planning process ensures best use of resources and keeps the operation moving in a positive direction to its conclusion.

FEMA makes a range of planning forms available to assist in developing an IAP, but make no mistake, the IAP is not about filling out forms: It is about synchronizing resources and harmonizing efforts to provide clear direction to those responsible for bringing the incident to successful conclusion. The IAP includes a comprehensive listing of the tactics, resources, and support needed to accomplish identified objectives within a specified time.

The IAP is a living resource, cyclical in nature, with a new plan developed for each operational period. At the outset of an incident, the situation may be chaotic and information may be incomplete or difficult to obtain. For this reason, the initial plan developed by an IC is typically a simple one, communicated concisely through oral briefings, and it may change rapidly. As additional management resources arrive, information becomes available, and specialized response units are engaged, the IAP may develop more thoroughly.

Whether written or oral, the processes used in determining tactics and developing an IAP are relatively constant. FEMA has developed a mnemonic resource called the Planning P to aid ICs in following a consistent pattern of planning (**FIGURE 4-10**).

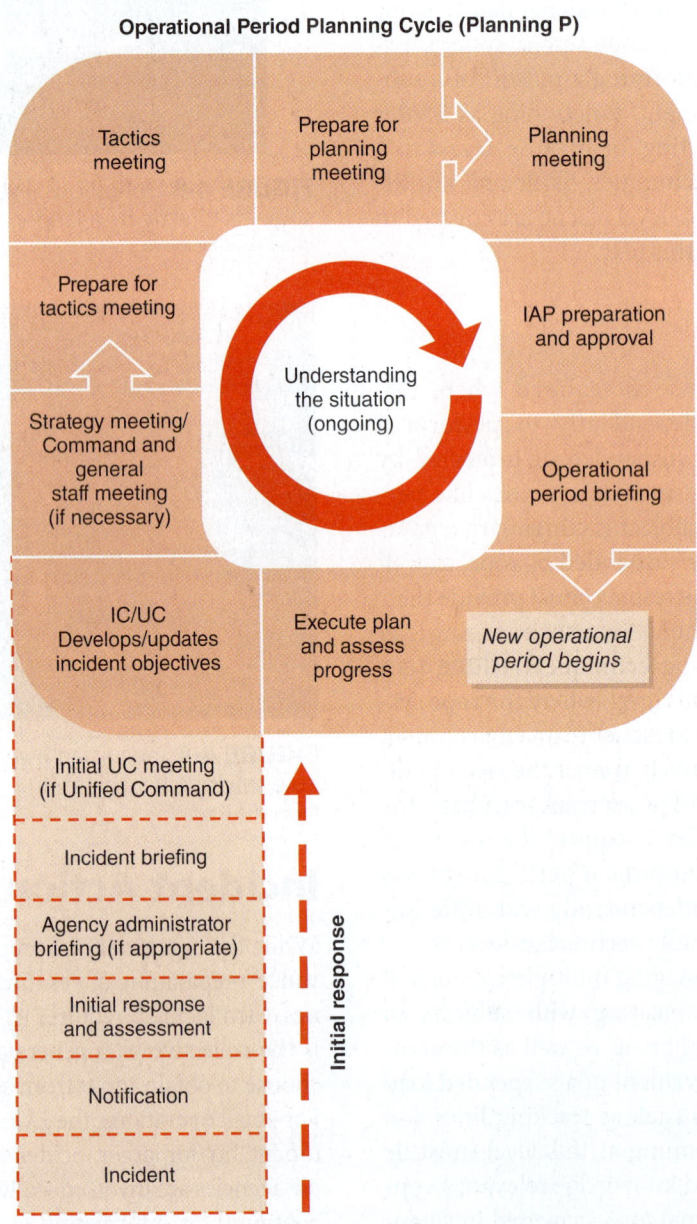

FIGURE 4-10 FEMA's Planning P.
Reproduced from FEMA. (n.d.). *IS-201: Forms used for the development of the incident action plan.* https://emilms.fema.gov/IS201/ICS0102summary.html

After the initial response phase, each operational period begins with an **operational period briefing** where members of the Command and general staff present incident objectives, review the current situation, and share information related to communications or safety. Following the briefing, supervisors brief their assigned personnel on their respective assignments.

Incident Planning Forms

All levels of rescuers should be familiar with the types and availability of forms disseminated by FEMA. These forms have been designed with input from emergency response organizations across the country to assist emergency responders in implementing ICS, creating IAPs, and documenting the progression of an incident.

These forms are best utilized by personnel who have adequate training and/or experience in effective use of ICS concepts. Although some agencies have their own versions of planning forms, the forms in this text are taken directly from FEMA resources; adherence to the standardized ICS forms promulgated by FEMA helps to promote consistency in the management and documentation of incidents, and to facilitate effective use of mutual aid. The NIMS ICS forms typically used in an IAP are listed in **TABLE 4-1**.

Of course, the ultimate purpose of any preplan is to facilitate the execution of a successful rescue. In all this planning, do not forget that in an actual incident, the focus must be on the person who is in need of rescuing. The processes and resources in this chapter are intended to enable that goal, not overshadow it.

LAST

The IAP should follow the LAST mnemonic detailing the phases of the incident for planning the consecutive elements of a search and rescue incident. There is no preestablished amount of time or resources required to achieve each phase; it is totally situation-specific:

- Locate
- Access
- Stabilize
- Transport

Locate

Locating a subject is sometimes easy, as when their location has been clearly communicated by the reporting party or where they are in plain sight of emergency services personnel on arrival. At other times, extensive search may be required. Search can occur in any environment: urban, wilderness, industrial, cave, confined space, etc. The methods used for search will vary depending on environment and other factors.

Rescuers involved in search or rescue in any environment should be competent and experienced in the environment where they are expected to work, even where environments may seem similar. This is primarily for the safety of the rescuers and subjects, but it is also true that experienced rescuers are more likely to generate a better outcome. Rescuers trained to the awareness level may be used at this, or any other, area of response, as long as what they are being asked to do is consistent with their skills and abilities. Follow local standard operating procedures.

Access

Accessing the subject is the next priority after the location of a subject is known. Access may be simple,

TABLE 4-1 NIMS ICS Forms Used to Create an IAP

ICS Form #	Form Title	Typically Prepared By
ICS 201	Incident Briefing	Incident commander
ICS 202	Incident Objectives	Planning section chief
ICS 203	Organization Assignment List	Resources unit leader
ICS 204	Assignment List	Resources unit leader & operations section chief
ICS 205	Incident Radio Communications Plan	Communications unit leader
ICS 205A	Communications List	Communications unit leader
ICS 206	Medical Plan	Medical unit leader (reviewed by safety officer)
ICS 208	Safety Message/Plan	Safety officer

via an established pathway, or it may be made more challenging if the subject is in a location not normally accessed by people. When access requires special tools or equipment, such as rope, care should be taken to prevent those access methods from creating additional hazards or causing further harm. Rescuers who are assigned to access the subject to render medical aid or assist in transport should access the patient cautiously and in harmony with one another, while remaining rescuers should immediately set about preparing for the next phases of the operation.

Stabilize

Stabilizing the subject may be as simple as calming them down, or as complex as needing to secure them to a rope safety system while providing trauma intervention. The first priority should be preventing further harm, whether this be securing them so they do not fall (farther), removing them from moving water, or isolating them from other hazards (**FIGURE 4-11**). Once they are secure, medical and other physical needs can be attended to. If the patient is far from definitive care, more attention may need to be given to peripheral or comfort needs.

Transport

Transport is the final step in the LAST mnemonic. This step includes all movement of the subject, whether technical evacuation, trail carry, or vehicle transport. As with the other steps, it might be as quick as handing the subject off to another prehospital provider (i.e., loading them into an aeromedical helicopter), or it may take a very long time.

The entire LAST process may take less than an hour for some incident responses, while in other responses it may take several operational periods. An operational period is typically (but not always) defined as 8 hours, emulating what many consider to be a reasonable work shift. Incidents that encompass multiple operational periods require more resources than short-duration incidents, as well as effective communication and briefing during transitions.

After Action Review

The ability of a rescue organization to achieve its mission is directly related to the performance of its members working together as a team. In the interest of aligning these, every incident (and training) should be followed by a review process. An effective debrief process will help with the following:

- Identify and affirm good practices
- Evaluate operational efficiency
- Identify opportunities for individual and team learning
- Maintain team and individual focus on continuous improvement
- Motivate rescuers toward self-improvement
- Prepare for the next response

A good review process consists of at least two parts: a **hot debrief** that takes place immediately following the incident, and opportunity for a **structured debrief**, which may occur later.

Hot Debrief

A hot debrief should be standard operating procedure for every incident, not just those where a perceived problem or concern is noted, and effort should be made to involve all individuals and agencies who took part. Conducting the hot debrief immediately after the subjects are released to the hospital or other care and while rescuers are still on scene affords the best opportunity to analyze and evaluate tactics, hazards, risks, and outcome while the operation is still fresh in everyone's minds.

The hot debrief should focus on *what*, not *who*, and should be conducted in an open spirit of constructive analysis. Any one of a number of methods can be used

FIGURE 4-11 Stabilizing the subject includes moving the subject to a place of safety.
Courtesy of Jason Williams.

to debrief an incident, but most address at least the following key points:

- What was the mission?
- How was it achieved?
- Where could improvements have been made?

Structured Debrief

The structured debrief differs from the hot debrief in that it only occurs when observation is made of important learning points that are deemed by leadership to warrant further review. These may be safety related or simply organized for the purpose of exploring opportunities for improvement. These sorts of debriefs typically involve some preparation and are expected to delve more deeply into specific issues, so they can be expected to take longer. Still, they should be performed as soon as practicable after the incident in question.

If the structured debrief involves a serious safety issue or if there are sensitive matters to be discussed, it is sometimes helpful to invite a facilitator who was not part of the event. It should not become a forum for airing personal grievances, nor a platform for pushing an agenda. The goal of this type of debrief is to address specific areas of risk or concern and to identify action items toward mitigating them.

Threat and Error Management System

In the aviation industry, an approach known as **Threat and Error Management System (TEMS)** is used to help maintain such perspective. Originally developed for flight deck operations, TEMS is a larger approach to big-picture safety management within an organization; however, the concepts can be effectively applied at any level. Using a TEMS approach during both hot debrief and structured debrief can help keep attention focused on critical issues.

The basic underlying presumption in TEMS is that threats and errors are part of everyday operations, and these threats and errors present the potential to generate unsafe states and undesired outcomes. Knowing this, efforts are directed toward recognizing and managing these. Using the principles of threats, errors, and undesired states can help maintain appropriate focus during a debrief. The components of a TEMS Framework include the following:

- **Threats**: These are defined as potential hazards that are beyond the influence of personnel. Examples may include environmental hazards, bystanders, or traffic.
- **Errors**: These are actions (or lack thereof) taken by responders that may result in unexpected or undesired outcomes. Examples may include a deviation from protocol, inappropriate reactions, or a failure to perform as expected.
- **Undesired states**: These are operational conditions wherein safety is reduced or compromised, often the last stage before an incident or accident. Such conditions are usually the result of ineffective threat and/or error management.

Five Why's

To help maintain focus on matters that should be addressed or managed more appropriately in future operations, the "Where could improvements be made?" portion of the discussion should be grounded in the following questions:

- What undesired states were created by threats that were present during the operation?
- What undesired states were created by errors made during the operation?

From here, employing the **Five Why's** approach can help identify the root cause so that appropriate corrective action can be implemented. This is a process popularized by Taiichi Ohno, visionary Toyota executive and an architect of the Toyota Production System. Ohno encouraged his staff to investigate the problem at the source and to ask "why" five times for every issue. It is a good technique to help make sure that you are really getting to the root cause of an issue. What truly caused that to happen? Consider the following example of implementing this method in the rope rescue environment:

- ERROR: The rescue pack was missing some equipment.
    - WHY? The pack was not resupplied following the last incident.
    - WHY? The pack was left in a vehicle.
    - WHY? There is no formal equipment check-in process for responders to follow after returning from a response.
    - WHY? Such a practice was not necessary in the past.
    - WHY? The team was smaller and the informal procedures were adequate.

Addressing opportunities for improvement within the context of threats, errors, and undesired states, and then utilizing the Five Why's, can help depersonalize the debrief and reduce the tendency for participants to become focused on their own opinions or preferences.

After-Action REVIEW

IN SUMMARY

- A well-written preplan is a useful tool for guiding the development of an incident action plan (IAP).
- The incident-specific IAP begins to form with the first responders to arrive at an incident.
- Information provided by the awareness-level rescuer in the initial phases of a response helps to ensure that resources deployed to a rope rescue incident are adequately prepared and adept in functioning in the environment where the incident occurred.
- Methods for activating the appropriate emergency response system may vary by jurisdiction, but generally involve reaching out to dispatch to request appropriate resources.
- The command structure reflects the three principles of the National Incident Management System (NIMS): flexibility, standardization, and unity of effort.
- The incident command system (ICS) is able to adapt to match the scope of an emergency, from single subject to multiple subjects.
- One of the most useful things awareness-level rescuers can do is to establish and maintain control of the scene and perimeter areas.
- Identifying and marking a controlled hazard zone helps to preserve safety of the subject and can also keep well-intentioned bystanders from being exposed to hazards.
- Controlling and limiting access to the rescue site and traffic around it is a big part of hazard mitigation; for the awareness-level rescuer, the primary recommended method for hazard mitigation is to stay away from hazards and keep others from them as well.
- Personal protective equipment (PPE) used for rope rescue operations will differ from that required in a fire-ground environment unless the rope rescue is occurring in the fire-ground environment. In most cases, environment dictates what is appropriate PPE for rope rescue.
- Awareness-level rescuers may be incorporated into incident response, but only at the level to which they are adequately trained and prepared.
- Operations-level rescuers are able to perform rigging and related operations for the movement of rescuers or the subject from one stable location to another.
- A technician-level response is required where rescuers need to work independently to perform a rescue related task while suspended or protected by a rope system.
- In the IAP, incident objectives for a given operational period are stated and a plan is formed to support these objectives using available resources.
- Every incident (and training) should be followed by a review process, whether it be a concise hot debrief or a detailed structured debrief.

KEY TERMS

Body substance isolation The use of protective barriers to reduce the risk of transmission of infectious agents to healthcare personnel.

Command and Coordination A NIMS term describing the structural interaction among leadership roles, processes, and recommended organizational parameters for incident management.

Command staff Positions that provide support and report directly to the incident commander, including public information officer, safety officer, and liaison officer.

Communications and Information Management Procedures used to establish and maintain a common operating picture and systems interoperability during an incident.

Emergency operations center (EOC) Location from which staff provide support to incident management with information and resourcing.

Federal Emergency Management Agency (FEMA) An agency of the United States Department of Homeland Security intended to build, sustain, and improve our capability to prepare for, protect against, respond to, recover from, and mitigate all hazards.

Five Why's A technique used to explore fundamental cause-and-effect relationships of an occurrence.

General staff Chiefs of each of the sections of ICS (Operations, Planning, Logistics, and Finance/Administration) who report directly to the incident commander.

Hot debrief An open, collective constructive analysis of an incident that takes place with all responders as a group immediately following completion of the operation.

Incident commander (IC) The individual who has overall management responsibility for an incident.

Multiagency coordination (MAC) groups Groups of representatives from different organizations who have authorization to commit resources from their respective agencies to a collaborative agreement.

National Incident Management System (NIMS) A comprehensive, national approach to incident management promulgated by FEMA for application at all jurisdictional levels and across functional disciplines.

Operational period briefing A meeting held at the beginning of each operational period with Command and general staff, along with other stakeholders, to share the incident action plan for the upcoming period

Public information officer (PIO) Individual assigned with communications relative to both press and members of the public to provide a consistent, clear, and accurate message about the incident.

Resource Management A term used in NIMS to describe coordination and oversight of personnel, tools, processes, and systems used during an incident.

Safety officer (SO) The person responsible for monitoring current and projected hazards associated with dangerous conditions affecting rescuers and others.

Span of control The ratio of subordinate personnel to one supervisor during an emergency incident. The optimum span of control is considered to be 5:1; this may be increased to a maximum of 7:1.

Structured debrief An in-depth review of a response that considers a specific area of concern; usually takes place a short time after the incident, so that participants can have time to research.

Threat and Error Management System (TEMS) An approach to safety management developed by the aviation industry that is grounded in the assumption that risk happens, and can be managed in a way so as to not impair safety.

Unified command Management of an emergency incident through the incident command system in which multiple agencies or jurisdictions are involved, each one having an incident commander assigned to the operation.

On Scene

1. What sorts of functional capabilities should a responder have who is trained at the awareness level in rope rescue?
2. What is the relationship between a preplan and an IAP?
3. Describe the parts of a Planning P.
4. What is the Incident Command System, and why is it important?
5. What does the mnemonic LAST stand for?
6. How does the TEMS system contribute to safety?

Chapter Opener: © Jones & Bartlett Learning. Courtesy of Loui McCurley; On Scene siren: © Bildgigant/Shutterstock.

CHAPTER 5

Awareness

Supporting the Operations- or Technician-Level Rescue Incident

KNOWLEDGE OBJECTIVES

After studying this chapter, you should be able to:

- Discuss the impact of the incident's location on a rope rescue. (p. 66)
- Describe the role of local standard operating procedures in determining how awareness-level rescuers may support operations- and technician-level operations. (pp. 66–67)
- Identify the basic personal protective equipment used at rope rescue incidents. (**NFPA 1006: 5.1.5**, pp. 67–73)
- Provide an overview of the principles of mechanical advantage in hauling systems. (**NFPA 1006: 5.1.1**, pp. 73–74)
- Describe the role of the awareness-level rescuer in hauling operations. (**NFPA 1006: 5.1.1**, p. 74)
- Identify safety concerns during hauling operations. (**NFPA 1006: 5.1.1**, p. 74)
- Recognize common operational commands utilized during hauling operations. (**NFPA 1006: 5.1.1**, pp. 74–76)
- Describe the role of the awareness-level rescuer in supporting the incident action plan (IAP). (**NFPA 1006: 5.1.5**, pp. 74, 76)
- Identify common facilities and stations at a rope rescue incident. (p. 76)
- Describe awareness-level support functions at a rope rescue incident in regards to supporting personnel rehabilitation. (**NFPA 1006: 5.1.1**, **5.1.5**, p. 78)
- Describe the best practices to help ensure rescuer accountability. (p. 76)

SKILL OBJECTIVES

After studying this chapter, you should be able to:

- Analyze local standard operating procedures to determine the role of the awareness-level rescuer in supporting operations- and technician-level tasks. (**NFPA 1006: 5.1.5**, pp. 66–67)
- Select personal protective equipment for rope rescue. (**NFPA 1006: 5.1.5**, pp. 67–73)
- Recognize common operational commands utilized during hauling operations. (**NFPA 1006: 5.1.1**, pp. 74–76)
- Analyze the incident action plan to determine the role of the awareness-level rescuer in supporting operations- and technician-level tasks. (**NFPA 1006: 5.1.5**, pp. 74, 76)
- Describe how progress is reported to the incident commander or supervisors. (**NFPA 1006: 5.1.5**, pp. 77–78)

You Are the Rescuer

Your agency is called to a tower site where an antenna was being lifted 225 feet (69 m) onto a tower by a crane and installed by a two-man crew. During the lift, the crane struck and bent the tower, leaving the two crewmembers hanging from their fall protection with no way to self-rescue. One states that he is fine but cannot reach any structural members to establish a self-rescue. The other is unresponsive.

1. What questions do you have about safety as you enter the site?
2. What training/skills/experience do you want rescuers to have?
3. What concerns will you have about the safety of rescuers during the course of the rescue?

Access Navigate for more practice activities.

Introduction

At the awareness level, personnel may be fielded under controlled conditions for the purpose of helping to support the operations- or technician-level rescue incident. This means that awareness-level rescuers must have foundational knowledge to maintain their own safety and carry out an assignment as directed, be capable of reporting progress to a supervisor and assisting with rehabilitation, and generally support the incident action plan (IAP). This includes not operating where skills, knowledge, and or abilities above the awareness level are required.

Any responder who will be present at a scene should have enough knowledge to manage their own safety during the incident. In any rescue operation, rescuer safety comes first, before that of the subject. A rescuer is of no use to the operation if they become injured or incapacitated. Some basic rules apply:

1. Never create additional subjects.
2. Do not create additional injuries.
3. Do not make the situation worse than you found it.
4. Do not cut corners on safety.
5. Gravity is incessant and unforgiving.

As a rescuer, remember the following:

1. Your own safety first
2. Safety of fellow rescuers second
3. Safety of the subject third
4. Safety of bystanders fourth

Incident Location Concerns

As discussed, rope rescue typically occurs in a specific environment with specific challenges, from confined space to tower rescue. For this reason, knowledge of rope rescue alone is not sufficient to ensure safety during a rope rescue response. Rescuers must also be familiar with the environment in which the rescue will take place. Of the 20 subdisciplines covered by NFPA 1006, 15 of these are related to the environment where the rescue takes place (e.g., water, confined space, tower, wilderness, cave, etc.). Rope rescue methods and techniques are unique in that they may be applied during other technical rescues, whether an urban building, an industrial facility, wilderness or cave environment, tower, or other high-angle environment (**FIGURE 5-1**). Where hazards or rescue types are combined, it is important for responders to understand all of the rescue types involved, as well as the environments.

Guidance on Providing Support

With training, awareness-level rescuers should be capable of following instructions to carry out an assignment in a support role. To be able to carry out an assignment, rescuers must understand the standard operating procedures of the authority having jurisdiction (AHJ), be capable of selecting and utilizing appropriate personal protective equipment (PPE) for the situation, and assist with equipment appropriate to the task.

Organizations will develop **standard operating procedures (SOPs)** based on the types of incidents that are likely to occur in their area. In some cases, the SOPs may be very generic in nature to allow rescuers latitude to choose the most appropriate method of operation for any given rescue. Training at the local level, then, is typically in accordance with these SOPs.

FIGURE 5-1 The environments that rope rescues occur can vary widely. **A.** Office building. **B.** Cave.
(A) © AerialPerspective Images/Moment/Getty Images; (B) © Theo Lawrence/500px Prime/Getty Images.

Local SOPs for rope rescue may vary depending on the location of the incident. For example, thought processes, equipment, and methods used for a rope rescue operation in a cave will likely differ from the thought processes, equipment, and methods used for a rope rescue operation on a tower or building. Awareness-level rescuers should be cognizant of this fact and refrain from utilizing SOPs from different locations or technical rescue subdisciplines in a given rescue unless directed to do so.

Awareness-level rescuers in a support role should perform specific tasks as directed. Technical rope rescue is a highly skilled discipline and requires a combination of training and experience to master good decision-making and rigging skills. The ability to explicitly follow instructions is infinitely important when serving in a support role.

Personal Protective Equipment Selection

National Fire Protection Association Standards

For decades, life safety rope and equipment that was tested and compliant with NFPA 1983, *Standard on Life Safety Rope and Equipment* was marked NFPA 1983 (2017). Since this standard has been absorbed into the new NFPA 2500, *Standards for Operations and Training for Technical Search and Rescue Incidents and Life Safety Rope and Equipment for Emergency Services*, markings will be adjusted to read NFPA 2500 (2022)—or, for a period of time, NFPA 2500 (NFPA 1983, 2002).

At the awareness level, rescuers are expected to be capable of selecting appropriate resources from a rescue tool kit to effectively support a rope rescue incident. This requires at least a foundational understanding of basic rope rescue equipment, including harnesses. Many agencies require that equipment be manufactured, tested, and marked in accordance with an appropriate, relevant equipment manufacturing standard such as NFPA 2500 (NFPA 1983), but there is equipment available that is not covered by this standard. The AHJ has final say over what is—and what is not—appropriate equipment for rescue with regard to the existing hazards.

A good reference for rescuers to use to learn more about choosing equipment for rescue is NFPA 2500, which contains material from the standard previously known as NFPA 1858, *Selection Care and Maintenance of Life Safety Rope and Equipment*. Whereas NFPA 2500 also contains the manufacturer's requirements formerly found in NFPA 1983, the content from NFPA 1858 is designed to help guide field personnel in the selection, inspection, cleaning, decontamination, repair, storage, and retirement of some components of rescue equipment, including life safety rope, harnesses, hardware, and victim extrication devices that are compliant with NFPA 2500 (1983).

> **TIP**
>
> Standards applicable to harnesses include the following:
> - NFPA 2500 (1983), *Standards for Operations and Training for Technical Search and Rescue Incidents and Life Safety Rope and Equipment for Emergency Services*
> - ASTM F1772, *Standard Specification for Harnesses for Rescue and Sport Activities*
> - ANSI Z359.11, *Safety Requirements for Full Body Harnesses*
> - EN 361, *Personal Protective Equipment Against Falls from a Height—Full Body Harnesses*

Common Protective Equipment at Rope Rescues

Rescuers should be familiar with and capable of using protective equipment that is appropriate to the tasks that they might be expected to perform during rope rescue incidents. At a foundational level, this includes appropriate clothing and footwear for the environment that they will be working in. It is important to emphasize that rope rescue might happen in any number of locations, and the personal protective equipment (PPE) for a particular operation will vary accordingly.

Turnout and bunker gear may be appropriate for a rescuer operating in a fire-ground environment, but this would not be the best choice for a wilderness environment, water-based rescue, or even structural rescue in an urban environment. Clothing items should be appropriate for the location in which the rescuer is working. If the rescuer is uncertain which equipment to choose, they should ask a supervisor for guidance. The general rule is to "dress for the sport you're playing."

Fire fighters will be familiar with some of the specialized equipment that may be found at a technical rescue incident, such as the following:

- Supplied air breathing apparatus (SABA), self-contained breathing apparatus (SCBA), and supplied air respirator (SAR), all of which should meet the requirements of the Occupational Health and Safety Administration's 29 CFR 1910.146, Permit-Required Confined Spaces.
- Personal alert safety system (PASS), which should meet the requirements of the NFPA 1500, *Standard on Fire Department Occupational Safety and Health Program*, and NFPA 1982, *Standard on Personal Alert Safety Systems (PASS)*
- Life safety ropes and system components, which should meet the requirements of NFPA 1500, *Standard on Fire Department Occupational Safety and Health Program*, and NFPA 2500, *Standards for Operations and Training for Technical Search and Rescue Incidents and Life Safety Rope and Equipment for Emergency Services*
- Communications equipment, which should meet the requirements of OSHA 29 CFR 1910.146
- Lighting equipment (e.g., flashlights, helmet-mounted lamps), which in some cases

FIGURE 5-2 Head lamp.
© NorthernExposure/Alamy Stock Photo.

may need to be intrinsically safe or explosion proof as defined by OSHA 29 CFR 1910.146. (**FIGURE 5-2**)
- Chemical protective clothing as defined by OSHA 29 CFR 1910.120

Although familiar to most fire fighters, some of this equipment may not be utilized at a rope rescue incident. It is dependent on the location of the incident and the local SOPs. PPE that is likely to be found at a rope rescue operation might include items such as harnesses, gloves, and helmets.

Harnesses

With respect to harnesses, the AHJ should specify what type of harness should be used for a given rescue operation. Harnesses are available in a variety of forms, fits, features, connection points, and performance specifications. Harness selection should be based on the intended use, including the following elements:

- **Personal escape**: An escape harness is typically worn during work at height operations explicitly for the purpose of enabling the rescuer to escape, if needed, to a lower position of safety. It is typically used with an emergency descent system.
- **Rescue**: Rescuers will normally wear a rescue harness to provide suspension during rescue operations (**FIGURE 5-3**).
- **Restraint**: Harnesses or belts worn for restraint are intended merely to prevent a rescuer from reaching a position from which they might fall.

FIGURE 5-3 An example of a rescue harness.

FIGURE 5-4 An example of a fall arrest harness.
© John99/Shutterstock.

- **Fall arrest**: A fall arrest harness is intended to catch a person who has fallen, and to be used as an adjunct to retrieval (**FIGURE 5-4**).

Awareness-level responders may be trained to wear a harness for their own personal safety, for example near an edge, as they perform tasks to support a rescue. However, rescuers with this limited level of training are not expected to take on operational roles and should not be put into (or accept) a role beyond their level of knowledge.

NFPA classifies harnesses into two main categories. **Class II harnesses** are composed of webbing primarily around the waist and thighs, and may also be known as seat harnesses. Class II harnesses offer good mobility and freedom of movement. **Class III harnesses**, also called full-body harnesses, incorporate additional webbing around the torso, upper body, and shoulders (**FIGURE 5-5**). While both types of harnesses offer security and are tested with a head-down orientation, the Class III harness offers dorsal and sternal connection points, which are more commonly used in rope access, industrial fall protection, confined space, and helicopter hoist operations. The front waist connection is generally preferred for litter tending, pickoffs, and other applications where mobility is key.

FIGURE 5-5 Class II harness.

It is especially important to note that some agencies may find it useful to utilize non-NFPA harnesses for certain specialized applications, such as cave rescue, lead climbing, or other situations. If the harness will be exposed to heat, flame, chemicals, or water, materials and performance criteria should be considered accordingly. Some harnesses are designed to be integrated with bunker gear; special care should be taken to ensure good fit and that the harness does not compromise the integrity of the protective garment.

The following features should be present in a rescue harness:

- Webbing width. Generally speaking, the wider the straps, the more the force is distributed over a greater area, resulting in more comfort. This should be balanced with specific harness design and weight implications. Typically, webbing should be at least 2 inches wide at critical points such as waist and thighs.
- Padding. This is a matter of preference. Some people prefer thick, cushiony padding, while others prefer a thinner, form-fitting design.
- Stitching. Critical points of the harness should be sewn with contrasting stitching for easier inspection.
- Ease of donning and doffing. The harness should be easy to put on and to adjust. This may vary by body type and personal preference.
- Comfort. While a subjective factor, this is important. The more comfortable the harness, the greater likelihood that it will be worn properly. It should not slip down or loosen when worn.
- Balance. Rescuers should be able to maintain a correct center of gravity when working in the harness. After a fall, the harness should allow the rescuer to easily return upright.
- Attachment points. Some harnesses have more attachment points than those required by the Class II and Class III specifications. Use caution and ensure that these attachment points are used in a manner consistent with their design and intended use. Misuse of a harness could result in a fall, rescuer injury, or other accident. Also note that some harnesses may include gear loops from which to hang spare carabiners and other equipment. Gear loops should NEVER be used as attachments to the rescue rope system; they are not engineered to resist the forces that a rescue rope system can sustain. Rescuers operating in close spaces, like wind turbines, confined spaces, or cave rescue, may choose to forego gear loops altogether and choose a harness without extraneous straps that might snag on protrusions.

Gloves

Gloves are worn in the rope rescue environment to protect the hands against the weather and, in the case of rope operations, against burns and abrasions from a running rope. They shield the hands and prevent discomfort that may cause the rescuer to lose control of the rope. Gloves designed to protect against weather and cold may not be robust enough for rope work. Those gloves designed for structural and wildland firefighting, and for utility use, are typically not suited for ropework because they do not provide sufficient dexterity, they are often quite bulky, and are not designed to provide protection of the hand across the path where the rope will run. In any case, gloves used for ropework should be reserved solely for that task. Note that ropes shared between functions run the risk of contaminating the rope with oil, soot, and debris from other uses.

While NFPA has standards for structural firefighting gloves (NFPA 1971) and utility gloves (NFPA 1951), there is no NFPA standard for rope rescue–specific gloves. Glove selection is something of a personal choice due to the wide range of hand sizes, performance requirements, and personal preferences of users. Some users feel that a snug fit provides greater dexterity, while others prefer a relaxed and loose fit. Where there is a likelihood of debris getting into the glove during use, a tight cuff with a Velcro closure can provide additional security. However, in hot environments, an open cuff offers ventilation. Some users prefer their gloves to also offer warmth, while others prefer that gloves allow the hands to remain cool. Some users require more thermal protection and/or extra cut resistance, while other users do not. Most glove users prefer the extra protection offered by full finger coverage, but an increasing number of rescuers find that, for rope rescue operations, fingerless gloves offer the best of all worlds. Again, the AHJ has the final say, but any of these will suffice as long as the glove provides adequate protection for the work to be performed.

In addition to providing protection, gloves must allow the hands to retain a sense of feeling so that the fingers can manipulate equipment. Do not mistake "fast-rope gloves" for rope handling gloves. Fast-roping is a special technique used for tactical helicopter insertions, wherein the team member descends a very thick-diameter rope using only gloved hands for braking. The gloves designed for this use are particularly thick and offer little dexterity. If a rescuer needs gloves with such extra heat protection (more than a few layers of leather), the rescuer is rappelling too fast for rescue ropework.

Gloves that offer special protection from other contaminants, such as chemical protective gloves or medical gloves for body substance isolation, may not be sufficient for rope work. In such cases, rope handling gloves may be worn over top of these types of gloves as needed. Every effort should be made to avoid

> ### TIP
> Be sure your gloves protect the part of the hand that YOU typically use for rope handling!
> Are you a (see **FIGURE 5-6**):
> - Palm gripper
> - Finger pincher
> - Thumb squeezer
>
> Reproduced from NFPA 1584. (2003). *Recommended practice on the rehabilitation of members operating at incident scene operations and training exercises* (2003 ed.). National Fire Protection Association. All Rights Reserved.

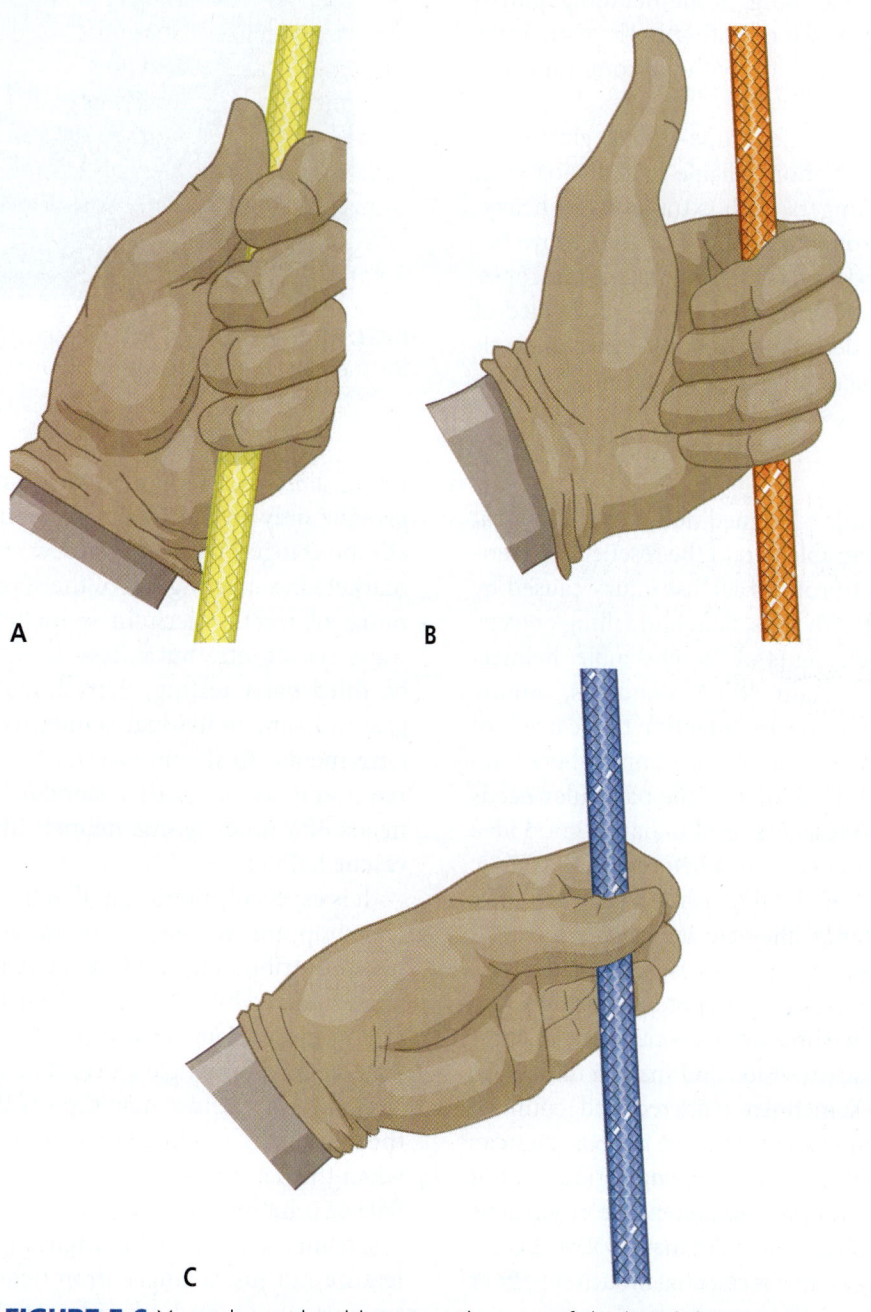

FIGURE 5-6 Your gloves should protect the part of the hand that you use for handling the rope. **A.** Palm gripper. **B.** Finger pincher. **C.** Thumb squeezer.

handling rescue ropes with contaminated chemical protective gloves. Contaminated ropes should be handled in the same manner as other contaminated PPE: either disinfected or discarded.

The best gloves for rope work are generally made of leather or synthetic leather with extra layers to protect the hand from heat and abrasion from a running rope. Heavy leathers, such as cowhide, offer greater durability and heat resistance, while soft leathers, such as goatskin, offer more finger dexterity. Leather is a super-durable and abrasion-resistant material that insulates against rope burn, yet is still pliable enough to allow the user a good amount of "feel" for tying knots, rigging, and rope handling. Rope-handling gloves with an extra layer or two of leather in the palm offer even greater heat protection without compromising dexterity.

Many purpose-designed rope-rescue gloves offer a combination of more pliable leather—or even synthetics—in the fingers, with extra layers of heavy-duty protection through the palm and the groove between the thumb and fingers for handling moving ropes, such as through a brake device. Gloves constructed of soft leather, such as deerskin or goatskin, offer an excellent compromise in comfort and performance.

Helmet

A well-fitting helmet designed for work/rescue at height is an indispensable part of the rescuer's PPE ensemble. The helmet protects against injury caused by impacts associated with slips, falls, and falling objects.

As with other elements of the ensemble, helmets are addressed by certain NFPA standards, among them NFPA 1951, but specific attention to the needs of technical rescuers is absent. At this writing, there is no one standard that fully addresses the particular needs of rope rescuers, so rescuers need to have a good idea of what they are looking for in a helmet.

While structural firefighting helmets generally offer good impact resistance, they are inadequate for most technical rescue operations. This is because they are typically quite heavy and they do not have a chin strap/retention system that works well for rope rescuers. These helmets tend to obstruct vision and make it difficult to maneuver in the sometimes cluttered and confined high-angle environment. For the same reasons, they can be very uncomfortable to wear for long periods and/or in hot weather. Rescue operations often take a long time, and rescuers are in hazard zones for many hours. A comfortable, well-fitting helmet is essential in such situations because the helmet must be worn for long periods.

Rescuers should use only helmets specifically designed for the activity and location (**FIGURE 5-7**).

FIGURE 5-7 Only use helmets specifically designed for the activity and location you are working in.
© Sementer/Shutterstock.

Using a helmet designed for something else may provide only an illusion of protection and can actually be dangerous to the wearer. Some helmets are marketed as meeting "all of the standards" for a wide range of rescue disciplines, including rope rescue, water rescue, etc., but unless these claims are backed by third-party testing, they should be taken with a grain of salt. Individual standards have specific requirements, so if one standard is met, the specific requirements in another standard will not be. The needs of a water rescue helmet differ from a trench rescue helmet.

It is especially important that helmets have a secure chinstrap and retention system. A ratcheting headband contributes to good fit and comfort, but must be supplemented by a chinstrap to keep the helmet properly in place. Helmets with elastic chinstraps are not suitable for high-angle work. As the elastic chinstrap stretches, the helmet may flip off the head and leave the wearer without head protection. This often occurs when the helmet is stressed, such as when the wearer falls or is hit by falling objects.

Helmets designed for high-angle work typically feature not just a single strap beneath the chin, but support points on both sides of the helmet as well as at the back of the helmet to help prevent the helmet from falling forward over the eyes. Such helmet retention systems are sometimes referred to as three-point retention systems. (The use of the term four-point is largely a marketing term, presumably stemming from the idea that if three is good, four must be better. However, this is something of a misnomer—the number of rivets or attachment points inside the helmet is less important than the fact that the helmet is anchored at three sides of your head: left, right, and rear.) None of the NFPA standards addresses this three-point retention concept, nor does the ANSI Z89.1 standard. The concept is addressed in two different European standards (EN 397 standard for industrial helmets and EN 12492 standard for mountaineering helmets), but with very different approaches. The main difference between the chinstrap requirements in these two is that the industrial requirement mandates that the chinstrap releases relatively easily so as to avoid strangulation in the event that the helmet is wedged into a crack with the rescuer's head attached, while the sport standard requires the chinstrap to hold at a higher strength to help ensure that the helmet stays on the head in a tumbling fall.

If the task warrants use of a helmet that is designed to release easily—for example, should the user's head become wedged during an operation—by all means, choose a helmet with a quick-release chinstrap feature. Otherwise, a good rope rescue helmet should have a chinstrap that requires much greater force to cause release; this helps keep the helmet on the rescuer's head in a tumbling fall.

The shells of helmets used in rope rescue are constructed of materials such as plastic, fiberglass, Kevlar, or composite materials. The shell should have some rigidity to resist penetration by sharp objects, but at the same time it should have enough flexibility to absorb some of the blow that otherwise would be directly transmitted to the skull and spine. The design of the helmet should protect the head against objects falling from above and hitting from the side.

The inside suspension of the helmet may hold the shell away from the skull, or the helmet may be lined with an impact-resistant Styrofoam-like material. While either is acceptable, consideration should be given to air circulation and comfort, particularly during hot weather or sweat-inducing labor, as well as to durability of the helmet during carrying and storage. A slight brim helps prevent rainwater or spray from dripping into the rescuer's face. Coverage on the sides, back, and front should provide adequate protection while still allowing a good upward field of vision and to ensure that the back of the helmet does not catch on equipment carried backpack style, such as rope bags or rescue kits.

Hauling Systems Overview

Rescuers use **hauling systems** for two basic reasons: (1) to make raising a rescue load easier and more convenient, and (2) to make it safer. The rigging of rope systems to achieve the task of hauling is performed by operations- or technician-level rescuers. Awareness-level rescuers may assist operations- or technician-level rescuers in raising loads by being on the **haul team**, the group of people who provide the power to raise the load. Always follow local standard operating procedures and training.

Hauling systems, especially those that also provide mechanical advantage, make the job of raising a load easier for the rescue team. The team needs less force to pull the load, though the task may take longer and the team will need to pull the rope a greater distance to raise the load. Rescuers who can construct hauling systems in varying sites and use equipment properly can set a place for raising that is both convenient and safe for both subject and rescuers.

For example, hauling systems could be established for any of the following criteria:

- Closer to vehicles and roadways
- Away from rockfall and other falling hazards
- Away from potential hostile activity
- Away from bystanders
- At a shorter drop

The most basic hauling system is a direct pull. For example, a rope has been let down a drop of 10 feet (3 m) and is connected to a load that weighs 100 pounds (444.8 N). The rope is directly on the load, and it is a straight haul. Therefore, bringing the 100-pound (444.8-N) load to the top will require 100 pounds (444.8 N) of force, plus some extra force to overcome the friction of the rope on the edge rollers. It also will require 10 feet (3 m) of rope.

Although this is considered a hauling system, it does not use any less force to move the load than if the rope had not been used. This is because the force pulling the rope (input) must be equal to or greater than the force on the load (output) to start the load moving. Adding a mechanical advantage pulley system to a basic hauling system allows less force on the input side of the system to move a greater load on the output side of the system.

How Hauling Systems Work

Pulleys are essentially wheels on axles with grooves in the wheels to guide the rope. When rigged with rope, a pulley becomes a simple machine, just as a lever or a ramp is a machine. The force required to move a load can be greatly reduced by using pulleys and rope to create mechanical advantage. With a mechanical advantage system, a relatively heavy load can be moved with minimal force. Depending on the configuration of pulleys and rope, varying amounts of **mechanical advantage** can be produced. The amount of mechanical advantage derived from a hauling system depends greatly on haul system efficiency, which includes how the system is rigged, and on the efficiency of the components (e.g., pulleys) used to build it.

The Haul Team

Haul team members should not be selected on the basis of brute force ability, but instead according to the following criteria:

- Ability to follow commands
- Ability to react quickly
- Sensitivity to the feel of the haul rope

Hauling systems create enormous forces on the rope rescue system. Unnoticed problems quickly can result in catastrophic system failure. Personnel constantly must monitor for potential problems, including, but not limited to, the following:

- Knots on moving rope that jam in cracks
- Broken gear that causes system failure
- Systems reaching their limit
- Pinned arms and legs

Good communication must be established between team captains and the haul team. Also, the haul team must keep in mind that what seems a normal speed for them seems very fast to those being hauled (i.e., the rescue subject and the litter attendant(s)). Therefore the haul team must pull slowly unless told to do otherwise, and must be prepared to stop at a moment's notice.

Communications for Hauling Operations

Awareness-level rescuers should understand the basic communications commands used during hauling operations (**TABLE 5-1**).

Supporting the Incident Action Plan

Awareness-level rescuers work primarily in a support role to an incident. This also includes the initial response to a technical rescue incident where operations- and technician-level rescuers are not yet on scene. In this situation, it is imperative that awareness-level rescuers do not engage in activities above their level of training and certification. This can have deadly results. The role of the awareness-level rescuer is to follow the directives of command staff to support and carry out the tasks outlined in the IAP. Awareness-level rescuers will often work under the Operations section, which is the "boots on the ground" part of the Incident Command

TABLE 5-1 Communications for Hauling

Stage of Operation	Command	Response
Roll Call	"Roll call!"	(Gets everyone's attention. Operations are about to begin. This is a sort of a "quiet on the scene.")
	"Belay ready?"	"Ready on belay!" (Device unlocked and in hand), or "Standby!" (Give indication of time needed.)
	"Main line ready?"	"Ready on main line!" (Haul team is in position) or "Standby!" (Give indication of time needed.)
	"Edge tender ready?"	"Ready at the edge!"
	"Rescuer ready?"	"Rescuer on belay and ready for raise!"
	"Safety ready?"	"Safety ready!"

Stage of Operation	Command	Response
Positioning the Load	"Position the load!"	(Action: Slide the package below the edge or in the work area.)
	"Slack on main line (or belay)!"	(Action: If needed to adjust position of the load, as necessary)
	"Tension on main line (or belay)!"	(Action: Pull any excess rope out of the system to prepare for haul, but do not raise the package at this time.)
Tensioning the System and Movement	"Set slack!"	(To be determined and used by haul team. DO NOT REDO ROLL CALL with each reset of the rope system. Operation carries on.)
	"Load the system!"	"Haul or raise on main line!" "Haul! (slowly)" (Action: Main line is initially tensioned to feel resistance.)
	*Rescuer needs: Commands should be specific to the needs of the tender, such as speed, belay, and indication of progress (e.g., "Haul on main line slow" or "Raise on main line fast").	
Termination	"Load at the edge!"	(Action: Stop, set, and reset for final haul, as necessary.)
	"Load on the ground!"	(No reply needed, but all rescuers remain at devices until told "off main" and "off belay.")
	"Off main line!"	(Load is no longer on main line and can be disconnected.)
	"Off belay!"	(Load is free and clear of the drop zone and can be disconnected from the belay.)
Additional Commands	"Stop!"	(Action: Command can be given by anyone; it means "freeze.")
	"Stop, stop, why stop?"	(A question asked to the group when an unexpected long pause has occurred.)
	"Tension!"	(Action: Too much slack exists. Take up excess rope.)
	"Slack!"	(Action: Too much tension. Need more rope. Give amount when appropriate, e.g., "Slack belay 2 feet.")
	"Prepare to change over! Rig for lower!"	(Action: Initiate change from raise to lower.)
	"Reverse haul!"	(Action: Command used to initiate the reversal of the mechanical advantage system to lower the load.)
	"Rock!"	(Safety issue: Given when any object is falling or dislodged.)

System (ICS). The Operations section of an incident is unique in that it expands from the bottom up. At the beginning or an operation, there may be only a few resources available, all of whom may initially report to the incident commander (IC). As the incident grows and additional resources are deployed, levels of supervision are added as needed. Limiting span of control is especially important in the Operations section because this is the area where the most hazardous activities are often carried out.

In following the directives they are given, awareness-level rescuers will contribute to the overall objectives of the IAP, including helping to ensure the safety of responders, subjects, and bystanders; helping to achieve tactical objectives, and contributing to the efficient use of resources.

Incident Facilities and Stations

Awareness-level rescuers must be familiar with the various types of facilities and stations commonly established in during a technical rescue incident:

- Command post: The location where primary command functions for an incident take place. Usually somewhere in the vicinity of the incident; there is only one command post at a given incident.
- Base: Often (but not always) co-located with the command post, where the ICS is coordinated for an incident. There is usually only one base per incident, although from time to time it may be necessary to have separate bases for different purposes (e.g., air base, search base, etc.).
- Staging area: A defined location where resources are collected or gathered while awaiting an assignment
- Rehabilitation: A defined location where incident personnel may rest and recover. Awareness-level rescuers may support rehabilitation operations by setting up the location or stocking the location with supplies.

Accountability

All responders to an incident should begin their shift with check-in procedures as established by the AHJ (**FIGURE 5-8**). Checking in officially records a responder as present at an incident and shows that they are available to receive an assignment. Using an established check-in process helps supervisors to track resources and ensure accountability, make personnel assignments, track personnel in an emergency, and account for all personnel during demobilization.

When working at an operation, it is important to follow the appropriate chain of command. In an ICS chain of command, there is an established line of authority wherein a given group of personnel will report to an individual, who is in turn subordinate to higher-level personnel. Unless otherwise instructed, when there is a question or issue with an assignment or something related to the incident, typically an individual should follow chain of command when requesting or transmitting information.

FIGURE 5-8 Before beginning your shift, follow your local check-in procedures to ensure accurate accountability.

FIGURE 5-9 A briefing is essential to ensure that all rescuers understand their roles.
Courtesy of Randy Tisor/IMCOM.

Briefings

Once checked in, resources are typically assigned to a point of contact for an initial briefing (**FIGURE 5-9**). It is good for rescuers to be prepared in advance for this briefing, know what they are listening for, and ensure that they obtain the information necessary to help them to be the best resource possible. Writing information down, if possible, will be helpful in the event that information needs to be passed on to someone else.

Briefings will typically consist of the following:

- Current situation, assessment, and objectives
- Specific job responsibilities
- Location of work area
- Procedural instructions for obtaining additional resources
- Safety hazards, required safety procedures, and PPE
- Identification of operational period work shifts

Reporting Progress

The transfer of essential information between rescuers during the course of a rope rescue is integral to a successful operation. Information is constantly being relayed over the duration of an operation, but there are certain stages at which information relay becomes critical. This is especially true in rope rescue operations where there may be a series of operational sites or task groups performing different, but related, roles. For example, one task group may be tasked with setting anchors and building a haul system, while another is responsible for rigging a high directional near the edge. Coordination of information about these interrelated tasks is essential to ensure that they are aligned properly and will function effectively together. Likewise, when a subject comes off a rope rescue system and is handed over to ground personnel to be carried or wheeled to a waiting ambulance, a succinct but thorough download of patient care information is required. Critical information relay might take place between segmented operational sites, between shifts, or when handing off the subject from one caregiver to another. The goal of any such transition is to ensure safety in the evacuation, continuity of care, and effective transfer of important information that may impact the outcome of the rescue.

Failed communication happens under the following conditions: a lack of a systematic process or information, poor recordkeeping, cultural differences, human factors, or differing expectations. Although written documentation is admittedly not always practical in field operations, it is arguably the best method for ensuring adequate transfer of information. Rescuers should practice using the closed-loop communications methods for delivering updates or reporting progress to a supervisor. Closed-loop communication is discussed in Chapter 3, *Hazards Associated with Rope Rescue*.

Whether verbal, written, or some combination thereof, information relay must provide the following:

- Critical information about the subject(s)
- Important big-picture details about the rescue operation
  - Goals of the operation
  - Tactical objectives
  - The progress of groups performing assigned tasks
  - Relevant intelligence that may influence future steps
- Resource needs
  - Logistics
  - Rehabilitation needs
  - Communications and information technology needs
- Any factors that may impact subject or rescuer safety

When sharing information among rescuers, each should take into consideration the level of expertise and the perspectives of the other persons involved.

Personnel Rehabilitation

Rescue is taxing, physically and emotionally. Locating, accessing, extricating, and transporting a subject from a dire situation can be an arduous task, often requiring extended periods of exertion with little opportunity to rest. The impact of stress illness and overexertion injury is well established in this field, but the detrimental effects on rescuers can be drastically reduced with a bit of care and attention.

The degree of physical fitness required of rescuers should not be underestimated. Ropes, carabiners, and other rescue hardware are heavy, and carrying them into a scene only to then have to carry them and the subject back out again is not an easy task. Rescuers should maintain a good level of fitness and be prepared to work in a variety of temperature ranges; dressing appropriately can also help prevent heat, cold, exhaustion, and other stress injuries.

The AHJ should implement appropriate procedures for personnel rehabilitation during incidents. The goal of rehabilitation is to offer rescuers opportunity to rest, replace fluids, eat, control temperature, and receive medical attention as necessary throughout their operational period.

Rehabilitation facilities may not be set up at operations expected to have a short duration, but when to implement rehabilitation, and what to set up, is at the discretion of the IC. This determination is typically made based on the type and location of incident, weather, tasks, and number of available personnel. Establishing and staffing the rehabilitation area is a task likely to fall to awareness-level responders. The rehabilitation area should be set up in a protected area outside but still near to the area of operations, and should provide adequate protection from the environment.

The awareness-level responder can assist with the care and rehabilitation of other rescuers by monitoring how they appear to be doing, providing assistance in the rehabilitation area, supplying water, and setting up cooling or warming stations.

After-Action REVIEW

IN SUMMARY

- To be able to carry out an assignment, rescuers must understand the standard operating procedures of the authority having jurisdiction (AHJ), be capable of selecting and utilizing appropriate personal protective equipment (PPE) for the situation, and assist with equipment appropriate to the task.
- AHJs will develop standard operating procedures based on the types of incidents that are likely to occur in their area.
- Common PPE at rope rescues includes harnesses, rope rescue gloves, and technical rescue helmets.
- Rescuers use hauling systems for two basic reasons: (1) to make raising a rescue load easier and more convenient, and (2) to make it safer.
- The rigging of rope systems to achieve the task of hauling is performed by operations- or technician-level rescuers. Awareness-level rescuers may assist operations- or technician-level rescuers in raising loads by being on the haul team.
- Rope rescue utilizes mechanical advantage systems to move heavy loads with minimal force.
- Awareness-level rescuers work primarily in a support role to an incident under the Operations section of the Incident Command System (ICS).
- To assist command in accountability, all responders to an incident should begin their shift with check-in procedures as established by the AHJ. Using an established check-in process helps supervisors to track resources and ensure accountability, make personnel assignments, track personnel in an emergency, and account for all personnel during demobilization.
- Whether verbal, written, or a combination, information relay must provide the following:
  - Critical information about the subject(s)
  - Important big-picture details about the rescue operation
  - Resource needs
  - Any factors that may impact subject or rescuer safety
- The AHJ should implement appropriate procedures for personnel rehabilitation during incidents. The goal of rehabilitation is to offer rescuers opportunity to rest, replace fluids, eat, control temperature, and receive medical attention as necessary throughout their operational period.
- The awareness-level responder can assist with the care and rehabilitation of other rescuers by monitoring how they appear to be doing, providing assistance in the rehabilitation area, supplying water, and setting up cooling or warming stations.

KEY TERMS

Class II harnesses Assemblies composed of webbing and worn primarily around the waist and thighs.

Class III harnesses Assemblies composed of webbing and worn around the torso, upper body, and shoulders.

Fall arrest The act of stopping a fall once it has started.

Hauling systems Rope systems generally constructed from life safety rope, pulleys, and other rope-rescue system components capable of lifting or moving a load across a given area.

Haul team The group of individuals who provide the power to raise the load.

Mechanical advantage The ratio of the output force produced by a machine to the applied input.

Personal escape An escape harness typically worn during work-at-height operations explicitly for the purpose of enabling the rescuer to escape if needed to a lower position of safety; typically used with an emergency descent system.

Restraint A system that prevents a person at height from reaching an edge where a fall might occur.

Standard operating procedures Written organizational directives that establish or prescribe specific operational or administrative methods to be followed routinely for the performance of designated operations or actions (NFPA 1521).

On Scene

1. Which of the technical rescue disciplines addressed by NFPA 1006 are most likely to overlap with rope rescue?

2. What PPE is an awareness-level responder most likely to use at a rope rescue incident? Why?

3. What is the importance of accountability in the ICS system?

4. What considerations might you take into account when deciding where a rehabilitation camp will be located?

SECTION 2

Operations

CHAPTER 6 **Developing the Incident Action Plan**

CHAPTER 7 **Hazard-Specific Personal Protective Equipment**

CHAPTER 8 **Rescue Equipment**

CHAPTER 9 **Ropes, Knots, Bends, and Hitches**

CHAPTER 10 **Principles of Rigging**

CHAPTER 11 **Anchorages**

CHAPTER 12 **Fall Protection and Belay Operations**

CHAPTER 13 **Patient Evacuation**

CHAPTER 14 **Lowering Systems**

CHAPTER 15 **Mechanical Advantage Systems**

CHAPTER 16 **Working in Suspension**

Operations

CHAPTER 6

Developing the Incident Action Plan

KNOWLEDGE OBJECTIVES

- Review the purpose of an incident action plan (IAP) and the role of size-up in generating the IAP. (**NFPA 1006: 5.2.1**, pp. 84–89)
- Describe how to utilize an incident objectives form when generating an IAP. (**NFPA 1006: 5.2.1**, pp. 85–89)
- Differentiate among objectives, strategy, and tactics.
- Identify the topics to be covered in IAP briefing.
- Explain how the IAP is coordinated with an incident command structure.
- Describe the process of personnel accountability reporting. (**NFPA 1006: 5.2.24**, p. 90)
- Identify rope rescue communications considerations.
- Identify the tasks that must be completed during demobilization. (**NFPA 1006: 5.2.24**, pp. 92–93)
- Identify the considerations for terminating command. (**NFPA 1006: 5.2.24**, p. 93)
- Identify the steps necessary to return all resources to a ready state. (**NFPA 1006: 5.2.24**, p. 93)
- Explain the documentation considerations associated with the termination of an incident. (**NFPA 1006: 5.2.24**, p. 94)
- Review the purpose of a postincident analysis. (**NFPA 1006: 5.2.24**, p. 94)

SKILL OBJECTIVES

- Communicate information during a rope rescue operation. (**NFPA 1006: 5.2.1**, pp. 90–92)
- Terminate a rope rescue operation. (**NFPA 1006: 5.2.24**, pp. 92–94)

You Are the Rescuer

You are among the first on scene at a stadium collapse incident. You see several bruised and battered people walking around in apparent shock, a few lying in the rubble, and still others stranded higher up on portions of the structure still standing. Several other agencies have been dispatched to assist, and it is clear that this will be a large, complex operation. While you will not be ultimately responsible for the operation, as first on scene, your role is crucial to setting the stage for an effective organizational structure, seamless communications, and safe management that involves a safe operation for all involved.

1. How do you expand the organization to deal with increasing numbers of personnel?
2. How many personnel can each person effectively manage?
3. How should rescue activities be overseen in the management structure?

 Access Navigate for more practice activities.

Introduction

Rescue teams can enhance their efficiency and safety by following established procedures. When team members know in advance the actions expected of them and can anticipate a team leader's instructions, they are better prepared mentally and physically to respond. When there are no established operating procedures in place, field personnel must make significantly more operational decisions, which in turn reduces efficiency and increases the likelihood of an incorrect decision.

Rescue groups should make a conscious determination as to whether or not their procedures should adhere to appropriate national standards, and—if so—to which ones. A fire rescue organization, for instance, may choose to adhere to National Fire Protection Association (NFPA) standards, whereas a mountain rescue team may adopt ASTM standards. Or, based on operational need, an organization may decide to follow a combination of relevant standards, generally with modifications relevant to their own application.

Rescue is not a one-size-fits-all proposition; while standards developers in national organizations often provide useful big-picture guidelines, no one understands local needs better than the responders who work there. Whatever national standards an organization may or may not decide to follow, a concerted effort should be made by knowledgeable, experienced, local personnel to develop standard operating procedures (SOPs) that are relevant and realistic to their response area.

Incident Action Plan

Armed with a solid baseline of SOPs and information from a thorough size-up, a specific strategy for moving toward resolution of the incident will begin to become apparent (**FIGURE 6-1**). Particularly for larger or more complex incidents, strategic plans should be recorded, reviewed, and updated in an incident action plan (IAP), which remains a living document until the conclusion of the incident. The incident commander (IC) oversees the development of the IAP, which may be verbal or, for incidents of longer duration, in writing. The IAP establishes the overall strategic decisions and assigned tactical activities for the incident, such as a decision to raise or lower a subject and is crafted with information determined during a scene size-up. Scene size-up is discussed in Chapter 2, *Size-Up*.

An IAP does the following:

- Unites the efforts of command and rescuers
- Clearly defines how resources fit into the operation
- Provides an objective touchstone for command decisions
- Clearly identifies tasks and assignments

FIGURE 6-1 A thorough size-up is critical in the development of an effective IAP.
Courtesy of Trask Bradbury.

- Establishes a framework for evaluating progression/success
- Helps to maintain a big picture perspective
- Lends to consistent and informed communications

In a large operation, IAP development might involve planning and operations command staff, while in smaller operations this task may fall to the IC at the outset, and perhaps even throughout the course of the incident. If ropes are required for safety or evacuation, personnel knowledgeable in their use should be in charge of planning and executing that part of the operation.

The IAP will encompass foundational strategies related to an incident, including identification of objectives, assignment of resources, establishing communications protocols, and providing for rapid intervention in the event that a rescuer is injured. The IAP is developed and updated on an ongoing basis by the IC with input from all general staff.

Incident Objectives Form

A good strategic IAP begins with identification of specific objectives, priorities, and safety considerations for the present and at least one subsequent operational period. While in a large or ongoing operation this information should be recorded on an incident objectives form (the Federal Emergency Management Agency's [FEMA's] ICS 202 form), the same thought processes will apply even to real-time, spontaneous planning at the outset of an operation. If a written IAP is used, the incident objectives form may be used as the front page of the IAP.

The complete IAP will encompass all operational periods for an incident but may be broken down into different segments relative to individual operational periods and/or different areas of command under the jurisdiction of the incident. There is no set requirement for how long an operational period lasts, but often these range from 6 to 10 hours depending on a range of factors. Provisions for rest, recovery, and rehabilitation of rescuers is one of the prime considerations in establishing length of operational periods. If an IAP is segmented for operational periods, the incident objectives form may be used as a separator between operational periods, each having its own smaller objectives relating to the overall plan. In any case, it should be made clear at the outset what the anticipated duration is for a given operational period.

Planning P

In addition to the incident objectives form, FEMA's Planning P concept may be used in generating an IAP. As previously discussed, the planning process is grounded in the following phases:

- Phase 1: Understanding the situation
- Phase 2: Establishing incident objectives
- Phase 3: Developing the plan
- Phase 4: Preparing and disseminating the plan
- Phase 5: Executing, evaluating, and revising the plan

These steps are represented in FEMA's Planning P, shown in **FIGURE 6-2**. The Planning P is a mnemonic that was developed early in the days of ICS to emphasize the critical importance of adequate planning in incident management. It had been observed that mismanagement of resources, inadequate tactics and strategies, safety issues, and poor incident outcomes were often attributable to lack of planning. The Planning P differentiates between initial and ongoing planning and provides a framework for methodically revisiting key areas for consideration.

While the Planning P in its expanded form can be used to manage an incident over a long period of time, rope rescue incidents are generally of a more urgent and immediate nature and can be resolved in a shorter period of time. Regardless of duration, the concepts detailed in the phases of the Planning P can be a useful tool for any rescue operation when crafting an IAP.

Phase 1: Understanding the Situation

While important at the outset, the steps in Phase 1 only apply to the initial phases of an operation and may be performed by awareness-level rescuers. These activities form the vertical leg of the Planning P. Once Phase 1 is completed, the activities outlined in the vertical leg are not repeated, and the Planning P becomes what FEMA refers to as an Operation O, represented by the circular portion of the P. The steps outlined in the Operation O are composed of the ongoing development and updating of incident objectives; establishing tactics to meet the objectives; effective dissemination of the plan; and execution, evaluation, and revision of the plan. These steps recycle and repeat themselves in an ongoing fashion with every operational period until conclusion of the operation. The information required for Phase 1 of the IAP has already been discussed in Chapters 2, *Size-Up* and 4, *Initiating a Response*.

Command Emphasis. Every operational period (or Operation O) is identified by a command emphasis that is at least in part influenced by the prioritization of objectives. Command emphasis is a section within the ICS 202 form that describes expected outcomes or

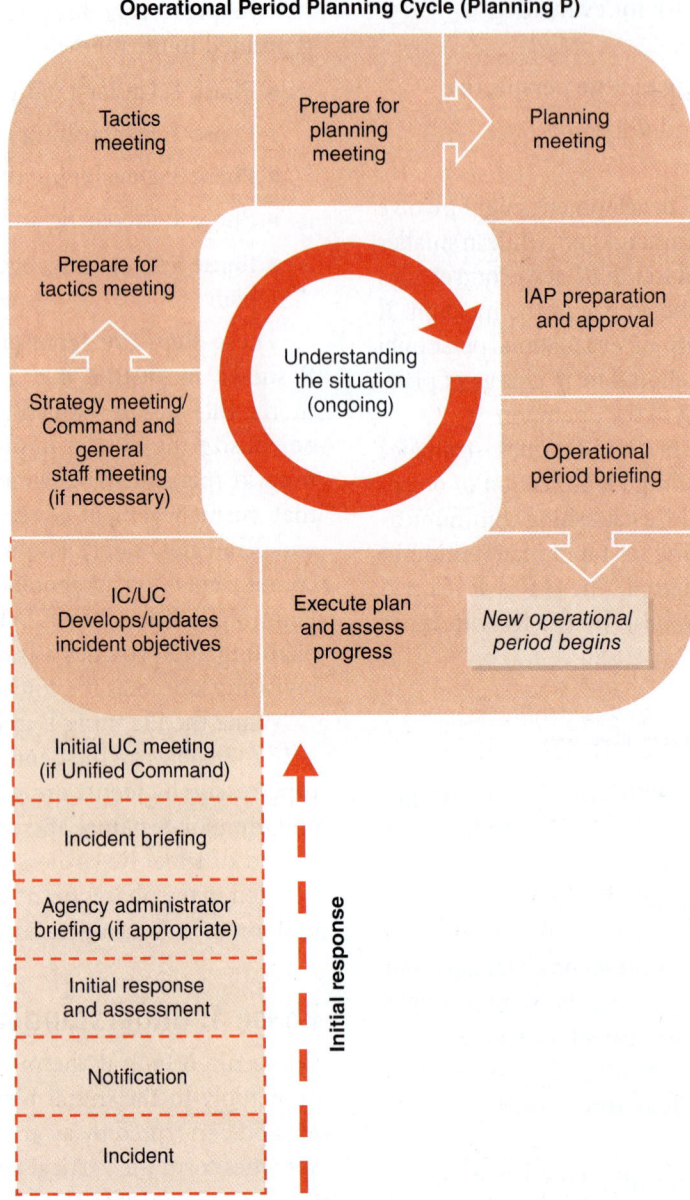

FIGURE 6-2 FEMA's Planning P tool.
Reproduced from FEMA. (n.d.). *IS-201: Forms used for the development of the incident action plan.* https://emilms.fema.gov/IS201/ICS0102summary.html

milestones for the operational period, a list of priorities, or the key message(s) that underpin the effort for that operational period.

Safety should be a key component of command emphasis observations, as probability and consequence of hazards will rise and fall with progression of the incident. Safety aspects of command emphasis may be related to weather, environment, darkness, state and stage of the operation, or other factors. The safety component of a command emphasis statement is a good place to highlight specific concerns identified by the safety officer in an incident, and in all cases should be consistent with the incident safety plan.

Command emphasis will also generally include a statement related to sequence of events or targeted goals to achieve for that particular period, such as a segment of the rescue to complete or a stage to accomplish. Often an incident will involve more than just a technical rope rescue phase; there will also be a need to locate, access, stabilize, and transport the subject. In some incidents, all of these phases may be able to be accomplished in one operational period, while other incidents may require several operational periods to fulfill all of the objectives. These are identified in the command emphasis for a given operational period.

Phase 2: Establishing Incident Objectives

Crafting Incident Objectives in the IAP.
Incidents are most effectively managed when specific objectives are identified in the IAP and used to drive the actions of personnel in every operational period and location. At the big-picture level, objectives are simply general goals or statements of what must be accomplished. On a more practical level, the incident objectives will ultimately become specific tactics for achieving the stated goals in Phase 3 and may be assigned to one or more operational periods. For example, broad-scale objectives for a stadium collapse incident might be to locate, extricate, and transport all impacted persons as needed. More specifically, objectives for an operational period might include a directive for a company to remove and document all ambulatory victims, and another for the Special Operations team to locate all buried victims within a given segment with 80% confidence.

If a form is used, and when briefings are given, the incident objectives should be listed in order of priority, with the most important being first. Inclusion of ALL the objectives for the entire incident response, as well as for the present operational period, will help rescuers to maintain a big-picture perspective even while striving toward shorter-term goals.

Incident objectives in the IAP should follow the S-M-A-R-T acronym model:

- **Specific:** Each objective should be very specific, with precise and unambiguous wording.
- **Measurable:** Each objective should be measurable by its outcome.
- **Action-oriented:** Each objective should be described in terms of actions, using verbs, to describe the anticipated achievement.
- **Realistic:** Each objective should be realistically achievable, particularly in light of the operational period and the available resources.
- **Time-oriented:** Each objective should state the targeted time frame in which it is expected to be accomplished.

Short, clear, and concise objectives should drive the allocation and utilization of critical resources available within the established operational period to achieve the overarching strategic goals of the operation, while at the same time remaining flexible enough to allow for strategic and tactical alternatives. If, for example, the Special Operations team suddenly decides they need assistance from Company B personnel, then Company B may be reallocated.

Incident objectives are intended to establish direction for what must be accomplished, but do not

FIGURE 6-3 Incident objectives are intended to establish direction for what must be accomplished, but do not specify how it will be achieved. In this case, the incident objective was to remove the subject from suspension within 1 hour.
Courtesy of Pigeon Mountain Industries.

specify how it will be achieved. For example, for an incident in which a window cleaner is hanging from his fall protection harness on the side of a building, the objectives might include the following (**FIGURE 6-3**):

- Establish and clear a hazard zone below immediately
- Evacuate affected people on impacted floors
- Access the subject within 20 minutes
- Relieve immediate threats to subject's safety
- Remove the subject from suspension within 1 hour
- Refer the subject to medical care

Clearly defined objectives form the foundation for managing an incident; tactical direction is derived from such objectives. Using incident objectives as a starting point helps ICs and management staff to maintain focus and direction while establishing a progressive pace and systematic approach to completing the goals.

Incident objectives and command emphasis should be communicated to all supervisory personnel at the Section, Branch, Division/Group, and Unit levels, either in the written form or as a verbal briefing.

Phase 3: Developing the Plan

Operational Tactics. In Phase 3, the rubber, as they say, meets the road. Here is where strategic plans are made and tactics formed to achieve the goal of rescuing the injured or stranded subjects. This will likely involve attention to medical needs as well as evacuation needs. The subject's medical condition should be evaluated as soon as possible, even as other rescuers prepare the rope rescue system. An effective way to accomplish this is to divide the job into specific tasks. Each task is assigned to a subgroup headed by individuals responsible to the group leader. Among subgroup tasks would be jobs such as rigging, anchoring, preparation of the litter, evacuation, crowd control, communications, safety equipment control, and rope management. The operations officer makes sure that all the actions of the various subgroups work together to achieve the ultimate goal: the rescue. This is an ongoing and ever-changing process and must be updated as progress is made.

Strategy differs from tactics in that strategy identifies an approach to achieving the objective, while tactics state the steps to achieve the strategy. For example, in the previously described incident, a strategy for achieving the objective of physically accessing the subject within 20 minutes might be to lower a rescuer(s) by rope from above.

Strategies are not tactics, but they do form the framework for developing tactics. A wise operations officer will be constantly forming and re-forming strategic plans, including backup plans, from the time they learn of the incident objectives. Strategic plans must be considered in light of available resources, incident parameters (including urgency and hazards), and likelihood of success.

Strategy should not be developed in a vacuum; rather, it is best formed in a collaborative effort with other stakeholders and experienced personnel to help ensure that relevant factors are taken into consideration and that the plan is achievable, appropriate, and a wise use of resources.

From the identified strategy, tactics are derived. Tactics involved in implementing the strategy of lowering a rescuer by rope from above might include as the following actions:

- Transport a rope rescue kit to the roof by way of the building elevator.
- Establish roof anchors.
- Establish a lowering system with a mirrored belay.

This may be when resources are assigned, either in a general or a specific sense. Tactics may be broken down into specific tasks, with each task assigned to a subgroup headed by individuals responsible to the group leader. The operations officer makes sure that all the actions of the various subgroups work together to achieve the ultimate goal: the rescue.

Distinctive identification (e.g., helmet, vest, clip-on tag) helps make operational assignments known as additional emergency personnel arrive. Assignments should also be identified on the radio to enhance incident-wide understanding of who is in command.

Understanding available resources and their capabilities is an important part of planning. In smaller incidents, this may occur "real-time," with very little time gap between Phase 3 tactics and Phase 5 execution. In larger incidents, there may be a more extended time period for Phase 4, the preparation and dissemination of the plan.

Phase 4: Preparing and Disseminating the Plan

Although incident command system (ICS) forms from FEMA are available for capturing and providing information as part of the IAP, these may or may not be used on all incidents. Regardless of whether or not ICS forms are used, a briefing should occur to inform rescuers of the incident objectives, strategic plan, and tactics. Topics to address in the briefing include the following:

- Incident objectives: Including information on what has already been achieved, what still needs to be achieved, and what is expected to be achieved in the present operational period
- Incident communications plan: Including all forms of communications, whether phone, radio, or otherwise. In larger operations, a telephone communications plan might include a contact list of key staff and staff agencies. Most rope rescue operations will be well served by radio alone, and in these cases, the briefing should identify assigned radio frequencies, trunked radio systems, talk group assignments, etc.
- Command staff information: Rescuers must be aware of who is in charge; who they report to; and assignments for other leaders, supervisors, and managers as appropriate.
- Assignments: Personnel should know their own assignments and should also be aware of what other teams or operational groups are doing.
- Incident map: A map is useful in showing where key incident facilities are located, identifying operational boundaries, and helping personnel to visualize how the strategic plan will come together.

- **Medical plan**: The medical plan includes relevant information pertaining to the subjects of the rescue, as well as to rescuers themselves. Information on medical responsibilities onsite, as well as nearby facilities, are relevant here.
- **Safety message**: Rescuers should be aware of relevant risks or threats and should be informed of the general safety message (which may evolve as circumstances change over the course of the operation).
- Other information, as needed: Where appropriate, rescuers should be aware of other units operating on the incident, traffic, weather, or other relevant information, plans for air support, etc.

An **operations briefing** should be held at the beginning of each operational period to ensure that all stakeholders are kept apprised of relevant operational information. In this briefing, the IAP for the upcoming operational period is reviewed.

In a larger, more extensive operation, briefings may be limited to those in leadership positions such as branch directors, division and group supervisors, task force leaders, team leaders, crew leaders, squad leaders, and incident support staff. However, where possible and especially on smaller incidents, it is generally advisable to include all available rescuers in the briefing so that information is transmitted most efficiently and consistently among personnel.

Phase 5: Executing, Evaluating, and Revising the Plan

An operation is effective only as long as those executing the operation are able to maintain dialogue with those in charge of it, and as long as plans continue to adapt to the changing needs of the incident. Information is exchanged through field reports, supervisory visits, and ongoing dialogue. If the task group assigned to access our window cleaner who is suspended in his harness on the side of a building finds that the elevator is out of commission, they might want to let Command staff know that they will be slogging gear up 20 stories before they can implement the plan. Once on the roof, if they observe storm clouds headed their way, this, too, should be communicated, as should operational progress, any obstacles, the subject's condition, and other information relevant to the operation. Ongoing communication is key to maintaining effectiveness of the ongoing IAP.

As the ongoing plan is evaluated and reviewed, it will continue to change until such time as the incident is concluded. An active and intentional approach to reviewing the progress of the incident will naturally lend itself to this process continuing to repeat itself from Phase 2 until the response activities have ended.

TIP

Maintaining an Organized Rescue

The following steps from the *National Park Service Technical Rescue Manual* can help keep a rescue operation organized and on target:

- Initiate a quick size-up of the incident, to verify the initial report and scope of the incident, by sending the closest available resource directly to the scene.
- Organize an immediate initial response to reach and stabilize the subject. Medically stabilize the subject by treating any life-threatening injuries; physically stabilize the individual so that he or she cannot fall farther; and psychologically stabilize the person with professional reassurance.
- Use ICS and identify positions (verbally on the radio and through the use of vests).
- Establish an accessible staging area for equipment that does not block or interfere with the area used for operational rigging.
- Limit communications with the rescuer or rescuers in technical terrain to the edge manager or the operations chief. Working in the vertical realm is an awkward task. Rescuers' efficiency is compromised if they are overtasked with too many instructions.
- Stay ahead of the logistics curve. Plan and act now. Be prepared for a rescue to take longer than expected. Request additional resources, supplies, and equipment well in advance so that efficiency does not suffer. It is better to request and cancel than to wait and wish you had called earlier.
- Keep rescue systems simple and safe. An overly complex system may compromise efficiency.

Data from Phillips, K. (2005). *National park service technical review manual* (10th ed.). Author. http://kristinandjerry.name/cmru/rescue_info/National%20Park%20Service/Basic%20Technical%20Rescue%20-%20Phillips.pdf.

Implementing the IAP

An IAP works best when it is coordinated within the ICS structure. One of the most important advantages of ICS is its flexibility. ICS is not a fixed system, but is designed to be flexible enough to expand or contract to the needs of any size incident. This is particularly important when the response to an incident grows, involving many more people. The organization adapts so that no one is overtasked and the organization remains functional and efficient. ICS is outlined in Chapter 4, *Initiating the Response*.

Span of Control

When utilizing an ICS framework to implement an IAP, it is best to view the structure in terms of need. All incidents will have a Command function (IC) and almost all will have an Operations function (operations officer). Other Command positions are often performed by the IC and delegated out to additional personnel only as needed. No incident needs to be burdened with "too many chiefs," but it is important to strike a balance between undertasking and overtasking leaders. The optimal span of control is a ratio of one supervisor to five subordinate personnel, and never more than seven. As the number of rescuers increases, all of the responders should be divided into smaller, tactical-level management components, such as teams, groups, or divisions, each with its own leader.

Supervisors must maintain a constant awareness of the location and function of all personnel assigned to them during a response. The fire service employs a **personnel accountability reporting (PAR)** system on the fire ground for this purpose. Personnel accountability may be accomplished through tactical worksheets, command boards, or apparatus riding lists (**FIGURE 6-4**). This tracking of resources also provides for efficient use of rescuers and helps to facilitate crew rotation for rest and rehabilitation during long-duration incidents.

FIGURE 6-4 A personnel accountability reporting system.
© Ted Pendergast/Shutterstock.

Chain of Command

Every individual responding to an incident should have one boss—and only one—at any given time. Every rescuer should know to whom they report, and every supervisor must be very clear on who works for them. This is for both safety and efficiency.

Roles and responsibilities should be assigned based on skills and abilities, along with the needs of the incident, rather than seniority or job titles. This may mean that a more senior person reports to a person lower in the management structure at times. Mature rescuers will maintain focus on the good of the whole, putting their egos aside at times like this and deferring to the person(s) assigned to command roles as appropriate.

Transfer of command responsibility should be explicit and direct, with a briefing held to announce the change so that there is no confusion.

Multijurisdictional/Multiagency Operations

Rescue operations that involve more than one jurisdiction or agency can be a breeding ground for conflicts over who is in charge, which complicates the development of unified response. There is provision within the IMS to resolve these potential conflicts through a system of **unified command**.

Under unified command, each agency or jurisdiction involved has an IC assigned to the operation (**FIGURE 6-5**). The IC jointly decides the incident objectives and appoints a single operations officer to carry them out. The key to unified command is the use of a single operations officer. The different agencies involved may easily maintain the integrity of their established teams by keeping their personnel together in functional units—such as branches, groups, or divisions—without any loss of team efficiency.

Communications

In the high-angle rope rescue environment, effective communication strategies are extremely important for effective incident management and personnel safety. By nature, conditions in the high-angle environment can often make normal communication very difficult.

Standard voice communications are often used in rope work. In many high-angle situations, however, the ability to hear voices is severely restricted by distance, wind, traffic, crowd noise, and other factors.

While the idea of radio communications may seem to be a simple solution to such challenges, conventional radio systems are subject to disruption by physical conditions at the rescue site, including confined

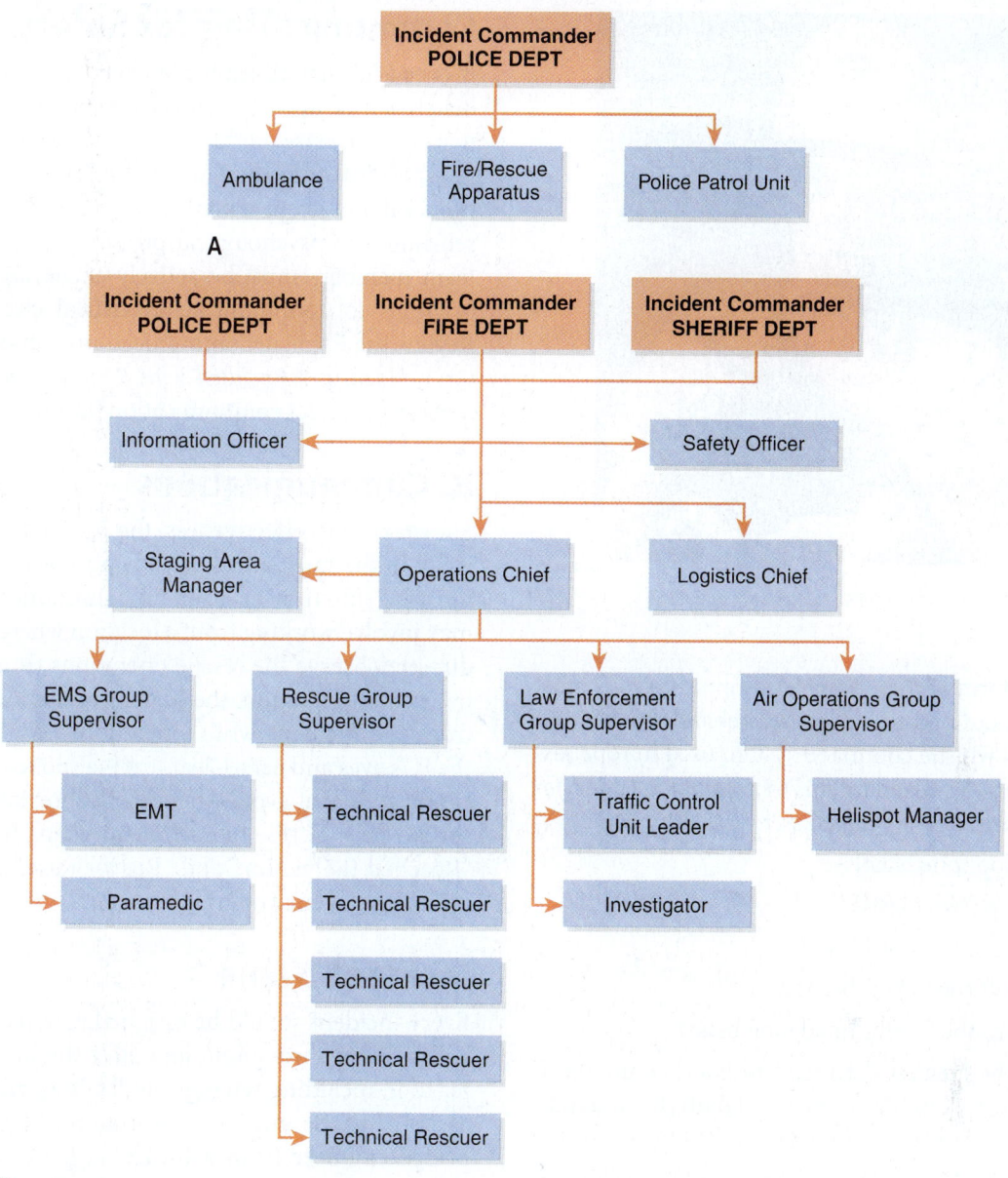

FIGURE 6-5 A. A simple ICS structure. **B.** An expanded ICS structure reflecting unified command by multiple jurisdictions.

spaces, terrain features, structures, and electromagnetic interference. In some cases, the situation can be improved by having all team members switch from repeater channels to simplex communication and/or by deploying a human radio relay at a strategic location.

If the team will be working with more than one frequency, there must be a systematic way to keep track of radio channels. Without a system for channels, confusion arises as individuals try to find a usable channel. If every radio has the same frequency setup, channel numbers can be used to indicate a frequency (e.g., "All units switch to channel 11."). If the frequency setup varies among the radios, the frequency name must be used (e.g., "All units switch to state law enforcement mutual aid."). A good strategy is either to label the channels on the radio or to keep a frequency sheet with the radio.

Another communications challenge can arise when rescuers are engaged in activities that require the use of both hands for other activities. In this case, a radio chest harness (**FIGURE 6-6**) may help to keep the radio protected and close at hand, requiring only a simple, short hand motion to key the mike. The addition of a remote hand microphone, throat microphone, or integrated radio headset with noise-canceling feature can also be useful.

Avoiding communication failures involves contingency planning. Have at least two forms of communication available, so that if a primary system fails, a backup system is available. One nonelectronic communication technique involves whistle blasts. These are limited in the amount of information that can be communicated, but they often are audible where nothing else works.

FIGURE 6-6 A radio chest harness.
Courtesy of Ken Phillips.

The exact form of these whistle communications must be worked out and practiced beforehand. **SUDOT** is a recognized whistle command system used in rope rescue (see ASTM Standard F1768, *Standard Guide for Using Whistles During Rope Rescue Operations*):

- **S**—Stop (one blast)
- **U**—Up (two blasts)
- **D**—Down (three blasts)
- **O**—Off rope (four blasts)
- **T**—Trouble! (continued long blast)

Often the greatest detriment to good communication is not electronic or mechanical failure, but the communicators themselves. In high-angle situations, it is imperative to communicate in a clear, concise, and specific manner. For example, does "right" or "left" mean as you face the structure or as you face away from it?

Here, borrowing from the natural or recreational world can be useful. On a cliff face, directions are often referred to in context of a climber facing the rock. Therefore, to the right of the rescuer would be face right and, to the left of the rescuer, face left. In river and stream operations, river right is on the rescuer's right as they face downstream, and river left is on the left facing downstream.

It is important to reduce the command vocabulary to as few words as possible and to use only words that are clear, concise, and have few syllables. Also, the same words should be used for specific actions. In lowering and hauling systems, the only word for cessation of action is "Stop!" Another word should never be substituted; "whoa," for example, could easily be mistaken for "slow," or even worse, for "go." For clarity in commands, any team member can say "Stop!", but only the team leader will give the command to proceed.

Communicating for Safety

It is crucial that all team members feel free to immediately speak up with critical information. Assuming that someone else on the team sees a hazard may result in a needless tragedy. This level of open communications will not occur naturally. To overcome the natural self-imposed psychological pressure not to speak up, team members must be actively encouraged through briefings to communicate in critical circumstances. Review the Explicit Communications section in Chapter 3, *Hazards Associated with Rope Rescue*, for a discussion on direct communication for safety.

IC Communications

For maximum effectiveness, the IC needs to be physically in the most advantageous position possible and must be able to hear all radio communications. This may involve working from a location where the IC can directly observe the rescue operations from a vehicle; in a remote situation, the IC may be stationed further from the incident while operations officer becomes the IC's eyes and ears. Clear text—a spoken communication style that avoids the use of abbreviated codes—should be used for radio transmissions (e.g., "Stop!" "Reached the victim," and "Rig for raise."). This optimizes interagency communication.

Incident Name

Every incident should be assigned a name. This name will be used throughout and after the incident to refer to it, including reference to facilities and resources assigned to the incident. For example, the rescue of a bridge engineer from a bridge support on Lashland Creek might be referred to as the "Lashland Bridge Rescue." If such incidents occur frequently in a similar location, additional identifying information, such as a date or an incident number, can be added for clarity. Naming an incident helps with coordination efforts and lends to clear communication, particularly where multiple operations might be occurring simultaneously or where resources are sourced from different agencies.

Demobilization

An important part of operational readiness is ensuring that resources are returned to a ready state as soon as practicable after the previous response. Leadership at any level should not leave their assigned post until all resources under their command have left those facilities. The IC should ensure that a demobilization plan is prepared and communicated to all resources, and that the plan is implemented.

The demobilization plan should provide for the return of all resources to a ready-state, including a timeline and process for releasing each resource from the incident, recordkeeping, and an anticipated timeline in which resources will once again become available. Considerations for logistics and resource demobilization include the following:

- Personnel
  - On-site debriefing (**FIGURE 6-7**)
  - Review/appraisal of performance as appropriate
  - Recording check-in, check-out, and other relevant information
  - Plans for rest, rehabilitation, and sanitation
  - Anticipated time to return to service
- Equipment
  - Inventory of rope rescue equipment and other items
  - Inspection and maintenance as required
  - Cleaning/drying of ropes and other equipment before storage
  - Return of equipment to the cache
  - Replacement of lost or damaged equipment
  - Resupply of expendable items
  - Documentation of use or disposition
- Vehicles
  - Refueling
  - Maintenance functions as required
  - Cleaning as needed
  - Restocking
- Specialty resources
  - Medical supplies
  - Air-support resources

Effective demobilization and stand-down of resources helps to ensure that resources are placed back into service and ready to go again as soon as practicable after the close of an incident.

FIGURE 6-7 A debriefing.
Courtesy of Dave Pope.

Manage Rescuer Risk and Site Safety

Rescuer safety is paramount in any response, including rope rescue incidents. It is the responsibility of every rescuer to maintain a vigilant watch for hazards and risks throughout the course of the operation, with attention to environmental, physical, social, and cultural factors that might pose a threat to rescuers and/or the operation. Experienced rescuers, especially, should take care to not become complacent. As an incident reaches its conclusion, both command staff and field personnel should continue to follow good accountability practices, including check-in and check-out procedures, span of control, and resource tracking as discussed earlier in this chapter. Command should not shut down until every resource (personnel and equipment) is accounted for and released from the incident. Hazard identification and risk management strategies are discussed in Chapter 3, *Hazards Associated with Rope Rescue*.

Scene Security and Custody Transfer

An initial survey of site safety will have been made during incident size-up, but maintaining a secure scene throughout the entirety of the operation is an important part of operational success and continues until all responders have cleared the scene. This includes continuing to limit traffic into and out of the general area surrounding the operation (within about 300 feet [91 m] of the incident, unless otherwise designated by command staff) as well as monitoring the vicinity for developing hazards. Barriers may be useful for helping to keep bystanders at a distance.

Where risks cannot be mitigated, areas of high risk (sometimes called a hot zone) can be marked off with marker tape, rope, or other visible means to alert rescuers to the need for special personal protective equipment (PPE) within that area. If sufficient personnel are available, assigning a person to monitor barriers can be helpful to prevent unauthorized entry and to ensure use of appropriate PPE and other protective measures.

Before an incident can be terminated, verification must be made that all rescuers are accounted for, that hazards created by the incident are not lingering, and that custody of the subject has been transferred to a responsible person. Local protocol will more specifically determine what constitutes appropriate transfer of custody, but where medical care is needed, generally this means that the subject has been handed off to more definitive care. In some cases, a subject may be released to self-care, a family member, or even law enforcement.

Recordkeeping and Documentation

Depending on the duration and complexity of the incident, there may be many documents associated with the response, or only a very few. In either case, all documents that were generated as part of the incident should be maintained in a secure location in accordance with protocols established by the AHJ. This should include enough accurate and complete information to reconstruct the major events and outcomes involved with the incident for legal, analytical, and historical purposes. As the old adage goes, "If it isn't written down, it didn't happen."

Documentation should include a complete record of what transpired during the effort to resolve the incident from start to finish, including ICS forms from FEMA as well as other relevant documents and materials. If more than one agency was involved in the response, the duplication services may be required to ensure that necessary copies are provided to all agencies and incident personnel as necessary.

Data Collection and Management Systems

The authority having jurisdiction (AHJ) will establish criteria by which information is filed, stored, maintained, and managed for further reference. In some cases, files may be electronic, while others continue to rely on paper documents in steel cabinets with drawers. Those responsible for maintaining records should be familiar with the requirements of their respective AHJ and should follow them.

It may be useful for the AHJ to also establish practices for collecting data relative to incidents that occur within their response area. Collecting and tabulating data in an organized structure can help to identify trends in incident types, shortfalls in preplans, or areas ripe for advancement in training and equipment. There is no right or wrong way of doing this; each agency should determine a method that works for them to provide relevant information.

Conduct Postincident Analysis

Termination of an event concludes with a process of reviewing site operations, hazards faced, and lessons learned from the incident. Postincident analysis should include a hot debrief, which occurs immediately following the incident while all (or most) resources are still on scene, and a structured debrief, which may occur some time later. Postincident analysis is addressed in Chapter 4, *Initiating the Response*.

A good postincident analysis process will include both of these types of debriefs, with the goal of developing action items toward the ultimate goal of continuous improvement. The TEMS framework, along with the "5-Why" analysis process, can yield very useful information and outcomes. Both of these methods are outlined in Chapter 4, *Initiating the Response*.

The goal of the postincident analysis is not to find fault or lay blame, but is an opportunity for honest reflection on oneself and the team, for the purpose of learning and striving toward excellence.

After-Action REVIEW

IN SUMMARY

- The incident action plan (IAP) begins to form with the first arriving responders; it lays the groundwork for the entire operation.
- The IAP is developed and updated on an ongoing basis by the incident commander with input from all general staff.
- FEMA's Planning P may be used to help generate an IAP. The five phases of an incident under Planning P are:
  - Phase 1: Understanding the situation
  - Phase 2: Establishing incident objectives
  - Phase 3: Developing the plan
  - Phase 4: Preparing and disseminating the plan
  - Phase 5: Executing, evaluating, and revising the plan
- An IAP works best when it is coordinated within the Incident Command System.
- In rope rescue, communication strategies include voice, whistle, radio, and hand signals.
- Terminating an incident effectively includes ensuring that all resources are accounted for and closed out, compiling relevant documentation, and readying for another incident.

CHAPTER 6 Developing the Incident Action Plan

KEY TERMS

Command emphasis A section within the Federal Emergency Management Agency (FEMA) form ICS 202 that describes expected outcomes or milestones for the operational period, a list of priorities, or the key message(s) that underpin the effort for that operational period.

Incident communications plan The Federal Emergency Management Agency (FEMA) form ICS 205 that enables rescuers to collect contact information for all personnel assigned to the incident including phone numbers, pager numbers, radio frequencies, call signs, etc., and functions as an incident directory.

Incident map A graphical representation, sketch, photograph, or actual map that shows the total area of operations, the incident site/area, impacted and threatened areas, or other graphics depicting situational status and resource assignment.

Incident objectives form The Federal Emergency Management Agency (FEMA) form ICS 202 enables rescuers to describe the basic incident strategy, incident objectives, command emphasis/priorities, and safety considerations for use during an operational period.

Medical plan The Federal Emergency Management Agency (FEMA) form ICS 206 enables rescuers to collect information on incident medical aid stations, transportation services, hospitals, and medical emergency procedures.

Operation O The repetitive cycle of planning and operations outlined by the oval part of the Planning P, which continues and is repeated each operational period.

Operations briefing A meeting conducted at the beginning of each operational period to inform supervisory personnel within the operations section of the incident action plan (IAP) for the upcoming period.

Personnel accountability reporting (PAR) A radio-based roll-call system initiated by command at predetermined intervals to ensure that all personnel are safe and accounted for.

Safety message The Federal Emergency Management Agency (FEMA) form ICS 208 is an optional form that may be included and completed as part of the incident action plan (IAP) to expound or emphasize key safety information regarding safety hazards and specific precautions to be observed during a given operational period.

S-M-A-R-T acronym A tool used in goal-setting to help guide planners toward setting objectives that are Specific, Measurable, Achievable, Realistic, and Timely.

Strategy An approach to achieving the objective.

SUDOT A standardized whistle-command protocol used in rope rescue where different numbers/types of whistle blasts each has a specific meaning.

Tactics The steps used to execute the strategy.

Unified command Management of an emergency incident through the incident command system in which multiple agencies or jurisdictions are involved, each one having an incident commander assigned to the operation.

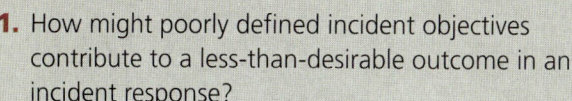

1. How might poorly defined incident objectives contribute to a less-than-desirable outcome in an incident response?

2. Think back to a recent response and use the S-M-A-R-T acronym to define the following:
   - Two objectives that might have been appropriate at the initiation of the incident
   - Two objectives that might have been appropriate when commencing evacuation
   - Two objectives that might have been appropriate during incident termination time

3. Considering your own response agency, list the resources that must be returned to ready-state at the conclusion of an incident, and what this involves.

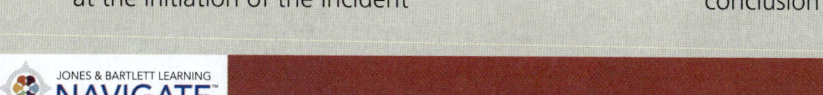

Section and Chapter Opener: © Jones & Bartlett Learning. Courtesy of Loui McCurley; On Scene siren: © Bildgigant/Shutterstock.

CHAPTER 7

Operations

Hazard-Specific Personal Protective Equipment

KNOWLEDGE OBJECTIVES

After studying this chapter, you should be able to:

- Differentiate personal protective equipment from rescue equipment. (p. 98)
- Define hazard-specific personal protective equipment. (pp. 98–103)
- Identify the standards associated with rope rescue personal protective equipment. (p. 98)
- Explain the considerations for selecting personal protective equipment for rope rescue operations. (pp. 98–103)
- Explain the differences among harness classes and what each is used for. (pp. 103–107)
- Identify the considerations for selecting a harness. (pp. 103–107)
- Identify at least two methods of constructing an emergency harnesses. (pp. 105–106)
- Identify the personal tools and equipment a rope rescuer might carry. (pp. 107–108)
- Describe how to clean personal protective equipment. (**NFPA 1006: 5.2.2**, p. 109)
- Describe inspection of personal protective equipment for damage, defects, or wear. (**NFPA 1006: 5.2.2**, pp. 108–109)
- Identify the documentation and recordkeeping recommendations for rope rescue personal protective equipment. (**NFPA 1006: 5.2.2**, pp. 108–109)

SKILL OBJECTIVES

After studying this chapter, you should be able to:

- Create an emergency hasty harness. (pp. 105–106)
- Clean rope rescue personal protective equipment. (**NFPA 1006: 5.2.2**, p. 109)
- Inspect rope rescue personal protective equipment for damage, defects, or wear. (**NFPA 1006: 5.2.2**, pp. 108–109)

You Are the Rescuer

You are being called to what is being described as a "need for high-angle rescue" in a densely populated urban area, but with little additional information. There are no steep or vertical natural features in the area, but you can see several warehouses, billboards, and assorted towers. You spot a small crowd of people in a parking lot between a large building and a cell tower. A unit from a different jurisdiction is already parked next to the building. The sun is going down and you can already feel the chill from the cold front that was predicted to bring a 30 percent chance of rain before midnight.

1. What site hazards are you looking for as you pull in and park?
2. What questions will you have for the first-arriving unit?
3. What kinds of personal protective gear do you expect to need?

 Access Navigate for more practice activities.

Introduction

The equipment used by rescuers generally fits into one of two categories: **personal protective equipment (PPE)** or rescue equipment. In this chapter we will focus on PPE, the items typically *worn by a rescuer* to provide protection against recognized hazards (**FIGURE 7-1**). Conversely, rescue equipment—which consists of gear used to perform a rescue task—will be addressed in Chapter 8, *Rescue Equipment*.

In the context of rope rescue, PPE might include such items as helmets, gloves, safety harnesses, and lanyards, while rescue equipment might include ropes, pulleys, braking devices, and carabiners. While clothing is not generally considered PPE unless it is necessary to protect against a specific hazard, clothing is important for the function it provides. Choosing appropriate clothing and personal equipment for the high-angle environment can give you, as a rescuer, a greater margin of safety, add to your comfort, and enable you to perform your job more effectively and efficiently.

The items rescuers choose to wear will be influenced by some combination of regulatory requirements, the requirements of the authority having jurisdiction (AHJ), the environment in which rescuers are working, and/or personal preference. There are wide variations among the clothing and equipment used by rope rescuers in different regions; some of this may be related to local protocol, temperature variations, and available resources. There will also be some variation based on the specific hazards presented in the incident. The emphasis of this chapter will be on **hazard-specific PPE**, which we will define as PPE particularly suited for protection against the hazards common to rope rescue, and those additionally inherent in rescue disciplines addressed by NFPA 1006, *Standard for Technical Rescue Personnel Professional Qualifications* and NFPA 2500, *Standards for Operations and Training for Technical Search and Rescue Incidents and Life Safety Rope and Equipment for Emergency Services*, where rope rescue is also likely to be a factor.

In addition to these criteria, also consider the particular needs in light of hazards imposed by the workplace, including electrical hazards, the fire-ground environment, chemicals, and nature.

Rope Rescue Personal Protective Equipment

Most firefighters will already be equipped with some range of hazard PPE. Although the hazard PPE may meet appropriate regulations and safety requirements,

FIGURE 7-1 The PPE worn by rescuers will depend on the hazards.
© Napa Valley Register/ZUMA Press, Inc./Alamy Stock Photo.

simply meeting an NFPA standard for PPE does not necessarily mean that the equipment is adequate for rope rescue purposes, or for a given rescue.

A classic example from the fire service would be PPE that meets the NFPA 1971, *Standard on Protective Ensembles for Structural Fire Fighting and Proximity Fire Fighting*. This standard is often used by fire departments when selecting garments for fire fighting. Equipment meeting this standard is subjected to test criteria that measure heat, flame, and chemical resistance. However, these criteria are generally not important factors for rope rescue operations. Because NFPA 1971 does not measure parameters that are specifically relevant to rope rescue operations, it is generally not an appropriate reference or specification for rope rescuers.

It is difficult to develop a universal standard for clothing appropriate to rope rescue operations, largely because rope rescue can occur in virtually any environment and the protection required will vary with accompanying hazards. PPE that is appropriate for a car-over-the-edge rescue may not be appropriate for a high-angle rescue occurring in a wilderness or cave environment, or for rescue from a transmission tower (**FIGURE 7-2**). What constitutes whether or not equipment is "appropriate" depends upon the environment and hazards present.

The AHJ may require that rope rescuers follow the PPE requirements listed in NFPA standards ranging from NFPA 2500 or previous editions of NFPA 1983, *Life Safety Ropes and Equipment*. While adherence to appropriate standards is important, this text will prioritize the maximum protection of the rescuer over a simple compliance mindset. Although relevant standards and regulatory requirements will be discussed, these are considered to be only a part of the equation, not the complete formula.

Equipment selection depends upon meeting local standard operating procedures, including regulatory compliance, performance specifications, size, color, and other measurable factors. In addition, factors such as comfort, fit, value, durability, and other personal preferences impact the selection of PPE.

Protective Garments

Garments worn by rope rescuers should fit well and be suited to the rescue environment. At this writing, there are at least 11 different NFPA standards that pertain to clothing worn by rescuers (**TABLE 7-1**); however, none of these is specific to high-angle rope work. Instead, these standards are focused on protection from specific hazards.

A

B

FIGURE 7-2 The protection required will vary with accompanying hazards. **A.** Motor vehicle crash. **B.** Transmission tower.
(A) © faboi/Shutterstock.

Even the document that seems most likely to contain detailed test criteria for rope rescue operations—NFPA 1951, *Standard on Protective Ensembles for Technical Rescue Incidents*—does not. NFPA 1855, *Standard on Selection Care and Maintenance of Protective Ensembles for Technical Rescue Incidents* is the user-guide companion to NFPA 1951. The parameters outlined in this document are focused on protection from elements such as surface abrasion, thermal conditions,

TABLE 7-1 NFPA Standards Addressing Personal Protective Equipment

NFPA 2112	Standard on Flame-Resistant Clothing for Protection of Industrial Personnel Against Short-Duration Thermal Exposures from Fire
NFPA 1971	Standard on Protective Ensembles for Structural Fire Fighting and Proximity Fire Fighting
NFPA 1976	Standard on Protective Ensemble for Proximity Fire Fighting
NFPA 1951	Standard on Protective Ensembles for Technical Rescue Incidents; Three Levels of Certification - (1)Utility, (2)Rescue & Recovery, and (3)CBRN
NFPA 1952	Standard on Surface Water Operations Protective Clothing and Equipment
NFPA 1977	Standard on Protective Clothing and Equipment for Wildland Fire Fighting
NFPA 1990	Standard for Protective Ensembles for Hazardous Material and Emergency Medical Operations
NFPA 1991	Standard on Vapor-Protective Ensembles for Hazardous Materials Emergencies and CBRN Terrorism Incidents
NFPA 1992	Standard on Liquid Splash-Protective Ensembles and Clothing for Hazardous Materials Emergencies
NFPA 1994	Standard on Protective Ensembles for First Responders to Hazardous Materials Emergencies and CBRN Terrorism Incidents
NFPA 1999	Standard on Protective Clothing and Ensembles for Emergency Medical Operations

chemical contamination, and even comfort and ergonomics, but none of these is specific to rope rescue. Perhaps a better resource is the US Fire Authority (USFA) report, *Protective Clothing and Equipment Need of Emergency Responders to US&R Missions.* Although this document is focused on Urban Search & Rescue (US&R), it does acknowledge that one of the most significant risks faced by technical rescuers is a need for mobility and comfort, and that incidents often occur over an extended period. Keep in mind that in fire service definitions, the term *technical rescue* most often refers to building/structural collapse, vehicle/machinery extrication, confined space entry, trench rescue, and water rescue.

The USFA report relies heavily on the requirements of NFPA 1977, *Standard on Protective Clothing and Equipment for Wildland Fire Fighting*, for the recommendations it provides. While this report—especially the elements related to fire protection—still may not be relevant to many rope rescue incidents, it does provide a good starting point.

A key concept to remember when selecting garments for rope rescue operations is that the weight and bulk of typical fire fighting garments may not afford the necessary comfort, mobility, and dexterity for rope rescue. Typically, the hazards faced by rope rescuers are of a physical nature. Garments should be form-fitting but not too tight, with no loose edges or wayward straps that may become caught on obstructions. It should provide adequate coverage and durability against abrasion and tears, in addition to being comfortable enough to wear for extended periods (**FIGURE 7-3**). Thermal protection should be considered relative to the environment, whether hot, cold, wet, or dry.

The principle of clothing protection is to provide protection against direct contact and against the elements. At a basic level, this includes appropriate abrasion protection and the ability to maintain warmth when needed, and also to allow the body to lose warmth before becoming overheated. Dressing in layers works well for rope rescuers. Each layer chosen should be specially selected for its ability to achieve a specific purpose. Typically, the purpose of a base layer is to wick moisture; then, one or more insulative layers may provide warmth, and finally, if needed, a top layer may be used to protect the rescuer from the elements.

FIGURE 7-3 Garments worn by rope rescuers should be form-fitting, but not tight.
© MartinFredy/iStock Editorial/Getty Images Plus/Getty Images.

A base layer that wicks moisture away from the skin is useful, especially for rope rescuers who are likely to work up a sweat. Some of the best base layers to use for wicking moisture (sweat) away from the skin are wicking synthetics such as polypropylene. These are arguably less comfortable than cotton, but aside from drawing moisture away from the skin they also dry quickly and retain insulative powers even when wet.

However, it should be noted that many of these synthetics are plastic-based and can melt, shrink, or stick to the skin when exposed to heat and flame. These may not be appropriate for individuals working in helicopters, in the fire-ground environment, or in other situations in which flash fire is a possibility. Under intense heat, such as that produced by a flash fire, the synthetic material may melt into the skin, complicating a burn injury. In these environments, a wool blend or cotton may be a better, and safer, choice.

Next to the base layer should be an insulating layer (if needed). This provides warmth, but it must also protect the wearer against chilling even when wet, either from precipitation or perspiration. The key for insulation value is the ability to retain air in the fibers and not get wet from outside or inside the garment. Thick, bulky insulative layers that restrict movement can be a detriment in ropework. Choose layers that offer high insulating value with low bulk, such as pile fleece or down. Vests of these materials are an especially good choice, as they retain body heat yet leave the arms free and unrestricted.

The final layer (if needed) is for environmental protection. This protection may be against weather conditions, or against contaminants. Again, bulky garments should be avoided for the reasons stated previously. Similarly, garments that are too loose or floppy should also be avoided as they can get in the way or become entangled in rigging.

The rope rescuer should present a compact and well-contained profile. Shirts and pants must be sized so that they do not bind when the arms are extended above the head or when the legs are raised. Clothing should protect the rope rescuer against adverse environmental conditions and provide maximum comfort during the anticipated activity.

Many departments utilize ensembles or jumpsuits made of Nomex, aramid, or FR cotton, but these may not be adequate for teams operating in a wilderness environment. Where a variety of hazards exist, priorities must be carefully balanced and protection suited to the conditions and exposures that pose the greatest risk and consequence to the rescuer.

Footwear

Among the requirements for footwear are comfort, protection, and adhesion. Although boots increasingly are partly or completely fabricated of materials such as Gore-Tex or plastic, leather is still the material with the qualities most needed in a multipurpose boot for rope rescue. Boots should provide support to the ankles and protect the feet from scrapes, cuts, and bruises, yet they should be pliable enough to be comfortable after hours of standing or walking. The soles should not be slick, like the soles of street shoes, but rather should have adhesion to help the wearer maintain balance against the surfaces found in the rope rescue environment. Rubber boots, such as those commonly used in the fire service, are not appropriate for rope rescue operations. They do not have the needed foot protection, and they impede foot and leg movement.

Lug soles, particularly those made by Vibram, have been very much in fashion, but they may not be required as long as the boot sole provides adhesion. Furthermore, in some cases a lug sole may actually be a disadvantage. Some types of lug soles may become dangerously slick when wet or caked with mud. For specialized rock climbing, technical climbing boots may be used. Technical climbing boots have soles constructed of special rubber compounds that may adhere to the rock better (**FIGURE 7-4**). Many of these climbing boots have little or no welt, and therefore may be better for certain climbing techniques, such as "edging" along the rock. However, because they are specifically made for climbing, they are not comfortable for walking or standing for long periods.

A good choice of socks is important for warmth, comfort, and prevention of injury such as blisters.

FIGURE 7-4 Rope rescuer footwear should provide traction.
© Perekotyp/Shutterstock.

FIGURE 7-5 NFPA 1951–compliant tech rescue utility gloves.
Courtesy of Pigeon Mountain Industries.

FIGURE 7-6 An appropriate and secure helmet is essential for rope rescuers.
Courtesy of Pigeon Mountain Industries.

A two-sock combination can reduce friction on the skin that causes blisters. An inner sock made of a synthetic, such as polypropylene, wicks moisture away from the foot so that it stays dry. A thick outer sock, made of a material such as wool, increases warmth and provides some protection for the foot.

Gloves

Gloves are worn to protect rope rescuers' hands against burns and abrasions and to prevent discomfort that may cause the rope rescuer to lose control of the rope (**FIGURE 7-5**). Heavy leathers, such as cowhide, offer greater durability and heat resistance, while soft leathers, such as goatskin, offer more finger dexterity. Extra layers and padding in the body of the glove offer protection from friction while increased flexibility in the fingers enables rescuers to retain their dexterity.

Do not mistake "fast-rope gloves" for rope handling gloves. Fast-roping is a special technique used for tactical helicopter insertions, wherein the team member descends a very thick diameter rope using only gloved hands for braking. The gloves designed for this use are particularly thick and offer little dexterity. If you need gloves with such extra heat protection (more than a few layers of leather), you are rappelling too fast for rescue ropework.

Helmet

A well-fitting **helmet** designed for work/rescue at height is an indispensable part of a rope rescuer's PPE ensemble (**FIGURE 7-6**). A helmet can protect the wearer from injury and can reduce the severity of injury from falling objects such as rocks or climbing hardware. It also can reduce the severity of brain injury should you fall and hit your head.

While structural fire fighting helmets generally offer good impact resistance, they are inadequate for rope rescue operations. This is because they are typically quite heavy, and they do not have a chin strap/retention system that works well for rope rescuers. These helmets tend to obstruct vision and make it difficult to maneuver in the sometimes cluttered and confined rope rescue environment. For the same reasons, they can be very uncomfortable to wear for long periods and/or in hot weather. Rescue operations often take a long time, and rescuers are in hazard zones for many hours. A comfortable, well-fitting helmet is essential in such situations because the helmet must be worn for long periods.

Rope rescue teams should only purchase helmets specifically designed for rope rescue activities. Other types of helmets may provide only an illusion of protection and can actually be dangerous to the wearer. It is especially important that helmets have a secure chin strap and retention system. Helmets with elastic chin straps are not suitable for rope rescue work. As the elastic chin strap stretches, the helmet may flip off the head and leave the wearer without head protection. This often occurs when the helmet is stressed, such as when the wearer falls or is hit by falling objects.

Any helmet used in rope rescue work should also have a three-point retention system. This means that in addition to support points on both sides of the helmet, there is also a third retention point positioned at the back of the helmet. Note that the term "three-point" refers to the three locations (left side, right side, rear) of security relative to the head, not the number of rivets in the helmet itself.

The suspension point in the back helps prevent the helmet from falling forward over the eyes. None of the NFPA standards address this three-point retention concept, nor does the ANSI Z89.1 standard. The concept is addressed in two different European standards (EN 397 standard for industrial helmets, and EN 12492 standard for mountaineering helmets) but with very different approaches. The main difference between the chin strap requirements in these two is that the industrial requirement mandates that the chin strap release relatively easily so as to avoid strangulation, while the sport standard requires the chin strap to hold at a higher strength to help ensure that the helmet stays on the head during a fall.

If the task warrants the use of a helmet that is designed to release easily, for example should the user's head become wedged during an operation, a helmet with a quick-release chin strap feature may be selected. Otherwise, a good high-angle helmet should have a chin strap that requires much greater force to cause release; this helps keep the helmet on the user's head if the user falls.

The shells of helmets used in high-angle rope rescue are constructed of materials such as plastic, fiberglass, Kevlar, or composite materials. The shell should have some rigidity to resist penetration by sharp objects, but at the same time, it should have enough flexibility to absorb some of the blow that otherwise would be directly transmitted to the skull and spine. The design of the helmet should protect the head against objects falling from above and hitting from the side.

The inside suspension of the helmet may hold the shell away from the skull, or the helmet may be lined with an impact-resistant Styrofoam-like material. While either is acceptable, consideration should be given to air circulation and comfort, particularly during hot weather or sweat-inducing labor, as well as to durability of the helmet during carrying and storage. A slight brim will help to deflect dropped objects and, in a natural environment, helps prevent rainwater or spray from dripping into the face. Coverage on the sides, back, and front should provide adequate protection while still allowing a good upward field of vision and to ensure that the back of the helmet does not catch on equipment carried backpack style, such as rope bags or rescue kits.

Harnesses for Rescue

Many people wonder whether it is safe to use a recreational climbing harness for rescue, or vice-versa. While the answer is a provisional "yes," it is important to understand the differences in design, and to select a harness that most closely meets the performance requirements associated with the application. Most climbing harnesses are made to be lighter weight, are cut for ease of movement, and are made for recovering from leader falls. Harnesses specifically designed for rescue usually are heavier and bulkier because they have wider webbing and padding for comfort for rescuers who must sit in suspension for longer periods. The attachment point on a climbing harness typically consists of a reinforced webbing loop that sits a bit lower than on a rescue harness, allowing the body to more fully rotate around the center of gravity in a lead-climbing fall, thereby absorbing force. The attachment point on a rescue harness is typically a bit higher, more suited to prolonged suspension, and may even have a metal attachment point into which carabiners are clipped. Climbing harnesses have a reinforced webbing loop to which a rope may be tied directly, or a carabiner clipped.

Another type of harness sometimes used in rescue is the caving harness. Caving harnesses are more similar to a climbing harness than a rescue harness, but the main attachment point sits even lower, for better functionality with ascending systems. These also often have abrasion guards on the seat and/or leg loops and may be paired with an independent **chest harness** to support a more efficient ascending system.

FIGURE 7-7 A seat harness.
Courtesy of Pigeon Mountain Industries.

FIGURE 7-8 A full-body harness.
Courtesy of Pigeon Mountain Industries.

Again, however, the straps tend to be narrow, not designed for extended periods of sitting in suspension.

When considering which harness to use, the question is generally not one of strength. Most harnesses made by a reputable manufacturer and certified to an appropriate standard are likely to withstand much greater force than that provided by the person wearing it. Instead, it is the features and benefits that set harnesses apart. Nowadays, many harnesses have begun to incorporate features from multiple user-groups to facilitate a wider range of application.

NFPA 2500 (1983) classifies harnesses into two groups:

- Class II: A **seat harness** meant for heavy-duty work by one person or in rescue situations in which another person's weight may be added in the course of the rescue (**FIGURE 7-7**)
- Class III: A **full-body harness** meant for fall protection and for use in rescues in which inversion may occur (**FIGURE 7-8**)

Considerations for a secure, comfortable seat harness include the following:

TIP

Previous editions of NFPA 1983 included a Class I harness, but this harness type was removed for the 2012 edition, and does not exist in NFPA 2500. Instead, the term BELT is now used.

- The harness should have waist, leg, and thigh supports to distribute body weight (or the force of a fall) over a relatively wide area.
- Where support features contact the body, such as at the waist and thighs, the webbing should be wide enough to provide comfort and not constrict blood flow. Padding can make the webbing even more comfortable.
- Stitching should be securely and evenly sewn and of contrasting color so that abrasion and wear can be detected.
- The harness should allow freedom of movement both when the rescuer is hanging in it and when wearing it on the ground.
- The harness should be easy to put on and to adjust.
- The harness should not slip down when the rescuer walks around.
- When you fall and are caught by the harness, it should allow you to easily return yourself upright.
- The harness must not allow rescuers to fall out when upside down.
- The harness should have a front tie-in point designed so that rescuers maintain a correct

center of gravity during whatever activity is being performed.

- The stress points, such as the tie-in, should be faced with extra webbing and/or use heavy metal connectors.
- Depending on how it is to be used, the harness should be certified as meeting appropriate standards, such as those established by the NFPA, ASTM, or ANSI.

Because anatomy varies from person to person and because the rope rescue environment requires different kinds of activities, no one seat harness design is suitable for everyone. Before selecting a particular seat harness and investing the money, rescuers should try several different designs to see which one is best. In certain circumstances, a Class III full-body harness may be preferable to a Class II seat harness. Such situations may include the following:

- When a person is involved in a dangerous activity that requires that they constantly be held upright. For example, a rescuer may be entering a confined space and is equipped with a harness attached to a retrieval line.
- When equipment is worn that makes a person top heavy, such as a breathing apparatus. The chest portion of the harness helps distribute the extra weight of the apparatus on the body.
- When a person is of greater than average weight. The higher tie-in point helps the person stay upright.
- During certain climbing or mountaineering activities when it is necessary to be held upright should a fall occur; for example, while wearing a heavy pack that would make it difficult to right oneself
- For placement on a subject in certain rescue situations, see the discussion on pickoff rescue techniques in Chapter 19, *Pickoff and Litter Management*.

In other rescue situations, some individuals may find the full-body harness too constraining, preventing the range of motion needed for rescue activities. For greater adaptability, a Class III harness, with separate seat-and-chest components, is preferred by many rescuers because it allows the wearer to choose options based on the specific conditions.

Whether choosing to work in a Class II or a Class III harness, this piece of equipment is perhaps the most important piece of personal gear for ropework.

An unsuitable rescue harness or one that is badly fitted can result in such severe discomfort that it can prevent the wearer from performing a task. Even worse, it can be dangerous to the wearer.

The portions of NFPA 2500 derived from NFPA 1858 provide guidance for the selection of harnesses for rescue and are a good reference. This standard advocates for the selection of harnesses that meet the minimum requirements of the equipment testing portions of NFPA 2500, as absorbed from NFPA 1983. NFPA 2500 also warns the user to consider the materials a harness is constructed of, particularly if it will be exposed to heat, flame, chemicals, or water. Fibers such as aramids or para-aramids may seem desirable for their heat resistance and high strength characteristics, but they also have lower impact force absorption capabilities and are not impervious to heat. Other features and accessories, such as gear loops, pockets, or methods for holding the loose ends of webbing, may enhance the functionality of the harness, but care should be taken to ensure that such accessories do not impede proper function of the harness. Finally, when a harness is integrated with bunker gear ensemble, it should not compromise the integrity of the protective gear as outlined in NFPA 1971 and vice versa.

Victim Extrication Device

NFPA 2500 makes provision for Class II and Class III victim extrication devices, which is a fancy term for harnesses intended to be placed on a subject. These harnesses are subjected to some of the same testing requirements as Class II and Class III rescuer harnesses, but may be easier to don and have fewer provisions for comfort.

Emergency Harnesses

A tied **emergency seat harness** ("hasty harness") may be constructed in the field to protect either a rescuer or a subject. Such harnesses are not addressed by NFPA, but it is a good skill for a rescuer to know how to secure a person in the vertical environment using just a piece of rope or webbing. Two methods of this are described here, but these should only be considered last-resort emergency approaches.

Diaper Emergency Harness

The diaper-style emergency harness is constructed using a piece of webbing or rope, usually around 10 to 12 feet (3 to 3.7 m) in length (**FIGURE 7-9**). To create

FIGURE 7-9 Diaper emergency harness.

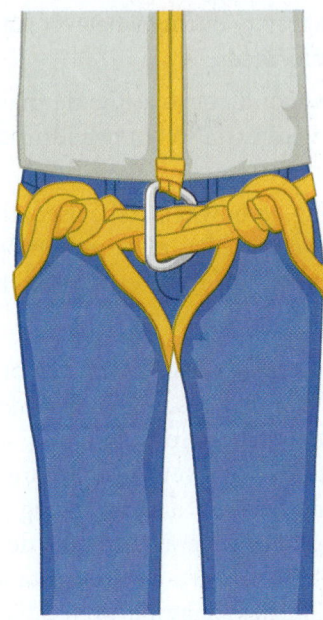

FIGURE 7-10 Swiss-seat emergency harness.

a diaper-style emergency harness, follow the steps in **SKILL DRILL 7-1**:

1. Tie the ends of a 10- to 12-foot (3- to 3.7-m) sling together to form a loop.
2. Grasp one strand of the loop and wrap it around the waist, parallel to the ground, with the knot in front.
3. The harness wearer should reach between the legs and grab the lower loop from behind.
4. The loop should be pulled up between the thighs, as a diaper, and the three angles of the triangle should be clipped together.

Clearly there are limitations to this type of harness, including a notable lack of security, low comfort, and minimal adjustability. This type of harness also does not stay in place well when not under tension.

Swiss-Seat Emergency Harness

A slightly more adjustable harness, the Swiss-seat emergency harness takes a bit longer to tie but is arguably more secure (**FIGURE 7-10**). To create a Swiss seat emergency harness, follow the steps in **SKILL DRILL 7-2**:

1. Using a 10- to 12-foot (3- to 3.7-m) piece of rope or webbing, find the center of the line and hold it against the left hip.
2. Wrap the rope behind the wearer and bring both lines together in front so that the length of line on the left side is slightly shorter than that on the right.
3. Make an overhand with two wraps, leaving the ends of the line hanging down in front of the legs.
4. Pull these ends securely between the wearer's legs, front to back, and snug each with a half hitch around the waist loop.
5. Bring both ends back to front and tie a square knot on the side with the shortest remaining length, adding an overhand on each side as backup.
6. To secure to the seat, clip through both strands of line at the waist.

These emergency harness ties may be used in a pinch, but they are no substitute for a well-designed, sewn, manufactured harness. A manufactured harness should be selected as soon as possible with respect to safety, security, and comfort. A well-engineered harness should support the pelvic girdle so that a person's weight does not create pressure points on the nerves and arteries in the groin and back. In a high-angle rescue situation, a person may have to be hanging in the harness for a relatively long period. Twenty minutes is not an unusual time, and periods longer than an hour are a possibility. Although no harness is totally comfortable in these conditions, the tied seat harness tends to cause greater discomfort and possibly circulatory problems.

Pickoff Seat (Rescue Triangle)

One type of commercially available victim harness that is not covered by NFPA standards is called a **pickoff seat** (or rescue triangle). Constructed of a combination of webbing straps and durable material, these are generally in the shape of an oversized isosceles triangle that is configured to hold a victim securely in its folds when wrapped and clipped properly together. Some pickoff seats are more complex than this, with arm holes and curved shapes.

Pickoff seats can be an efficient and quick solution for urgent rescue. The lighter and simpler the seat, the more likely it is to be available and transported easily to the site. However, the seat must be large enough to ensure that the subject sits deeply in the seat with their center of gravity well below the connection/suspension point.

Escape Belts

One type of emergency belt that is addressed by NFPA 2500 is classified as an **escape belt**. These belts are intended to provide emergency escape capability to a fire fighter from a life-threatening emergency at height. An escape belt differs from a ladder belt in that the escape belt is subjected to additional force testing because it may be used for suspension, whereas the ladder belt is intended only for restraint.

Although some belts are commercially sewn and even certified as meeting the minimal requirements of NFPA 2500 (1983), they are arguably less secure and possibly riskier to use than improvised tied emergency seats. Escape belts do not have leg loops to hold the belt down when the wearer is suspended, and they are generally constructed of narrow fabric, such as 2-inch (51-mm) webbing. As such, they do not provide the support of a harness and pose a high danger of creating a constrictive horse-collar arrangement around the torso, a configuration that can be painful at best, and even potentially incapacitating to the user.

A Note on Suspension Trauma

There has been much ado in recent years about the dangers of **suspension trauma** (also referred to as *suspension intolerance, suspension syndrome,* and other terms), a potentially hazardous condition that can occur when a person hangs motionless in a seat harness for a long period.

The sense of urgency that has sprung up around the idea of quickly rescuing any worker who is suspended in a harness is not necessarily a bad thing, but rope rescuers should note that the medical community is divided on the pathophysiology, urgency, and treatment for this condition. It has even been suggested that the mechanism of injury may vary between incidents.

The important takeaway for the rescue community is this: Treatment for the condition was previously believed to be to keep the subject upright, in a seated position, even after the subject is on the ground. However, more current protocols suggest that treating the subject for shock, including providing high-flow oxygen and fluid replacement, is the preferred approach.

It is still prudent to make prompt rescue of any person who is suspended in a harness, particularly one who is unconscious or unable to move. Meanwhile, rescuers can increase their own comfort when it is necessary for them to be suspended in a seat harness for long periods by using foot loops, stirrups, or an **etrier** (a short ladder made of webbing) attached to their harness or on the rope above their attachment point.

> **TIP**
>
> Life belts, **ladder belts**, pompier belts, and similar safety belts with support only around the waist must not be used as the single point of support in high-angle activities. These types of equipment are designed only as a safety element to help prevent falls from ladders or other elevated positions.
>
> When a person hangs free in them, the belts can constrict the waist and rib cage, impairing breathing and possibly damaging internal organs. They also can slip up under the armpits, potentially causing permanent damage to the nerves and resulting in permanent paralysis of the arms. Their use, in place of a seat harness, as the single support can result in injury, permanent disability, or death.

Other Considerations

Preventing Dropped Objects

Pieces of high-angle hardware, such as carabiners, are easily lost during rescue work in the rope rescue environment. Even worse, hardware can easily be dropped and can injure a person who happens to be in the path of its fall. When rescuers are working in the rope rescue environment, all equipment should be kept attached to something secure. A convenient place is the equipment loops that many harnesses have. Tool lanyards are also a useful resource for helping to prevent dropped objects when working at height. There are several commercially available tool lanyards for different sizes and weights of tools. These are adequate for small items, but large tools weighing 12 pounds (5.4 kg) or more are better secured by a separate line.

Light Sources

Rope rescue operations often take place at night or in enclosed areas with no light. Consequently, all personnel should carry a reliable source of light. Rescuers will need to have both hands free during rope rescue operations; therefore, these light sources should be in form of a hands-free device, such as a light that mounts to a

FIGURE 7-11 A head lamp.
© NorthernExposure/Alamy Stock Photo.

helmet or a head lamp. A head lamp follows the movement of your head, usually placing the light where you are looking.

Choose a head lamp that is easily adjustable and field serviceable. Always carry extra batteries and a spare bulb. Many head lamps have battery packs to the rear of the head, which helps balance the weight on your helmet (**FIGURE 7-11**). If you choose this kind of head lamp, you must be sure that it will remain stable on your head and not fall off easily.

The battery type that has the longest shelf life and is resistant to the effects of cold is the lithium cell. However, it is also the most expensive, and older models have been known to cause explosions by venting gas. After the lithium cell, the alkaline battery is the next most desirable in terms of long life and operating temperature, and it is not as expensive as the lithium battery. A secure switch that prevents the light from being turned on accidentally when in storage will also help preserve battery life.

While halogen bulbs provide amazingly bright light, they also drain batteries very quickly. The use of light-emitting diode (LED) technology has made for dramatic improvement in headlamps in recent decades. LED lights are not actually filament lightbulbs, but rather solid-state diodes that give off light when charged with a current. Newer, high-intensity models make for extremely lightweight head lamps with incredibly long burn times. For those who remain committed to conventional or halogen bulbs, an LED/conventional bulb combination head lamp provides the best of long-duration light from the LED and a strong beam when brightness is needed for distance.

Individuals entering potentially explosive atmospheres must use light sources that are intrinsically safe for the particular conditions. This means that the design of the lighting equipment safeguards it against ignition in a hazardous atmosphere. There are several levels or classes of hazards, and lamps that are certified for one level may not be safe in others.

Knives

There are few topics so divisive as that of carrying knives in the high-angle environment. A basic expectation of some, the very concept causes others to break out in a cold sweat. Careless use of knives can have terrible consequences. In addition to presenting the real danger of personal injury, knives in the high-angle environment are a threat to life because of the ease with which they can destroy fiber-based life-safety equipment. An exposed knife blade is particularly dangerous to loaded webbing, and even ropes. When stretched, as they are when supporting weight, webbing and rope are more susceptible to being cut by a sharp object.

A common argument in favor of carrying a knife revolves around the urgency presented when a person's shirt or hair has gotten drawn into a rappel device, causing them to become stranded. However, in such a situation, the person is likely to be under stress, possibly in pain, and may have limited freedom of movement. Realistically, it would be very difficult to cut a person's hair away out without touching the knife edge to a loaded fiber member.

One alternative to a knife is the tool used by emergency services personnel to cut seat belts in motor vehicle crashes. This instrument has a recessed blade so that it does not accidentally cut a lifeline as easily as a naked knife blade would. Another option is to use trauma shears, which are robust enough to cut through tough material but also require intentional action to apply.

Even better than cutting is the use of self-rescue skills and optional equipment to extricate oneself from such situations. Much can be accomplished with just a Prusik knot or ascender and a footloop!

Inspection, Maintenance, and Recordkeeping

Always read manufacturer's instructions and obtain proper training on new equipment prior to the equipment being placed into service. Equipment may be marked with etching, engraving, permanent marker, or some other method so that it can be tracked and monitored over time (**FIGURE 7-12**). The date the equipment is placed into service should be recorded, and instructions for that piece of equipment should be retained and stored for reference as needed. The name

FIGURE 7-12 An example of a tagged rope.
Courtesy of Pigeon Mountain Industries.

of the product, manufacturer, date purchased, date in service, unique identifier, and other relevant information may be logged in a recordbook or spreadsheet.

A competent rescuer should inspect equipment before each use. This is not a detailed inspection, but may simply involve a cursory once-over and an abbreviated function check—just enough to ensure that the equipment is in good working order and ready for use. Any equipment found to not be in good working condition should not be used. Preuse inspections do not necessarily require formal documentation.

In addition to preuse checks, all equipment must regularly undergo a more thorough, periodic inspection by a competent person. Periodic inspection is a more in-depth process, where form and function are closely considered, purchase and in-service dates are reviewed, and wear and tear is considered. These inspections are typically documented. Periodic inspection should be performed at regular intervals—at least annually, although more frequent checks (3 to 6 months) may be more appropriate for busy teams. Criteria for inspecting equipment may generally be found in manufacturer's instructions.

Signs of Damage

Hardware items, such as carabiners, descenders, pulleys, etc., are arguably easier to inspect than their textile counterparts. For this reason, a mnemonic that is sometimes used as something of a checklist for inspecting hard goods is ACADEMIC, with each letter representing a specific consideration:

- A: Alignment. Is the item properly aligned with itself, as manufactured?
- C: Cracks. Are there visible hairline cracks, especially at connecting points?
- A: Action. Does the item function as intended, without sticking or jamming?
- D: Deformation. Are there any deformities in the body of the item?
- E: Edges. Are there sharp or excessively worn edges, especially at rope paths?
- Mi: Missing. Is any subcomponent missing or loose?
- C: Corrosion. Do you observe corrosion, especially at joints?

Soft goods, such as ropes, harnesses, slings, and other textile items, can be more susceptible to abrasion and wear than hardware, and can suffer damage that is more difficult to see, such as loss of elasticity or strength. A mnemonic to help remember what to look for when inspecting textile goods is T-CHAPS, again with each letter representing something specific to look for:

- T: Thermal. Thermal damage typically appears as glazed, charred, or hardened fabric.
- C: Contamination. Clues might include discoloration, stiff or soft spots, or even odor.
- H: History. A check of the usage log is an especially important part of soft goods inspection.
- A: Age. Fibers deteriorate with age—a du Pont study suggests 10 years for nylon.
- P: Physical damage. Cuts, abrasion, tears, or even fuzzy spots should get a second look.
- S: Soiling. While dirt is not necessarily caustic, a very dirty rope is harder to inspect and may signal contamination.

To inspect soft goods using the T-CHAPS mnemonic, follow the steps in **SKILL DRILL 7-3**:

1. Select a rope or other soft good, and inspect using the T-CHAPS mnemonic.
2. Look and feel for glazing or hardness that could indicate thermal damage.
3. Look, feel, and sniff for discoloration, inconsistencies, or odors that might indicate the presence of contaminants.
4. Check the history/use log of the item to ensure that it has not been subjected to an extraordinarily hard life. Ropes with harder use history are candidates for early retirement.
5. Check the manufacturing/in service date to ensure that the soft good is within its expected lifespan.
6. Look and feel for inconsistencies in the material; when in doubt, throw it out!
7. Determine whether the item is excessively dirty, and wash if necessary.

Life safety equipment is designed to be used time and time again. With proper use, care, and storage, equipment may last up to several years. Treat your gear with utmost respect. Your life depends on it.

Maintenance

Maintaining hazard PPE is an important step in keeping it ready for use and to protect its lifespan. Consult the manufacturers' guides on cleaning and storing PPE and follow their recommendations.

After-Action REVIEW

IN SUMMARY

- Personal protective equipment (PPE) encompasses all equipment (from clothing outward) used directly to protect oneself, while rescue equipment is equipment used to effect a rescue.
- Within the context of rope rescue, hazard-specific PPE refers to specialized equipment needed for rope rescue, and the environments where rope rescue takes place.
- While most firefighters are equipped with personal protective equipment, what is appropriate for rope rescue depends upon the environment of the incident and the hazards present.
- Equipment selection depends upon meeting local standard operating procedures, including regulatory compliance, performance specifications, size, color, and other measurable factors. In addition, factors such as comfort, fit, value, durability, and other personal preferences impact the selection of PPE.
- The weight and bulk of typical fire fighting garments may not afford the necessary comfort, mobility, and dexterity for rope rescue. Garments should be form-fitting but not too tight, with no loose edges or wayward straps that may become caught on obstructions.
- Among the requirements for footwear are comfort, protection, and adhesion.
- Gloves for rope rescue should protect rope rescuers' hands against burns and abrasions.
- A helmet can protect the wearer from injury for falling debris or falls.
- NFPA 2500 (1983) classifies harnesses into two groups: Class II (seat harness) and Class III (full-body harness).
- An unsuitable rescue harness or one that is badly fitted can result in such severe discomfort that it can prevent the wearer from performing a task.
- Follow the manufacturer's guidelines when inspecting, maintaining, and documenting equipment.

KEY TERMS

Chest harness A type of harness worn around the chest for upper body support. In the high-angle environment, it should never be used as the only source of support; it should always be used in conjunction with a seat harness.

Emergency seat harness A temporary tied harness that is used when a manufactured, sewn seat harness is not available; also known as a *hasty harness*.

Etrier A short ladder made of webbing which is attached to a harness or on the rope above the attachment point to provide additional stability.

Full-body harness A type of harness that offers pelvic and upper body support as one unit.

Hazard-specific PPE Personal protective equipment particularly suited for protection against hazards common to rope rescue, and those additionally inherent in rescue disciplines addressed by NFPA 1006 and NFPA 2500, where rope rescue is also likely to be a factor.

Helmet A head covering that protects against head injury both from falling objects and from head impact.

Ladder belts Devices that fasten around the waist and are intended for use as a positioning device for a person on a ladder. They should never be used as the sole means of suspension.

Personal protective equipment (PPE) Those items of equipment that can be worn or used directly by rescuers to protect themselves against a recognized hazard.

Pickoff seat A harness constructed of a combination of webbing straps and escape belt; also called a *rescue triangle*.

Rescue equipment Equipment or gear used to perform a rescue task.

Seat harness A system of nylon or polyester webbing that wraps and supports the pelvic region to attach the wearer to the rope or other protection in the high-angle environment. There are two classes of NFPA harnesses: Class II (meant for heavy-duty work by one person or in rescue situations in which another person's weight may be added in the course of the rescue) and Class III (a full-body harness meant for fall protection and rescue where inversion may occur).

Suspension trauma A life-threatening condition where venous blood flow from the extremities to the right side of the heart is reduced due to lack of movement and harness strap compression.

On Scene

1. What criteria can you use to determine whether a given harness is appropriate for use in your rescue squad?

2. You are on a rescue scene performing preuse inspection on gear being deployed for a vertical evacuation, and find the webbing of one of the two available harnesses to be discolored. What do you suggest to the rescuer?

3. You are called to the scene of a tower rescue, and workers on scene offer you a set of twin lanyards for your team to use. How do you determine whether it is acceptable for use, and what the limitations might be?

4. You are asked to inspect a harness. What do you look for?

Operations

Rescue Equipment

KNOWLEDGE OBJECTIVES
- Describe the components of a rope rescue equipment program. (p. 115)
- List the considerations in selecting rope rescue equipment to respond safely to an incident. (pp. 115–117)
- Identify documentation and recordkeeping recommendations for rope rescue equipment. (**NFPA 1006: 5.2.3**, p. 143)
- Explain the strength ratings used for rope rescue equipment. (pp. 117–119)
- Define and describe load ratios. (pp. 117–119)
- Describe the purpose of carabiners in rope rescue systems. (pp. 119–126)
- Describe the purpose of screw links in rope rescue systems. (pp. 126–127)
- Describe the purpose of braking devices and descenders in rope rescue systems. (pp. 127–131)
- Describe the difference between autolocking and nonautolocking braking devices. (pp. 127–131)
- Describe the purpose of belay devices in rope rescue systems. (pp. 136–137)
- Describe the purpose of brake bar racks in rope rescue systems. (pp. 131–135)
- Describe the purpose of escape devices in rope rescue systems. (pp. 135–136)
- Describe the purpose of rope grabs and ascenders in rope rescue systems. (pp. 137–139)
- Describe the purpose of anchorage connectors in rope rescue systems. (pp. 139–140)
- Describe the purpose of pulleys in a rope rescue system. (pp. 140–141)
- Describe how to clean rope rescue equipment. (**NFPA 1006: 5.2.3**, p. 125)
- Describe how to inspect rope rescue equipment. (**NFPA 1006: 5.2.3**, pp. 143–145)
- Identify the signs of damage, defects, or wear in rope rescue equipment. (**NFPA 1006: 5.2.3**, p. 144)

SKILL OBJECTIVES
- Clean rope rescue equipment. (**NFPA 1006: 5.2.3**, p. 125)
- Inspect rope rescue equipment for damage, defects, or wear. (**NFPA 1006: 5.2.3, 5.2.7**, pp. 143–145)
- Document rope rescue equipment inspection and maintenance. (**NFPA 1006: 5.2.3**, p. 143)

You Are the Rescuer

You arrive at the scene of a grain entrapment incident at a local brewery. A worker who had entered a grain hopper is trapped up to his chest, and another coworker has entered the bin to try to assist. The only access is to climb a ladder on the side of the bin and drop through a roof hatch. Both workers will need to be evacuated from the bin, and one will also have to be released from the grain. You have been assigned the task of assembling rope rescue equipment for raising the workers out of the bin through a hatch in the top, and for lowering them to the ground.

1. How will you determine which pieces of hardware are safe to use?
2. How will you know which pieces of hardware adhere to the standards your team follows?
3. Which pieces of hardware can be used in hauling systems?

 Access Navigate for more practice activities.

Introduction

Selection of equipment for rope rescue operations is a topic that is intertwined with operational preplanning, hazard analysis, staffing, and training. The equipment your organization chooses is based on a variety of factors, including what kinds of incidents might be anticipated within the jurisdiction, the types of hazards likely to be present during a rope rescue, the number and qualifications of rescuers, defined standard operating procedures, and the type and frequency of training rescuers receive.

Once equipment is selected and placed into service, consideration should be given to organizing equipment for optimal, streamlined deployment at a rescue scene. This requires thoughtful consideration and attention to detail. Labeling, color-coding, and pre-configuring gear will all help to improve operational readiness, as will a well-planned approach to packing and storing equipment (**FIGURE 8-1**).

Any agency or organization expecting to maintain an operations- or technician-level response capability for rope rescue should also develop and implement a program for the selection, care, and maintenance of appropriate life safety rope and related equipment. A useful reference document to assist in this process is NFPA 2500, *Standards for Operations and Training for Technical Search and Rescue Incidents and Life Safety Rope and Equipment for Emergency Services*. This document incorporates information previously found in NFPA 1858, *Standard on Selection, Care, and Maintenance of Life Safety Rope and Equipment for Emergency Services*, a user-oriented standard designed to help users establish basic criteria for the selection, inspection, and care of life safety rope and associated equipment. It focuses largely on equipment that is compliant with NFPA 2500 and its predecessor NFPA 1983, *Standard on Life Safety Rope and Equipment for Emergency Services*, to which much of the equipment used in technical rescue is certified.

FIGURE 8-1 Labeling, color-coding, and preconfiguring gear will all help to improve operational readiness, as will a well-planned approach to packing and storing equipment.

Rope Rescue Equipment Program

A systematic approach to rope rescue equipment selection, care, and maintenance will help with the following:

- Ensure the availability of appropriate equipment that is suitable for the intended use
- Maintain life safety rope and equipment in a safe, usable condition
- Remove equipment that is unfit for service

A rope rescue equipment program should have standard operating procedures (SOPs), a document listing the roles and responsibilities of both the organization and of responsible personnel, and equipment records. This system should include guidelines for the following:

- Equipment selection
- Equipment inspection
- Cleaning and decontamination
- Guidelines and policies for repair
- Storage
- Retirement and disposition

There is a saying that, "If it isn't written down, it didn't happen." Ensuring that a program is well-documented both in terms of how it works as well as the equipment records themselves will help to facilitate training and provide a reference when information is needed about a specific piece of equipment.

The following details should be recorded and maintained for each item of life safety rope and rescue equipment in an equipment record:

1. Equipment identification
2. A copy of all manufacturer guides
3. Date of purchase
4. Date placed in service
5. Service location (if applicable)
6. Manufacturer and model number
7. Month and year of manufacture
8. Dates of use, including how the equipment was used, weather conditions, potential damage, and other circumstances relating to use
9. Dates of cleaning and inspection
10. Removal from service, as applicable

Recordkeeping is facilitated by a good, well-thought-out system of labeling and color coding. The authority having jurisdiction (AHJ) should determine which markings/criteria are important to the organization, including ownership, date placed in service, identification of what cache/location an item belongs to, duty assignment, assigned vehicle, or other criteria (**FIGURE 8-2**).

FIGURE 8-2 Equipment that is designated for use in a particular location, or stored on a certain vehicle, should be marked accordingly.

Equipment should be organized and stored on responding vehicles in a way that lends itself to the necessary sequence of deployment to increase operational efficiency at the scene. This may include gear kits organized by functionality (for example, an anchor kit, lead climbing kit, or patient restraint kit) or pre-rigged into rapid deployment solutions for lowering, raising, or other technical operations.

If equipment is staged in kits or collections, or if it is pre-rigged into a hasty deployment solution, marking and labeling bags or containers with contents, weight, or other information may also be helpful.

Equipment Selection Considerations

Selection of rope rescue equipment begins with a clear understanding of the types of incidents within the jurisdiction where rope rescue operations might be required, including the following:

- Environmental factors
- Incident locations
- Steep to vertical terrain
- Frequency of incidents
- Frequency of training
- Cooperative agreements with other agencies

The types of incidents that might require rope rescue capabilities might include the following:

- Industrial accidents
- Municipal incidents

- Utilities incidents
- Climbing/caving accidents
- Confined space incidents
- Car-over-the-edge accidents
- Farm machinery incidents
- Remote/wilderness accidents
- Water incidents
- Snow/ice responses

In many cases, the same types of rope rescue equipment can be used across multiple environments; in other cases, the unique circumstances of incident types may dictate the need for specialized equipment. The AHJ has final authority over which equipment is used, with consideration to the appropriateness for intended uses, whether or not it meets relevant standards, and the training/skill levels of rescuers.

A competent person should take into consideration at least the following factors when evaluating rope rescue equipment:

- The appropriateness of the available equipment to the anticipated need
- The rope diameters used by the organization and others with whom they are likely to work
- The compatibility of hardware (descenders, rope grabs, pulleys, etc.) with the rope diameters used
- Whether equipment is, or needs to be, compliant with NFPA 2500 (1983) or other relevant standards
- The minimum breaking strength (MBS) of equipment in its intended rigging configuration
- Desired MBS of the system
- Potential fall distances, swing fall potential, and other hazards
- Skills/training of available personnel with the intended equipment

Where applicable, organizations should prioritize equipment that is certified as being compliant with the current edition of NFPA 2500 (1983), which establishes baseline performance requirements for various components of gear. NFPA 2500 contains *Standards for Operations and Training for Technical Search and Rescue Incidents and Life Safety Rope and Equipment for Emergency Services*, including NFPA 1670, NFPA 1983, and NFPA 1858, all of which were previously standalone documents.

Although the National Fire Protection Association (NFPA) does not directly evaluate or approve products or equipment, selecting equipment that is affirmed by a third-party organization as being compliant with NFPA 2500 (1983) can give the AHJ confidence that it meets minimum performance criteria for the intended use.

The AHJ may sometimes choose to allow or acknowledge equipment that does not meet NFPA standards. The NFPA 2500 (1983) equipment standard addresses only the most commonly used components in urban rescue operations. There is a great deal more equipment out there that is designed and used for other disciplines, such as recreational climbing, rope access, caving, fall protection, etc. The fact that this equipment is not marked as meeting NFPA standards does not negate its value.

That said, it is imperative that rescuers understand the performance capabilities and limitations of all equipment before putting it into use. One of the hazards of a standards-based selection philosophy is that sometimes users equate a mark or designation as inherently safe, to the point of being completely unaware of what that mark means in terms of strength and performance. Although a piece of equipment may meet some standard other than NFPA 2500 (1983), this should not be presumed as an "equivalency." NFPA 2500 (1983) boasts some of the most robust requirements in the life safety equipment standards world, including extremely high performance criteria, third-party testing, and a manufacturer surveillance program to ensure quality and consistency that goes beyond that of any similar equipment standards promoted by the American National Standards Institute (ANSI), Committee for European Normalization (CEN), ASTM International (ASTM), or the Union of International Alpine Associations (UIAA). A manufacturer's claim that a piece of equipment meets the requirements of NFPA 2500 (1983) holds no sway unless that piece of equipment is truly tested, marked, and listed by a qualified testing laboratory.

Compatibility between components is another key consideration, as various items may be designed for different diameters of rope, different intended uses and applications, and different environments. Every rescuer who might use a piece of equipment should be trained in its capabilities and limitations, including matters of compatibility.

When equipment is delivered to the organization, prior to placing it into service, it should be inspected by a competent person to ensure that it arrived as ordered, that it meets the organization's specifications, and that it was not damaged during shipment. The AHJ should also ensure that rescuers have been appropriately trained on the equipment prior to it being placed into service (**FIGURE 8-3**).

Rope, as the foundation of any rope rescue system, is a big topic. The components of the equipment addressed here should be considered in context of the

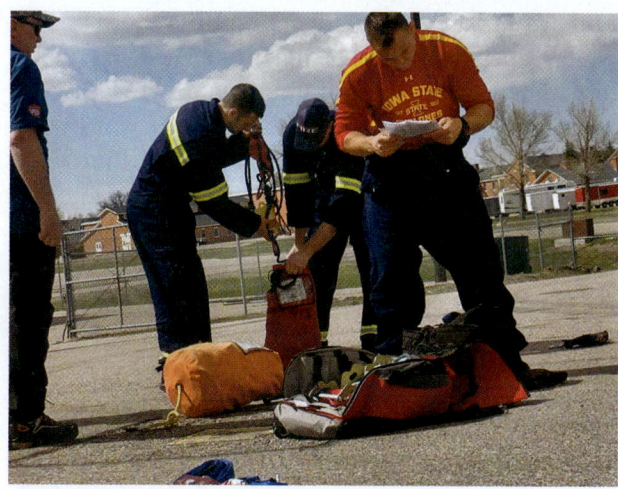

FIGURE 8-3 A competent equipment inspector should verify viability of equipment.

rope chosen and used by the responding agency. See Chapter 9, *Rope, Knots, Bends, and Hitches,* for a detailed discussion on rope.

Equipment Strength and Safety Factors

Aside from being compatible with other components of the system, all equipment should be of sufficient strength to offer appropriate performance under whatever load might be applied, plus offer a sufficient factor of safety above and beyond that. In rescue applications, loads greater than a single body weight are common, and forces are likely to be intensified by rigging angles and friction. NFPA 2500 (1983) classifies equipment into three categories:

- NFPA 2500 (1983) – G (General). G-rated equipment is intended for general use; this category most closely replicates the historical requirements of NFPA 1983. It is grounded in a philosophy that if each and every component is highly durable and very strong, even moderately trained rescuers are unlikely to create systems that will overload it.
- NFPA 2500 (1983) – T (Technical). T-rated equipment is the present moniker for a classification of equipment that was introduced into NFPA 1983 in the 1990s, acknowledging that more highly skilled technicians who are trained to think in terms of system safety can build and analyze systems that are sufficiently strong and appropriate even with components of lower strength. The T rating allows for equipment with lower performance requirements than G-rated gear, but higher than E-rated gear.
- NFPA 2500 (1983) – E (Escape). E-rated equipment has lower test requirements than either of the other classes, largely because it is expected to be used only in a limited capacity, for personal escape. The lower requirements allow for lighter, smaller, more portable gear that, while perhaps not as robust as its T- and G-rated counterparts, may be more compact and conducive to being carried on by rescuers and thus readily available for life-threatening emergencies.

There is a lot of misunderstanding about these classifications, as rescuers have tried to extrapolate user information from this manufacturer's standard over the years. To further complicate matters, in different versions of the standard, the T classification has changed several times. It was first introduced as "P," which stood for Personal Use, and then to "L," which stood for Light Use, and now it is "T," which stands for "Technical Use."

The actual performance requirements are not the same for all components—hard goods (like carabiners) tend to have somewhat lower strength requirements than soft goods (like ropes) due to their lesser susceptibility to degradation through wear and tear. While the G, T, and E classifications at least give manufacturers a way of classifying gear, these classifications only differentiate the absolute minimum performance levels. Many qualifying pieces of equipment actually exceed the minimum performance requirements, offering experienced users the potential for even greater flexibility and margin of safety. Rescuers should also be aware of the actual strength and other ratings marked on the products to facilitate well-informed decisions based on their respective level(s) of training and anticipated forces applied by their systems.

Load Ratios

One area where knowing the actual strength ratings is important is in determining the load ratios used for safety factors. The simple classification of equipment into E, T, and G ratings is less useful when rescuers are inclined to think in terms of the forces placed on their system as the forces relate to the strength of various points in that system. In fact, in the case of T-rated equipment, the NFPA standard clearly expresses that users are expected to understand system forces in context of specific equipment strength and performance criteria. When equipment meeting standards other than NFPA 2500 (1983) is selected for use, this becomes even more important, because other standards are less robust than NFPA 2500 (1983).

For example, European standards assume a rescuer weight significantly lighter than the 300-pound

(136-kg) test weight assumed by the NFPA tests, not to mention lighter overall loads, shorter-distance evacuations, lower potential forces, and more controlled environments. While these lower test requirements are convenient for improved repeatability in highly refined laboratory tests, they do not translate well to actual application. The NFPA 2500 (1983) tests strike a more realistic balance between laboratory test repeatability and actual field use.

When equipment is manufactured, it helps if the maker has some target or desired load that it should be capable of withstanding. As an example, most European standards assume a use-load of 80 kg (176 lb) as the weight of an equipped person, while U.S. standards (NFPA, ANSI) tend to assume the weight of an equipped person to weigh nearer 300–310 pounds (136–140 kg). Of course, to have confidence that the equipment can consistently support this use-load, manufacturers incorporate a load factor so that the equipment is somewhat stronger than the expected use. There is no consistent formula for what load factor is used by a given manufacturer. The load for which a product is designed is known as the **design load**.

In the case of technical rescue equipment, a value known as **minimum breaking strength (MBS)** is derived from the design load. MBS is the force at which tested samples of equipment actually fail during testing. Life safety equipment standards typically state a required MBS for compliance.

As defined by NFPA and other life safety standards, MBS is calculated by a testing laboratory as three standard deviations below the mean breaking strength of a minimum number of samples tested in manner of use. This testing/reporting criteria helps to ensure a high level of confidence in the cited MBS of the product, providing a good baseline from which to determine the **system safety factor (SSF)**.

Rescuers in the field use MBS information in conjunction with their knowledge of forces applied by rescue systems to ensure sufficient system strength. By choosing life safety equipment with a known MBS, a rescuer has a foundational tool for analyzing the strength of that component against the anticipated load. The ratio between the MBS and an equipment component is known as a **component safety factor**. For example, if a rope rated to 9000 pounds (4082 kg) is used for a rescue system with a mass of 600 pounds (272 kg), this might be said to be a 15:1 component safety factor.

This concept of component safety factors has long created misunderstanding in fire-rescue communities because design loads were used from the earliest versions of NFPA 1983 to establish standardized component MBS requirements. The original versions of that standard built on a concept first born from a white-paper called *Line to Safety*, published by the International Association of Fire Fighters (IAFF) in response to the accidental death of two New York fire fighters. This white-paper asserted that while a 7:1 or 10:1 safety factor might be adequate for industrial safety, it was not appropriate for critical rescue use in emergencies. It went on to say that a two-person load should be assumed to be 600 pounds (272 kg), and that a 15:1 safety factor should be applied to rescue rope.

Using this paper as a guide, the first edition of NFPA 1983 mirrored the concept that one-person life safety rope would be designed to have a maximum working load of no less than 300 pounds (136 kg) and that two-person life safety rope would have a maximum working load of no less than 600 pounds (272 kg). That edition of the standard went on to say that the minimum breaking strength requirement for two-person rope would 40 kN, or approximately 9000 lbf, with a corresponding MBS requirement for single-person rope. Dividing that 9000-lbf rope strength by the 600-lb rescue load, we get 15:1.

In 2001, the NFPA 1983 document removed references to load ratios, numbers of persons, and safety factors, but still the misinformation about 15:1 safety factors (and a related concept of 10:1 safety factors) remains among trainers and rescuers. Although fire fighter Jim Kovach from Ohio addressed this topic at the International Technical Rescue Symposium as early as 2005, many rescuers still believe that NFPA standards specify a 15:1 safety factor. They do not. In fact, determination of an appropriate safety factor is left to the AHJ and should be based on situationally relevant information including hazards, risk, levels of training, and probability and consequence of failure, to name a few.

The factor of safety with which the AHJ should be most concerned is NOT the component safety factor at all, but the SSF. An SSF is determined by analyzing the strength of the entire system, as each component is interconnected with other components in the system and comparing that against the anticipated force or load at that point. The ratio between the least strong point in the system and the potential load at that point is the SSF. A static SSF is analyzed as the ratio during normal static loading, while a dynamic SSF is analyzed as the ratio during worst case dynamic loading.

Knowing the expected breaking strength of each component is essential in determining a SSF, but knowing the strength of those components as they interact together is even more important. This knowledge comes only with experience and devoted study. This is the reason that seasoned, experienced rescuers lean toward equipment that is marked with minimum breaking strength (MBS) ratings. Markings to safe working loads (SWLs) or Working Load Limits (WLLs), which are commonly found on equipment used for lifting and industrial rigging in non-life-safety applications, does not provide the user with sufficient information for determining even a proximal SSF, leaving the true level of protection offered by the system in question.

Carabiners

Carabiners are a type of self-closing connector that link together different elements of the high-angle system (FIGURE 8-4). They are also sometimes called biners or snap links. In Europe, the United Kingdom, and some parts of Canada, carabiner is spelled karabiner.

NFPA 2500 (1983) classifies carabiners into two categories. There is no separate classification for escape carabiners:

- NFPA 2500 (1983) – G. Tested to a rated strength of 40 kN (9000 lbf), this classification of carabiner is intended for use by any trained organization performing rescues at a level where higher strength may be required to achieve the desired level of safety.
- NFPA 2500 (1983) – T. Tested to a rated strength of 22 kN (5000 lbf), this classification of carabiner is intended to be used by highly trained rescuers who might benefit from lighter weight and/or smaller profile equipment, and who are capable of maintaining an acceptable level of safety despite a lower strength rating.

Aside from the strength criteria of these classifications, there are several other features that should be considered when selecting carabiners, in light of the risk assessment and the organization's needs, training, and capabilities.

Material

Carabiners are most often constructed of either aluminum alloy or steel. Advantages of aluminum alloy are that it is significantly lighter weight and less expensive than steel, and it does not rust (although it can undergo some forms of corrosion). However, the locking mechanism on some aluminum designs eventually may wear out and severe shock loading may cause permanent damage. Additionally, some aluminum carabiners are not as strong as comparable steel designs.

Because of its hardness, the locking mechanism on steel carabiners tends to be less susceptible to wear and tear, and steel carabiners may also hold up better under severe shock loading. Steel is heavier than aluminum (an important concern when more than a few carabiners must be carried for any distance) and is typically more expensive. Unless plated, steel is also susceptible to rust, and as a result, steel carabiners generally require more maintenance than the aluminum type.

Some steel carabiners are plated to resist rusting. If you plan to buy plated carabiners, make sure they have stainless steel hinge springs. Unless they are stainless steel, the springs could rust and break. Consequently, the gates would not swing closed automatically, making the carabiner dangerous to use.

If you work in particularly harsh high-angle environments exposed to corrosive agents such as salt spray or chemicals, choose carabiners constructed of all stainless steel. Stainless steel is more resistant to corrosion, but be sure to choose high-quality equipment whose springs, pins, and other components are also made of stainless steel. Corrosion-resistant gear is no substitute for proper care and maintenance, so do not neglect that important part of your equipment program.

Carabiner Components

The basic parts of a carabiner are the frame and the gate. The gate is secured to the frame by way of a hinge, while a latch at the nose helps to maintain the integrity of the unit when the gate is closed (FIGURE 8-5).

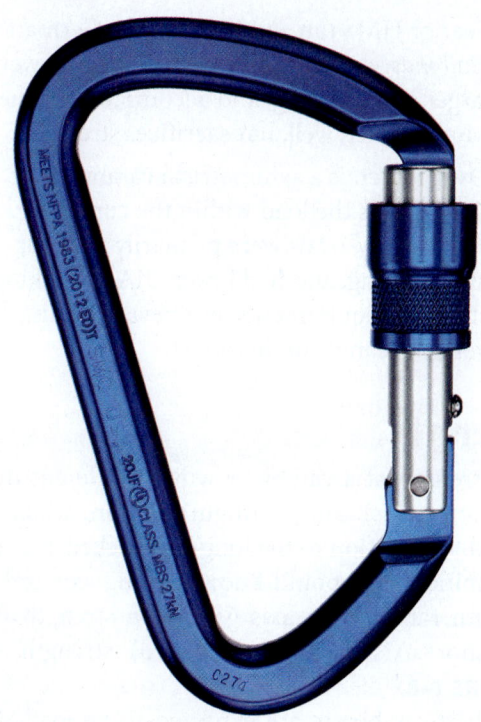

FIGURE 8-4 NFPA Carabiners.

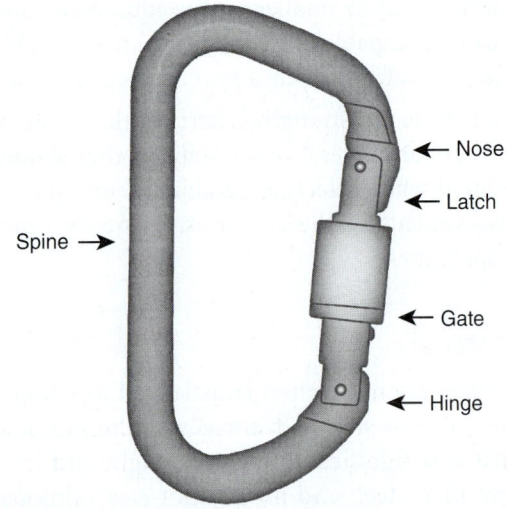

FIGURE 8-5 Basic parts of a carabiner.

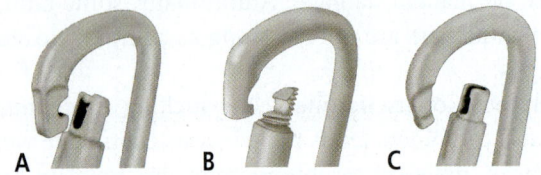

FIGURE 8-6 Different types of gate latch mechanisms. **A.** Pin and slot. **B.** Claw and slot. **C.** Keylock.

Many users prefer a smooth latching mechanism; while notched latch mechanisms are arguably more secure, they can snag during use. Some carabiners also feature a locking mechanism that helps to secure the opening between the gate and frame when locked.

A carabiner is typically strongest along its long axis, especially when the load is focused along the part known as its spine; a secure gate is integral to maintaining the overall rated strength of the carabiner. If the gate is opened, whether intentionally or inadvertently, the strength of the carabiner is compromised (**FIGURE 8-6**).

Basic Carabiner Shapes

Carabiners are manufactured in a wide variety of shapes and sizes. Originally all carabiners were of a simple oval shape, but over the years different features and configurations have been developed to maximize strength, gate opening, ease of handling, and other characteristics. Commonly used carabiner shapes include the following (**FIGURE 8-7**):

- D, which helps to shift the load to a more optimal position along the spine
- Modified D, which makes for a slightly larger gate opening (at the cost of a small amount of strength)

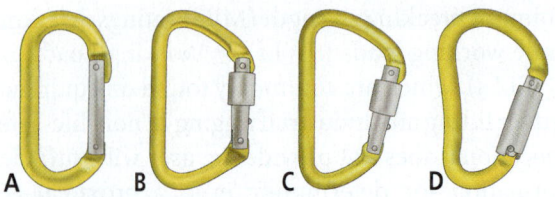

FIGURE 8-7 Carabiner designs. **A.** Oval. **B.** D shape. **C.** Modified D shape. **D.** HMS.

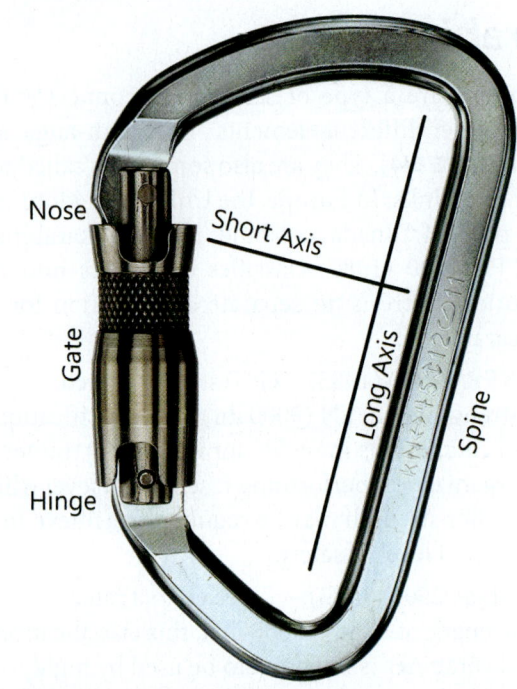

FIGURE 8-8 Long and short axis of a carabiner.

- Pear or HMS (an acronym for the German word *Halbmastwurfsicherung*), which has an even larger gate opening, and accommodates the Munter hitch well, but sacrifices strength
- Oval, which is a symmetrical carabiner that centers the load within the connector. Historically, ovals were primarily used for aid climbing, and had lower UIAA breaking strength requirements, but newer locking versions can be quite robust.

Strength

The strength of a carabiner will be different in each direction (axis) and configuration in which it is loaded. In addition to the long-axis locked strength of a carabiner, you should know its long-axis unlocked strength rating, long-axis gate-open strength rating, and short-axis (gate cross-loaded) strength rating (**FIGURE 8-8**).

When carabiners are expected to be loaded with heavy loads, or with dimensionally larger materials,

FIGURE 8-10 National Fire Protection Association (NFPA) markings for carabiners.

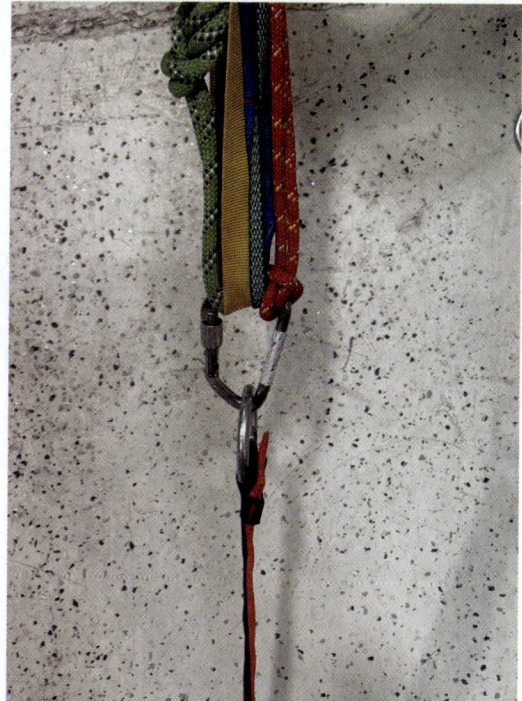

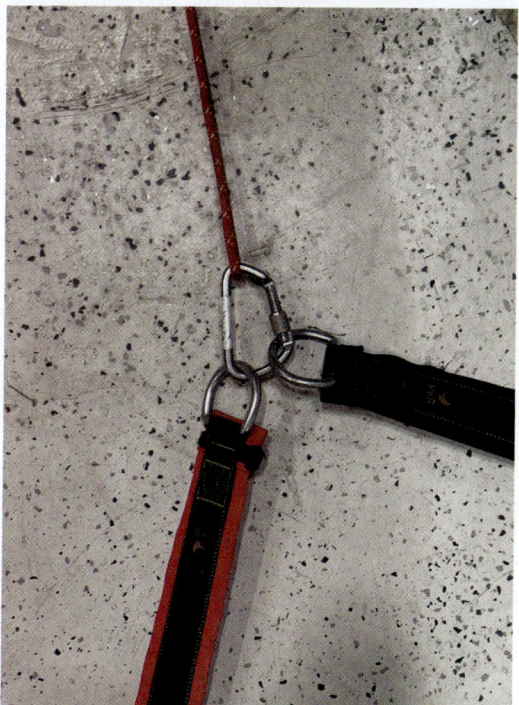

FIGURE 8-9 Overstuffing a carabiner can compromise its strength.

choosing a larger size will help to facilitate proper loading of the carabiner. As noted previously, carabiners are strongest along the spine, so when material is stuffed into the carabiner to the point that it approaches the gate side of the carabiner, strength is compromised (**FIGURE 8-9**).

Carabiners are typically marked in some way to show the strength or performance rating of the item.

These markings are typically derived from the standard(s) to which the item is marked. Carabiners that meet NFPA 2500 (1983) will be marked accordingly (**FIGURE 8-10**). In addition to reference to the standard itself, you should also find the name or logo of the manufacturer, along with the label of the third party that certifies that the product meets the standard.

NFPA-certified carabiners also have either a "T" or a "G" stamped into the frame. The letter "T" indicates that the carabiner is for "Technical Use" and designed for experienced users, having a major axis minimum breaking strength, with gate closed, of 22 kN (5000 lbf). The letter "G" indicates that the carabiner is for "General Use", with a more forgiving minimum major axis gate-closed breaking strength of 40 kN (9000 lbf). Carabiners marked to older editions of NFPA 1983 (prior to 2012) may be marked with an "L" instead of a "T."

Testing

In some cases, it may also be fine for carabiners marked to other standards to be used for rope rescue applications, even within the fire service. This is the prerogative of the AHJ, with the main priority being that the determination has been made by a competent person qualified to do so. Other relevant standards to which rescuers may find carabiners marked are shown in **TABLE 8-1**.

Keep in mind that not all of these standards utilize the same three-sigma method of calculating MBS; some require only that each sample tested by the test lab exceed the requirements of the standard.

According to NFPA 2500 (1983) test method requirements, a minimum of five samples are tested, and then a MBS is calculated by subtracting three times the standard deviation from the mean (average) of the test result numbers. That result, or lower, is what the manufacturer is required to advertise.

NFPA's approach to using a three-sigma calculated MBS results in a reported value that is statistically significant. In this case, a carabiner tested to a standard that does not require three-sigma analysis might be advertised with a strength of 5600 lbf (25 kN), while the highest strength MBS that could be advertised

TABLE 8-1 Comparison of Standards Requirements for Carabiners

Standard	Major Axis Requirement	Minor Axis Requirement
NFPA 2500 (1983) (General Use Rescue)	40 kN (900 lbf)	7 kN (1500 lbf)
NFPA 2500 (1983) (Technical Use Rescue)	22 kN (5000 lbf)	7 kN (1500 lbf)
ASTM F1956 (Light Use Rescue)	27 kN (6000 lbf)	7 kN (1500 lbf)
ANSI Z359.12 (Industrial Fall Protection)	22 kN (5000 lbf)	16 kN (3500 lbf)
EN 362 (European Industrial PPE)	20 kN (4500 lbf)	7 kN (1500 lbf)
UIAA/CE 12275 (climbing and mountaineering)	20 kN (4500 lbf)	7 kN (1500 lbf)
ASTM 1774 (climbing and mountaineering)	20 kN (4500 lbf)	7 kN (1500 lbf)

Abbreviations: PPE, personal protective equipment

according to NFPA's three-sigma method would only be 5100 lbf (23 kN). The smaller the numbers get (as with cross-load strengths) the more significant this concern becomes. Chapter 9, *Rope, Knots, Bends, and Hitches*, discusses the three-sigma method in detail.

It is also worth knowing that NFPA is the only rescue equipment standard that mandates a certified Quality Assurance program, periodic sample testing, and annual retesting of certified equipment. Other standards permit up to several years between retesting, or even have no retest requirements at all.

Some carabiners carry no labeling and may not be tested to any standard. If the carabiner is not labeled in accordance with a relevant standard (NFPA, ANSI, etc.), you may not be able to be sure about factors such as open gate strength, the strength of minor and major axes, and other qualities. In any case, if the actual rated breaking strength is not stamped on the item, it is best to check the referenced standard to find out just what the stamp means. Fire fighters working in an industrial environment may find carabiners marked to the ANSI Z359.12 standard, some of them even designated for coworker-assisted rescue. The major difference between ANSI Z359.12 carabiners and the others mentioned here is their gate loading strength. The gate of an ANSI Z359.12 carabiner must withstand a test load of 3600 lbf (16 kN) in all directions.

Even among carabiners with similar specifications, there is a balance to be struck between mass, weight, size, and strength. For heavy-duty rescue use, carabiners of solid aluminum or steel are a better choice than lightweight hollow aluminum models. The AHJ should determine which types of carabiners are most relevant to their intended use. In some cases, different carabiners may be appropriate for uses in different applications within the operation.

Strength Rating for Carabiners

Strength ratings for carabiners usually represent the ideal situation; in a D-shaped carabiner, for example, when material pulls on only a small area of the carabiner next to the spine. If material pulls across a wide area of the carabiner top or bottom (such as wide webbing does), greater stress could be placed on the gate side of the carabiner. Consequently, the carabiner may fail at a load lower than its rated strength.

Carabiner strengths are typically marked using scientific notation, with the unit of measurement being kilonewtons (kN). One newton (1 N) equals about 0.225 lbf, or about the weight of an apple. One thousand newtons would be referred to as a kilonewton (1 kN), which equals about 225 lbf.

Gate Openings

Another important consideration in carabiner selection is that of gate opening dimensions. When the gate is opened to its maximum, the dimensional gap will determine the size and bulk of material that can be inserted into the carabiner. While a smaller opening can help prevent overstuffing, for some activities the carabiner gate opening must be larger. For example, a carabiner that may have to be clipped over a litter rail, or onto a ladder rung, may need to have a larger opening.

Locking Mechanisms

The main job of a carabiner is to maintain its link with the other elements of the rope rescue system. To do this, the carabiner gate must remain securely closed. If it does not remain closed, the connecting elements

CHAPTER 8 Rescue Equipment   123

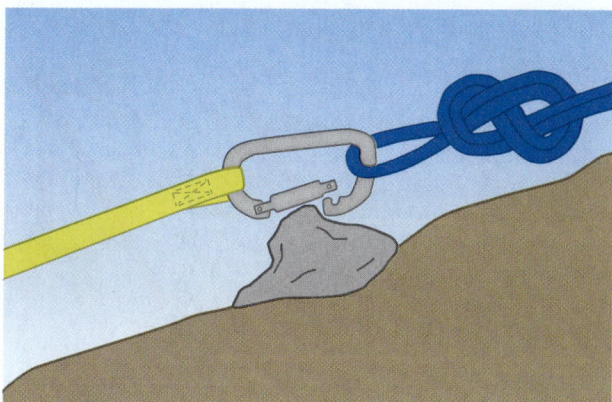

FIGURE 8-11 Gate opened by rock edge.

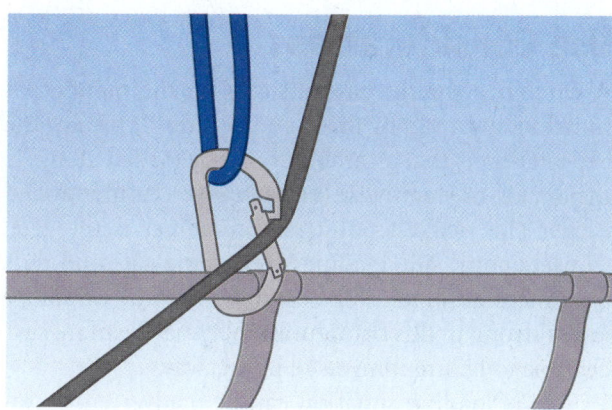

FIGURE 8-12 Gate opened by rope or webbing.

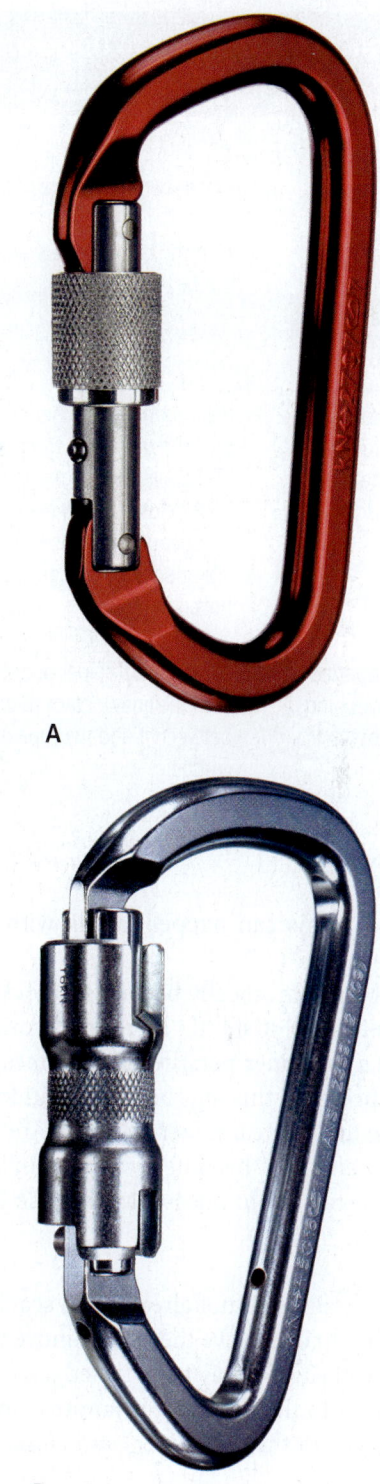

FIGURE 8-13 A. Manual locking carabiner. **B.** Autolocking carabiner.

can come unlinked and the system will fail. Carabiner gates can come open accidentally in several ways. Among the most common are as follows:

- The carabiner is pressed against the edge of a wall or rock, forcing the gate open (**FIGURE 8-11**)
- A rope or piece of webbing is pulled across the carabiner gate, forcing it open (**FIGURE 8-12**)

To prevent accidental gate opening, locking carabiners should be used for rescue.

Although specific designs vary with the manufacturer, locking carabiners usually fall into the following categories: manual and autolocking. A manual-locking carabiner (screw-lock) is a locking sleeve that manually screws over threads to ensure closure. The sleeve may be mounted on the gate and extend over the nose of the carabiner as it is screwed closed, or the sleeve may be mounted on the frame of the carabiner and extend over the gate as it is screwed closed. Autolocking carabiners (autolock) have a spring-loaded sleeve that automatically snaps into place as the carabiner closes, with no separate action required by the rescuer. Unlocking these may simply require a twist of the sleeve, or some additional safety mechanism may be present that necessitates an additional action to open (**FIGURE 8-13**).

Whatever type of carabiner is used, those using it are responsible for ensuring that the carabiner gate stays closed during use. Locking carabiners can and do come open after being locked. **TABLE 8-2** lists

TABLE 8-2 Countermeasures to Unlocking Carabiners

Unlocking Situation	Countermeasure
The carabiner gate unlocks by rubbing against a wall or cliff face.	Turn the carabiner away from the face.
With some carabiner designs, vibration can cause the locking sleeve to unscrew.	Position the carabiner such that the gate is at the bottom and gravity keeps the locking screw closed. (This may not be an adequate solution in high-vibration environments, such as helicopters.)
The gate is opened by rope or webbing running across it.	Move the carabiner out of contact with the line or place padding between it and the rope or webbing.

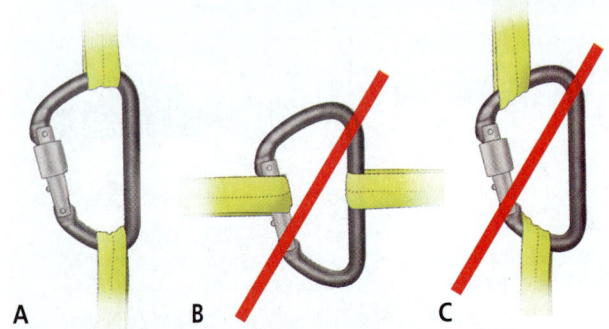

FIGURE 8-14 A. Acceptable loading. **B.** Unacceptable loading, nonlocking carabiner. **C.** Unacceptable loading, locking carabiner.

common ways this can happen, along with possible countermeasures.

If a carabiner chronically becomes unlocked without apparent cause, it should be retired from service. Likewise, if a carabiner persistently becomes jammed in a locked position, this may be an indication of overuse or wear, and the carabiner should be retired. If a screw-lock carabiner becomes jammed in a locked position, you can try to release it using the following procedure:

1. If the carabiner is not already on a seat harness, attach it to one. Have the wearer move to a secure position, away from the edge of any drop.
2. Connect to the carabiner by another means (sling or another carabiner) and connect to an anchor.
3. Reload the carabiner by leaning into it, or sitting down with it attached to the anchor point.
4. Check to see if the locking nut then can be loosened easily while under tension.
5. If it still cannot be loosened, tightly wrap a short piece of webbing around the locking nut to gain leverage.
6. If the previous step does not work, the only option may be careful use of a pair of pliers, after which the carabiner should be considered for retirement.

Use Considerations

A carabiner should be used only in the **manner of function** approved by the manufacturer. Typically, the expectation is that a carabiner will be loaded along its major axis, or lengthwise (**FIGURE 8-14**). As mentioned earlier, the weakest point of a carabiner is the gate. Consequently, side loading stresses the gate and puts an unnatural force on the carabiner. As the carabiner is less strong in this configuration, it may be more susceptible to failure if misused in this way.

A side-loading situation can be inadvertently created when a carabiner is rigged so that it cannot rotate with the forces placed on it. This twisting force can cause the carabiner to break. This is especially a cause for concern when a carabiner is hard linked to one or more points. Some examples of hard linking are as follows:

- A carabiner clipped into eyebolts on a wall
- A carabiner clipped into a second carabiner that cannot rotate
- A carabiner clipped directly into a vehicle tow hook

To avoid the pitfalls of hard linking, a soft link may be clipped between the hard links. A soft link is really nothing more than a short section of rope, webbing, or cordage whose primary purpose is to provide flexibility and strength, especially when connecting two hard components to one another. A soft link may be created simply by inserting a length or loop of cordage or webbing between the hard points. If you intend to add a soft link to your system, be sure you know the performance specifications in the configuration in which you plan to use it, whether end-to-end, looped, girth hitched, etc. A swivel may also be used to help mitigate the effects of twisting during use, but as swivels are also made of metal, they can be less forgiving of certain loading configurations (**FIGURE 8-15**).

CHAPTER 8 Rescue Equipment 125

damaged, has sustained significant impact loading, or has been dropped. Always inspect carabiners prior to use. Perform a visual inspection to ensure that there are no cracks, sharp edges, corrosion, burrs, or excessive wear. Check for proper gate function, and look for signs of excessive loading, such as frame deformation, damaged hinge pins, or dented latches. The body of the carabiner should not be gouged deeply, dented, corroded, or otherwise damaged. Locking mechanisms should function smoothly and properly.

Clean carabiners function best. Light dust and dirt may be brushed or blown from the frame and hinge, while dirtier or sticky gates may be washed in warm, soapy water or a citrus-based cleaner. It is especially important to clean and lube carabiners after contact with saltwater, salt air, or other corrosive environments. Be sure to dry carabiners thoroughly before placing them back in storage. If necessary, carabiners may be lubricated with dry graphite or any dry, waxed-based lubricant, focusing on critical areas such as hinge pins, springs, latches, and locking knobs. This also helps to help prevent corrosion. Wipe off excess lubricant before storing the carabiners. **TABLE 8-3** presents solutions to various carabiner problems.

FIGURE 8-15 A. A soft link may be clipped between hard points to provide flexibility. **B.** A swivel can help mitigate the effects of twisting.

TIP

All equipment used in the high-angle environment is designed to be used in a specific manner of function. This is particularly true of carabiners. Any equipment used in a manner other than that which it was designed may result in failure of the equipment and severe injury or death.

Inspection and Maintenance

Carabiners are an important link in the life safety chain, and as such should be used only if in good condition. Discontinue use of any carabiner that is

TABLE 8-3 Troubleshooting Carabiner Problems

Problem	Solution
Gates stick or close sluggishly.	This problem may be caused by contamination with dirt or by corrosion. Pressurized air can be used to clean the mechanism. Do not use oil or grease-based lubricants because they attract more dirt and grit. Lubricate the gate mechanism with a lubricant.
The carabiner gate has a broken spring.	The carabiner should be discarded.
The latch mechanism is broken.	The carabiner should be discarded.
The carabiner is bent (usually the result of overloading or of loading the carabiner without closing the gate completely).	The carabiner should be discarded.

Screw Links

Triangular or semicircular screw links are another type of connector that is used, especially where smaller dimensions are desirable. Instead of the swing-open gate found on carabiners, screw links have a screw locking sleeve that must be manually closed to secure the opening (**FIGURE 8-16**). As with carabiners, users of screw links should avoid side loading, which stresses the gate, puts an excessive force on the screw link, and severely reduces its strength, possibly causing failure.

These types of connectors must be manually screwed completely closed, which requires numerous turns to the sleeve. Failing to securely close a screw link, or it working itself open during use, are very real potential hazards. For this reason, when screw links are used, they should be integrally connected (that is, using a tool) before work begins, and left in place throughout, rather than opening and closing repeatedly.

NFPA 2500 (1983) life safety equipment standards do not address screw links. One of the challenging things about screw links is that not all are intended for life safety applications. The only standard that addresses screw links for this purpose is the European standard, *EN362 Personal protective equipment against falls from a height — Connectors*. Hardware store screw links are not tested to this standard, and are quite a bit less expensive than those tested and marked to EN362. Only those explicitly tested and marked in accordance with EN362 should be used for life safety applications.

Like a carabiner, a screw link is strongest when loaded on its primary axis, with the sleeve locked. Loading a screw link in any other way is dangerous, and can result in failure at lower loads. EN 362 calls for a screw link to be rated to 25 kN (5600 lbf) along the major axis (gate closed and locked) and 10 kN (2250 lbf) along the minor axis (gate closed). The idea that a screw link is as strong in a short axis as it is in its primary axis is a myth. Another important consideration is that—unlike carabiners—screw links are not tested in a gate open configuration. This is important because they are not self-closing and their integrity is completely dependent on their being screwed closed properly. Therefore, they are particularly susceptible to human error, as when not screwed fully closed.

Because they are often used in lifting, screw links are typically marked with a WLL that is one-fifth of the rated breaking strength, rather than with the breaking strength itself. This WLL approach is typical of the cranes/lifting industry, but should be used

FIGURE 8-16 Screw links. **A.** Triangular screw link. **B.** Oval screw link. **C.** Semicircular screw link.
Courtesy of Pigeon Mountain Industries.

with caution in rescue rigging. WLL essentially dictates a component safety factor, but the figures used are not derived from three-sigma calculations for statistical significance. Most rescuers prefer to work with minimum breaking strength information so that they can more accurately assess system safety factors. This is discussed further in Chapter 10, *Principles of Rigging*.

Although at face value the 25 kN (5620 lbf) strength required by EN 362 for the major axis of a screw link may seem stronger than the 22 kN (5000 lbf) requirement of NFPA T-rated carabiners, the difference may not be so significant when you consider the difference in how those figures are attained. Specifically, the EN requires only that the samples tested exceed the required minimum, as noted above, whereas the NFPA standard requires statistical analysis of at least five samples tested by the test lab. This approach is designed to protect the user—and, as was already discussed earlier in this chapter—without statistical analysis, data can be misleading.

Looking back at the sample test results in Table 8-2, you will note that the outcome of these test results would be a 25 kN (5620 lbf) rating according to the European standard, but only a 22 kN (5000 lbf) rating according to the NFPA standard. The same principle would hold true for any equipment tested, including screw links.

This is not to say that screw links are inappropriate—only that they should be used with care, and with full understanding of their capabilities and limitations. A key advantage of screw links is that they are diminutive in size, and can shorten the length of an otherwise cumbersome system, as when used with a footloop in an ascending system. Because it is so important that they be screwed completely closed before loading, screw links are best used where they can be integrally mounted into a system—that is, screwed closed and tightened with a tool so that they cannot be unintentionally released in the field.

Braking Devices and Descenders

A **braking device**, also called descender or descent control, is a piece of hardware that is used to help manage the rate at which a load is lowered in a system or at which a rescuer descends. Most braking devices rely on friction to perform their function, so rescuers must be attentive to the potential for heat buildup, which is a natural result of friction. Heat buildup is increased by heavier weights and/or faster speeds. In extreme cases, heat buildup in descenders can damage rope or other equipment, cause second-degree burns to bare skin, and have other adverse effects. Choosing a descender designed to dissipate heat, and not running rope too quickly through the device, will help control heat buildup during use.

A braking device may be used in a stationary position, fixed to an anchorage with the rope running through it (in which case it is most accurately referred to as a braking device), or may be used in a manner in which it moves along the rope, such as in a rappel (in which case it may also be referred to as a descender) (**FIGURE 8-17**).

Braking devices offer different features and specifications for different applications. An ideal braking device used for a heavy rescue litter lower, for example, may not be the best choice for emergency personal escape. A good understanding of braking device features and performance specifications will help the AHJ select the best option(s) for each application.

Common uses for braking devices in a rescue organization include the following:

1. Nonemergency rappel or single-person descent
2. Emergency rappel or bailout
3. For lowering a rescuer, a litter, or both
4. As a belay device
5. Any combination of the above

Selecting the appropriate braking device for a given application is a matter of taking into consideration not just rope diameters and compatibility, but also the intended use, rescue methods used by the organization, weight considerations, multifunctionality, and training of personnel. Descenders that are reeved through the descender in a linear fashion, rather than requiring the rope to wrap or twist, are best for rope management and care.

FIGURE 8-17 A braking device being attached to a rope.
Courtesy of Steve Hudson.

Some descenders are intended for use with a specified diameter (or diameter range) of rope. Experienced rope users will find that even braking devices that are not designated by the manufacturer as being rope-specific will perform a bit differently with different ropes. Generally speaking, the following are true:

- Smaller-diameter ropes will run more quickly through a braking device than larger-diameter ropes through the same descender.
- Polyester ropes will run more quickly through a braking device than nylon ropes.
- Dry treated ropes will run more quickly through a braking device than standard ropes.
- Harder/stiffer ropes will run more quickly through a braking device than softer/flimsier ropes.
- The more rope weight hanging below the device, the faster it will run.

Autolocking and Manual Devices

One of the greatest challenges to rescuers these days is that both autolocking and manual devices are used in the field—and each has its own set of advantages and disadvantages, depending on application. An autolocking device requires the operator to engage some sort of override to permit the rope to slide through, and will automatically stop the rope in the event that the operator completely lets go. While completely letting go of a device is not considered normal operation, rescuers who are accustomed to the additional measure of security this feature offers may be inclined to develop habits that rely on this feature. Conversely, many experienced rescuers prefer a manual device that requires the operator to consistently maintain an active grip on the rope to perform a braking function. Should the rescuer let go of this type of device, the rope will run freely through it (unless stopped by a secondary safety system.) Manual devices tend to be simpler to operate, have fewer moving parts, and be less sensitive to rope diameter than autolocking devices. Each type has its own unique advantages and disadvantages, but in either case adequate training and experience is imperative.

Specifications to consider when selecting a braking device:

- Whether an autolocking device is required
- Whether a panic-lock feature is required
- Ability to adjust friction
- Size and weight of the device
- Compatibility with the organization's life safety ropes for rappel or belay
- Compatibility with the organization's escape rope or webbing
- Material of construction
- Ability to dissipate heat
- Levels of training required for personnel competency in desired skills

Additional Considerations

Variable friction is an important consideration during lowering or descending operations where the weight of the load is likely to change, or when the weight of the rope beneath person descending varies (such as on a long rappel.) Such a device goes beyond a simple stop/go function to include ability to wrap, clip, or add friction components as necessary.

Desired performance specifications of a braking device must be balanced with the size, shape, and weight of the equipment. For lowering and raising operations next to a road, a truck-mounted capstan may be a desirable choice, but where rescuers must carry equipment into the field, size and weight are considerations.

Some devices are able to be used with a wide range of rope and/or webbing, while other devices are quite dependent on being used with a particular construction, material, or diameter. Responding organizations usually change out their ropes and soft goods more frequently than hard goods such as descenders, and so this is a consideration. Most hard goods manufacturers do not also actually manufacture their own rope (even if it is so labeled), so descenders that function with a range of ropes are generally a safer option. The best descenders are not acutely rope specific, although compatibility with a narrow range of diameters is not unusual. Rescuers will notice a marked difference in the performance of their braking device even with ropes of the same make and model, depending on the age and condition of the rope. It is important to train on the same kinds of ropes as are deployed on actual operations.

The material of which a descender is made will have some influence on its weight, friction imparted, ability to dissipate heat, durability, resistance to corrosion, susceptibility to wear, and other performance considerations. While rescuers will appreciate the durability and smooth operation of steel, they will be less keen on carrying it into the field. Aluminum is lighter, but more susceptible to wear. Rescuers should be trained to carefully inspect rope paths for wear, and retire any excessively worn descenders.

Friction and Heat

Friction is the means by which most braking devices perform their function, and a natural by-product of friction is heat. The heavier the load, the greater the friction, and the longer the descent, the more heat buildup there is. A very hot braking device can pose a burn hazard to rescuers' skin as well as to ropes and soft goods with which it comes into contact. The strength of the device itself may also be diminished as a result of overheating. When selecting and operating a braking device, rescuers should give heed to the intended use—the weight of anticipated loads and the distance(s) these loads will normally be lowered. Heat dissipation derives from greater surface area and an open design. If a device does become overheated during use, dousing it with water can help to cool it through convection.

Guidelines for Selecting a Braking Device

1. Make sure the descender is sized properly for the diameter of rope you will be using.
2. The descender should create enough friction to give you absolute control over the descent without using brute strength. You should be able to go as slowly as you like and be able to stop at any time.
3. Keep in mind that hardware and equipment make you significantly heavier than your "street weight." You may also be wearing gloves.
4. Be sure that you know how to lock-off the device before you use it, so that you can secure yourself and remain stopped on the rope with hands off the device.
5. Be sure the descender is strong enough to have an adequate safety factor for the intended use.
6. The descender must be adequate for the length of descent needed.
7. Be sure the device is adequate for use in your intended application (i.e., lowering or descending.)

Types of Braking Devices

Response organizations should choose braking devices with consideration to the purpose(s) for which they will be used; desired performance capabilities; and the frequency, level, and type of training that will be available to users. At a minimum, braking devices (whether used for lowering or descending) should meeting the following criteria:

- Provide suitable control over the speed of descent
- Offer smooth braking, without bouncing or undue shock loads
- Not cause abrasion, plucking, milking, or stripping of the rope sheath
- Not become accidentally detached from the rope

An organization may choose a range of braking devices to be used for different purposes—for example, lowering, descending, and personal escape—and then when that equipment finds its way to the field, rescuers must have sufficient knowledge to determine which device is best for a given operation with consideration to weight of load, length of lower, environmental factors, available rope, auxiliary equipment, etc.

NFPA classifies descenders into the same three categories it does other equipment: E (escape), T (technical), and G (general). The biggest difference among these three classifications is the load and use for which they are intended. In addition to echoing certain test methods and requirements from the ISO standard, NFPA also requires deformation testing. Braking devices with a G rating must withstand at least 11 kN (2470 lbf) without deformation, while T-rated devices must withstand at least 5 kN (1124 lbf) without deformation. Escape devices are subjected to a dynamic test.

Braking Devices for Rescue

Autolocking

With rope rescue capabilities becoming more commonplace among responding agencies, and the proliferation of rope access methods being used in professional rescue, the fail-safe advantages of autolocking devices have increased in popularity. Although arguably less versatile, devices with autolocking and panic-lock features do offer peace of mind for a broader spectrum of users. These devices tend to be very rope-specific, and capable of being used only with a very narrow range of diameters (**FIGURE 8-18**).

FIGURE 8-18 An example of an autolocking device, the Petzl Maestro.

Using an autolocking descender with a diameter of rope other than that for which it is designed can create additional hazards, including insufficient friction, jamming, or even accidental disengagement.

Most autolocking devices function by means of wrapping the rope, either fully or part-way, around a moving bollard, or block of metal. Tension applied to the working end of the rope pulls the bollard in such a way that the rope is compressed or pinched against a block, and progress stopped. In many cases, the bollard is enclosed by a shell, which opens to allow insertion of the rope, and closes to keep it in place. Most of these devices can be used not just as a descender, but also as a progress capturing pulley for raising operations. Typically, these devices are fitted with a handle by which the braking mechanism can be rotated, releasing pressure on the rope. Feeding rope through the device is a two-handed operation, with one hand operating the handle and the other serving to apply friction on the tail of the rope.

Climbing Technology Sparrow

The climbing technology sparrow is a self-braking descender designed for descent and ascent. An external brake-spur permits rapid addition of a modicum of friction while the automatic-returning brake lever offers speed control. When the lever is pulled too far, the speed is reduced, but not stopped altogether. The climbing technology sparrow is designed for use with an 11-mm rope.

CMC MPD

Although the form of the CMC MPD differs from the other bollard-type devices in this list, it is included here because it, too, incorporates a bollard and is designed to serve as both a braking mechanism and a pulley. Like the other devices in this list, it permits switching from lowering to raising without any change in hardware. This device is available in two models, one for 11-mm rope and the other for 12.5-mm rope.

Harken Clutch

A recent addition to this class of products, the Harken clutch also features a stainless steel rotating sheave. The sheave, stationary in the locking direction, is designed to make an audible clicking noise in the other. It offers both an autolocking feature as well as panic-lock, and has a unique force limiting feature to prevent overload. The Harken clutch features a becket, which is a hole for attaching a carabiner, so that it may be integrated into haul systems, and rope can be installed without removing the device from its anchorage.

ISC D4/D5

The ISC D4/D5 Work Rescue Descenders are designed for 11-mm and 12-mm rope, respectively. The offset bollards of these double-stop devices are unique for the range of control that they offer. With both autolocking and anti-panic features, the device is designed to slip between 4 and 6 kN (890 and 1349 lbf), so as to be able to be used as a fall-arrest/belay device. Designed for heavier loads without extra friction or redirect, this descender can be rigged to the rope without removing it from its anchorage connection. It features a release mechanism that is designed to reduce the risk of accidental opening but is easily operated with gloves.

Petzl ID

One of the earliest bollard-type devices to be designed for the professional market, this standby product is advertised as a self-braking descender/braking device. With different handle positions for descent, panic stop, and storage, there are two versions of this device: one for 10- to 11.5-mm rope (maximum load 150 kg [330 lb]) and the other for 11.5- to 13-mm rope (maximum load 250 kg [550 lb]).

The ID is known for having a relatively narrow sweet spot, which can be problematic for lighter loads or low slopes. Because the handle adjustment position at which rope can travel freely through the device is so narrow, it can be difficult to operate when forces on the device are lower. In older models, a button at the end of the handle can override the cam; in newer versions manufactured after 2019, the operator can manipulate the cam with their thumb.

Petzl Maestro

The Petzl Maestro is one of the more recent additions to this class of descenders. It is unique in that the bollard is really a large-diameter sheave, like a pulley. It is fixed in one direction, but rotates in the other, making it quite efficient for raising. The outside shell features a protruding horn, allowing the user to temporarily increase the amount of friction applied by a moderate amount, as long as the brake hand holds friction on the rope.

When loaded, the device automatically locks off. The braking mechanism is released by pulling the handle; the more the handle is pulled, the more the moving brake plate disengages to let the rope run. This device does not have a panic-lock; if the handle is pulled too far, the rope can run freely through the device—possibly causing loss of control. This device should be used with care, especially by less experienced persons.

The Maestro is available in two versions, one for use with ropes from 10.5 mm to 11.5 mm in diameter (maximum working load 250 kg [550 lb]), the other for use with ropes from 12.5 mm to 13 mm (maximum working load 280 kg [617 lb]).

Skylotec Lory

The Skylotec robust descender and belay device is designed for professional rescue, work positioning, restraint, and anchorage, in addition to standard belay, lowering, and raising operations. It is especially adept at feeding and taking rope, and blocks automatically in case of a fall. The Skylotec Lory features both autolock and panic stop elements and is equipped with a descent handle that ensures maximum security and control. It is designed solely for use with 11-mm rope.

Skylotec Sirius

The Skylotec Sirius is designed to be an ergonomic, durable, ultra-compact device that prevents twisting, kinking, and excessive wear on the rope. Its autoreturning lever has a minimal range of motion for maximum efficiency and safety, and a small hole at the top can be rigged with cordage for remote operation. A secure attachment point permits installation of rope without removal from the anchor point and a becket accommodates integration into haul systems. According to the manufacturer, the Sirius has been submitted for NFPA testing for a G rating with both 11-mm and 12.5-mm rope, with the same device.

SMC Spider

The SMC Spider is the only device in this list tested and certified to function effectively with a wide range of rope diameters. While most other devices offer different versions to function with very specific diameters of rope, this one works with rope diameters of 10 mm to 12.5 mm. It functions best for lowering, descent, and positioning, and is designed such that the rope may be installed or removed without having to unclip from its anchorage. The SMC Spider is autolocking in a hands-free mode and has a very wide sweet spot, which is the adjustment at which the rope travels freely through the device, before the handle finally reaches a panic-lock position.

Precautionary Notes

Most autolocking devices function with only a very narrow range of rope diameters; it is extremely important to use the correct rope diameter in these devices. The very narrow range also means that ropes of different hands, or hardnesses, will also perform somewhat differently in them.

If an organization that uses a certain combination of rope and device plans to change either rope or device, field evaluation of compatibility should be performed prior to that change, and personnel should be provided ample training time to become accustomed to the differences prior to a possible response.

It should also be noted that wet, icy, or muddy ropes will perform differently in a braking device than dry ones. This difference is much more notable with autolocking devices than with nonautolocking devices. If such use is likely or expected, rescuers should practice under these conditions to become accustomed to the difference in friction and feel. In addition, it is important to ensure that debris does not become lodged in the shell of the device, as this could impede proper function or even damage the rope.

Finally, care should be taken when using autolocking devices for heavy loads, long lowers, or extended descents. Due to the limited range and capacity of these devices, they may not (by themselves) offer sufficient friction for heavy loads and/or may jam up when there is a great deal of rope weight beneath them, as on a long rappel. Further, the limited rope path and smaller mass of most of these devices means that they do not dissipate heat as well as, for example, a brake rack.

Nonautolocking

Nonautolocking braking devices should be used only by those with specialized training and skill. Although these devices may offer unique performance benefits, including versatility, capability of being used with a wider range of rope diameters, adjustable friction, and/or optimization for especially long descents, the fact that they do not automatically lock when the user lets go makes them less appropriate for inexperienced users. These devices should be used only by very experienced users, when the benefits they offer outweigh the drawbacks, and when potential hazards are mitigated.

Nonautolocking devices can be retrofitted to achieve the benefits of an autolocking device by utilizing a self-jamming friction hitch to create an autoblock, in the system. This is a rigging solution, not a device, so is further discussed in Chapter 18, *Personal Vertical Skills*.

Brake Bar Racks

While not manufacturer specific, a **brake bar rack** is most conducive to adjusting friction over a wide range, even while under load (**FIGURE 8-19**). SMC Gear

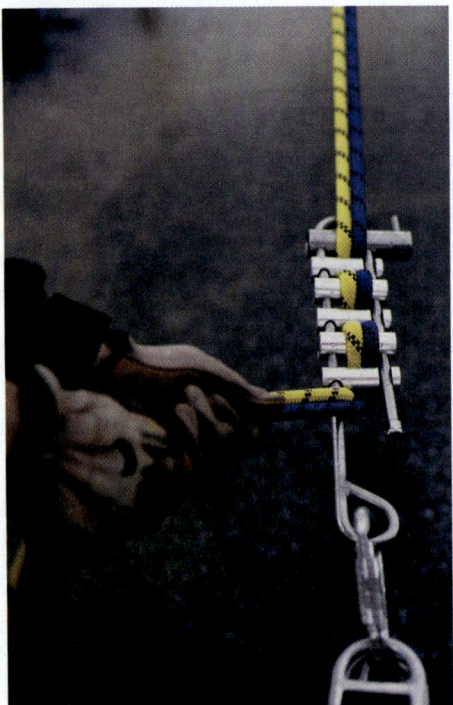

FIGURE 8-19 NFPA-compliant brake bar rack.

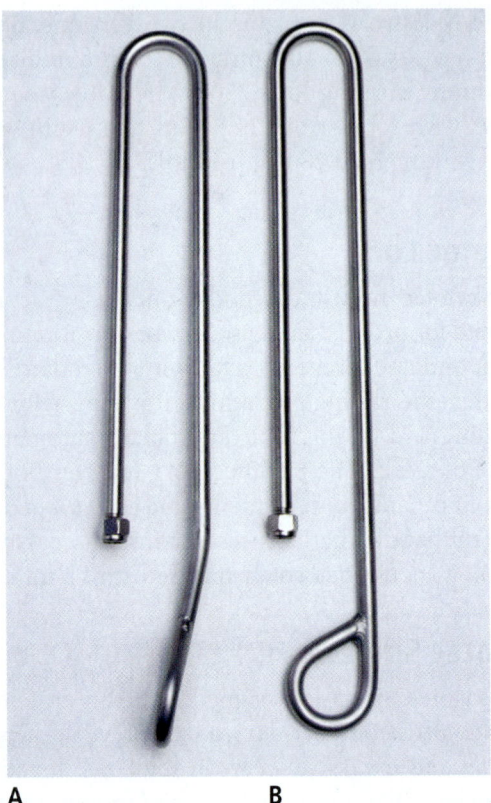

FIGURE 8-20 J-rack. **A.** Eye with 90-degree twist. **B.** Straight eye.
Courtesy of Pigeon Mountain Industries.

offers a brake bar rack assembly that meets the NFPA 2500 (1983) equipment standard.

The brake rack (also called Cole Rack or rappel rack) offers a great deal of control and allows the user to adjust the amount of friction easily during use. It excels in very long rappels, and also works well for lowering heavier loads such as those encountered during rescue operations. However, it is free-running and does not autolock. The brake bar rack consists of two primary elements:

1. A frame. Most frames are made of stainless steel. Less common are titanium versions, which are lightweight but more expensive.
2. Bars. These are designed to fit over the frame, with a hole drilled in one end so that the bars slide freely along one side of the frame and a notch on the opposite end to clip on and off of the other side of the frame. Larger bars tend to create greater friction than small-diameter bars. Most brake bars are made of hollow steel bars, although aluminum bars are used in some older designs.

There are two common frame configurations, the J-rack and the U-rack. The J-rack is an open-ended frame with one leg longer than the other (**FIGURE 8-20**). The user clips to the device by means of an eye on the leg of the longer bar at the open end. For rappelling, the J brake bar rack works most efficiently with the open side of the rack toward the ground. Some seat harnesses have a horizontal D ring as a clip-in point; this positions the open side of the rack facing the rappeler's side. To compensate for this, some versions of the rack have a 90-degree twist in the eye so that the rack remains with the open side toward the ground. The other rack design, known as the U-rack, has legs of equal length and is operated with the U in the usual position (**FIGURE 8-21**). When used, a carabiner is clipped across the bend at the bottom part of the U.

With either frame configuration, the rope is woven through the bars to create an optimum amount of friction. Under tension, the rope keeps the bars in place on the frame. Friction is controlled by (1) applying pressure with the brake hand on the rope below the rack, (2) by varying the bar spacing, and/or (3) by varying the number of bars engaged on the rope. The bars must be loaded in the correct sequence for safe operation of the device. Longer frames accommodate more bars, making them an excellent choice for long drops or situations where a wide range of friction adjustment is desirable (**TABLE 8-4**).

When a brake rack is used correctly, the bars are the only elements that wear out. They must be replaced from time to time. Depending on the bars'

CHAPTER 8 Rescue Equipment

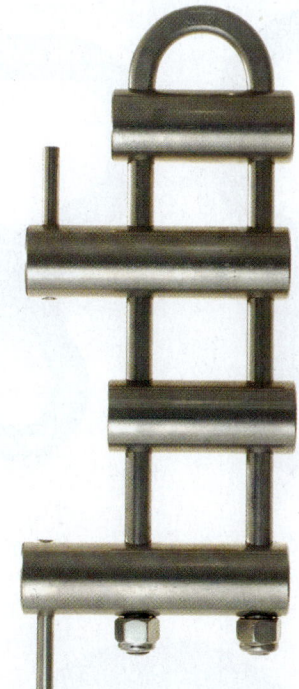

FIGURE 8-21 U-rack.
Courtesy of Pigeon Mountain Industries.

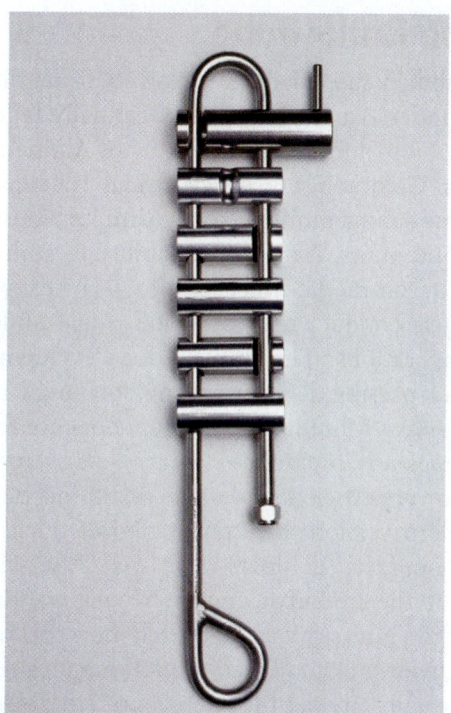

FIGURE 8-22 Brake rack with hyper-bar.
Courtesy of Pigeon Mountain Industries.

TABLE 8-4 J-Style Versus U-Style Brake Bar Racks

J Style	Advantage	Has greater flexibility in changing friction by changing a specific number of bars while the rack is engaged with rope
	Disadvantage	Tends to pull at a slight angle when loaded, so it may be more difficult to keep rope centered on upper bars
U Style	Advantage	Rope tends to remain centered when the rack is loaded
	Disadvantage	More difficult to change the number of bars safely when the rack is loaded

requirements for friction, they can be purchased in a variety of sizes. In the most common configuration, the rack is arranged with a 1-inch (2.5-cm) diameter top bar. On J-racks, this top bar may be grooved to keep the rope in the middle of the bars as it runs through the device. On U-racks, the second bar down is often grooved. The remainder of the rack usually is filled out with five aluminum bars, each $^7/_8$ inch (2.2 cm) in diameter.

Some brake racks also have a tie-off bar (also known as a hyper-bar) that is of a larger diameter and extends out to one side (or both sides in some designs) to make it easy to add friction or to tie-off the rope (**FIGURE 8-22**). A pin at the end of the bar helps help keep the rope from slipping off the bar. On the six-bar rack, a tie-off bar can be placed at the top; some short versions of the U-rack have two tie-off bars, with the second one as the bottom bar.

Scarab

This unique and compact version of a bar-type device is very small compared to a typical rack and offers similar performance characteristics, such as not twisting the rope and ability to vary friction. Of course, simply due to its smaller mass the amount of friction that can be added is limited, and not sufficient for a very wide range, and it does also not dissipate heat as well as a larger device.

While the scarab offers neither autolocking nor panic-lock features, it may be used with a wide range of rope (9–13 mm), it can be rigged in single rope or dual-rope configurations, and it is very strong. The frame and crossbar of the scarab should be monitored closely, and the unit retired when the crossbar is worn more than 0.030 in. or the frame is worn more than 0.090 in.

Alpine Brake Tube

This tubular shaped device evolved from the Forrest Wonder Bar in the 1970s. Originally fabricated by Tom Fiore and then Phil Leuthy of Alpine Rescue Team in Colorado, the Alpine Brake Tube excels in accommodating multiple ropes simultaneously and in passing knots. It also offers ability to adjust friction to accommodate very light to very heavy loads, making it conducive to both low-angle and high-angle operations. The impetus for development of this knot-passing device was long, low-angle rescues in the Rocky Mountains where anchors are few, but long ropes are impractical to carry. As it turns out, it is also very efficient and offers additional benefits.

Modern versions are constructed of thick aluminum tubing that dissipates heat well. Risers help to keep it off the ground during lowering operations, and protruding pins serve as tie-off retainers. Brake tubes are relatively bulky, and heavy, but function effectively with virtually any size life safety rope. Where the need to pass knots and/or run multiple ropes through a single device, the brake tube is an excellent choice.

Figure-8 Descender

A Figure-8 descender (also called simply a Figure 8) is shaped rather like the numeral 8, but having rings of unequal size. The smaller ring (or lower one when in use) is clipped into a seat harness or anchor with a carabiner or screw link. The rope passes through the larger ring (or upper one when in use) to create friction. Figure 8s are made of either steel or anodized aluminum alloy.

Figure 8s are a free-running type of device, with the user controlling the speed of the load with a brake-hand; the amount of grip or tension on the rope determines speed. These devices may be used with any type and diameter of rope, including doubled ropes, and they dissipate heat well. While the Figure 8 does not offer autolocking or panic-locking features, its open shape is quite versatile and offers several rigging options.

An inherent problem with the venerable 8 is that it has a propensity to tangle ropes. These devices also offer a limited amount of friction for descending, can be difficult to take rope through while belaying, and are generally not considered a good choice for heavier loads. In fact, Figure 8s have become all but obsolete, even for recreational use.

Conventional Figure-8 descenders are mostly found in small sizes and typically have a rounded or slightly squared large ring (**FIGURE 8-23**). They range in size from the very small Escape 8 to larger

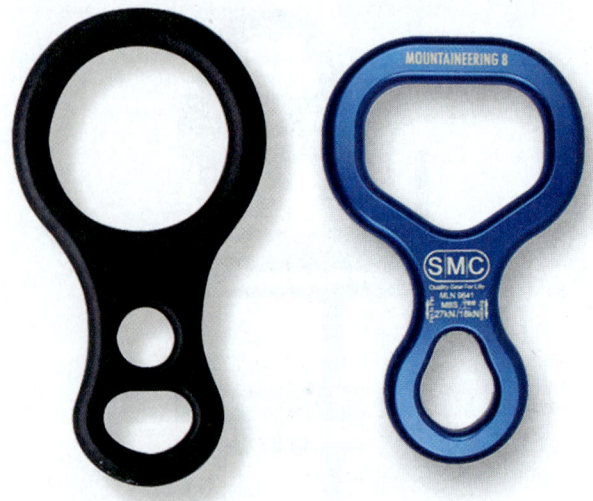

FIGURE 8-23 Conventional Figure 8 descender.
Courtesy of Pigeon Mountain Industries.

FIGURE 8-24 Figure 8 with ears.
Courtesy of Pigeon Mountain Industries.

styles used in recreational climbing. Some Figure-8 descenders have protruding "ears" fabricated into the large ring (**FIGURE 8-24**). Known as Rescue 8s, these are specifically designed so that the rope contours better around the large ring and is less susceptible to accidental girth hitching. This type of Figure-8 is typically of a larger size than recreational 8s, offering greater heat dissipation and facilitating use with larger diameter ropes.

The most commonly used configuration, the standard rig offers a modicum of friction and is easy to rig (**FIGURE 8-25**). This mode does, however, induce twist into the rope, and requires disconnection from the harness to de-rig. This higher friction method

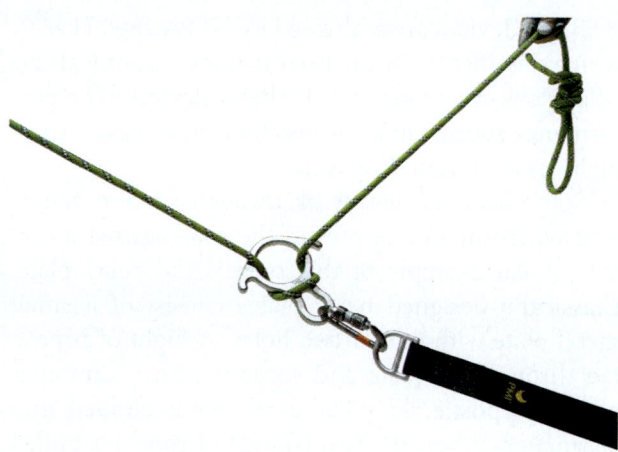

FIGURE 8-25 A standard rig.

of rigging offers additional control for heavier loads, but likewise imparts a twist into the rope and requires complete disconnection from the harness to install or de-rig. An alternative method of increasing friction for heavier loads, care should be taken when using this method to avoid accidentally unlocking the host carabiner. Note, also, that the host carabiner should never be opened while under load.

This advanced method of rigging a Figure 8 is especially for use in environments where the user might want to escape the system rapidly. This advantage, however, is likewise a potential disadvantage in that it could inadvertently come off the rope when unweighted. While not considered an autolocking device, a Figure 8 rigged in this manner will autolock when the tail of the rope is trapped between the tensioned part and the device. To release it, simply pull down on the small ring of the 8 with one hand while continuing to control speed with the brake hand. This method works better on descent than in lowering mode.

Your choice of wraps will depend on your skill level, weight of the load, type of rope, conditions, and device dimensions.

Escape Devices

Some emergencies in the rope rescue environment may require an immediate descent; for this reason, it is wise to carry equipment that is either specifically designed for this purpose, or that can be improvised to accomplish this end. **Escape devices** are designed to be more compact, and suited to a smaller diameter of rope (usually 7–9 mm) so that they may be easily carried by a rescuer in a pocket or small pouch, ready to be deployed in a life-threatening emergency.

Some escape devices are autolocking, and some are not. There is a body of thought in the field that because escape ropes tend to be so thin, descent may be difficult to control and therefore the descender should be both autolocking and panic-locking. There is another body of thought that suggests that the idea of a real emergency escape is to evacuate rapidly, and autolocking features increase the risk of becoming inadvertently stuck on rope. It is left to the AHJ to weigh the risks, to determine which of these is of greater concern, and to purchase equipment and train personnel accordingly.

Escape Devices: Autolocking
Petzl EXO

The Petzl EXO is an autolocking descent system that evolved from the Petzl GriGri recreational belay device. It offers a self-braking system as well as anti-panic function, permitting the user to control descent on a slim escape rope. The device is also designed so that it can be held open manually as the user moves across a horizontal surface. As with all devices, the Petzl EXO requires special training before use.

Sterling F4

Simply designed with smooth, beveled holes, the F4 features a lever that allows rope to feed when activated, and to lock when released. The rescuer operates the device by activating the device with one hand while controlling the free end of the rope with the other. When the device is released, descent stops; in the event that a person is injured or incapacitated while descending they would simply stop, suspended on the rope.

Rigging the F4 is simply a matter of reeving a rope through the holes to the extent necessary for the anticipated load. The number of holes reeved will depend on the type and diameter of rope, and the weight of the user.

Escape Devices: Nonautolocking

For those who subscribe to the idea of free-running emergency escape systems, there are a number of options. The caveat that comes with such equipment is that there is no room for error: If a user lets go of the brake hand, the device will not lock off. The advantage, of course, is that these devices are easier to feed and adjust quickly, making them less likely to jam in the heat of an emergency.

Escape 8

A smaller version of the previously mentioned Figure 8 works well for this purpose. Designed as a personal escape descender for fire fighters in bailout situations, these small devices work well for either single or double ropes. Unlike autolocking escape devices, which tend to be very rope specific, the Escape 8 can be used for ropes from 7.5 mm in diameter all the way up to 12 mm in diameter. When needed, particularly with smaller diameter ropes, an additional wrap can help add friction. The compact size makes it lightweight and easy to carry, and the squared-off shape also contributes to better friction.

PMW Hook

The PMW Hook is designed with a variety of functions in mind, including as a personal escape descender, to make a rescue lower, or as an anchorage connector. The only device in this list designed with such versatility in mind, the PMW Hook is available in two sizes. For lowering or descent, the device is rigged by taking a bight of rope through the large hole and looping it over the neck of the device, similar to the manner in which a Figure 8 is rigged. Friction may be easily adjusted by taking additional wraps around the top part of the hook.

Belay Devices

Belaying refers to the process of protecting a person from falling by controlling an unloaded rope (the belay rope) in a way that secures the person on the rope in case the individual's main line rope or support fails. Belaying a rescue load is different from belaying one person. Commonly used recreational climbing devices and techniques that work well for a single body weight may not be adequate for belaying a rescue load.

Historically, climbers have belayed one another by attaching the **belay rope** directly to both the belayer and to the person being belayed. The oldest belay technique is the body belay, which involves running the belay rope around the body of the belayer (usually around the waist). In this way, the rope can be brought tight if the person on the end of the rope (the climber, rappeler, worker, or rescuer) falls. The rope is held by friction around the belayer's body. Body belaying has significant disadvantages:

- The force of the fall may cause the belayer to lose control of the rope and drop the climber.
- The force of the fall can easily injure the belayer.
- The belayer can become entangled in the rope. If the belayer does catch the climber, they must hold the individual until the climber can become secure.

Because of these problems, alternatives to body belaying have been developed and are preferable.

Belay devices are addressed by NFPA 2500 (1983), which classifies them into two categories: technical-use belay devices and general-use belay devices. The performance specifications for each of these class ratings includes a relevant drop test.

Some belay devices work through friction generated when the device presses the rope against a carabiner. An example of this type is the belay plate. Classically designed belay plates consist of a small metal plate with one or two holes. A bight of rope is fed through the plate and secured with a carabiner on the opposite side. The carabiner is clipped into an anchor. When the two strands of rope are pulled apart, a high degree of friction is created on the rope. This stops the fall of the climber. Although commercial versions of these are available, the small hole of a Figure 8 descender may also be used to create a belay plate for personal loads (**FIGURE 8-26**).

By poking a bight of rope through the small hole of the Figure 8 and clipping it with a carabiner, the device may be used for belaying. This method allows rope to be played out quickly when belaying, and is the only use of the Figure 8 that does not induce twisting. Another example of a personal belay device that works through friction is an Air Traffic Controller (ATC) by Black Diamond (Salt Lake City, Utah). The ATC can be used with ropes ranging from 8.5 to 11 mm in diameter. Some personal belay devices work through a camming action on the rope. An example of this type is the Beal Birdie.

Personal belay devices are not covered by NFPA 2500 (1983), but may be a valuable tool for rescue personnel to carry in the event that they may need to belay another rescuer across a leading edge or while lead climbing, such as while accessing a tower or other structure in an industrial environment.

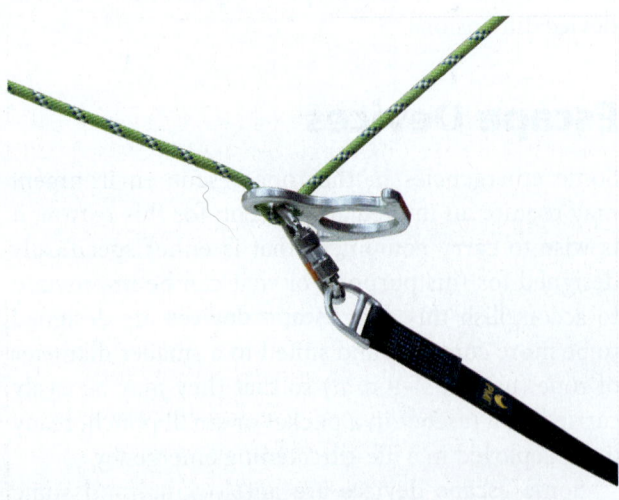

FIGURE 8-26 Figure 8 belay mode.

> **TIP**
>
> A personal belay device may not be the most appropriate device for catching loads of more than one person's body weight. Before using any device or system to belay a rescue load, you should test it under conditions similar to those you will encounter in an actual rescue situation. This ensures that you will be able to catch the load when it falls.

Some of the braking devices noted earlier in the chapter are also approved by the manufacturer for belaying of a rescue load. Always follow manufacturer instructions and guidelines when using a braking device for belay, as some devices are not well suited for dynamic impact loads. That said, the commonly used twin tensioned rescue belay method calls for the use of two descenders rather than a belay device. See Chapter 14, *Lowering Systems*, for more on the twin tensioned rescue belay method.

The Münter hitch (or Italian hitch) requires only a rope, an anchor, and a carabiner. Properly tied and operated Münter hitches can work as personal belay systems for falls with a low fall factor. The Münter hitch is less effective with heavy loads, but may be suitable if there is sufficient friction in the system. The ability to catch a load with a Münter hitch varies, depending on the rigging situation. For example, it may be easier to catch a load with a Münter hitch if the rope runs across an edge or face or through directional pulleys. These elements in a high-angle system add friction and thus help absorb the force of a falling load. The Münter hitch is covered in more detail in Chapter 9, *Ropes, Knots, Bends, and Hitches*.

No belay device or technique is perfect for all rescue loads and in every rescue environment. Great caution must be exercised in choosing any belay device for rescue loads. Before you put the device to use, test it under realistic conditions that you will encounter in your own rescues.

In selecting devices and methods for belaying, the rescuer should follow the guidelines set forth by the AHJ with respect to the following:

- The maximum potential impact load and arrest distance for the load
- The anticipated static load to which the device (and system) might be subjected, considering system configuration and methods used
- Possible approaches to mitigating potential impact force through reduction of mass, reduction of fall distance, additional means of energy absorption, or some other manner.
- Operational capabilities and training levels of the users
- Operational conditions, such as weight and environment

Rope Grabs and Ascenders

Rope grabs, which are devices that grip the rope, have a variety of uses in rope rescue. The term encompasses a broad spectrum of devices, usually metallic in nature with a camming device that is designed to grip the rope firmly. Friction hitches of rope or cordage that also grip the rope are not considered "devices" and will be addressed in Chapter 9, *Ropes, Knots, Bends, and Hitches*.

When selecting ascenders for use, ask the following questions:

- What size rope are they designed for?
- What is their strength rating and how is it determined?
- Can they be operated easily (placed on and taken off the rope) with one hand?
- Do they have a secure safety catch to prevent them from accidentally coming off the rope?
- Are they comfortable in the hand?
- Can they be used easily while wearing gloves?

Rope grabs typically slide freely in one direction and lock off in the other. They provide adequate safety only when attached to run in the proper direction. Here we will address three types of rope grabs: those designed for fall arrest, those designed for rigging, and those designed for personal ascent.

Fall Arrest

Rescuers may come into contact with fall arrest rope grabs when working in an industrial environment or on towers. Some fall arrest rope grabs are intended for use on rope, while others are designed for cable. They are typically of a design that is not easily removed from the rope. Fall arrest rope grabs are intended to be used with a fall arrest lanyard and, in some cases, a force absorbing lanyard. They do not function properly when they are grabbed during a fall, nor when attached to the wrong size rope. Fall arrest rope grabs are generally quite specific with regard to the type of rope or cable upon which they are designed to be used. Some fall arrest rope grabs, such as the Petzl ASAP, are approved by the manufacturer for rescue belay. Always follow manufacturer's instructions for such use.

A fall arrest rope grab should always be attached to the rope in the proper direction. Many have an arrow on them, which should point along the lifeline towards the anchor point. A firm tug in the direction of potential fall will help to ensure that it is properly attached. During use, the device should be kept as high as feasible above the rescuer, to minimize potential fall distance. Some fall arrest rope grabs are designed to self-trail while others must be manually adjusted. Care should be taken to not grip or squeeze the activation feature of the rope grab during a fall as this could prevent it from working properly. Some rope grabs have a "parking feature" that can be activated to lock the device at any desired point on the lifeline; take care to release this feature before climbing.

Rigging

Rope grabs designed for rigging, such as the PMI/SMC Grip, may be marked to NFPA General Use or Technical Use ratings. NFPA 2500 (1983) specifies that a G-rated rope grab must hold 11 kN (2500 lbf) without permanent damage to the device or the rope upon which it is tested, while a T-rated rope grab should hold 5 kN (1124 lbf) without permanent damage to the device or the host rope. Rope grabs designed for rigging should not be used for fall arrest.

These rope grabs may be used to grip rope for progress capture or other purposes in rescue rigging. Most rigging rope grabs do not have handles. Typically, they will have an enclosed, clamshell-type body, and must be taken apart to be placed on rope. While perhaps more tedious to put onto the rope, once they have been assembled they are not likely to come off until intentionally removed.

The camming device on many rope grabs designed for rigging is often designed with a ridged cam that makes contact over a maximum amount of surface area to grip a rope firmly while imparting a minimal amount of damage to the host rope (**FIGURE 8-27**). The ridged cam is less likely to clog up with ice or snow than a toothed cam. The rope grab may also have a notch in the shell where the cam meets the body to improve grip and reduce likelihood of damage. That said, under very high or shock loads, even a rope grab such as this may damage rope.

Although ascenders may simply be considered as one type of rope grab, in common use the primary differentiator between the two is that a rigging rope grab is generally of a clamshell or similar design that requires two hands to install and remove from rope (and is thus less likely to become inadvertently disconnected), while an ascender has a more open design that can be readily applied and removed from rope, often one-handed.

FIGURE 8-27 A rope grab cam compared to an ascender cam.

Personal Handled Ascender

A personal handled ascender, such as the CT Quick Roll, works on the same principle as other rope grabs: Used properly, it is designed to slide freely in one direction (up) and to lock in place under a downward force. While there are several types of ascending systems, for the system to work effectively at least two devices are required. The rescuer ascends rope by sliding one ascender up the rope while supported by the other, then alternating the action to progress up rope.

Handled ascenders like these are typically used in conjunction with a personal lanyard for safety, and an etrier or footloop. A hole or two at the bottom end of the device will help facilitate different ascending styles, and features like a built-in pulley can accommodate climb-assist systems. Look for an ascender that fits your hand comfortably, that does not come off the rope accidentally, and that has a sufficient working load for your needs.

A slotted cam will help prevent accumulation of mud or ice. Usually the cam of personal ascenders will have very fine, sharp teeth angled in a downward direction to provide a firm bite on the rope. While useful for recreational climbing and mountaineering, these teeth can snag and jam with heavier rescue loads, and may be more likely to inflict damage and increased rope wear.

Not all ascenders have handles. Some people prefer to use ascenders by wrapping their fingers around the upper shell rather than using the handle anyway, and elimination of the handled portion of the device makes it lighter and more compact.

A chest ascender is a type of handleless ascender that is mounted onto the harness at chest level to provide progress capture as the rescuer climbs. Traditionally, these have been designed with a small twist in the frame, as with the Petzl Croll, to help it lie more ergonomically in a vertical position and parallel to the body of the user when rigged between the sternal and the ventral attachment. Others, like the Beal Hold Up, incorporate a specially designed vertical attachment point that achieves the same result.

Personal ascenders should not be used in hauling systems and other rescue rigging involving loads much greater than one person's body weight, nor as belay devices, because they are more likely to damage the rope under high weight loading and shock loading.

Anchorage Connectors

Making an appropriate connection to a sufficient anchorage is a foundational part of rope rescue. Discussion here will be limited to equipment used to connect to an anchorage. Specific guidance for creating and using anchorages is provide in Chapter 11, *Anchorages*.

Anchorages used in rescue may be comprised of an existing structure or natural element, or they may be installed as necessary (**FIGURE 8-28**).

Regardless of the anchorage selected, the means and method used for attaching to it will at least in part determine its performance. The connecting components should be of sufficient strength to achieve the desired goal, and should be attached in such a way so as to prevent unwanted movement or disengagement of the rescue system from the anchorage.

Anchor Slings

Anchor slings are made of webbing, rope, or cable, typically with metal D rings or sewn loops at each end where a carabiner can be clipped. These straps can be a quick way of setting reliable anchors. Anchor straps come in two basic types, as outlined in NFPA 2500 (1983). The heavy-duty ones (with the NFPA G designation) typically have an end-to-end breaking strength of about 8000 pounds (35.6 kN) while lighter-weight versions (NFPA rated T) typically may have an end-to-end breaking strength of 4945 lbf (22 kN). Strap-breaking strengths may be higher when the straps are rigged to form a basket. Slings with a larger ring at one end may be used to form a choker.

Some anchor straps, such as the PMI Self-Padded Choker Sling, are fitted with a presewn pad to help protect the load bearing part of the strap. Others may have a heavy-duty buckle so that the strap can be adjusted to various lengths; however, these buckles may slip with a force less than the rated breaking strength of the strap. Some presewn slings are simply sewn into a circular loop, without specific terminations for connection. These are very versatile, and may be also used as an anchor sling, among other things.

Before using a presewn sling, ensure that it has adequate tensile strength to achieve the desired safety factor. When abrasion or cutting is a concern, steel anchor strops are an excellent choice that offer additional abrasion resistance and cut protection. Typically constructed of varying lengths of steel cable, terminated with swaged loops at each end, these are most often used in fall protection. For this reason, strength ratings generally hover around the 5000-pound (2268-kg) mark, but by using multiples, an adequate rescue anchorage can be constructed.

When selecting anchor slings, consider the AHJ's requirements in respect to the following:

- Length
- Width
- Weight
- Terminations
- Material
- Adjustability
- Color

Beam Clamps

The angular steel structures found in many industrial plants and urban environments can be particularly challenging to anchor to. Not only can the edges of these beams be destructive to anchor slings, they are also often quite large, and may or may not be conducive to wrapping. One solution is the portable beam clamp. The typical use for these is industrial fall protection, so rescuers may find the 5000-pound

FIGURE 8-28 A structural anchorage for rescue.

(2268-kg) rated breaking strength to be inadequate. However, utilization of multiple anchor points can result in a sufficiently strong anchor for rescue.

Bolt Anchors

Industrial sites and urban structures often will have preengineered fall protection anchors over commonly accessed areas, such as confined space entry points, tanks, vaults, and other work surfaces. While using preinstalled anchorages is perfectly acceptable for rescue, before using any installed bolt anchorage point be sure that it is rated for life safety use and find out what it's strength is. Most fall arrest anchorages are rated to only 5000 lbf, with positioning anchorages rated to 3000 lbf. It may be necessary to use two or more existing anchorages to achieve your desired strength.

Some bolt anchors are fixed, or stationary, while others may swivel (**FIGURE 8-29**). Be sure that the anchorage connector is adequately positioned for the intended direction of pull. Some bolt anchors are designed to be removable, used either in concrete or in a steel structure. These vary widely in strength and performance, and are highly dependent on the quality and nature of material into which they are inserted. Again, it is incumbent on the rescuer, under the authority of the AHJ, to ensure that the anchorages used are adequate and appropriate for rescue loads.

Portable Anchors

In some situations, tripods—or their advanced cousin the multipod—provide an excellent solution. Known as portable anchors, these devices require extensive training, but adept use can expand the capabilities of a rescue team dramatically. The need for such a device should be considered based on a risk assessment, training, and the organization's response capabilities. When considering such a device, important factors include number and type of head attachments, height and strength of the unit, and optional accessories.

FIGURE 8-29 Removable bolt.
Courtesy of Steve Hudson.

Packaging, storing, and transporting a tripod or multipod can be a particular challenge due to its size. It is particularly important to keep components together and ready for quick assembly and for rescuers to be adequately trained in prompt deployment.

Adaptations on the tripod concept, such as the Arizona Vortex and the TerrAdaptor, are advanced concepts in high rigging that can be rigged in as simple, or as complex, a manner as needed. The TerrAdaptor is unique in that it is designed to be used as a monopod, bipod, tripod, quadpod or even in a davit configuration, and may be coupled with additional units in an endless range of configurations. It is adjustable from 4 feet (1.2 m) to upwards of 12 feet (3.7 m) in height, and available with a variety of feet and attachment options to adapt to any terrain. It may be used as an anchorage, as a change of direction, or simply as a high point for lifting a rope system off the ground.

Pickets

A picket system is an alternative in a natural area where no anchors are available. Although a picket system can work very well when correctly rigged, establishing it properly usually takes a great deal of time. In addition, not all soil types can hold pickets securely. Loose, sandy, or muddy soil, or snow, may not hold well regardless of the number of pickets used. Pickets should be made of appropriate material for the specific use or environment, with good shear strengths. Most picket anchor systems consist of several rows of pickets.

Pulleys

Although a **pulley** is designed primarily to reduce rope friction, this capability makes it useful in a number of functions in the rope rescue environment (**FIGURE 8-30**). Pulleys also can be used as follows:

- Change the direction of a running rope
- Position a rope more conveniently, such as to an area where people using the rope will be less exposed to falling, dropped objects, or where rescuers may have more room
- Reduce abrasion on a rope (e.g., a pulley could be used to hold a rope up from an edge or to bring it away from other rope or webbing)
- Develop a mechanical advantage in hauling systems

As with other auxiliary equipment, NFPA 2500 (1983) classifies pulleys either as "T" (technical) or

CHAPTER 8 Rescue Equipment

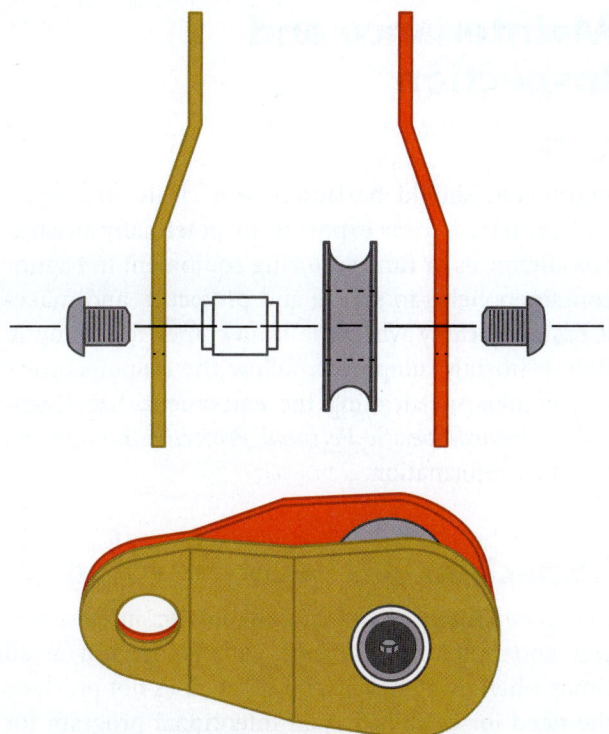

FIGURE 8-30 Parts of a pulley.

"G" (general). When selecting pulleys for a rescue organization, the AHJ should consider the following:

- Efficiency: Ball-bearing pulleys are more efficient than bronze-bushing pulleys, but not quite as strong. They also do not take stress, such as sudden blows, as well as the bronze bushing.
- Single or double: Double pulleys work better than two pulleys side-by-side where two lines follow a common rope path; however, double pulleys should not be used in a single-pulley configuration.
- Ratchet: If a Prusik will be used as a progress capture, a Prusik minding pulley should be selected.
- Size and overall dimensions: Consider the bulk, weight, and number of pulleys to be carried, balancing this with selecting a large enough diameter to optimize efficiency.
- Sheave width: Rescue pulleys are manufactured in ½-in. (12.5-mm) and 5/8-in. (16-mm) widths.
- Sheave diameter: Selecting pulleys with a sheave diameter of at least four times the diameter of rope helps to optimize system efficiency
- Strength: The side plates generally are the weakest part of a pulley. Pulley strength should correspond at least in part to the strength of the rope with which it is intended for use. Pound for pound, steel side plates are stronger than those made of aluminum.
- Compatibility with rope: It is okay to use smaller ropes on a larger sheave, as long as the rope is not so small that it will get caught between the pulley wheel and the sideplate during use. It is not okay to use a larger rope on a smaller sheave.
- Side plates: Movable side plates allow the pulley to be placed on the rope anywhere along its length; side plates that extend beyond the edge of the sheave help to protect the rope.

Specialized Pulleys

Some pulleys are designed for specific tasks. For example, the large sheave of the knot passing pulley is designed so that knots connecting lengths of rope can pass over it easily. This type of pulley also travels well across multiple ropes simultaneously, making it useful for highlines. Fitted with locking pins, when the sheave of this pulley is locked into place it makes an excellent high strength tie-off.

Prusik minding pulleys are designed with squared off sideplates, which allow a Prusik hitch to push up against it without jamming or being sucked into the pulley. In fact, the squared off sideplates do such a good job holding the hitch open this type of pulley has come to be known as a Prusik minding pulley.

Other Hardware

Rigging Plates

Rigging plates serve as collection points for multiple anchors and/or rigging points (**FIGURE 8-31**). They help keep anchor rigging organized, can save time in setting anchors, and can add a degree of safety by helping rigging personnel quickly visualize and understand their rigging.

Rigging plates commonly are used where multiple lines come together at a common point—for example at an anchor, at the master attachment point on a litter, or where multiple anchor lines are collected at one point.

Look for rigging plates that are strong (NFPA rated G) and that have contoured edges that are less likely to damage rope and carabiners. The holes should be large enough to accept large locking carabiners easily.

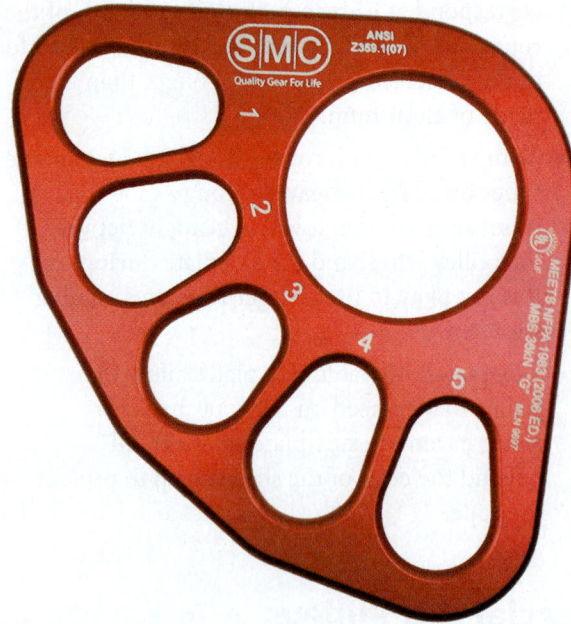

FIGURE 8-31 Rigging plate.
Courtesy of Pigeon Mountain Industries.

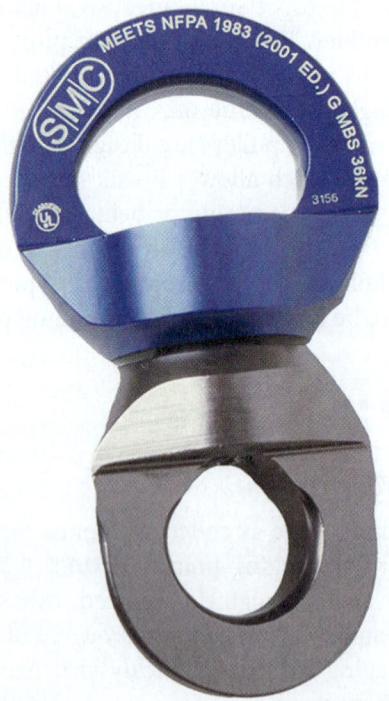

FIGURE 8-32 Swivel.
Courtesy of Pigeon Mountain Industries.

Swivels

Swivels may be used in rope systems to help prevent rope and equipment entanglements (**FIGURE 8-32**). They are particularly useful for reducing the potential for torque, or to reduce spin.

Maintenance and Inspection

Care

Equipment should be stored in a clean, dry place, and protected from exposure to potentially hazardous chemicals or fumes. Storing equipment in bags or containers helps to secure and protect it, and makes it easier to carry when the time comes to deploy it. Before storing equipment, follow the manufacturer's instructions on cleaning the equipment. See Chapter 7, *Hazard-Specific Personal Protective Equipment*, for more information.

Inspection and Recordkeeping

Rescuers should be on the lookout for inconsistencies and potential problems with equipment at all times when using it. This, however, does not preclude the need for establishing an intentional program for inspection of gear at regular intervals. Equipment should be inspected as follows:

- Before placing it into service
- Before any use
- At regularly scheduled intervals

Equipment should be first inspected prior to placing it into service. This is also a good time to remove packaging, mark the equipment for future traceability, make an equipment log form, and examine the item in detail to ensure that there are no signs of manufacturing defects or shipping damage. This inspection should be quite thorough, including detailed examination of any subcomponents, buckles, and moving parts.

Before use inspection is a bit less detailed, but no less important. This inspection includes both visual and tactile inspection—that is, looking at and touching the item to affirm that it is in good working order. Many agencies choose to do a before-use inspection as they are putting gear away from its previous use, but this does not preclude the need for additional inspection as the gear comes out of storage for the next operation. It is important to ensure that no critters have chewed on textile parts, pieces have not been damaged by being pinched or slammed in compartment doors, and the gear is safe for use.

In addition to these inspections, provision should be made for a thorough inspection at regular intervals. There is no set "right" time to do these, but intervals between thorough inspections should not be more than 1 year. Most organizations choose to do a thorough inspection at least twice a year, and very busy organizations may do them as often as every month.

A thorough inspection is a time-consuming process, with a significant amount of time spent on each and every item inspected. This is the time for in-depth visual and tactile examination of the item itself along with any records that might accompany it.

How to Inspect

In addition to following industry best practice and information provided in manufacturer's instructions, equipment inspectors should take into consideration each piece's particular use and any specific hazards or contaminants to which the equipment might be exposed in the environment. For example, perhaps there is nothing in the manufacturer's instructions about radiation exposure, animal excrement, or by-products of certain processes that exist in your environment. If information about the potential effects of something on a piece of gear is not readily available, an agency may need to do some research of their own to determine potential effects. Some damage is visible to the eye, but some is not.

While a rescuer can be trained in what to look for, using the information gained to make an informed decision about continued usage and retirement is more difficult. Such decisions are rather subjective in nature and expertise is attained only through a combination of training and experience, including tutelage under an experienced competent person.

When performing a thorough inspection of equipment, always take into consideration known information about the product, including age, purchase date, and usage history. These points of information should be taken into consideration in context of the results of visual and tactile inspection to make a final decision. Lives are saved by adhering to the maxim, "when in doubt, throw it out."

Soft Goods. When inspecting soft goods, the mnemonic T-CHAPS is a good reminder of what to check for (**FIGURE 8-33**):

- T: Thermal. Thermal damage typically presents as glazed, charred, or hardened fibers. Look for shiny, slick, or hard areas of rope, and feel for overly smooth or stiff areas.
- C: Contamination. Visual clues might include discoloration, while tactile indicators may be spots that are overly stiff or soft to the touch. Odor may also be an indicator of contamination.
- H: History. A check of the usage log is an especially important part of soft goods inspection. Ropes that have record of unusually high frequency of use, heavier loads than

FIGURE 8-33 An example of soft good damage.

FIGURE 8-34 An example of hard goods with physical damage.

normal, or use in harsh conditions may warrant earlier retirement.

- A: Age. All fibers deteriorate with age; an experienced inspector should factor this together with rope history and visual/tactile inspection to make a determination about life expectancy.
- P: Physical damage. Physical damage might be visible as hourglassed or bulging sections of rope, or as cuts, abrasion, tears, or even fuzzy spots. It is important to both look and feel for physical damage.
- S: Soiling. While dirt is not necessarily caustic, a very dirty rope is harder to inspect and may signal contamination.

Hard Goods. Hardware items, such as carabiners, descenders, pulleys, etc., can be inspected using the mnemonic ACADEMIC, with each letter representing a specific consideration (**FIGURE 8-34**):

- A: Alignment. Is the item properly aligned with itself, as manufactured?

- C: Cracks. Are there visible hairline cracks, especially at connecting points?
- A: Action. Does the item function as intended, without sticking or jamming?
- D: Deformation. Are there any deformities in the body of the item?
- E: Edges. Are there sharp or excessively worn edges, especially at rope paths?
- Mi: Missing. Is any subcomponent missing or loose?
- C: Corrosion. Do you observe corrosion, especially at joints?

While still a factor, age is less of a concern in hard goods than in soft goods because metallic items do not decompose as readily as textile products. See Chapter 7, *Hazard-Specific Personal Protective Equipment*, for more information on inspecting soft and hard goods.

Damaged Equipment

Equipment that does not pass inspection even after cleaning should be removed from service. If the manufacturer (or a relevant standard) has cited a maximum lifespan or obsolescence date, equipment should not be used beyond that date. Equipment should be destroyed or rendered inoperable on retirement to prevent further use.

In most cases, damaged equipment should be discarded immediately in a manner that precludes it being used again by anyone for life safety purposes. In some cases, it may be acceptable to modify a piece of equipment at the point of damage as long as that modification does not become a factor in use. For example, if a rope sustains specific damage at a certain point but is known to be otherwise in great condition, a decision to cut the damaged part out of the line and place the resulting shorter lengths back into service may be perfectly acceptable. However, in this case the equipment placed back into service must be unquestionably sound.

After-Action REVIEW

IN SUMMARY

- An established rope rescue equipment program will help ensure that the right equipment is in working order and ready for rescuers to utilize in an emergency.
- A rope rescue equipment program contains guidelines on equipment selection, inspection, maintenance, and documentation.
- The selection of rope rescue equipment begins with the identification of the type of emergency to which rescuers are responding.
- Rescuers should understand equipment specifications and how those relate to their intended use.
- NFPA 2500 (1983) establishes the baseline performance requirements for rescue gear and the AHJ has final say over the appropriateness of equipment.
- NFPA 2500 (1983) classifies equipment into three ranges: G (General), T (Technical), and E (Escape).
- Carabiners are a key piece of rope rescue equipment. NFPA 2500 (1983) classifies carabiners in to two categories: G and T.
- Carabiners are constructed from either aluminum alloy or steel. Commonly used carabiner shapes include D, modified D, and pear or HMS.
- The strength of a carabiner will be different in each direction and configuration in which it is loaded.
- Triangular or semicircular screw links are another type of connector that is used, especially where smaller dimensions are desirable.
- A braking device is a piece of hardware that is used to help manage the rate at which a load is lowered in a system or at which a rescuer descends.
- Braking devices include autolocking and nonautolocking and the NFPA classifies descenders into E (escape), T (technical), and G (general).
- Belaying is the process of protecting a person from falling by controlling an unloaded rope (the belay rope) in a way that secures the person on the rope in case the individual's main line rope or support fails.

- The term "rope grab" describes a broad spectrum of devices, usually metallic in nature with a camming device that is designed to grip the rope firmly.
- Anchorages used in rescue may be comprised of an existing structure or natural element, or they may be installed as necessary.
- Pulleys are primarily designed to reduce rope friction and also can be used as follows:
- Change the direction of a running rope
- Position a rope more conveniently, such as to an area where people using the rope will be less exposed to falling, dropped objects, or where rescuers may have more room
- Reduce abrasion on a rope (e.g., a pulley could be used to hold a rope up from an edge or to bring it away from other rope or webbing)
- Develop a mechanical advantage in hauling systems
- Additional hardware utilized in rope rescue systems includes rigging plates and swivels.
- In order to ensure that rope rescue equipment is ready to be utilized in an emergency, follow the manufacturer's instructions for cleaning, inspecting, and storing equipment.

KEY TERMS

Belaying To protect against falling by managing an unloaded rope (the belay rope) in a way that secures one or more individuals in case the mainline rope or support fails.

Belay rope The line attached to one or more individuals that provides protection from a fall.

Brake bar rack A descending device (also known as a rappel rack) that consists of a J- or U-shaped metal bar to which are attached several metal bars, which create friction on the rope. Some racks are limited to use in personal rappelling, whereas others may also be used to lower rescue loads.

Braking device A piece of hardware used to help manage the rate at which a load is lowered in a system or at which a rescuer descends.

Carabiners Load-bearing metal connectors with a self-closing gate used to link the elements of the high-angle system. Also sometimes called *biners*, *snap links*, or *krabs*.

Component safety factor The ratio between the minimum breaking strength and an equipment component.

Design load The weight for which a product is designed to manage.

Escape devices Rope rescue equipment designed to be more compact, and suited to a smaller diameter of rope (usually 7–9 mm) so that they may be easily carried by a rescuer in a pocket or small pouch, ready to be deployed in a life-threatening emergency.

Manner of function The method in which a particular piece of equipment was designed to be used.

Minimum breaking strength (MBS) The force at which tested samples of equipment actually fail during testing; calculated by a testing laboratory as three standard deviations below the mean breaking strength of a minimum number of samples tested in manner of use.

Pulley A device with a free-turning, grooved metal wheel (sheave) used to reduce rope friction; it also has side plates to which a carabiner may be attached.

Rope grabs Device that grips the rope.

System safety factor (SSF) The ratio between the least strong point in the rope rescue system and the potential load at that point.

On Scene

1. How might a braking device for personal use differ from a braking device used for a rescue load?

2. How can the NFPA 2500 (1983) designations of G, T, and E be useful? How might they be misused or misinterpreted?

3. What questions/concerns might you have about a load rating that you find stamped on a piece of equipment?

4. When inspecting a piece of life safety/rescue equipment, what kinds of things will you be looking for?

5. How would you respond to an associate from a neighboring agency who claims that they primarily use equipment meeting European EN standards because there are more lightweight options available?

6. How can you tell if a piece of equipment is third-party verified to a given standard?

CHAPTER 9

Operations

Ropes, Knots, Bends, and Hitches

KNOWLEDGE OBJECTIVES
- List the materials used to create rope. (**NFPA 1006: 5.2.4**, pp. 148–150)
- Describe the characteristics of rope construction and how it impacts performance. (**NFPA 1006: 5.2.4**, pp. 150–152)
- List the considerations to weigh when selecting rope for a task.
- Describe the anatomy of a knot. (**NFPA 1006: 5.2.4**, pp. 157–158)
- Identify the purpose of knots in a rope rescue system. (**NFPA 1006: 5.2.4**, p. 159)
- Describe the specific knots utilized in rope rescue systems. (**NFPA 1006: 5.2.4**, pp. 160–172)
- Describe the practices to follow for the proper storage of ropes.
- Describe the practices to follow for the inspection and maintenance of ropes. (**NFPA 1006: 5.2.3**, pp. 172–180)
- Describe how to identify and remove damaged rope from service. (**NFPA 1006: 5.2.3**, pp. 175–178)

SKILL OBJECTIVES
After studying this chapter, you should be able to:
- Select the appropriate rope for a rescue task. (**NFPA 1006: 5.2.4**, pp. 152–157)
- Tie an overhand knot. (**NFPA 1006: 5.2.4**, p. 160)
- Tie a barrel knot. (**NFPA 1006: 5.2.4**, p. 161)
- Tie a figure 8 stopper knot. (**NFPA 1006: 5.2.4**, p. 161)
- Tie a figure 8 on a bight. (**NFPA 1006: 5.2.4**, p. 162)
- Tie a figure 8 follow-through. (**NFPA 1006: 5.2.4**, p. 163)
- Tie a double figure 8 loop. (**NFPA 1006: 5.2.4**, p. 164)
- Tie a high-strength bowline. (**NFPA 1006: 5.2.4**, p. 165)
- Tie an interlocking long-tail bowline. (**NFPA 1006: 5.2.4**, p. 165)
- Tie an inline figure 8. (**NFPA 1006: 5.2.4**, p. 166)
- Tie a butterfly knot. (**NFPA 1006: 5.2.4**, p. 167)
- Tie a figure 8 bend knot. (**NFPA 1006: 5.2.4**, p. 168)
- Tie a grapevine bend. (**NFPA 1006: 5.2.4**, p. 169)
- Tie a ring bend. (**NFPA 1006: 5.2.4**, p. 170)
- Tie a Prusik hitch. (**NFPA 1006: 5.2.4**, p. 171)
- Tie a clove hitch. (**NFPA 1006: 5.2.4**, p. 172)
- Bag a rope. (**NFPA 1006: 5.2.3**, p. 175)
- Clean a rope. (**NFPA 1006: 5.2.3**, pp. 178–180)
- Dress rope ends. (**NFPA 1006: 5.2.3**, p. 180)

You Are the Rescuer

You arrive at the scene of a high-angle rope rescue involving the rescue of an individual who is 100 feet (30 m) up a communications tower. A family member on scene says that the individual is distraught and climbed up the tower structure trying to "get away" from people, but now is stranded. There is no ladder or obvious climb path. The reporting party says he does not believe that the individual is injured, but he is not certain. The individual on the tower appears to be conscious and alert. Another unit from an adjoining jurisdiction arrived just before you and is pulling rescue gear from their unit. They tell you not to bother with getting your own gear because they have everything needed for the rescue, including rope. Your team has never worked with them on a rescue and has never trained with their team. They say their rope "is approved by the NFPA."

1. How can you determine that the rope from the other responders will be safe to use?
2. On your vehicle, you have both dynamic climbing rope and low-stretch rope. Which should you use for this operation?
3. You have both 7/16-in. (11-mm) and ½-in. (12.5-mm) ropes. Which should you use?

JONES & BARTLETT LEARNING NAVIGATE Access Navigate for more practice activities.

Introduction

Rope is the foundation of any rope rescue activity. Many kinds of rope are used for rescue tasks. Each kind of rope contains different materials and is constructed in a specific manner according to its performance characteristics. The ideal performance characteristics to specify when choosing a rope depend on the specific need and environment for which the rope will be used. Before choosing a rope, you must decide whether it will be used for personal loads or rescue loads; whether you need it to excel in force absorption or stability for managing loads; how durable the sheath needs to be; how soft and easy to tie it should be; or whether it will be exposed to high heat or flame. Choosing the correct rope for a job will make the task easier and safer to accomplish, whereas a poorly chosen rope can result in severe problems or hazards for rescuers.

Fibers Used to Make Rope

Natural Fibers

For many years, ropes made of natural fibers, such as sisal, hemp, and manila, were standard (**FIGURE 9-1**). About the time of World War II, mass production of rope made of synthetic fibers, such as nylon or polyester, began.

Currently, synthetic fiber ropes are considered standard for situations in which the safety of a person is "on the line." National organizations such as the International Association of Fire Fighters, the International Society of Fire Service Instructors, and the National Fire Protection Association (NFPA) have all condemned the use of natural fiber rope in life safety applications because natural fiber ropes experience the following issues:

- Show low resistance to abrasion
- Have a limited ability to absorb shock loading
- Degrade in strength even with the best care
- Can rot without outward, visible signs
- Have lower breaking strengths than ropes of the same diameter made of synthetic fibers such as nylon or polyester
- Do not have strands that are continuous along the rope's entire length because natural fibers are never more than a few feet long

FIGURE 9-1 For many years, ropes made of natural fibers, such as sisal, hemp, and manila, were standard.
© Viacheslav Nikolaenko/Shutterstock.

Synthetic Fiber Ropes

Synthetic fiber ropes have several important advantages over natural fiber ropes. For example:

- Synthetic fiber ropes do not rot.
- Synthetic fiber ropes do not age as quickly.
- Synthetic fiber ropes can be made into more advanced rope designs than natural fibers.

Several different synthetic fibers are used to make ropes. Each fiber has distinct characteristics that make it suitable for certain uses and unsuitable for others.

Polyolefins (Polypropylene and Polyethylene)

Polypropylene or polyethylene ropes are common in recreational boating activities and in **commodity ropes**, such as hardware store utility rope, twine, clothesline, and other nonspecialty commercial products. Because they float, rescuers may come across polypropylene ropes in water rescue throw bags. These are among the least expensive of ropes, but because of their low tensile strength, low abrasion resistance, and low melting point, they should not be used for direct loading in life safety operations.

Advantages of **polyolefin** ropes include the following:

- Do not absorb water
- Float (specific gravity of 0.91); consequently, they are useful in activities on the water
- Have good chemical resistance (pH 2 to 12)

Disadvantages include the following:

- Have relatively low tensile strength (6 to 6.5 grams per denier [gpd] breaking tenacity)
- Have poor abrasion resistance
- Have low melting points (150°F to 200°F [65.6°C to 93.3°C])
- Have poor shock-absorbing (shock-loading) capability
- Have poor resistance to damage from sunlight

Polyester

Polyester fibers are found in a number of life safety applications. However, because polyester does not handle shock loading as well as nylon, it generally is not found in dynamic climbing ropes.

Advantages of polyester rope include the following:

- High tensile strength even when wet (7 to 10 gpd breaking tenacity)
- Good abrasion resistance
- Resistant to damage from acids (pH of 3.5 to 7.5)
- Lower elongation than nylon

Disadvantages of polyester rope include the following:

- Inability to handle shock loading as well as nylon (12 percent to 15 percent elongation at break)
- Susceptible to damage from alkalis
- Inability to float (specific gravity of 1.38)
- "Slipperiness" of material can make for fast descents

Nylon

Nylon is produced in several different types, the two most commonly used in life safety ropes being **nylon 6** and **nylon 6,6**. Of the two, nylon 6 is preferred for higher elongation properties (and lower cost) while nylon 6,6 is preferred where greater durability and lower elongation characteristics are desired.

Advantages of nylon rope include the following:

- Optimum breaking tenacity (7.8 to 10.4 gpd)
- Balanced shock-loading capability (15 percent to 28 percent elongation at break)
- Resistant to damage from alkalis (pH 6.5 to 10.5)

Disadvantages of nylon rope include the following:

- It may lose 10 percent to 15 percent of its strength when wet (regained when dry).
- Susceptible to certain strong acids
- Does not float (absorbs water)

UHMPE (Extended Chain, High-Modulus Polyethylene)

UHMPE is a polyethylene yarn, known by brand names such as Spectra and Dyneema. Lauded for its incredible strength-to-weight ratio, this material is said to be "10 times stronger than steel" and gained initial popularity in this market in the form of slings and runners for climbing. Now incorporated into ropes and other products, rescuers must take note that UHMPE has a very low melting point, does not absorb shock loads, and is a very "slippery" fiber that does not hold knots very well.

Advantages of HMPE ropes include the following:

- Have high tensile strength (30 to 35 gpd breaking tenacity)
- Float (0.97 specific gravity)
- Do not absorb water

Disadvantages of HMPE ropes include the following:

- Have a low melting point (about 150°F to 200°F [65.6°C to 93.3°C])
- Have poor shock-absorbing capability (2.7 percent to 3.5 percent elongation at break)
- Are very slippery; therefore, special knots may be required to hold when tied

> **TIP**
>
> Denier is a weight-per-unit-length measure of any linear material such as yarn. The measurement is a numeric representation of the weight in grams (g) of 9000 meters of the material. The smaller the number, the finer the yarn. The tensile strengths of yarns are often rated as grams per denier (gpd).
>
> While denier is still being used by some manufacturers, *tex* is a term increasingly being used. Tex is defined as the mass in grams per 1000 meters. The commonly used unit is decitex, abbreviated dtex, which is the mass in grams per 10,000 meters.

Aramids

An aramid is a particular type of polyamide that has been altered to increase heat resistance. Para-aramid fibers such as Kevlar, Twaron, and Technora excel in heat resistance (where UHMPE is lacking) while still offering exceptionally high strength-to-weight ratios, but they do not float, they absorb water, and they are easily damaged by repeated flexing.

Advantages of aramid ropes include the following:

- Are resistant to high temperatures (350°F [176.7°C] working limit)
- Have high tensile strength (18 to 26.5 gpd breaking tenacity)
- Are resistant to organic solvents

Disadvantages of aramid ropes include the following:

- Are easily damaged by continued small-radius flexing (as in knotting)
- Have poor shock-loading capability (1.5 percent to 3.6 percent elongation at break)
- Are sensitive to chlorine and some acids and bases

Rope Construction

The choice of a rope for a specific job depends not only on the fiber from which the rope is made, but also on the manner in which the rope is constructed.

Laid

Laid construction, also known as twisted or hawser lay, means that small fiber bundles of material are twisted and then combined in larger bundles, usually in groups of three, which are twisted around one another in the opposite direction (**FIGURE 9-2**). **Laid rope** construction resembles the designs of older types of rope made of natural fibers. (See the section, "Moderate Elongation Laid Life-Saving Rope," later in this chapter.)

When loaded, the fibers in laid rope tend to untwist slightly, causing inherent spin and kinking as a descender spirals down the twist of the strands. Laid rope tends to be very stretchy, and it also tends to kink unless handled carefully. Because the load-bearing fibers are not covered by a sheath, they are exposed to potential damage by abrasion during use.

Ropes of laid construction have been displaced in most high-angle work by other designs.

Plaited

Plaited rope usually consists of bundles of fibers plaited together (**FIGURE 9-3**). Plaited ropes tend to be soft and pliable, but they are prone to picking (snagging and pulling out of fiber bundles).

Braided

The two types of braid used in the construction of braided rope are the solid (single) braid and the hollow braid.

FIGURE 9-2 Typical laid rope construction.
Courtesy of Steve Hudson.

FIGURE 9-3 Plaited construction.
Courtesy of Steve Hudson.

In a solid (single) braid rope, the rope is constructed entirely of a single weave of three or more fiber bundles (**FIGURE 9-4**). The design is sometimes called clothesline braid. Because the load-supporting fiber bundles in single-braid construction are vulnerable to destruction when the rope is being used, single-braid ropes have limited use in critical safety operations.

A hollow-braid rope is similar in appearance, but sometimes has a filler, such as scrap yarn or filament plastic. It typically is found in inexpensive hardware store–type rope. Such rope can be quite similar in appearance to life safety rope, but does not have the same performance specifications.

Double Braid

A double-braid rope is essentially a rope constructed of a solid braid covered with a hollow braid (**FIGURE 9-5**). One braid acts as the rope core; the second braid is constructed around it to act as a sheath and to help protect the inner braid.

Double-braid ropes are typically woven loosely to maintain softness and flexibility, but this also increases their susceptibility to picking, abrasion, contamination, and sheath slippage.

Kernmantle

The term **kernmantle** comes from a compound German word, *kern* (meaning "core") and *mantle* (meaning "sheath" or "cover"). The kernmantle rope design consists of a central core of fibers that supports the major portion of the load on the rope. This core is covered by a woven sheath, which supports a lesser portion of the load. The amount of material in the core versus sheath is not directly proportional to the amount of load that is taken by each. Kernmantle ropes are prized for their balance, strength, ease of handling, and resistance to damage (**FIGURE 9-6**).

Kernmantle ropes used for life safety applications are typically made on a braiding machine that carries yarn-filled bobbins around and around a core material in a serpentine pattern to form a braided sheath around the strength-giving core. The amount and type of yarn committed to each part of the rope, the angle at which the braided yarns are wound on the sheath, and the tension at which the braid is applied are some of the factors that contribute to overall performance of the rope.

Static and Low-Stretch Kernmantle

When applied to kernmantle rope, the term **static** means a type of rope with very low stretch (no more

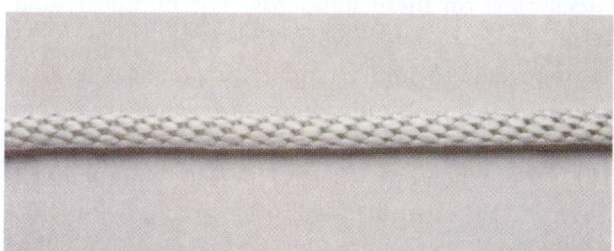

FIGURE 9-4 Solid-braid construction.
Courtesy of Steve Hudson.

FIGURE 9-5 Double-braid construction.
Courtesy of Steve Hudson.

A

B

FIGURE 9-6 Kernmantle rope. **A.** Dynamic kernmantle rope. **B.** Static or low-stretch kernmantle construction.
Courtesy of Steve Hudson.

than 6 percent elongation at 10 percent of its rated minimum breaking strength). A rope core of fiber bundles that are nearly parallel to one another creates this very low stretch. Some static ropes have so little stretch that what there is results largely from the inherent stretch of the core fiber.

The term **low stretch** refers to a type of kernmantle rope with a little more stretch than static kernmantle ropes (no less than 6 percent and no more than 10 percent elongation at 10 percent of its rated minimum breaking strength). Most low-stretch ropes have more twist in each core strand to provide additional mechanical elongation to the inherent stretch of the core fiber.

Because static or low-stretch ropes have so little stretch, they cause a more abrupt stop when catching a fall. This sudden stop subjects the climber's body, the equipment in the system, and the anchors to greater impact loading than would a dynamic rope.

Most static or low-stretch kernmantle ropes also have a thicker, tighter sheath than do dynamic kernmantle ropes. This thicker sheath helps protect the core from damage by abrasion and helps prevent dirt and grit from entering the core and damaging the inner fibers. However, the tighter sheath results in a rope that is stiffer than, and not as easy to handle as, a dynamic kernmantle rope with its thinner sheath.

Performance Characteristics

While material and construction are important, what is more important are the performance characteristics that the combination of these factors produce in the rope that you intend to use. The starting point for evaluating or classifying rope is the relevant standards that apply to your intended use. Examples include NFPA, American National Standards Institute (ANSI), ASTM International (ASTM), Union of International Alpine Associations (UIAA), or even EuroNorm (EN) standards.

For example, the best rope for belaying a rescuer who must climb above an anchor point (i.e., lead climb) is a rope meeting the requirements set out by the UIAA in the standard known as UIAA-101—or its corresponding EN document, EN 891. Although these standards are actually written with recreational climbing in mind, they are the only ones to address rope performance criteria with consideration to the effects and force dynamics of lead climbing falls. The primary limitation of dynamic rope for rescue applications is that it elongates significantly and somewhat erratically under load, making it very difficult to control effectively in lowering and raising operations.

Rope Selection

When choosing a rope, consider at least the following specifications:

- Diameter. Compatibility with other gear is especially important.
- Minimum breaking strength. Be sure it is rated in accordance with an appropriate standard test method.
- Elongation. Too much results in a bouncy rope, but too little can exacerbate forces.
- Abrasion resistance. More tightly woven ropes tend to be more abrasion resistant.
- Hand. A soft hand is great for tying knots, but this feature is juxtaposed with abrasion resistance.
- Color. Consider visibility in your field of work, and possible color coding schemes.
- Length. Needs will vary depending on your response area and type.

Of course, you can only compare specifications between ropes if they are measured according to the same test methods and reporting criterion. Imagine you are measuring diameter by taking the average of 12 caliper measurements over the cross section of a rope at 2-foot intervals, and your partner is measuring diameter by weighing 1 meter of material and calculating the result. You will have very different outcomes.

NFPA 2500 (1983) refers to test methods from Cordage Institute standard CI 1801, so this text will focus most on these standards. The European standard for low-stretch rope, EN 1891, differs. This is important to understand on a practical level, as the data are not similar enough to effectively compare breaking strengths (or other features) between two ropes when one is tested in accordance with EN 1891 and the other is tested in accordance with NFPA 2500 (1983).

Diameter

Knowing the actual diameter of your life safety rope as it is accurately measured and reported is essential, as many components of equipment are designed for use with very specific sizes of life safety ropes. Precision and accuracy is less important in nonspecialty rope used in household and commercial applications (commodity ropes), and the diameter of these is sometimes reported based on mass per unit length. Technically, this is known as nominal diameter, whereas life safety rope diameter that is measured more precisely is cited as actual diameter. Practically speaking, and in marketing materials, this distinction is rarely acknowledged.

Ropes certified to NFPA 2500 (1983) are measured for diameter using calipers at several points across the length of the rope. There is a limit to how much the values can vary. Using the correct diameter of rope for your auxiliary equipment is essential to ensure compatibility and proper performance.

Some rescuers cite the importance of diameter in making a rope more grippable for the user. However, with this class of equipment, rescuers should not be relying on barehand rope techniques. The grippability of even a 12.5-mm (1/2-in.) rope is not adequate to ensure safety for rescue purposes. Instead, rely on auxiliary equipment such as connectors, braking devices, rope grabs, and other gear to hold fast to the rope.

Most ropes used for rescue range in diameter from 9.5 mm to 13 mm, (3/8 in. to ½ in.) while personal escape ropes may be in the 7-mm to 9-mm (9/32-in. to 7/20-in.) range. According to the NFPA, ropes must be between 11 mm and 16 mm (7/16 in. and 5/8 in.) to be rated with a G-rating, between 9.5 mm and 12.5 mm (3/8 in. and ½ in.) to hold a T rating, and between 7.5 mm and 9.5 mm (19/64 in. and 3/8 in.) to carry the E rating for escape rope. See **TABLE 9-1** for the properties of several different types of rope.

TABLE 9-1 Types of Rope

	Diameter	Minimum Breaking Strength (MBS)	Manufactured Terminations	Elongation	Fiber Heat Resistance	Rope Heat Resistance
General Use Life Safety Rope	11–16 mm (7/16–5/8 in.)	40 kN (8992 lbf)	85% of MBS **OR** 40 kN (8992 lbf)	1–10% at 10% MBS	204°C (400°F)	NA
Technical Use Life Safety Rope	9.5–12.5 mm (3/8–1/2 in.)	20 kN (4496 lbf)	85% of MBS **OR** 20 kN (4496 lbf)	1–10% at 10% MBS	204°C (400°F)	NA
Moderate Elongation Lifesaving Rope	11–16 mm (7/16–5/8 in.)	40 kN (8992 lbf)	85% of MBS **OR** 40 kN (8992 lbf)	1-25% at 10% MBS	204°C (400°F)	NA
Throw Rope (water rescue)	7–9.5 mm (9/32–3/8 in.)	13 kN (2923 lbf)	85% of MBS **OR** 13 kN (2923 lbf)	NA	NA	NA
Escape Rope	7.5–9.5 mm (19/64–3/8 in.)	13.5 kN (3034 lbf)	85% of MBS **OR** 13.5 kN (3034 lbf)	1–10% at 10% MBS	204°C (400°F)	NA
Fire Escape Rope	7.5–9.5 mm (19/64–3/8 in.)	13.5 kN (3034 lbf)	85% of MBS **OR** 13.5 kN (3034 lbf)	1–10% at 10% MBS	204°C (400°F)	600°C (1112°F)
Escape Webbing	1 in. (flat)	13.5 kN (3034 lbf)	85% of MBS **OR** 13.5 kN (3034 lbf)	1–10% at 10% MBS	204°C (400°F)	NA
Fire Escape Webbing	1 in. (flat)	13.5 kN (3034 lbf)	85% of MBS **OR** 13.5 kN (3034 lbf)	1–10% at 10% MBS	204°C (400°F)	600°C (1112°F)

Abbreviations MBS, minimum breaking strength

The best choice for your application will depend first on the descenders, rope grabs, and other components with which it will be used. Beyond that, choosing a rope with a large enough diameter to provide adequate durability and an appropriate minimum breaking strength for your needs are important factors. It may be tempting to choose the largest diameter rope available, which might be as large as 5/8 inch (15.9 mm) for a rescue rope, but doing so could limit compatibility with other gear, as well as the overall additional weight when transporting gear to a rescue site.

> **TIP**
>
> NFPA now also permits webbing to be used in place of rope for escape systems. Webbing usually has a flat construction, rather than round, so diameter measurement is necessarily across the wide, flat surface. It should not be assumed that a 1-inch webbing really has 1 inch of thickness to protect against abrasion.

Most rescue teams select 7/16-inch (11.1-mm) or 1/2-inch (12.7-mm) static or low-stretch kernmantle ropes for their main rescue lines. However, each team should make its own decision based on such factors as rescue needs, environment, and the nature of their team.

Large-Diameter Ropes

The one obvious advantage to very-large-diameter ropes is their overall strength, but consider instead focusing on the safety factors relative to the entire system and how it is used as a more reliable means of maximizing safety.

> **TIP**
>
> **Beware the Pitfalls of Large-Diameter Ropes**
> - Higher cost (more material used in the rope)
> - Greater weight (more difficult to carry to a rescue site)
> - Handling problems (additional weight of hanging rope on the system)
> - Incompatibility with other equipment (most equipment is designed for 11- to 12.7-mm (7/16- to ½-in.) rope)

Breaking Strength

Breaking strength refers to the force at which an item will break. Of course, the only way to know the actual strength of any item is to break it, and no two samples of any item are going to break at exactly the same force, so the manufacturer is challenged to provide a reasonably accurate and reliable figure upon which to base load ratio calculations.

When publishing breaking strengths for life safety equipment, it is important that the manufacturer publish a figure in which the rescuer can have a high degree of confidence. The rescuer does not want to know the highest possible strength that a product may have, nor even the average strength of a product. The information that is of most value to the rescuer is the minimum, or lowest, strength at which equipment is reasonably likely to fail. This strength is generally referred to as minimum breaking strength (MBS).

To establish a common method for calculating a reliable and repeatable MBS rating, the NFPA and Cordage Institute life safety rope standards use a statistical method called **3-Sigma** to determine a reported MBS that gives rescuers a high probability of confidence in the performance of their rope **TABLE 9-2**. According to the standard test method, rope strength is measured by wrapping the rope several times around a bollard (or short post) and pulling to failure. A minimum of five samples are tested, and then the MBS is calculated by subtracting three times the standard deviation from the mean (average) of the test result numbers. That result, or lower, is what the manufacturer should advertise.

TABLE 9-2 MBS Calculation Example Using the 3-Sigma Method

	lbf	kN
Sample 1 test result	6510	28.96
Sample 2 test result	6520	29.00
Sample 3 test result	6600	29.36
Sample 4 test result	6750	30.02
Sample 5 test result	6680	29.71
Average (mean)	6612	29.41
Standard deviation of five samples	103.2957	0.4595
3 Sigma MBS = Mean less 3 × stnd	6302	28.03

As you can see in this example, the calculated MBS that will be reported is significantly lower than any of the actual test results. Using the 3-sigma method helps to ensure that some 99.87 percent of ropes made to this exact same design should have a breaking strength above the 3-sigma rating. The manufacturer's quality assurance program is what maintains that continuity.

When reviewing breaking strengths and comparing products, be sure you are comparing similar information. Some manufacturers do not use the 3-sigma method to determine the MBS. Some manufacturers that are not building and marketing gear to CI1801, ASTM F32, or NFPA 2500 (1983) standards may have other ways to determine breaking strength. For example, ropes that are marked as meeting EN 1891 are not subjected to a statistical analysis; in that case, it is only the actual test results that are reported, and in that case the strength requirement for all low-stretch ropes is only 15 kN (3372 lbf) for type A ropes and 12 kN (2697 lbf) for type B ropes. Some gear manufacturers do not even tell customers how they arrive at the breaking strength numbers in their catalogs. If you do not see what standard a company uses for breaking strength in their advertisements or presentations, always ask.

Because it is such an easy value to look at, there is often a tendency to gravitate toward the strongest possible rope for a particular diameter. Although strength is important, higher strength often is achieved by using materials that hardly stretch at all. This lack of stretch means poor energy absorption if an unexpected force is suddenly applied to the system. Special care must be taken when rigging with a very-low-stretch rope. Another way to increase strength in a kernmantle rope of a given diameter is to put less yarn in the sheath and more yarn in the core. Although this can result in a strong rope with a soft hand, it can also make a rope with little sheath to protect the core from cuts, dirt, chemicals, and abrasions.

NFPA 2500 (1983) classifies rope as being for General Use (G-Rated) if its calculated MBS value is at least 40 kN (8992 lbf), and for Technical Use (T-Rated) if its calculated MBS value is at least 20 kN (4496 lbf). In essence, this means that users who are using T-rated rope are expected to be more thoroughly trained and capable of ensuring that they maintain adequate safety margins while using the rope. Escape rope need be only 13.5 kN (3034 lbf) strong.

Elongation

Elongation and force absorption are related, but different. Ropes that offer high impact-force absorption, such as dynamic climbing rope, are typically also high in elongation, and vice-versa. NFPA 2500 (1983) does not specify force absorption characteristics of ropes, but it does call for measurement of elongation, which is required to be between 1 percent and 10 percent. Ropes that are lower in elongation are more stable for lowering and raising, but cause a more abrupt stop when catching a fall. This results in greater impact loading than would a dynamic climbing rope.

Rescue ropes are primarily intended to be used with a relatively low fall factor (**FIGURE 9-7**). With a

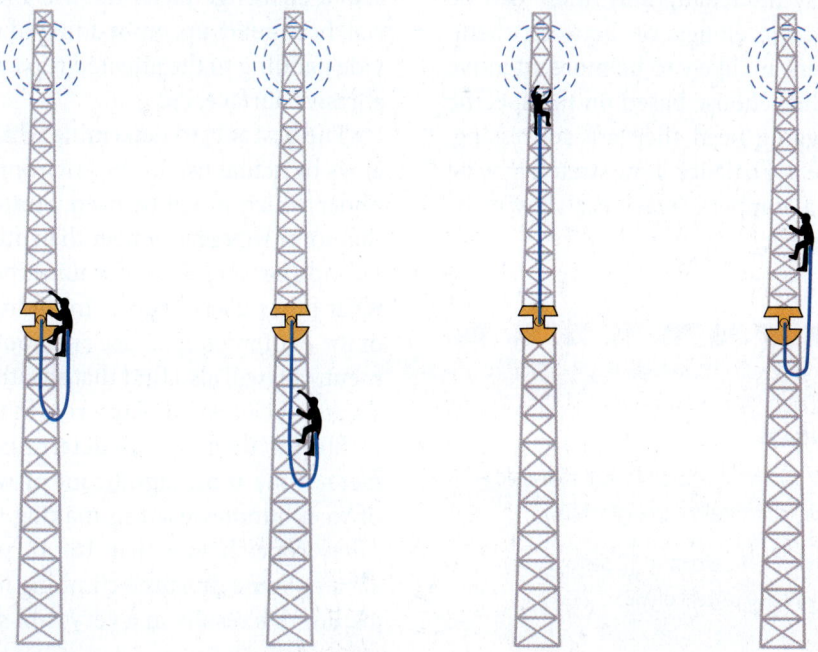

FIGURE 9-7 Comparison of fall factors.

TABLE 9-3 Reducing Rope Stretch

Method	Disadvantage
Increased proportion of core to sheath	Less abrasion resistant sheath
Polyester core	Less forgiveness in shock load
Tighter sheath:core relationship	Harder to tie knots
Para-aramid core	Weakens under repeated bending
UHMPE core	Lack of knotability

sufficient length of rope already in play, the limited range of falls that can be reasonably anticipated in rescue situations do not warrant a high force-absorption capability. More important in this case is stability for lowering and raising.

Ropes that are less dynamic and have less ability to elongate offer greater control during rescue operations. For example, imagine a litter being eased over a parapet wall at the top of a building. As the load suddenly transitions onto the rope, a stretchy rope could result in a sudden drop of several feet **TABLE 9-3**.

NFPA 2500 (1983) permits both static and low-stretch ropes. While these two types of ropes may be used relatively interchangeably, there can be a noticeable difference in elongation between them. Experienced rope users are likely to be more attentive toward which rope they choose based on the specific type of rescue or rigging need they are addressing. Again, however, there is a balance. Low stretch may be achieved in a rope in a number of ways, each having its own relative disadvantage.

> **TIP**
>
> **CI Elongation Definitions**
>
> Static = Elongation > 1 percent < 6 percent when measured at 10 percent of minimum breaking strength (MBS)
>
> Low stretch = Elongation > 6 percent < 10 percent when measured at 10 percent of MBS

NFPA 1983 also recognizes something known as moderate elongation life-saving rope. This rope is not classified as life safety rope by NFPA 1983 and is intended only for special-use applications where elongation between 10 percent is 25 percent are necessary in life-saving operations. A typical use for moderate elongation laid life-saving ropes would be a basic fire-ground rescue operation where a rescuer is being lowered from a roof for a pickoff, and there is a high possibility of an impact load from a person suddenly jumping onto the rescuer. They would not be used for other types of rope rescue such as lowering or hauling a litter, in mechanical advantage haul systems, or highlines.

Abrasion Resistance

Abrasion resistance is a difficult thing to quantify and is not presently measured in NFPA 2500 (1983). In fiber technology, abrasion resistance generally refers to the ability of a yarn to resist surface wear when rubbed against another yarn of the same material. In real-world use of life safety rope, that quality holds little sway. Yarns that do well in yarn-on-yarn abrasion tests typically do not fare as well when tested against the more abrasive surfaces found in field applications where life safety rope is used.

Although a number of test methods have been proposed to measure this rope characteristic, each has its flaws. Ropes that test well in a longitudinal test, where it is abraded up and down on a surface, may not perform as well in a test where abrasion occurs side-to-side, in a sawing fashion. The test surface is also a challenge in terms of choice of abrasive material, heat build-up, embedding of fibers that then provide padding to the affected rope, wearing down of the abrasive surface, etc.

The best way to determine which ropes outlast others is by actual use in the environment and conditions under which it will be used. Simply put, you will find that some ropes last longer than others as you use them. Do not be surprised if your experience differs from what other users report. In addition to the influence of the equipment you use and your particular environment, you will also find that sheath material, braid, and sheath thickness all play a role in the outcome.

Sheath thickness is determined by a number of factors, the most significant of which is the number of yarn bundles used in making the sheath. Most life safety ropes have either 16, 32, or 40 bundles in the sheath. Some dynamic climbing ropes may have up to 48, but this results in a very thin sheath.

During the manufacturing process, the piece of equipment that carries the bundles of sheath yarn around and around the core to form the outer braid is called a **carrier**; thus, we would refer to the robustness of a sheath by saying that these are 16-, 32-, or 40-carrier ropes. More bundles result in a smoother, more pliable, softer sheath, while fewer bundles contribute to toughness and durability.

Hand

How a rope feels in your hand is called the **hand** of the rope. It is influenced by such things as number of yarn bundles used to create the sheath, tightness of weave, and material. In general, the softer the hand of the rope, the less abrasion resistant it will be. While hand is somewhat subjective, the related measurement of knotability can provide a related, more objective analysis. **Knotability** is measured by simply tying an overhand knot in the rope and measuring the remaining gap with a tapered plug gauge.

The proportion of the gap to the diameter of the rope is called knotability. So, a rope with a knotability rating of .5 would be found to have a measured gap of half the diameter of the rope. A rope with knotability of 1.5 would be found to have a measured gap of one and a half times the diameter of the rope. The lower the number, the easier it is to tie knots and work with the rope. The higher the number, the more durable it will be.

Color

Ropes are available in a wide range of colors and patterns. Generally speaking, extruded fibers such as polyester will sport brighter colors than dyed filament fibers like nylon. Although many people choose a rope color for aesthetic reasons, color also can serve a functional purpose. For example, if several ropes are used together, the different colors can help distinguish one line from another so that the rescuer immediately knows which rope to haul or lower.

Some agencies choose to color code their ropes. This can provide a convenient way to remember for rescuers. For example, all red ropes go on Rescue Truck #1, while all yellow ropes go on Rescue Truck #2. Another approach is to code by length. For example, all red ropes are 150 feet, blue ropes are 300 feet, and yellow ropes are 600 feet. Color may also be used to indicate such things as diameter, year of purchase, storage location, duty cycle, etc.

It is left to the discretion of the authority having jurisdiction (AHJ) whether, and how, color coding might be used. If used, care should be taken to ensure thorough training so that all personnel understand the protocol.

Another method of color-coding ropes is to color code only the ends using a coating such as whip end dip. Perhaps the best method of managing rope information is to utilize end labels. These write-on labels have a light adhesive on the back and may be secured in place with a clear coat of whip end dip, or a length of shrink tubing.

Length

There is no one standard rope length that is right for every agency. The best length of rope for your needs will depend on your response area, the types of operations you run, and how your personnel are trained. Ropes may be measured in meters or in feet, so pay particular attention to the unit of measurement when purchasing rope, and also when working alongside another agency. Weight and training levels are also factors in choosing rope lengths. Longer ropes are heavier and more difficult to transport, so avoid erring toward very long ropes. On the other hand, if ropes are too short for an operation, personnel must be trained to perform knot passes. Of course, this is a skill that should be in the toolbox of every rescue agency, but it bears thinking about.

Anatomy of a Knot

Much of rope terminology comes from sailing and marine uses and can be a bit confusing when applied to land-based operations. For the purposes of rescue and rigging, it is best to keep things relatively simple. Having a common understanding of terms will help rescuers to effectively communicate with one another during an operation, and during training.

A rope in use is referred to as a **line**. Generally, we refer to the part of a rope where a knot is tied and a function underway as the working end of a rope, while the free end that goes over the edge and will touch the ground is known as the running end. The remaining length of the rope between the working end and the running end is known as the standing part. This term is often confused with the unused portion of rope remaining behind the working end—which, oddly enough, does not really have a name. Some refer to this as the **tail**.

Another pair of terms that many rescuers confuse is the difference between a loop and a bight. A bight is a U-shaped section of rope with parallel sides, and a loop is the portion of rope formed into a circle with the ends crossing each other. Notice the difference: A loop closes, while a bight remains open at one end.

The term *knot* is generally used to refer to any kind of fastening made by tying rope or webbing together in a prescribed way; however, there are two notable and unique categories of knots with which rescuers should be familiar. One is the bend, which is a class of knot that joins two ropes (or webbing pieces) together. The other is the hitch, which is a manner of tying that requires some other object to maintain its shape.

A couple of additional notable categories of knots include the backup, or safety knot, and the stopper knot. The safety knot is an extra tie—usually a simple overhand or double overhand—intended to provide secondary safety and prevent a primary knot from loosening, or to secure the tail of a rope. The stopper knot is just that: It is a bulky knot tied in the running end of the rope to prevent a device from accidentally running off the end.

Knot-tying instruction relies on knowledge of certain basic terminology (**FIGURE 9-8**):

- **Bight**: A U-shaped section of rope with parallel sides.
- **Hitch**: A knot that attaches to or wraps around an object or rope in such a way that when the object or rope is removed, the knot falls apart. Some hitches are adjustable along the object or rope they are attached to.
- **Loop**: A portion of rope formed into a circle with the ends crossing one another.
- **Running end**: The portion of rope used for lowering or hauling, or that will touch the bottom when rappelling.
- **Standing part**: The portion of rope between the running end and the working end.
- **Turn**: A single pass of a rope behind an object.
- **Working end**: The end of the rope that is used to tie and form the knot.

Manufacturer-Supplied Eye Termination

A **manufactured eye termination** is a permanent formed loop in the end of a rope created by the rope manufacturer, as an alternative to a knot (**FIGURE 9-9**). The inside of the eye may be protected by a layer of fabric, metal, or other protective material, and the sewn part is often protected by material such as plastic tubing. Such terminations make for quick rigging or facilitate hooking to connections such as the davits on some helicopters.

Some users specify their life safety ropes with permanent eye terminations supplied by the manufacturer. NFPA 2500 (1983) now has performance test standards for manufacturer-supplied eye terminations. Specifically, they are required to have a calculated MBS from a series of tests that is EITHER at least 85 percent of the certified rope's calculated minimum breaking strength OR at least equal to the rated strength requirement for the host rope (20 kN [4496 lbf] for T, 40 kN [8992 lbf] for G, and 13.5 kN [3034 lbf] for E.)

Some rescuers believe that a rope marked to NFPA 2500 (1983) must meet the requisite test methods with the termination. However, according to the standard, this is not true. What this means is that if you purchase an NFPA-marked rope that is rated to 9000 pounds (4800 kg), it could have an end termination

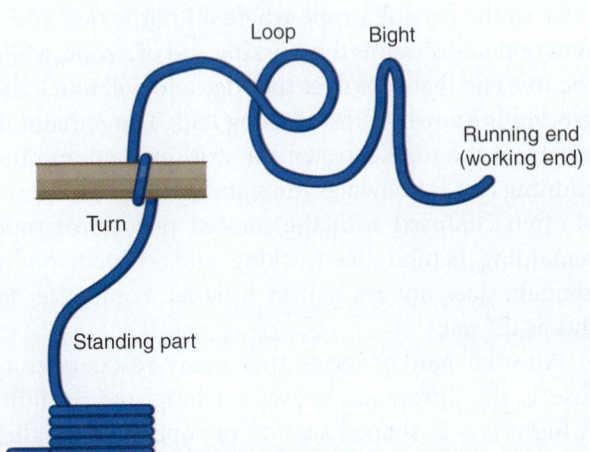

FIGURE 9-8 Standard rope terminology.

FIGURE 9-9 An example of a manufacturer-supplied eye termination.
© jocic/Shutterstock.

that is rated to only 7650 pounds (3500 kg). It is vital to know what the strength and performance characteristics of your rope are.

Knots

While elimination of knots is a lofty goal, anyone who works at height will quickly learn that **knots** are indispensable for joining together many elements in the rope rescue system. Among other functions, knots are used in the rope rescue environment for the following:

- Anchoring
- Tying ropes together
- Tying webbing together
- Tying loops in rope and webbing
- Tying people directly into ropes
- Creating certain belay systems
- Dealing with emergency situations (e.g., devising an emergency seat harness)
- Backing up other knots
- Keeping rope ends from pulling out of equipment
- Ensuring personal safety (e.g., preventing a rappeller from rappelling off the end of a rope)
- Creating emergency ascenders
- Tying safety lines
- Improvising when other elements of a system fail
- Extricating oneself from unexpected difficulties

The rope rescuer must be able to tie knots correctly, confidently, and without hesitation. They must be well versed in the ways these knots are used, and competent to select the appropriate knot for a given application. If you go into the rope rescue environment without these knot skills, you may be a danger to yourself, to a rescue subject, and to your fellow rescuers.

Proficiency requires practice. An aspiring rope rescuer should own at least two lengths of rope, each several feet (or a couple of meters) long, so that you can continually practice knot tying. Because many activities take place under severe environmental conditions, every operations-level rescuer should be able to tie knots under stress, in the dark, when cold, using only one hand, and with diminished physical ability.

Qualities of a Good Knot

Although knots vary in their specific use, all good knots have certain characteristics in common, including the following:

- They are relatively easy to tie.
- It is easy to determine whether they have been tied correctly.

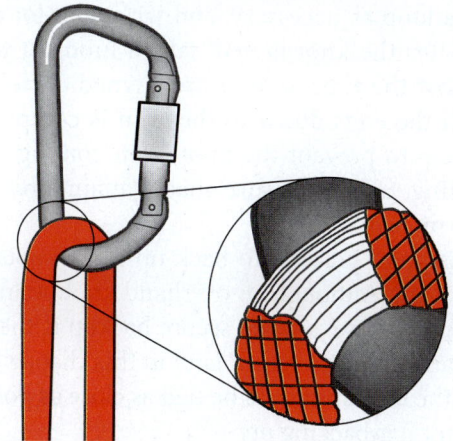

FIGURE 9-10 Effect of bending rope.

- Once tied correctly, they remain tied.
- They have a minimal effect on rope strength.
- They are relatively easy to untie after loading.

Every knot diminishes the strength of rope to some degree. This strength reduction is the result of sharp bends and pinching—generally speaking, the sharper the bends or the tighter the pinch, the less efficient is the knot (**FIGURE 9-10**).

Some knots, such as bowlines, have sharper bends, resulting in greater strength loss than knots that have more open bends, such as the figure 8. Ultimately, the strength of knots, along with other elements of a high-angle system, must be taken into consideration when deciding on a rope load ratio or a system safety factor for a rope.

> **TIP**
>
> An improperly tied knot, or incorrect application of a knot, could result in serious injury or death.

Removing Knots

Knots should be removed from a rope before the rope is put away and stored, for the following reasons:

- Leaving a knot loaded and tied in a rope over a long period may cause permanent strength loss in the rope.
- Knots left in a rope over a long period tend to set and become more difficult to untie.

Completing the Knot

Knots should be well-dressed. This is to say that the end result should be neat and clean, with only as large a loop as necessary, and without a lot of extra tail. After the knot is tied, take a moment to make sure that the rope strands are aligned correctly and to pull the ends down so the knot is compact. This will help to prevent the knot from coming apart or capsizing, and to ensure that it maintains its best performance.

It is good practice to back up most knots with a safety knot. Although an overhand knot often is used for this purpose, a more secure backup is the double overhand knot (discussed later in this chapter). When used, the backup should be tied as close as possible to the knot it is backing up.

> **TIP**
> For clarity, many knots in this manual are not shown with safety knots. However, it should be assumed that safety knots are an appropriate addition to most knots used for life safety applications.

Specific Knots for the Rope Rescue Environment

Overhand Knot

An overhand knot is one type of stopper knot. A stopper knot is a type of knot tied into the running end of a rope to prevent the line from feeding through a device. Usually, it is used in lowering or rappelling so that the rescuer does not inadvertently let the rope run completely through the device in the event that the rope length is insufficient. To tie an overhand, simply make a loop in the rope, tuck an end of the rope through the loop from behind, and pull taut, as shown in **SKILL DRILL 9-1**.

Another type of stopper knot is a barrel knot, which can be transformed from an overhand knot (**FIGURE 9-11**). The barrel knot is tied by following the first two steps of tying an overhand knot, then looping it two more times:

1. Form a bight and bring the working end under itself.

SKILL DRILL 9-1
Tying a Simple Overhand Knot

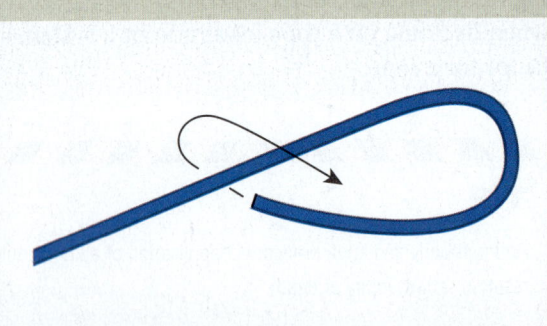

1. Form a bight and bring the working end under itself.

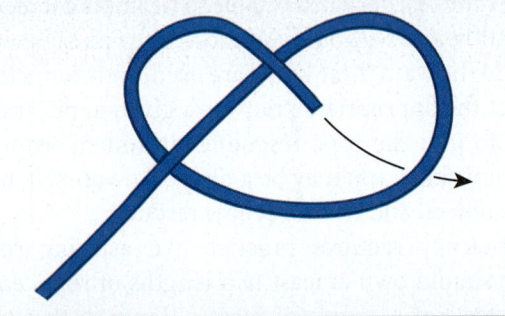

2. Bring the working end back into the bight and under itself.

3. Pull the working end and the standing part to tighten.

CHAPTER 9 Ropes, Knots, Bends, and Hitches 161

2. Bring the working end back into the bight and under itself for a total of three times.
3. Pull the working end and standing part to tighten.

Figure 8 Family of Knots

Many authorities experienced in the rope rescue environment prefer the figure 8 family of knots because for the following reasons:

- More likely to be tied correctly
- More likely to be remembered
- Easier to tell quickly if it is tied correctly
- Remains stable if loading on it comes from a direction different from that intended
- More likely to remain tied after repeated loading and unloading
- Less likely to invert and become untied when pulled across an obstruction or when the tail of the knot is pulled
- More efficient (stronger) than a bowline

Figure 8 Stopper Knot

A third option for a stopper knot is a figure 8 knot. To tie this knot, follow the steps in **SKILL DRILL 9-2**.

Finally, if two lines require a stopper knot (**FIGURE 9-12**), the figure 8 can be utilized to create a stopper knot by the following:

1. Tying a stopper knot in each line
2. Tying a stopper knot together into both lines

Figure 8 On a Bight

The figure 8 is also a good choice to form a loop in the end of a rope. The first step in tying this knot is to

FIGURE 9-11 Barrel knot.

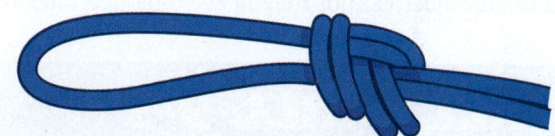

FIGURE 9-12 Stopper knot in two lines.

SKILL DRILL 9-2
Tying a Figure 8 Stopper Knot

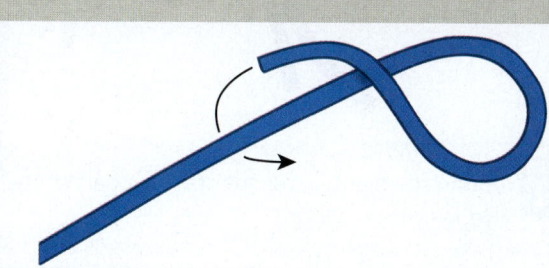

1 Form a bight with the working end over the working part of the rope.

2 Bring the working end under the working part (which will form a second bight) and into the first bight.

3 Pull the working end and working part in opposite directions to create the "8."

form a bight in the end of the rope. You will recall that a bight is simply the loop formed when the rope is doubled back on itself. The figure 8 on a bight knot is easy to remember if you visualize the bight as one strand of rope and tie it just as the simple figure 8 is tied. To tie a figure 8 on a bight, follow the steps in **SKILL DRILL 9-3**. A figure 8 on a bight is used as a secure loop in a rope for clipping into with safety lines, anchor lines, persons being lowered, or similar situations.

> **TIP**
> Be careful that you tie a figure 8 on a bight and not an overhand on a bight (**FIGURE 9-13**).

Figure 8 Follow-Through

When it is not possible to pass a simple loop over the end of an object, a knot may need to be tied after the rope is looped around something, such as a structural beam or a tree. In this case, a figure 8 follow-through (also called a figure 8 retrace) may be used. To tie a figure 8 follow-through knot, follow the steps in **SKILL DRILL 9-4**.

Double Figure 8 Loop

Another useful variation on the versatile figure 8 is the double figure 8 loop, also known as *bunny ears*. This is

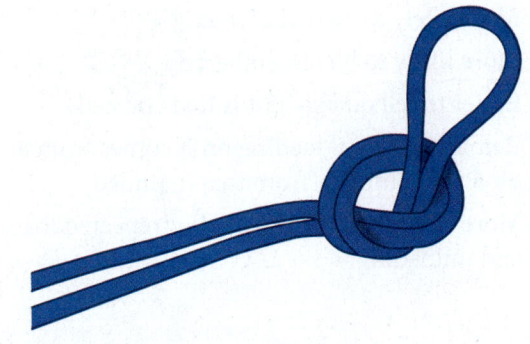

FIGURE 9-13 Overhand on a bight.

SKILL DRILL 9-3
Tying a Figure 8 on a Bight

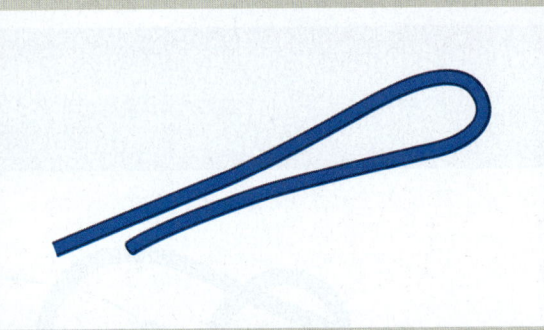

1 Form a bight. Visualize the closed end of the rope as the working end.

2 Bring the bight across the standing part of the rope to form a loop.

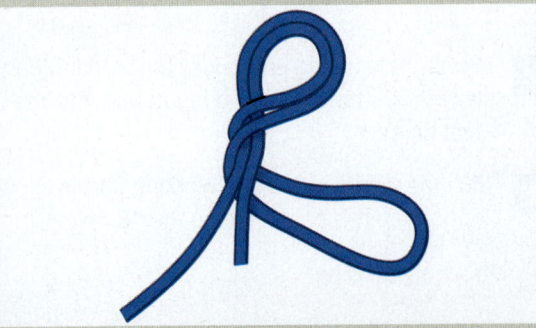

3 Bring the bight into and through the loop.

4 Pull the working end and working part in opposite directions to create the "8."

CHAPTER 9 Ropes, Knots, Bends, and Hitches 163

SKILL DRILL 9-4
Tying a Figure 8 Follow-Through Knot

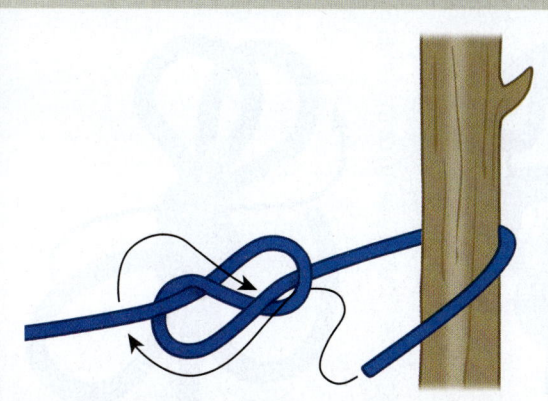

1 Begin by tying a loose figure 8 knot. Leave a few feet of tail at the working end of the rope.

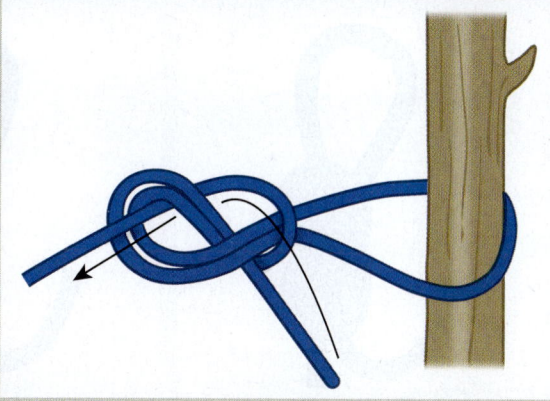

2 Bring the working end around the attachment point and back to the loose figure 8.

3 Retrace the working end completely through the loose figure 8.

4 Pull the knot tight and dress it to assure that all the rope strands are compact and do not cross one another.

TIP
Tying a Figure 8 Follow-Through Knot
1. Note that the figure 8 follow-through always begins with the tying of a simple figure 8 knot as a foundation well back from the end of the rope.
2. After the simple figure 8 has been tied, pass the end of the rope around the anchor point, then follow back through, parallel to the first knot. Follow every contour of the first knot with both rope ends going in the same direction.
3. Do not confuse this knot with the figure 8 bend.

a useful knot for sharing a load between two anchors, among other things. To tie a double figure 8 loop, follow the steps in **SKILL DRILL 9-5**.

Optional Approach: Bowline Knots
As we previously discussed, many authorities prefer the figure 8 family of knots. However, the bowline has a long tradition, so it is covered here (**FIGURE 9-14**). A high-strength bowline provides a more secure loop than a single bowline and is usually easier to untie after being loaded. A high-strength bowline, too, should be tied with a safety knot at the tail.

164 Rope Rescue: Principles and Practice

SKILL DRILL 9-5
Tying a Double Figure 8 Loop

1 Begin with a figure 8 on a bight knot with the bight positioned at the top of the knot. This bight should be fairly large.

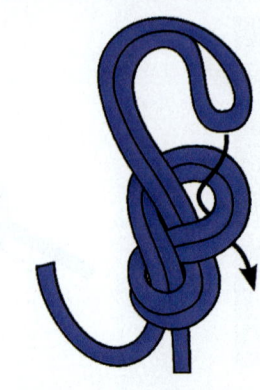

2 Bring the bight into the top part of the double 8.

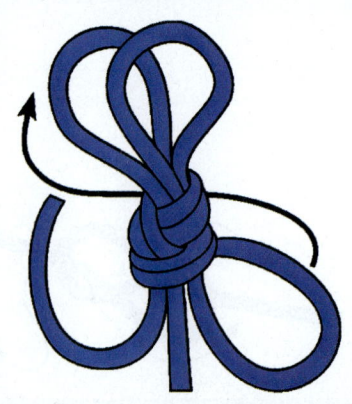

3 Continue threading this bight through so that you have twin ears at the top.

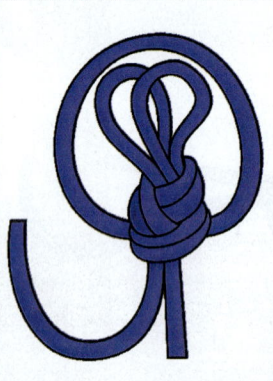

4 Bring the bight upward and over the twin ears.

5 Bring the bight down and under the main knot.

6 Now tighten the ears.

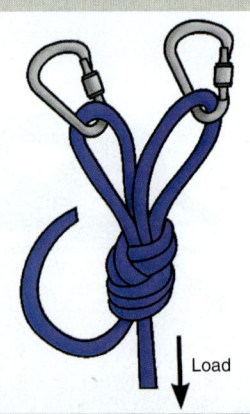

7 Clip a carabiner into each ear. You can adjust the knot by lengthening one ear as you shorten the other.

FIGURE 9-14 Bowline.

To tie a high-strength bowline, follow the steps in **SKILL DRILL 9-6**.

An interlocking long-tail bowline is an interesting variation on this theme. It is useful as a tie-in for vertical litter lowers and raising operations. Used in this fashion, it brings the main line and the belay line to a single point, such as the main attachment point on a litter bridle, at the same time provides tails to back up the subject and the litter tender. To tie an interlocking long-tail bowline, follow the steps in **SKILL DRILL 9-7**.

CHAPTER 9 Ropes, Knots, Bends, and Hitches   165

SKILL DRILL 9-6
Tying a High-Strength Bowline

1 Create two loops of equal size. Leave plenty of tail in the working end.

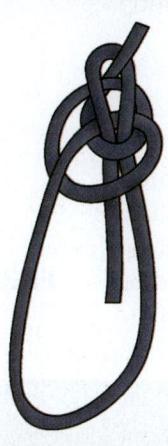

2 Thread the working end back through the two loops from below, around the standing part of the rope, and then back into the two loops from above.

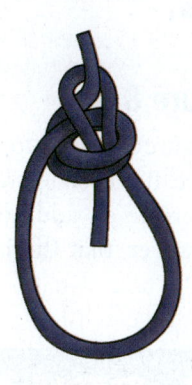

3 Tighten the elements of the rope and dress the knot.

SKILL DRILL 9-7
Tying an Interlocking Long-Tail Bowline

1 Tie a basic bowline with a small loop, leaving a long tail in the working end to attach to the litter and litter tender.

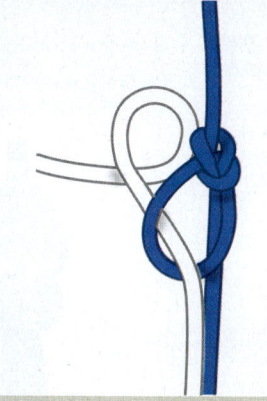

2 Thread the second rope up through the small loop of the first bowline and make a second loop.

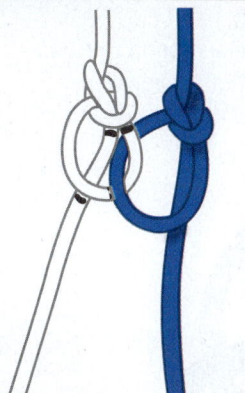

3 Create the second basic bowline with a long tail in the working end, making certain that both loops interlock.

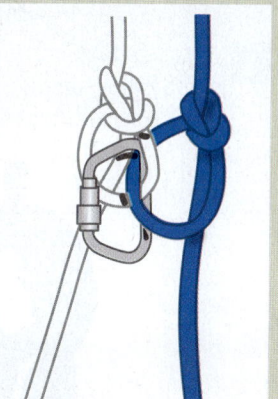

4 Clip a large locking carabiner across the loops of both bowlines.

Rope Rescue: Principles and Practice

The triple loop bowline is another variation on the bowline theme, with such uses as creating a multi-point anchor, improvising a harness, and separating load connections. Using Skill Drill 9-7 as a starting point, a triple loop bowline can be created by extending the tail to create a third loop (**FIGURE 9-15**).

Inline Figure 8

The inline figure 8 is used to create a loop in the middle of the rope. It is used to create handholds for a haul line and to create an additional anchor point. Keep in mind, however, that the inline figure 8 should be loaded in only one direction; otherwise, it can capsize and fail. To tie an inline figure 8, follow the steps in **SKILL DRILL 9-8**.

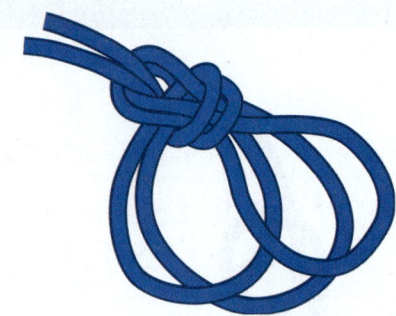

FIGURE 9-15 Triple loop bowline.

SKILL DRILL 9-8
Tying an Inline Figure 8

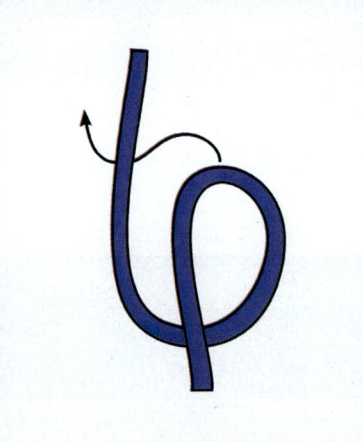

1 Make a loop in the rope and bring the loop under the standing part of the rope.

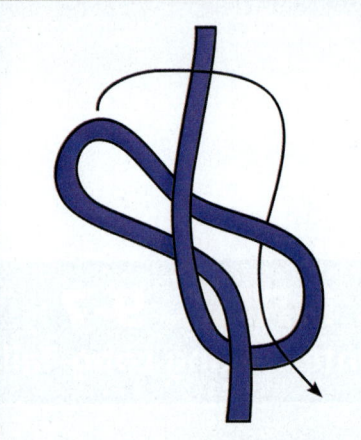

2 Pull on the loop to make it slightly larger and then bring it under and through the original loop.

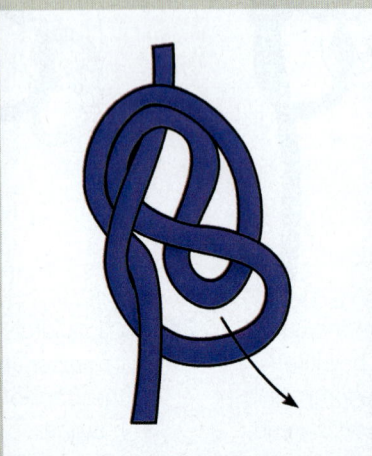

3 Pull it tight, making sure there are no loose elements in the knot.

4 Attach a carabiner. Make certain that the knot is loaded in the correct direction.

Butterfly Knot

A butterfly knot, also called a *lineman's knot*, provides a secure loop in the middle of a rope. Unlike some similar knots, such as the inline figure 8, the butterfly knot can be loaded in multiple directions. Be aware that mistying can result in a "false butterfly," an inferior knot prone to slipping. To tie a butterfly knot, follow the steps in **SKILL DRILL 9-9**.

SKILL DRILL 9-9
Tying a Butterfly knot

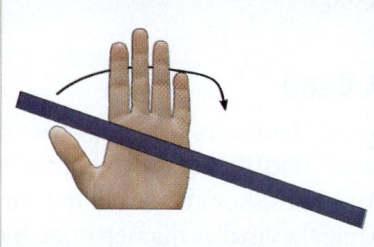

1 Hold your palm facing you. Lay a rope across the palm.

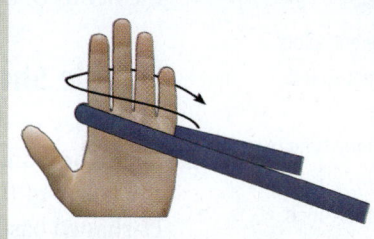

2 Bring the working end of the rope around the back of the hand.

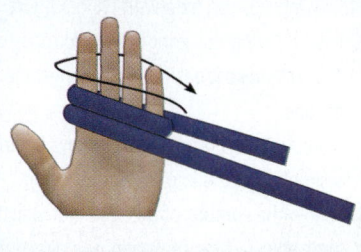

3 Bring the working end around the hand again with the second wrap above the first wrap.

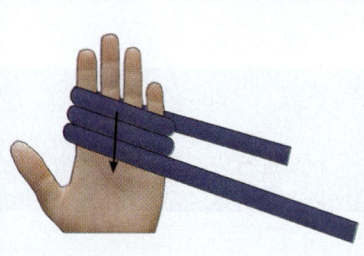

4 Bring the working end around a third time with the third wrap above the second wrap. The strand nearest to the fingers should be brought over just one strand to become the new middle strand.

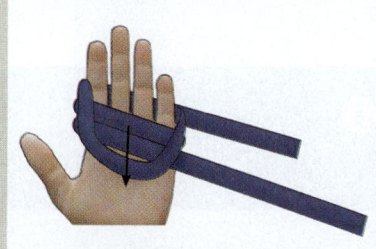

5 From the third wrap, pull some slack and form a shallow bight. Pull this bight down and below the first two wraps.

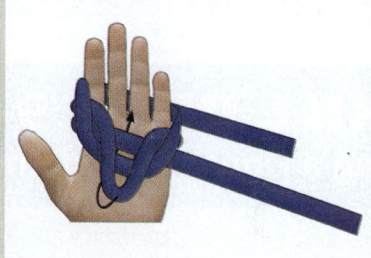

6 Pull the bight under the first two wraps.

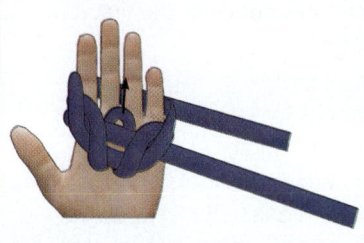

7 Pull the bight above the first two wraps.

8 Tighten by pulling both sides of the rope, making sure the knot holds its shape.

9 Check the rope to make certain it is compact and there are no loose elements inside it.

Figure 8 Bend Knot

The term **bend**, as applied to knots, refers to the joining of two ropes. A figure 8 bend is used to join two ropes or the two ends of one rope for the purposes of connecting two pieces of rope or creating a loop of rope by joining the two ends of one rope. To tie a figure 8 bend, follow the steps in **SKILL DRILL 9-10**.

> **TIP**
>
> The following tips can help with tying a figure 8 bend:
> 1. First, try tying this knot using two ropes of different colors. This will make it easier to distinguish the different strands of rope.
> 2. Note that the figure 8 bend always begins with the tying of a simple figure 8 knot as a foundation.
> 3. Follow the contour of the first knot exactly, with the rope ends approaching from opposite directions.

Optional Approach: Grapevine Bend

Another knot that can be used to join two rope ends securely to form a longer rope or to form a loop is the grapevine bend (also known as the double fisherman's knot). Loops of rope have a variety of uses, including the creation of Prusik hitches. This knot is very secure, but it may be more difficult to learn and to tell whether it is tied correctly. The grapevine bend can be difficult to untie after it is loaded, particularly in softer hand ropes. To tie a grapevine bend, follow the steps in **SKILL DRILL 9-11**.

The grapevine bend is best used to join rope ends of a similar diameter. It should not be used for webbing, for ropes of greatly unequal diameters, or for materials that may tend to untie or creep back through the bends of the knot.

Double Sheet Bend

For ropes of unequal diameter, a double sheet bend could be considered (**FIGURE 9-16**). For this bend, the larger of the two lines should be used to form the U-shaped base, while the smaller diameter line is used to form the wraps. This is one tie where an overhand safety is well-advised.

Ring Bend (Water Knot)

The ring bend (water knot), also known as the overhand bend, is used only for webbing. The ring bend is

SKILL DRILL 9-10
Tying a Figure 8 Bend Knot

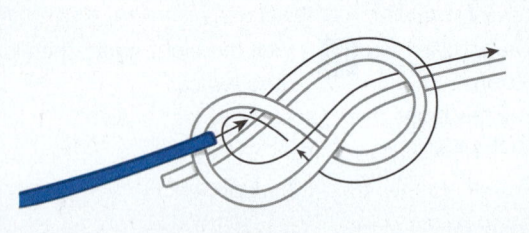

1 Begin by tying a loose figure 8 knot on one rope. Take the second rope and follow a reverse path with the other rope.

2 Make certain that the second rope stays parallel to and does not cross the first rope.

3 Pull the two ropes together. Make certain that the two ropes tighten parallel and symmetrically. While not necessarily required, some rope professionals choose to add a safety backup to each of the lines as they exit the figure 8 bend.

SKILL DRILL 9-11
Tying a Grapevine Bend

1. Lay the two rope ends parallel to one another. Wrap the end of one rope around both ropes two full turns in a clockwise manner. Then, thread it back through the inside of the two turns.

2. Wrap the end of the second rope two full turns around both ropes in the opposite direction in a counterclockwise manner and thread it back through both turns. Pull both ropes tight and dress the knot.

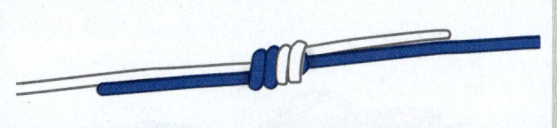

3. When tied correctly, the two turns from each half of the knot should lie parallel and flat against one another on one face of the knot. They should appear as a double "X" on the opposite face. The tail of each rope, when tied correctly, should end up on the side of the knot, opposite the side it entered.

FIGURE 9-16 A double sheet bend.

used to join two pieces of webbing or the two ends of one piece of webbing for the following purposes:

- Forming a longer piece (joining two pieces of webbing)
- Forming a loop (tying together the two ends of one piece of webbing)

Make sure the webbing follows flat through the knot. A twist in the webbing inside the knot will allow the knot to slip at relatively low loads. The ring bend is used only for webbing, never for rope. Because webbing is flat, it allows the strands to overlay itself with a smooth contour, which is a big part of what secures the bend in place. Because rope does not have this same quality, a ring bend in rope can come out much more easily.

When tying a ring bend, first try tying it using two pieces of webbing of different colors. This will make it easier to distinguish the different pieces of webbing as you tie. To tie a ring bend, follow the steps in **SKILL DRILL 9-12**.

After dressing the ring bend and pulling it tight, leave at least 2 inches (5.1 cm) of webbing in each webbing end. Although webbing contours well in a ring bend, it does tend to be slippery. For additional assurance, back up both sides of the knot with an overhand knot or sew loose ends down sufficiently to keep them from working through the knot. A ring bend in webbing should be inspected frequently to ensure that it has not worked loose over time.

Hitches

A **hitch** is a knot that attaches to or wraps around an object or rope in such a way that when the object or rope is removed, the knot falls apart. There are two primary hitches: the Prusik hitch and the clove hitch.

Prusik Hitch

The Prusik hitch is a sliding friction hitch that works well as a lightweight, soft rope grab for ascending,

SKILL DRILL 9-12
Tying a Ring Bend

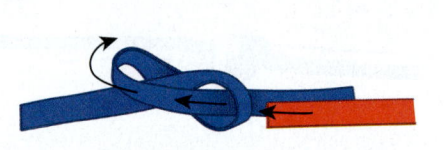

1. At about 6 inches (about 15.2 cm) from the end of the webbing, tie a loose overhand knot.

2. Take the second end of webbing and follow a reverse path through the first overhand knot.

3. Make certain that the second rope webbing stays parallel to and does not cross the first webbing.

4. Leave about 6 inches (about 15.2 cm) of webbing on the opposite end.

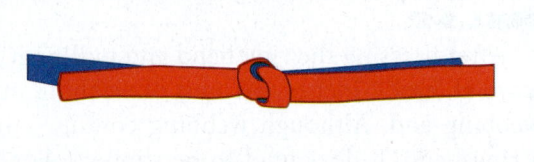

5. Pull both pieces of webbing tight, making certain the knot is symmetrical.

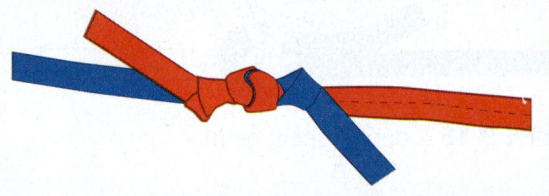

6. If slip is a concern, an overhand knot may be tied on each side of the ring bend as a safety.

belaying, and progress-capture. The hitch is made by first tying a length of cordage into a loop, and then wrapping that loop several times around another, larger, rope. This hitch, when loosened, can be slid up and down the rope—yet, when it is under tension, the hitch will not slide. To tie a Prusik hitch, follow the steps in **SKILL DRILL 9-13**.

Clove Hitch

A clove hitch is used for anchoring to rounded anchor points, such as litter rails. To tie a clove hitch, follow the steps in **SKILL DRILL 9-14**. This method assumes that you have use of the working end of the rope, so that you can pass the end around the object to which you are hitching.

SKILL DRILL 9-13
Tying a Prusik Hitch

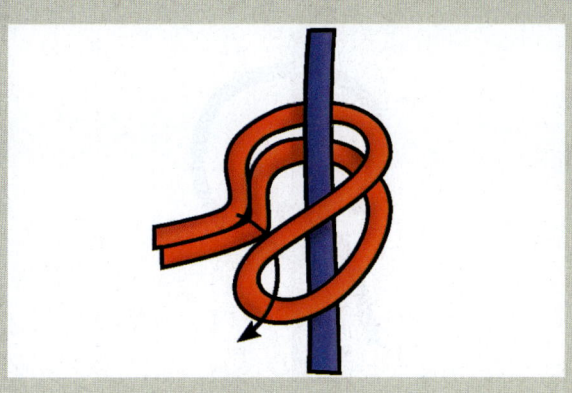

1 Anchor a kernmantle rope vertically. Make a continuous sling of smaller diameter cord. Make a bight in the sling in the loop (make this away from the connecting knot to prevent it from interfering with the Prusik hitch). Place this bight behind the main rope. Bring the larger side of the loop around the main-line rope toward you and pull it through the bight.

2 Bring the larger side of the loop around the side of the rope opposite you. It should look like the beginning of a girth hitch.

3 Bring the loop around the side of the rope nearest you and bring it into the loop. Make certain that the remainder of the cord goes inside the bight.

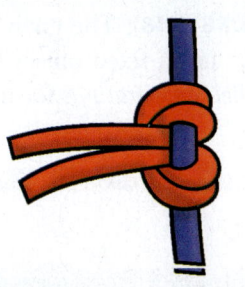

4 If you are making a two-wrap Prusik hitch, pull the sling tight, making certain that the hitch is contoured evenly.

5 If you are making a three-wrap Prusik, repeat step 3. Tighten and contour the hitch.

SKILL DRILL 9-14
Tying a Clove Hitch

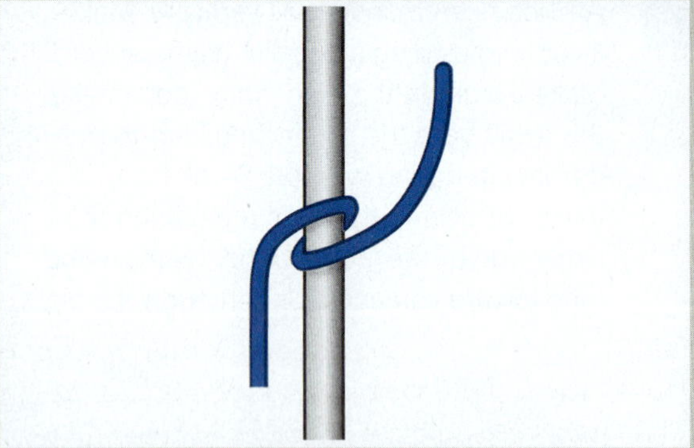

1 Wrap the running end of the rope around a round object.

2 Bring the running end over itself and around the object again.

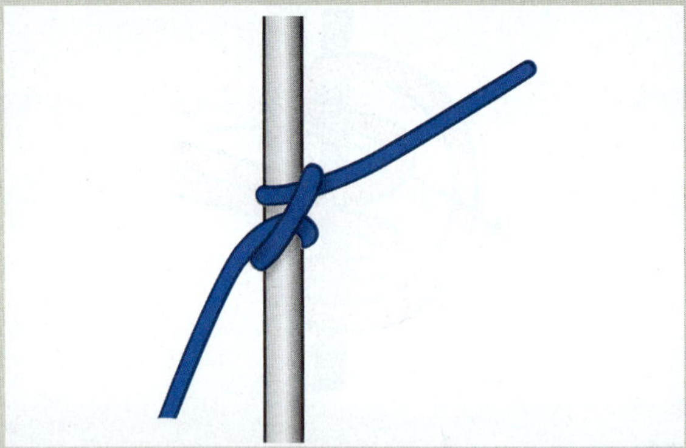

3 Bring the rope under itself.

4 Tighten the hitch, making certain that it is symmetrical and compact.

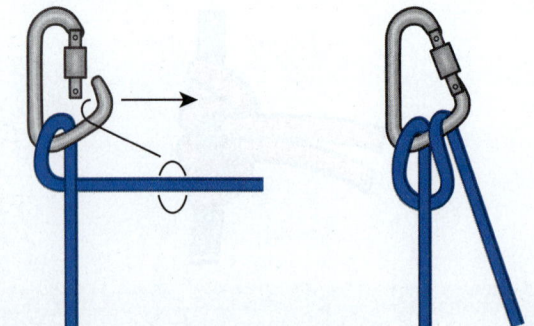

FIGURE 9-17 Münter hitch or Italian hitch.

Additional Hitches

The Münter hitch is used for one-person belays (**FIGURE 9-17**). The Truckers hitch is a handy hitch that every rescuer should know and can be used to increase tension (**FIGURE 9-18**). The girth hitch may be used with webbing and a fixed object (**FIGURE 9-19**). See Chapter 12, *Belay Operations* for more on the Münter hitch.

TIP

Many knots are known by different names depending on the region and the rescue organization. **TABLE 9-4** lists several common knots and alternate names for them.

Care of Ropes

The modern high-angle rescue rope is a marvel of design and engineering. However, a rope's performance, how long it lasts, and its safety still depend on how well it is cared for. The condition of a rope ultimately

CHAPTER 9 Ropes, Knots, Bends, and Hitches   173

depends on its history: the age of the rope, the conditions to which it has been subjected, and the care it has received.

If a rope is owned and used by only one person, that person probably knows the history of the rope. However, if more than one person is using the rope, there has to be a system for tracking the rope's history. The common way of tracking a rope's history is to keep a **rope history log**. Each rope should have its own log.

Keeping a Rope History Log

Each rope must have its own log card with pertinent information on the manufacturer, diameter, design, tensile strength, date of purchase, and critical data. The log card should have enough space to allow rope technicians to note each time the rope was used and for what activity. Specific entries must be made whenever the rope was subjected to abuse that could affect its performance or safety. An example of a rope history log is shown in **FIGURE 9-20**.

It is essential that entries for each rope be made every time ropes are returned to storage after use. Every user must follow this discipline; otherwise, the rope history is incomplete.

Identification and Marking of Ropes

Because most groups with high-angle rescue gear have ropes of similar color, length, and diameter, a means of distinguishing each individual rope must be found so that its history can be kept. Even if ropes are coded by color, as discussed earlier in this chapter, each rope should be marked with some distinct identification, such as a number or letter, that corresponds to its card. This identifying mark, known as a **rope ID**, must be placed on the rope so that it is unmistakable and cannot be eradicated or lost. Some examples of rope ID methods include the following (**FIGURE 9-21**):

- Hot stamping the end of the rope.
- Writing on a special paper or tape and securing it to the end of the rope with clear plastic tape, clear heat-shrink tubing, or a protective coating such as clear whip end dip.
- Adding RFID (radio frequency identification) tags. RFID tags are electronic chips increasingly used in commerce to track and manage inventory. RFID tags can be inserted in the end of ropes, assisting in tracking the rope history and use. To track ropes with RFID, you need the chips, a needle to insert the chips, a chip reader, and tracking software.

FIGURE 9-18 Truckers hitch.

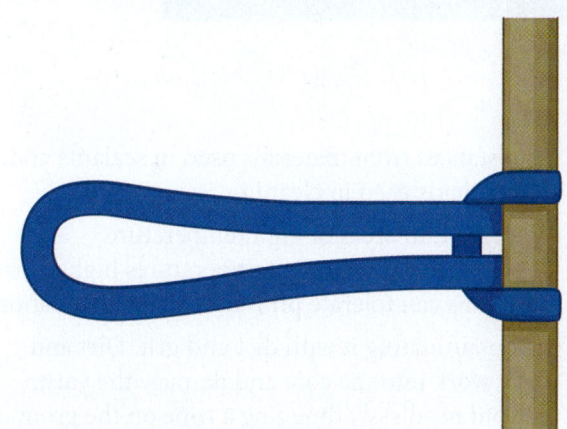

FIGURE 9-19 Girth hitch.

TABLE 9-4 Knot Names	
Knot Name	**Alternate Name**
Girth hitch	Lark's foot
Inline figure 8	Inline figure 8 loop
Double figure 8 loop	Bunny ears
Butterfly knot	Lineman's knot
Grapevine bend	Double fisherman's knot
Ring bend	Water knot
High-strength bowline	Double bowline

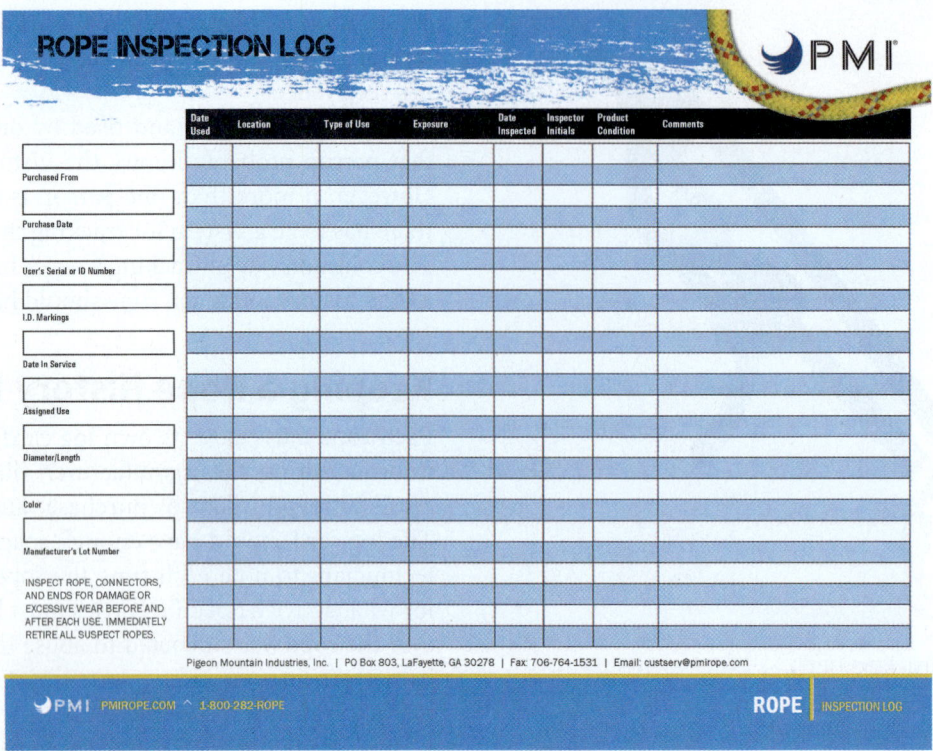

FIGURE 9-20 Rope history log.
Courtesy of Pigeon Mountain Industries.

FIGURE 9-21 Rope tagging example.
Courtesy of Pigeon Mountain Industries.

Storing Ropes

A life safety rope must be stored in a place of its own where it is protected from harm. Rope can be damaged in any number of ways, including the following:

- Age. All fibers used in life safety ropes degrade over time.
- Leaving it in sunlight. Although some research indicates minimal loss of strength of life safety rope due to sunlight exposure, this is likely the result of the core being protected from direct sunlight by the sheath. Despite the fact that most fibers used in life safety ropes have UV stabilizers in them, nylon, polyester, and other polymers degrade with prolonged exposure to sunlight.
- Exposing it to potentially harmful substances. Any chemical exposure, including vehicle exhaust, fumes or residues from storage batteries, and other substances, can be damaging to rope.
- Leaving it on the floor. Concrete floors leach alkaline, which is especially harmful to polyester, but they may also contain damaging substances from materials used in sealants and from acids used in cleaning.
- Storing it in areas of high temperature. Prolonged exposure to temperatures higher than humans can tolerate promotes rope degradation.
- Contaminating it with dirt and grit. Dirt and grit work into the core and damage the yarn. Avoid needlessly dragging a rope on the ground, and clean it periodically.

Bagging Ropes

One of the most convenient ways of storing, transporting, and protecting a rope is called bagging. Some of the advantages of bagging include the following:

- The bag helps protect the rope from damage while keeping it clean.
- You can usually flake rope into a bag quicker than you can coil it. **FIGURE 9-22** shows a fast technique for bagging a rope.
- A bag with a shoulder strap or pack straps is a convenient way to carry the rope.
- A bagged rope is easy to deploy. Simply secure the upper end of the rope and drop the bag over the edge. In most cases, the rope flakes out of the bag without tangles. Secure the bottom end of the rope to the bottom of the bag so that the bag is not lost when you drop it.

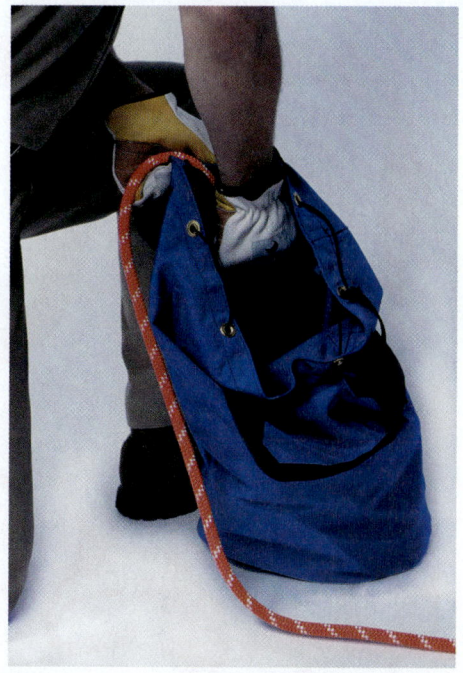
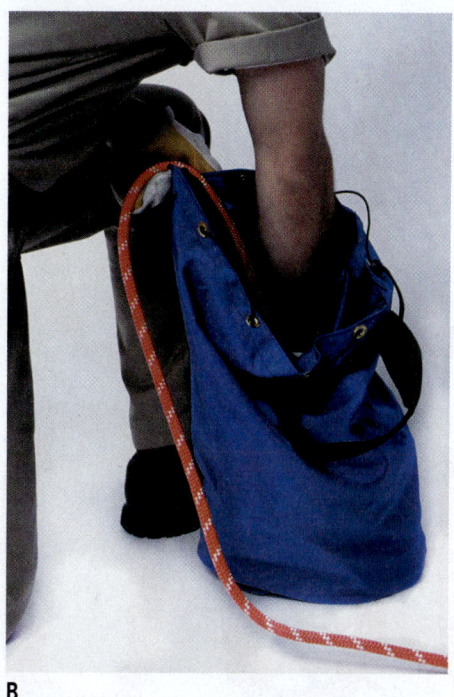

A B

FIGURE 9-22 Bagging a rope.
Courtesy of Steve Hudson.

To bag a rope, follow the steps in **SKILL DRILL 9-15**.

1. If you are wearing a seat harness, clip a carabiner into the harness and run the end of the rope through the carabiner and into the bag. If you are not wearing a harness, begin with the next step.
2. Grasp the top edge of the bag and hold the bag open and upright with your nondominant hand (e.g., left hand for right-handed individuals).
3. Lightly trap the rope between your thumb and index finger as it enters the bag.
4. With your dominant hand (right hand for right-handed individuals) below the other hand, grasp the rope and pull it into the bottom of the bag.
5. Slide the dominant hand back up to the other hand, take another length of rope, and pull it down into the bag.
6. Continue with these short strokes until the rope is bagged.

Coiling

Before bagging became common practice, coiling commonly was used for storing and transporting ropes. The specific type of coil depends on the circumstances or environment in which the rope is to be used. Some basic types of coils (i.e., mountaineer coil, caver's coil, and butterfly coil) are shown in **FIGURE 9-23**.

How Ropes Are Damaged

Harmful Substances

TABLE 9-5 includes some of the common substances that can destroy or cause deterioration in certain kinds of rope. In addition to those listed in the table, many strong chemicals can be damaging to both nylon and polyester rope. Avoid any contact with a chemical unless you know for sure it is harmless to rope fiber.

Overloading a Rope

Overloading a rope causes internal damage that could endanger those using the rope in the future. Damage from overloading usually occurs when a rope is used in activities for which it was not intended and when the load greatly exceeds the rope's safe working load. Some examples of overloading a rope include towing vehicles and lifting heavy objects.

A separate set of ropes, for utility use only, must be used for activities such as these two examples. Utility lines must be stored separately from life safety ropes and be distinctly marked, for example, "Utility line—not for life safety operations."

Damage from Falling Objects

Objects such as rocks or tools that fall on the rope, particularly when it is under load, can do serious damage. Any time you see heavy or sharp objects fall directly on a rope, retire the rope, even if damage is

FIGURE 9-23 A. Mountaineer coil, a traditional climber's coil, can be made quickly. **B.** Caver's coil is designed to be carried through caves without snagging. **C.** Butterfly coil is useful for paying out rope, such as dropping down a face. Also, the ends can be secured so that the coil can be carried on a person like a backpack.

TABLE 9-5 Substances Harmful to Rope

Rope Material	Harmful Substances
Nylon	Acids, particularly those found in storage batteries Bleaches
Polyester	Alkali (such as found in soot)

not readily apparent. When the rope has been used in a rock fall zone, you should inspect the rope for damage.

Abrasion

One of the most common means of destroying a rope or shortening its life is abrasion. This kind of damage usually is avoidable. Damage from abrasion commonly occurs when the rope is under tension and is lowered and raised across a rock or over the edge of a building. Abrasion often happens when a person is doing "bouncy" rappels or ascending, causing the rope to "saw" back and forth across a rock or hard object. Rope protection is discussed in Chapter 10, *Principles of Rigging*.

Thermal Damage

Thermal damage results when two pieces of synthetic material rub together. This is very destructive to rope and can cut a line as surely as a knife. Thermal damage usually occurs when one rope runs across another rope or across webbing, or when one line moves quickly across one spot in a second line that remains stationary. Thermal damage can occur in the following situations:

- Two ropes under tension with one remaining stationary while the other, being lowered, runs across the first
- A loaded rope running across an anchor rope or webbing also under load
- A rappeller holding a rope against the seat harness webbing while performing a rapid rappel

Thermal damage to ropes can happen quickly, without warning, and can be catastrophic. Everyone working in the high-angle environment must constantly be on the alert for thermal damage and take steps to avoid it.

Some ways to prevent thermal damage between ropes include the following:

- Rigging ropes so that they do not make contact and create heat fusion

- Holding ropes away from each other with pulleys or edge rollers
- Padding a stationary rope where another rope runs across it
- Making sure never to place a moving rope and a stationary rope in the same edge protection device, such as edge rollers (separate rollers or devices are used)

Thermal damage occurs when one rope is stationary and the other moves across it in one spot so that heat builds up. If both ropes are moving constantly so that one spot is not subjected to heat build-up, destruction is less likely. For example, in a Münter hitch, the rope is running across itself, but all surfaces are moving. Therefore, when used correctly, the Münter hitch is not likely to cause thermal damage.

Rope Damage Through "Flash" Rappels

All rappel devices operate through friction of the rope across the device; this results in heat build-up that increases with the speed of the rappel. Fast rappels must be avoided because they can damage rope through heat build-up. Such "flash" rappels also indicate poor technique or lack of control (or both) on the part of the rappeller.

Rotation of Ropes

In ropes in frequent use that are always anchored on the same end, the handling characteristics eventually change because of sheath milking at the lower end. This can be observed either as a "bunching" of the sheath over the core at the end of the rope, or as a flattening as the sheath protrudes beyond the core. When a rope is used for many rappels, the ends of the rope should be alternated as anchors to help prevent a change in handling characteristics.

Strength Loss Through Knots

All knots reduce the overall strength of a rope, but some knots cause a greater loss than others. The general rule is this: Knots with tight bends, such as bowlines, cause greater strength loss than knots with more open bends, such as the figure 8 family of knots.

Effects of Bending a Rope

Whenever a rope is placed under load in a sharp bend, some strength is lost (**FIGURE 9-24**). The rope fibers on the outside of the bend receive a greater share of the load, and those on the inside of the bend

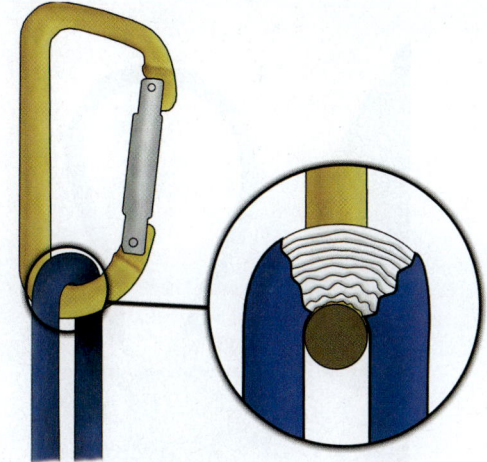

FIGURE 9-24 Effects of bending rope.

receive very little of the load or none at all. Common situations in which ropes undergo this kind of stress include ropes that have knots or kinks when they run over a sharp bend, such as in a carabiner or small-diameter pulley.

The 4:1 Rule

Rope users have traditionally estimated rope strength loss and system efficiency based on a D:d ratio. The D:d is the ratio of the diameter (D) of the bend divided by the overall diameter (d) of the rope. As a rule of thumb for modern nylon and polyester life safety ropes, when a bend in a rope becomes less than the diameter of the rope itself, strength begins to be significantly affected. This ratio is difficult for users initiated by wire rope and natural fiber ropes to comprehend. With those products, acceptable D:d ratios have historically been in the 6:1, 10:1, or even 25:1 range. Synthetic fiber ropes are a different matter altogether.

When it comes to system efficiency, higher ratios will yield better system performance, particularly in systems where rope is moving around a bend, such as in a pulley. Very small pulleys mean low efficiencies as the rope turns around them. As a standard guide, most rope users recommend a D:d ratio of at least 4:1 for optimum system efficiency.

To choose a pulley using the 4:1 rule, compare the diameter of the rope with the diameter of the pulley sheave. If the diameter of the rope is 1/2 inch (1.3 cm), the diameter of the pulley sheave should be at least 2 inches (5.1 cm).

Inspecting a Rope

Rope inspection is an ongoing process that is performed before, during, and after rope use. Rope is inspected in two ways: by look and by feel. After each

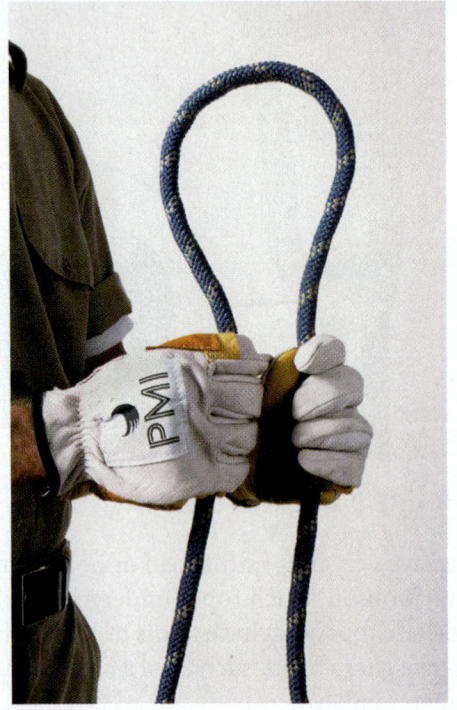

FIGURE 9-25 Rope inspection.
Courtesy of Steve Hudson.

use, inspect your rope thoroughly by looking and feeling along every inch of its length. When inspecting the rope, use the mnemonic T-CHAPS to help you remember what things you are looking and feeling for (**FIGURE 9-25**). See Chapters 7, *Hazard-Specific Personal Protective Equipment*, and 8, *Rescue Equipment*, for more information.

Establishing Responsibility for Life Safety Ropes

As with other life safety devices, such as breathing apparatus, ropes used by a team must be assigned a chain of responsibility. Someone must be responsible for knowing where they are, how they have been used, who has used them, and what condition they are in. Someone must be responsible for inspecting them after each use, for keeping a log for each rope, and, when appropriate, for removing them from service.

Retiring a Rope

Unfortunately, the only tests currently available for reliably measuring rope strength destroy the rope. Therefore, it is essential to be able to determine whether a rope should be retired. That ability is the result of education in rope use and construction combined with experience and good judgment.

Compared with many other types of equipment, rope is an inexpensive tool. The cost of replacing a rope is certainly less than that of a severe injury or loss of life.

Washing Rope

Ropes tend to become dirty with use. Using a rope when it is dirty shortens its life; therefore, one element of a rope inspection program is deciding when the rope needs a bath. A rope should be washed only if it is in obvious need. Overwashing can cause the rope to stiffen or shrink, or both.

Soiling obviously affects the appearance of the rope, but the most serious effect is hidden. Particles of grit and dirt can eventually work their way into the core of the rope and damage the load-supporting yarn as it stretches and flexes. Furthermore, dirt on the surface of a rope accelerates wear on hardware such as rappel devices, much as sandpaper would.

Aluminum particles also are damaging to rope. The metal particles are forced into the rope as it runs through metal hardware such as rappel devices. Correct washing of your rope can also remove many of these damaging particles.

Rope-Washing Devices

Commercial devices specifically designed for washing ropes are available. Some operate very much like the

FIGURE 9-26 Rope washer.
Courtesy of Pigeon Mountain Industries.

FIGURE 9-27 Chain coil.

hose-washing devices used by fire departments. As water jets spray into the center of the device, the rope is pulled slowly through it. One model has a built-in brush to help scrub away the surface dirt (**FIGURE 9-26**). It adjusts to various rope diameters up to 3/4 inch (19.1 mm).

Rope-washing devices are most effective against larger particles of dirt; for deeper cleaning, additional steps may be warranted.

Cleaning Ropes with a Washing Machine

Washing machines can thoroughly and effectively clean rope. Front-loading washing machines are the best choice, as the agitator in top-loading machines can become entangled with the rope, causing potential damage to both the rope and the machine. This hazard can be reduced somewhat by placing the rope in a mesh rope-washing bag. Close the bag securely before placing the bagged rope in the machine. If a mesh bag is not available, coiling the rope can help to prevent tangling. A commonly used coil for washing rope is the chain coil (**FIGURE 9-27**).

The following guidelines can be used in washing ropes:

- Use gentle, nondetergent soaps and follow package directions for their use. Specially fabricated low-pH rope soap, available from most rope manufacturers, is a good choice to gently remove dirt, oil, grease, soot, smoke, and grime. Otherwise, use a gentle cleaner, such as Woolite or Ivory Snow laundry soap, that is intended for use with synthetics. Do not use dishwashing liquid detergent, as the grease-cutting solvents degrade fiber treatments.
- Some ropes have had an additional chemical "dry" treatment to prevent the rope from picking up water. This is particularly useful in areas where wet ropes may freeze. Some of these treatments are easily removed with soapy water, and such ropes should be cleaned only with plain water.
- Do not use bleaches or bleach substitutes.
- Use the "cold water" setting.
- Rinse thoroughly to remove all traces of soap.
- Carefully dry the rope without heat. Hang the rope loosely out of direct sunlight and allow it to air dry.

Fabric Softeners

Some people use a small amount of fabric softener during the rinse cycle to give the rope sheath a soft feel. Ropes rinsed with a fabric softener solution that is mixed according to the fabric softener manufacturer's directions actually should perform better than ropes washed only with soap and water. Fabric softeners can make it easier to tie knots by delaying the rope stiffness that comes with age. The lubricants in fabric softeners can help replace the lubricants originally furnished on the yarns by the manufacturer, which are removed by repeated washing of the rope. Replacing the lubricants helps the yarns load more evenly, which may raise the tensile strength slightly in older, well-washed ropes. Always use fabric softener diluted according to manufacturer's instructions.

Special Cleaning Problems

Despite careful handling, ropes may become spotted with oil, grease, or mildew. There is no indication that any of these substances alone destroys rope fiber. However, they are unsightly and may stain clothing or high-angle gear. Petroleum substances may cause other contaminants to stick to the rope.

These substances often can be removed by soaking the rope in cool, soapy water and scrubbing the affected areas with a fingernail brush. Do not use strong, solvent-based cleaners. Many solvents that loosen grease and grime also dissolve nylon. Contact the rope's manufacturer for specific types of cleaning problems.

Dressing Rope Ends

Cut rope ends should always be carefully dressed. Frayed ends have a sloppy, unprofessional appearance, become snagged, and eventually grow in size.

FIGURE 9-28 Hot cutter.
Courtesy of Pigeon Mountain Industries.

The most effective method of cutting nylon or polyester rope is to use an electric hot cutter to melt the rope in two (**FIGURE 9-28**). Before the rope is cut, the spot where it is to be severed should be firmly taped. Electric hot cutters are less effective on aramid ropes, which do not melt. If you are cutting an aramid rope, or if you need to cut a nylon or polyester rope without a hot cutter, the following steps should be taken:

1. Firmly tape the spot to be cut to prevent fraying.
2. Cut down through the center of the tape.
3. Immediately fuse the cut ends. For fibers that melt, heat from a lighter or other small flame can be used. Alternatively, a coating material such as whip end dip may be used.
4. Taper the fused end slightly. It should not be the shape of a mushroom, because the rope end could get snagged when pulled through hardware or rock.

After-Action REVIEW

IN SUMMARY

- Rope is the foundation of any rope rescue activity. Choosing the correct rope for the task will make the task easier and safer to accomplish.
- Ropes may be made of natural or synthetic fibers. Synthetic fiber ropes are considered standard for situations in which the safety of a person is "on the line."
- Synthetic fiber ropes include polyolefins, polyester, nylon, UHMPE, and aramids. Each type of fiber has its advantages and disadvantages.
- The selection of a rope depended upon what the rope is made of and how it is made. Rope construction methods include laid, plaited, braided, and kernmantle.
- When choosing a rope, consider the rope's diameter, minimum breaking strength, range of elongation, abrasion resistance, the feel of the rope itself (hand), color, and length.

- Knot-tying instruction relies on knowledge of certain basic terminology, including bight, hitch, loop, running end, standing part, turn, and the working end.
- Knots are used in the rope rescue environment for tasks from anchoring to backing up other knots to creating emergency ascenders.
- It is good practice to back up most knots with a safety knot. When used, the backup should be tied as close as possible to the knot it is backing up.
- The lifespan of a rope depends upon the care, inspection, and maintenance of this vital piece of equipment and should be documented.
- A rope should be washed only if it shows dirt. Overwashing can cause the rope to stiffen or shrink, or both.
- Commercial devices specifically designed for washing ropes are available but may not get the surface of the rope completely clean.
- Washing machines can thoroughly and effectively clean rope. However, the machines must be used carefully, and certain specific precautions must be taken.
- Ropes rinsed with a fabric softener solution that is mixed according to the fabric softener manufacturer's directions should perform better than ropes washed only with soap and water.
- Do not use strong, solvent-based cleaners when cleaning rope. Many solvents that loosen grease and grime also dissolve nylon.
- The most effective method of cutting synthetic fiber rope is to use an electric hot cutter.

KEY TERMS

3-sigma A statistical method of calculating minimum breaking strength (MBS); this method results in a figure that is 99.87 percent reliable.

Bend A class of knot that joins two ropes or webbing pieces together.

Bight The open loop in a rope or piece of webbing formed when the rope is doubled back on itself.

Carrier The piece of equipment that carries the bundles of sheath yarn around and around the core to form the outer braid during the manufacturing process.

Commodity ropes Generic, nonspecialty ropes used in household and commercial applications.

Hand How a rope feels in a person's palm; influenced by the number of yarn bundles used to create the sheath, tightness of weave, and material.

Hitch A knot that attaches to or wraps around an object or rope in such a way that when the object or rope is removed, the knot falls apart.

Kernmantle A rope design consisting of two elements: an interior core (kern), which usually supports the major portion of the load on the rope, and an outer sheath (mantle), which serves primarily to protect the core but also may support a minor portion of the load.

Knots Fastenings made by tying rope or webbing together in a prescribed way. Knots include bights, bends, and hitches.

Laid rope made by twisting three or more strands together with the twist direction opposite that of the strands. Plain, or hawser, laid ropes have three strands, whereas shroud laid ropes have four strands.

Line A rope that is being utilized in a rope rescue system.

Loop A portion of rope formed into a circle with the ends crossing one another.

Low stretch A quality of a type of rope designed to be used in applications such as rescue, rappelling, and ascending in which high stretch would be a disadvantage and no falls, or only very short falls, are expected before the climber is caught by the rope; refers to ropes with slightly more elongation than the traditional static ropes.

Manufactured eye termination A permanent formed loop in the end of a rope created by the rope manufacturer, as an alternative to a knot.

Nylon 6 A type of nylon used in rope manufacturing. Because of its shock-absorbing qualities, nylon type 6 is found in most climbing ropes. One trade name for this type of nylon is Perlon.

Nylon 6,6 A type of nylon used in rope manufacturing. With its resistance to wear and reduced elongation under load, most static or low-stretch ropes are constructed of type 6,6. In North America it is manufactured by DuPont and the Monsanto Corporation.

Polyester A type of fiber used in some rope manufacturing. Also known by the trade name Dacron. Dacron is often found in the core of very low-stretch static ropes due to polyester's much lower elongation compared to nylon fiber.

Polyolefin A group of fiber types (e.g., polypropylene, polyethylene) used in the manufacture of ropes often used in water applications.

Running end The portion of rope used for lowering or hauling, or that will touch the bottom when rappelling.

Safety knot A second knot used to secure the tail of a primary knot; also known as a backup or keeper knot.

Standing part The portion of rope between the running end and the working end.

Static rope A type of rope designed to be used in applications such as rescue, rappelling, and ascending in which high or moderate stretch would be a disadvantage and in which no falls, or only very short falls, are expected before the climber is caught by the rope. Static ropes have slightly less elongation than low-stretch ropes built to the same standard.

Tail The unused portion of rope remaining behind the working end of a rope.

Turn A single pass of a rope behind an object.

Working end The end of the rope that is used to tie and form the knot.

REFERENCE

Gonzales, Laurence. 2014. *Deep survival: Who lives, who dies, and why.* New York, New York: W. W. Norton & Co.

On Scene

1. To reach a civilian caught on a bridge abutment, a rescue must lead climb beyond their anchor point. What type of rope should be used?

2. What are the important considerations for selecting a rope for a given task?

3. When might an aramid rope be desirable over a polyester rope?

4. Name at least one tie that can be used for each of the following:
   - Form a loop in the end of a line
   - Join two ropes
   - Form a safety knot
   - Create a friction hitch

CHAPTER 10

Operations

Principles of Rigging

KNOWLEDGE OBJECTIVES

After studying this chapter, you should be able to:

- Understand the principles of rigging. (**NFPA 1006: 5.2.5**, **5.2.7**, pp. 185–191)
- Explain the relationship of angles to forces. (**NFPA 1006: 5.2.6**, pp. 186–187)
- Describe the purpose and types of edge protection. (**NFPA 1006: 5.2.8**, pp. 188–191)
- Describe how to calculate expected loads in a rope rescue system. (**NFPA 1006: 5.2.5**, **5.2.6**, **5.2.22**, **5.2.23**, pp. 191–192)
- Describe the factors that influence the integrity of a rope rescue system. (**NFPA 1006: 5.2.6**, pp .192–194)
- Describe how static loads differ from dynamic loads. (**NFPA 1006: 5.2.6**, p. 192)
- Understand when a belay is appropriate, and why. (**NFPA 1006: 5.2.9**, pp. 192–193)
- Describe the purpose of dual-tensioned systems. (pp. 192–193)
- Understand the methods for reducing force to system components. (**NFPA 1006: 5.2.17**, p. 194)

SKILL OBJECTIVES

After studying this chapter, you should be able to:

- Identify rope rescue system priorities based on anticipated forces. (**NFPA 1006: 5.2.9**, p. 194)

You Are the Rescuer

You respond to an incident in which a BASE jumper has caught their chute on a bridge cap. They are suspended several feet below the horizontal support, which itself is 75 feet (29 m) below the lowest deck surface, and over 400 feet (122 m) above the ground. There is no place to anchor above the deck, and the river below precludes any possibility of lowering the rescue load to the ground. A rescuer will have to be lowered to extricate and secure the subject, and the two will then have to be raised together back to the deck.

1. What forces will be applied to the lowering system?
2. How much lifting power will rescuers need to raise the load?
3. At what points in the system are you likely to find friction, and what effect will it have on each of the systems used?

 Access Navigate for more practice activities.

Introduction

A good rescuer understands not just how to rig and set up commonly used systems, but also the concepts behind the rigging. Building good rescue systems requires an underlying ability to analyze the physical properties of equipment and the manner in which it is used together, and the influence that motion, energy, and other factors will have on forces during the operation of the system (**FIGURE 10-1**).

It is not necessary to overcomplicate these concepts or to use complex engineering terms. While the principles we will discuss here are commonly used and understood in engineering, one need not be an engineer to be a good rescue rigger. In fact, many rope rescuers have a more solid grasp of the fundamental, practical aspects of physics than they realize! People who have a natural proclivity toward ropes and rigging often have a very good grasp on physics concepts, even if they have never been formally trained in them. Those to whom these tendencies may not come quite as naturally can still develop an understanding of general principles that will help provide a strong foundation for choosing, creating, and evaluating rescue systems.

This chapter will focus on rigging concepts—not so much what to do, but how to think about rope-based rescue systems so that you can more effectively build systems, evaluate forces, and maximize safety. Equipment that is rigged together using the principles in this chapter and used to perform some function is generally referred to as a **rope system**. Rope systems used to effect a rescue are usually called a **rope rescue system**. The principles discussed in this chapter apply to all types of rescue systems discussed in this book (anchoring, lowering, mechanical advantage, personal skills, etc.) and more. Whether a rope rescue system is fixed or moving, it should be rigged a sufficient distance back from the edge so that the rescuer has a safe zone in which to operate.

FIGURE 10-1 The principles of physics are applied during rope rescue procedures.

TIP

Slow is Smooth; Smooth is Fast

This mantra from military marksmanship emphasizes that rushing in a reckless manner is much riskier than slow, careful, and deliberate actions. Being well organized, a rescue team can breed efficiency in its emergency response efforts. A disciplined team communicates effectively, and the team members accomplish their tasks without rushing or yelling. Team members know what to do and are trained to the level of competency that is required for their mission.

Principles

Force

Anyone using rope rescue systems will soon become familiar with the term **force**. A proper definition of force is the push or pull an object experiences as a result of its interaction with another object or external influence. Intuitively, we all know what force is when we experience it, but there is a key concept in this definition that we may not always think about. Specifically, that is the concept of force being relative to an interaction between two objects or between an object and an external influence such as gravity. Force can only exist in relationship; it exists only as a result of an interaction. As soon as the interaction stops, the force also stops. Force exists only as a result of an interaction, which we can usually recognize as pulling/pushing.

Those of us accustomed to thinking in imperial units of measure should keep in mind that weight and mass are not the same thing. The term **mass** refers to the amount of material in an object, while **weight** is the effect of gravity attracting two masses. Although mass and weight are not the same thing, we can somewhat use them interchangeably for objects near the Earth's surface. The confusion between mass and weight provides a strong reason to use metric units rather than imperial units when calculating forces.

Mass, being the amount of material in an object, is independent of gravity. For instance, in space, the mass is still present, but the weight that would be felt on Earth is not. However, for objects near the Earth's surface, the force due to gravity acting on a mass is measured as weight.

Gravity is, in and of itself, technically not a force. It is a phenomenon that pulls physical objects toward one another. It is only when an object (say, a rescuer) is pulled toward the Earth by this phenomenon that the concept of force becomes relevant. Gravity pulls an object toward the Earth at a constant rate. The force exerted on an object is a result of the attractive force between the Earth and that object.

The resulting force for objects near the Earth's surface can be calculated using the equation:

Gravity Force (or weight) = 9.8 Newtons (N) / kilogram (kg) × mass of an object (kg)

Although it is an invisible influence that acts on our rescue systems from a distance, the predictability of gravity makes it a concept that we can use to our advantage.

With gravity pulling objects toward the Earth at a constant rate, we can know that a rescue load dangling from the end of a rope will create predictable tension in the system unless some other force acts upon it to change it. Imagine: If the direction or magnitude of the force of gravity were constantly varying, we humans would never quite know which way is up, and a load at the end of the rope might be exerting extraordinary force one moment, and floating the next.

Other influences on our systems may come through direct contact—for example, a tensioned rope used to pull a load, or friction between objects. These are more variable influences and more difficult to predict, but no less important to our analysis of force on rescue systems. In fact, the forces that we intentionally apply to a system to move a load are generally of this type. The magnitude and direction of the applied force will determine if, how much, and where the rescue load moves. When we use a tensioned rope to pull a load upward, the constant force of gravity is pulling the load toward the center of the Earth. The load moves against the force of gravity in the direction exerted by the countering force of the rope. As we pull harder on the rope (with greater magnitude), the load will move faster.

TIP

Weight Versus Mass: Units of Measure

Precise measurements require a distinction between weight and mass. Pounds and kilograms are measurements of weight. An accurate measurement of force is expressed in newtons. A newton (N) is the unit of force required to accelerate a mass of 1 kilogram 1 meter per second. The impact loads on rope and the breaking strength of equipment are usually expressed in kilonewtons (kN); in the English system, this is expressed in pounds/force (lbf). Major organizations that set standards for high-angle rescue use the International System of Units (SI). Equivalents are as follows:

1 newton = 0.225 lbf
1 kN (1000 newtons) = 225 lbf

TIP

International System of Units (SI)

This text was written in the United States, where the English system of measurement (e.g., pounds, quarts) is commonly used. To make the information more accessible to readers outside the United States, equivalent measurements in the International System of Units (SI) have been included. In some cases, the SI units are approximate conversions from the English system.

Vector

The term **vector** is used to refer collectively to direction and magnitude of a given force. Direction is a familiar term, and magnitude is a term that refers to how strong the influence is. It is measured as a result of the angle of the force with respect to a directional reference and the amount of force. With this in mind, the force vector of a system pulling on a floor-mounted anchor might be measured as a result of its direction of pull relative to the floor and the magnitude of the pull. Together, the direction and magnitude of a given force are referred to as a vector.

Every force has magnitude and direction, so a vector is associated with *every* force. This is an important principle, because some rescuers tend toward using the term only in complex rigging, such as with high directionals. However, by grasping and using the concept of vectors in everyday rigging, the principle becomes clearer even as rigging becomes more complex.

For every rope rescue system, rescuers should always ask themselves, "What vectors (i.e., force magnitude and direction) are acting on the system at any given time?" By addressing this question directly, we can rig systems to intentionally incorporate forces that influence, mitigate, or even counteract other forces, to ultimately achieve our rescue goals. For example, a load may be transported across a chasm by attaching it to a horizontally rigged rope, known as a highline. In this system, the vector induced by gravity is countered by vectors induced by the rope on the two anchors opposite one another. For the system to be ultimately useful, however, we must induce yet another force vector, in the form of a haul line to pull the load across the highline.

Resultant

The term **resultant** has come into vogue in rescue circles to describe the concept of multiple vectors that occur as a result of, or as a part of, complex rigging. However, in truth, there is a resultant any time that multiple vectors interact with one another, even if it is in a straight line. Just as *every* force has direction and magnitude (i.e., *every* force is a vector) so also *every* combination of vectors has a resultant.

In a straight-line system, the net force exerted on an object is the result of adding all the force vectors acting upon it. The sum of all the forces (vectors) is called the resultant. This concept is illustrated by the equations in the illustration, where the direction of pull is indicated by arrows, and the magnitude is indicated by line length (and numbers) (**FIGURE 10-2**).

In this illustration, the forces indicated are acting in the same, or in directly opposed, directions. The load will remain stationary as long as the forces are

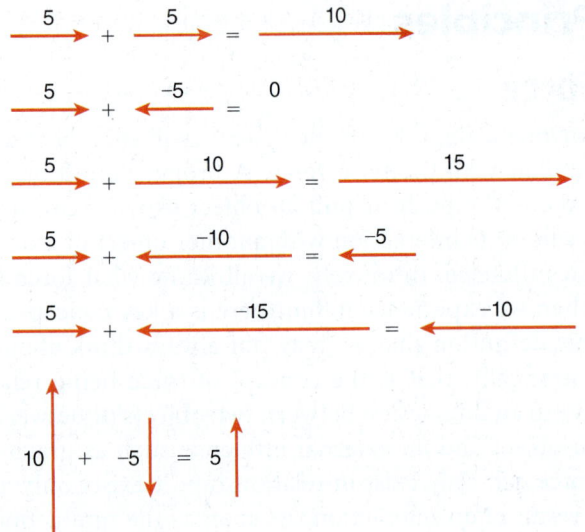

FIGURE 10-2 Vector addition.

equal and opposed, but if an imbalance occurs, the load will be attracted to the higher force. As the load is lowered by a rope rescue system, the relationship between the direction/magnitude of the force of gravity and the direction/magnitude of the force exerted by the rope system will determine its path of travel. If that path needs to vary, rescuers may need to introduce yet another rope-based vector to redirect the resultant of the first two.

Angles in Rescue Systems

An **angle** is the measure of the difference in direction between two lines as measured at the intersection of the two line segments. In rope rescue, we are most concerned with the point(s) at which two vectors meet. This occurs where a rope is terminated, such as in anchors, or bent, such as around a pulley (**FIGURE 10-3**). Although an angle is not, itself, a force (remember, a force must have both magnitude and direction), angles do change the direction in which a force is being exerted—which, in turn, creates a new vector. Wherever two vectors interact, a new resultant vector is created. Rescuers must be capable of analyzing the direction and magnitude of forces as they change and evolve during use of a rope rescue system. Properly managed, angles can be an asset in a rope rescue system as they are used to redirect the path of a rescue load, create mechanical advantage, distribute loads in an anchor system, etc.

When two vectors interact at an angle to one another, the direction and magnitude of each of those vectors will change. Also, simply by acting upon one another, these two vectors will create a new vector of some magnitude acting in some direction. Consider a lowering system where the rope path follows a straight

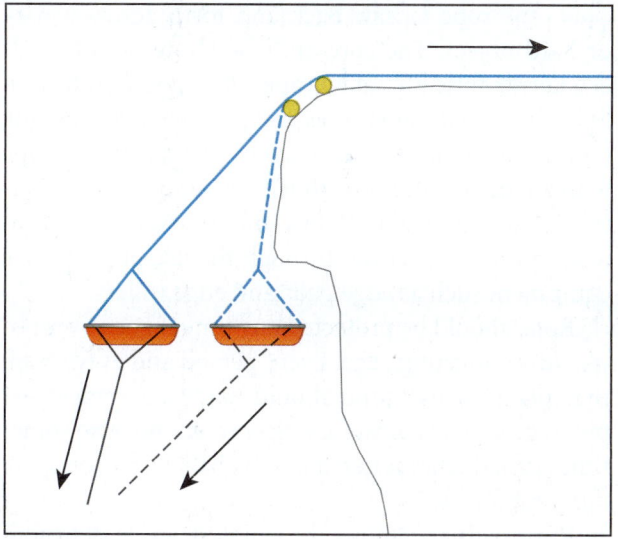

FIGURE 10-3 Forces occur when an object is pushed or pulled in a direction as a result of interaction with another object.

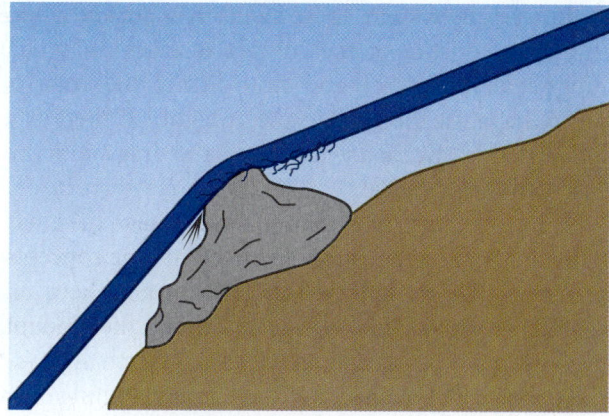

FIGURE 10-4 When a rope is dragged over the edge of a rocky cliff, there will be friction on the rope where it comes into contact with the rock, in a direction opposite the motion.

line from anchor to load. Should this straight-line path prove inconvenient or hazardous, it may become necessary to offset the direction of travel by inserting a pulley at some point part way down the line. The rope above the pulley, between it and the anchor, will have one vector; the rope below the pulley, between it and the load, will have a second vector; and the pull created by these two vectors upon the pulley itself will result in yet another vector. Each vector affects the other vector(s) in the same system, causing the direction and magnitude of the original vectors to change.

Tension

Tension is a term that we use in rope rescue to describe the stress in a system or system component due to transmitted forces that pull in equal and opposite directions. Tension is created in the rope or other system components by the forces at each end as they act in opposition to one another. The magnitude of tension is equal to the mass of the object multiplied by gravitational force, plus or minus any acceleration forces that are also acting on the object.

When an object is supported (at rest) by a rope, the tension in the rope is equal to the weight of the object. Likewise, in an anchored system, tension is applied equally and opposite longitudinally along the rope. For this reason, parts of a system such as rope, webbing, and other similar equipment are sometimes referred to as **tension members**. Tension may be applied in multiple directions by different systems at the same time. For example, when a load is being raised in a cross-haul system, there are typically two or more haul systems pulling simultaneously on the load to cause it

to move vertically and horizontally at the same time. See Chapter 16, *Working in Suspension*, for more on cross-haul systems.

The opposite of tension is **compression**. Compression, too, is a type of transmitted force, in this case referring to the inward or pushing forces that something exerts on another object or structure. Compression, like tension, is the result of opposing forces. While it may not seem at face value that we deal with many compressive forces in rescue, the fact is that compression does play a part in rescue systems. Compression may be found in a rope rescue system where a tripod anchorage is supported by the ground, or where an edge roller is pressed into an edge by a rope under tension.

Friction

Friction is an object's resistance to movement. For example, when a rope is dragged over the edge of a rocky cliff, there will be frictional force on the rope where it comes into contact with the rock, in a direction opposite the motion (**FIGURE 10-4**).

Friction is calculated by a mathematical formula that takes into account the force that is resisting the motion and the force that is pressing the objects together. The amount of force that it takes to overcome an object at rest and initiate movement determines the static coefficient of friction. The resistance to continued motion while the object is moving determines the dynamic (kinetic) coefficient of friction. For example, an inline spring scale can be attached to an object at rest on a table. As the scale is tensioned, the point at which the object begins to move, as measured on the scale, can be used to calculate its static coefficient of friction. The measurement on the scale as the object continues to move is typically a bit lower, and can be used to calculate its dynamic coefficient of friction.

In a rescue system, these values will depend upon the characteristics of the contact surfaces (material, roughness, moisture) and how firmly the rope is pressed against the surface (i.e., weight.) If there is a lot of friction, the load will be harder to drag and more susceptible to damage.

A rope rescuer may not need to calculate the coefficient of friction, but should understand the concepts and recognize the effects that friction may have on rigging systems. These effects go beyond the amount of force in play, and can include damage such as physical damage that may occur to soft goods running over sharp edges as well as potential damage from the heat that friction causes.

In rope rescue operations, friction occurs between the rescue load and all of the surfaces with which it comes in contact, between the rope and everything it touches, and between the rescuers and the surfaces with which they come into contact. In such cases, there is a combination of forces at play: friction, compression, and tension.

When planned for, the addition of friction to a system can contribute positively to an operation, such as by helping a rescuer to control the load. With enough friction, the force of a 600-pound (272-kg) load in a direct line lower might only impart 400 pounds (181 kg) at the braking device. When not planned for, friction can make the work more difficult or even create hazards and jeopardize the operation. With too much friction, the force of a 600-pound (272-kg) load in a direct line raise might impart as much as 2000 pounds (907 kg) on a haul system. Generally speaking, friction works for us when lowering a load and works against us when raising a load. As a rule of thumb, friction can be assumed to amplify the effects of a load by a factor of about 2.5. It is important to understand the implications of this on equipment and systems.

> **TIP**
>
> In terms of system efficiency, a smooth edge is generally better than a sharp edge, and a gentle bend is preferable to an extreme one—although the length of rope that comes into contact with the edge will also play a role.
>
> Reproduced from McCurley, L. (2013). *Falls from height: A guide to rescue planning*. John Wiley & Sons.

Edge Protection

Ropes are most susceptible to damage when they are under tension and run over a rock or over the edge of a building. Bouncy rappels or ascending can also cause the rope to saw back and forth across a rock or hard object. The core-and-sheath design of modern kernmantle life safety ropes offer good protection from frictional damage, as the primary load-bearing core is shielded from damage by the sheath. A rope with a tighter, stiffer sheath will generally provide better abrasion resistance than a soft one. Additional protection can be achieved through the use of protective equipment such as edge pads and edge rollers.

Rope should be protected any time it goes over potentially damaging edge. Every person and every team that plans to use rope should carry equipment for preventing edge abrasion. Edge protection, also sometimes called edge softeners, is available in a couple of different forms.

Rope pads are among the simplest and least expensive means of protecting rope from abrasion. They work best for fixed ropes, such as rappel lines. Among the commonly used types of pads are heavy-duty canvas pads. Natural fiber canvas, such as cotton, is preferred where there is a chance that the pad may be used with moving rope because synthetic fibers are prone to melting due to heat generated by frictional forces.

For greater protection and durability, a square of heavy-duty cotton canvas can be folded twice, stitched around the edge to prevent fraying, and then cross-stitched (**FIGURE 10-5**). For added convenience, large grommets can be set into two or more corners. The grommets can be used to anchor the pad so that it does not slide away as the rope is moving. A complete edge protection kit should include a variety of pads, ranging from approximately 2 × 3 feet (0.6 × 0.9 m) to 2 × 6 feet (0.6 × 1.8 m) or larger. Several commercially made canvas rope pads are also available.

Historically, sections of discarded fire hose were often converted into effective rope pads, even for moving ropes. In the past, most fire hoses were made of a cotton jacket over a rubber core. By slitting the fire hose lengthwise, a secure pad could be devised where the rubber interior could be laid directly against the surface to be covered, with limited slip, and the rope could be run over the cotton jacket side. However, in recent decades, synthetic fibers such as nylon and polyester have replaced cotton as the material of choice for fire hose jackets.

Still, even synthetic fire hose makes a decent rope sleeve for stationary rope. In this case, it is best to devise the hose length into a rope-sleeve. To make a fire-hose rope sleeve, follow these steps:

1. Split the hose down the center.
2. To prevent the rope from slipping out, secure the edges of the hose with a closure such as snaps or Velcro.

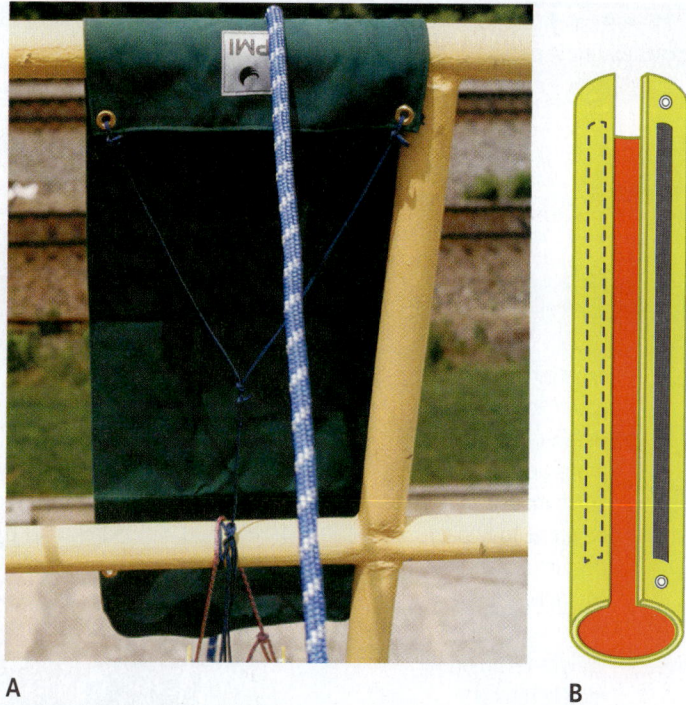

FIGURE 10-5 A. Canvas rope pad. **B.** Fire hose rope pad.
A. Courtesy of Pigeon Mountain Industries.

3. Set a hole or grommet in each end of the fire hose so that it can be anchored to prevent it from slipping down the rope.

This type of protector is also available commercially, often made of ballistic cloth or other fabric, or even metal. A rope sleeve is designed to wrap completely around the rope, and as such is primarily intended for use with stationary ropes (**FIGURE 10-6**). Those made of fabric are typically designed to be secured in place with a hook-and-loop or other closure, and may also have grommets and a small cord to anchor the sleeve to the host line. This type of protection is especially useful where a rope comes into contact with an abrasive or potentially damaging structure midline.

Another type of commercially available edge pad is a plastic version. While plastic edge protectors are not as flexible as fabric, they do tend to conform at least somewhat to the protected surface, and also provide more durable protection. Most plastic edge protectors may be used with either stationary or moving ropes. Some plastic edge protectors are designed to be modular, and allow the addition of an infinite number of units to create a rope-track for a rope to travel over a difficult surface.

By far the most efficient type of edge protector available is the aluminum edge guard. These are an excellent choice for either stationary or moving ropes

FIGURE 10-6 Supermantle on a rope.

and are available in various designs to either conform to a variable surface, or to effectively protect a 90-degree edge. Most have moving rollers along the rope path (although some are fixed), but in any case all edge guards must be affixed in place to prevent them from moving or from being dropped during a rescue.

TABLE 10-1 Advantages and Disadvantages of Commercial Rope Protectors

Commercial Rope Protectors	Advantages	Disadvantages
Metal devices	Afford greater protection than soft protection	Reduce friction, particularly in hauling Usually more expensive than soft protection Heavier than soft protection Some models may be prone to tipping over unless anchored carefully.
Soft protection	Less expensive than metal devices	Weighs less than metal protection May not give as much protection as metal devices Does not reduce friction as much as metal devices

FIGURE 10-7 Edge roller.
Courtesy of Steve Hudson.

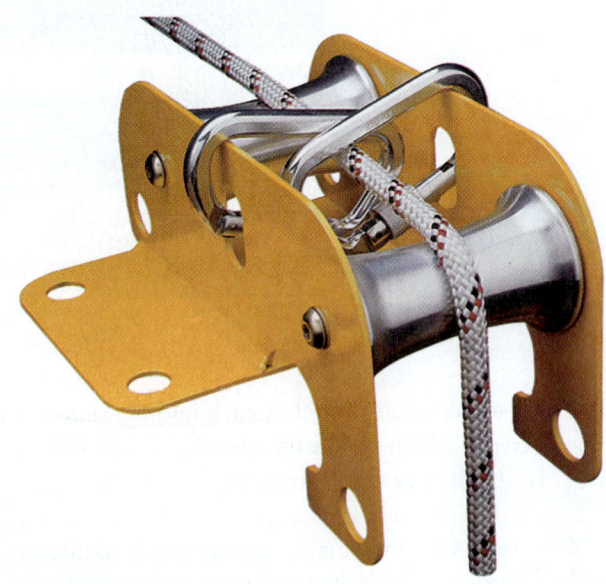

FIGURE 10-8 Roof roller.
Courtesy of Pigeon Mountain Industries.

The advantages and disadvantages of some different types of commercially available devices are listed in **TABLE 10-1**.

Metallic edge rollers are one of the most efficient and effective means of protecting ropes from abrasion. They usually are more expensive than soft protection, but they have the added advantage of greatly reducing the friction of rope over an edge. This is particularly important in hauling systems, in which edge friction makes raising more difficult and puts great stress on equipment. Some edge rollers may tip over if they are not anchored carefully. Three main types of edge rollers are available: single-unit rollers, roof rollers, and roll modules.

In single-unit rollers, two rollers are set into a frame (**FIGURE 10-7**). Two or more of these units usually are needed to provide adequate edge protection. They generally must be stabilized by anchoring them with their attachment points. When they are stabilized, these units perform well on irregular surfaces, such as cliffs and other natural conformations. One particularly versatile edge roller design is composed of three linked units consisting of hard aluminum side plates with two roller bars each. It can be linked together with the furnished screw links at sharp edges or separated for long, gradual breakovers.

A roof roller is specifically designed to excel in protecting a sharp edge, and consists of a unit of two rollers set in a 90-degree frame (**FIGURE 10-8**). They work best on buildings and other structures where 90-degree angles are present. A roll module has rollers on the sides as well as the bottom, so that if the module tips over, it still offers protection. The four modules are linked with screw links and can conform to a variety of surfaces.

Commercial, purpose-built rope protection devices are now the choice of many professionals. While either edge pads or rollers may be used for protecting nonmoving ropes, for systems where ropes are being dragged over an edge, rollers are the better choice.

If no premade edge pad is available, other materials may be pressed into service to protect the rope from abrasion. Some examples of improvised rope pads are as follows:

- Packs
- Turnout coats

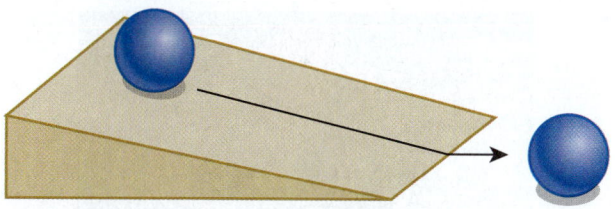

FIGURE 10-9 The fall line is the natural path a ball would take if rolled down a slope.

- Clothing
- Blankets
- Carpet squares (wool or natural fiber)

Artificial high directionals are another type of device that can help protect ropes from edge abrasion.

Fall Line

The **fall line** is one final concept that we will discuss as having an important influence on forces imparted on rescue systems. The fall line can be described as the path a ball would follow when rolling down an incline (**FIGURE 10-9**). It is influenced by factors such as gravity, friction, and obstacles, and it plays an important role in what directional forces are imparted on the rope rescue system. Learning to accurately assess and integrate the fall line into rope rescue rigging, whether in urban or natural surroundings, is an important skill. Because rolling an actual ball down a steep or vertical plane prior to rescue is less than practical, assessing the fall line is a bit of a soft skill that requires an experienced eye and awareness of the myriad of factors to predict. Successful rope rescue rigging takes into account the expected fall line when rigging anchors and planning the load path, as the direction of pull is such an important part of vectors influencing the system (as discussed earlier in this chapter.)

Putting It Together

The principles addressed in this chapter are foundational to analyzing and evaluating rope rescue systems. An operations-level rope rescuer should be capable of using these principles to analyze rescue systems at a glance, at least for obvious errors and concerns. An operations-level rope rescuer should also be capable of applying these principles to effectively manage a rope rescue system. Rescue system are discussed in Chapters 11–16.

Safety Factors

The concepts described earlier in this chapter regarding forces in a rope rescue system are essential building blocks in developing an understanding of safety factors. It is imperative that rope rescuers can identify the weakest point in a rope rescue system and analyze it against the maximum anticipated load at that point. The rope rescuer must be able to scrutinize force multipliers both within and on the system and adapt rigging methods accordingly—whether by using a different knot, changing direction of pull, using different equipment, etc.

The term **load ratio** is used to express the difference between the strength of something and the anticipated load that is expected to be placed on it. Looking now beyond components to the system level, the ratio between the breaking strength of the weakest point of a rope rescue system as compared with the maximum anticipated load at that point is called a **system safety factor**. A system safety factor may be calculated using the following formula:

$$S / L$$

Where S = Strength of the weakest point in the system and L = Maximum anticipated load

Load ratios were first discussed in Chapter 8, *Rescue Equipment*, relative to design load of rescue equipment components.

The load ratio that represents the system safety factor may be different from the load ratio that represents a component safety factor. As an example, consider that the strength of an NFPA 1983 G–rated rope must be at least 40 kN (8992 lbf). The difference between this and the commonly accepted **rescue load** of 600 lbf (2.66 kN) is a factor of about 15:1 (15 × 600 = 9000 lbf [2.66 = 40.03 kN]). We would refer to this as a component safety factor. However, when we tie a figure-eight knot in the rope, it can reduce the strength of the rope by as much as 30 percent, bringing the 9000-lbf (40.03 kN) rope strength down to around 6300 lbf (2.66 kN). If this is determined to be the weakest point in the system, the static system safety factor as measured against that 600-pound (272 kg) load would be more like 10.5:1.

Ideally, however, your safety factor should take into consideration the highest potential force on the system at that point, not just the weight of the load. During operation of the rope rescue system, the forces exerted can be significantly higher than the weight of the load, particularly in the event of a failure. When a rope rescue system is in motion, the load is no longer static or still, it is dynamic. Dynamic loads produce kinetic energy, which adds an additional force onto the system. Taking the previous system as an example, if a worst-case foreseeable failure might result in a worst-case scenario impact force of 2.5 times the load, or 1500 lb (680 kg), at the point in question, our safety factor would be likewise reduced to a value closer to 3:1 (4800 / 1500 [2177/680]).

The safety factor should be a prime consideration when constructing a rope rescue system. In order to ensure the safety factor, the rope rescue system should be constructed to absorb a force greater than anticipated.

Rope Systems

Equipment that is rigged together using the principles in this chapter and used to perform some function is generally referred to as a rope system. Rope systems used to affect a rescue are usually called a rope rescue system.

Rope rescue systems may be fixed or moving. A fixed rope rescue system is one in which the ropes do not move (or are stationary). In such cases, ropes are anchored in place and rescuers use the ropes for ascending or descending. A fixed rope rescue system should anchored in place, with the anchor a sufficient distance back from the edge so that the rescuer has a safe zone in which to access the rope for rigging and derigging their descending or ascending system. Only one user at a time should access a fixed rope system, but these types of systems can be an excellent choice for providing rescuers access to and from a rescue site.

Some rescuers use what is known as **single rope technique (SRT)**, where all attachments are made on a single rope. Others use **dual rope systems**, where access equipment is used on one rope and backup/safety equipment is attached to another (**FIGURE 10-10**). Both types of systems are viable, appropriate systems, and should be selected with consideration given to the level of training and skills of personnel, system goals/performance, as well as probability of equipment failure.

Moving rope systems are dynamic systems in which rope is used to lower or raise a load. This is the type of system most often used to effect a rescue, as subjects are generally unable to ascend/descend fixed rope as previously described. When using a moving rope system, a second system is usually employed as a belay. Sometimes a moving rope system is used as a belay for a person who is moving along the ground or on a fixed rope system.

Dual-Tensioned Rope Systems (DTRS)

A foundational concept of any rope rescue system is that *the failure of any one point does not result in catastrophic failure*. In other words, the system should be redundant. For this reason, systems that incorporate a primary system backed up by a belay system are nearly always preferred over single rope technique. Another way to create a redundant system is to use two separate

A

B

FIGURE 10-10 A. Single rope technique has all attachments on a single rope. **B.** Dual rope systems have access equipment on one rope and backup/safety equipment attached to another.

lowering lines, each sharing a portion of the load but at the same time each backing up the other. Known as a **dual-tension rope system (DTRS)**, this approach is lauded for offering lower impact force and fall distances in the event of a main line failure.

The standard approach, a **two-tensioned rescue system (TTRS)** (or twin-tensioned rescue system), in which two virtually identical systems are employed and simultaneously weighted, has become especially prevalent in recent years due to the advent of auto-locking braking devices that may be used for either lowering or backup (**FIGURE 10-11**).

Some TTRS rescue systems use two ropes running through one braking device, such as a brake tube or **braking/belay device** (**FIGURE 10-12**). Other systems use two brakes, each controlling a rope attached to the

CHAPTER 10 Principles of Rigging

FIGURE 10-11 A two-tensioned rescue system.

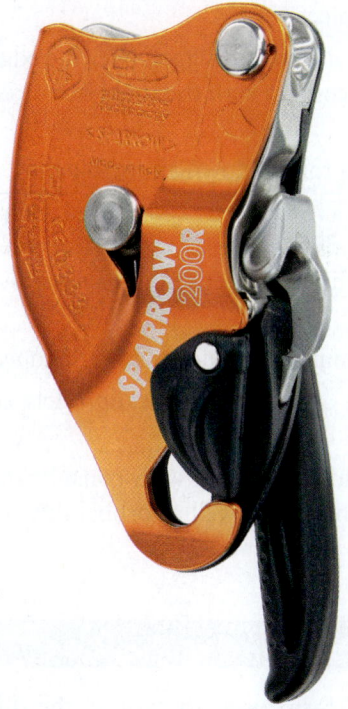

FIGURE 10-12 A descent/belay device.

rescue load. In such cases, the two-brake systems are designed so that each lowering device backs up the other. In essence, each braking device belays for the other one.

TTRSs offer several advantages in certain situations. For example, they do not require special equipment solely for the belay, and the roles of brake person and belayer require the same skills and can be interchanged easily. Also, TTRS usually do not accidentally engage and hang up, they do not require load-releasing hitches, they provide a soft catch, and they are predictable in the way they will perform.

However, in such a system the operation of the two lines must be well coordinated. If two ropes are run through a single device by a single operator, coordination is easier but redundancy is limited. If two ropes are run through two devices by a single operator, redundancy is somewhat improved, but having one person managing two devices at the same time can present challenges, particularly if things are not going smoothly. If two ropes are run through two devices by separate operators, the individuals operating the brakes must be well coordinated and alert for failure of the other system. With certain devices, there may also be difficulty with changing direction suddenly from lowering to hauling. As with all rescue operations, equipment and methods should be selected based on circumstances and need.

The TTRS may be operated using any descent device that is also capable of being used as a belay. In guiding and mountaineering, even tube-style stitch plate–type devices are sometimes used for this purpose. In general, considerations for evaluating a multifunction device include the following:

- Load range/maximum rated load
- Ease of use
- Smooth transition during initial release of heavy load
- Ease of transition from lower to raise (and vice versa)
- Whether additional friction is required for heavy loads
- Effective function as a progress capture

Specific equipment is discussed in Chapter 8, *Rescue Equipment*.

Optimizing Force

A rescuer can reduce and optimize forces in a rescue system by understanding the principles discussed in this chapter, and altering systems accordingly as needed. Sometimes, a rescuer will need to find a means by which to increase force—for example, when creating a mechanical advantage system. Other times, force may need to be reduced so as to maintain adequate safety factors. Always, the ongoing analytical assessment of force is a necessary part of every rescue system.

By understanding that force is a vector that has both direction and magnitude, and by understanding external influences such as friction, angles, fall line, and multiple vectors acting upon one another, the rescuer is better prepared to determine which of these should best be adjusted to achieve the desired result within the parameters of appropriate safety factors.

After-Action REVIEW

IN SUMMARY

- Rescuers should have a firm grasp on the concepts of forces and how they relate to one another in a rope system before trying to execute a rope rescue.
- Vector refers collectively to the direction and magnitude of a given force. Together, the direction and magnitude of a given force are referred to as a vector.
- Every force has magnitude and direction, so a vector is associated with *every* force.
- Just as *every* force has direction and magnitude (i.e., *every* force is a vector) so also *every* combination of vectors has a resultant.
- An angle is the measure of the difference in direction between two lines as measured at the intersection of the two line segments. In rope rescue, we are most concerned with the point(s) at which two vectors meet.
- Tension is created in the rope or other system components by the forces at each end as they act in opposition to one another. The magnitude of tension is equal to the mass of the object multiplied by gravitational force, plus or minus any acceleration forces that are also acting on the object.
- Friction is an object's resistance to movement. For example, when a rope is dragged over the edge of a rocky cliff, there will be frictional force on the rope where it comes into contact with the rock, in a direction opposite the motion.
- The fall line can be described as the path a ball would follow when rolling down an incline.
- It is imperative that rope rescuers can identify the weakest point in a rope rescue system and analyze it against the maximum anticipated load at that point. The rope rescuer must be able to scrutinize force multipliers both within and on the system and adapt rigging methods accordingly, whether by using a different knot, changing direction of pull, using different equipment, etc.
- Some rescuers use single rope technique (SRT), where all attachments are made on a single rope.
- Others use dual rope systems, where access equipment is used on one rope and backup/safety equipment is attached to another.
- Both types of systems are viable, appropriate systems, and should be selected with consideration given to the level of training and skills of personnel, system goals/performance, as well as probability of equipment failure.

KEY TERMS

Angle The intersection of two vectors.

Braking/belay device A device that can be used for braking or belay.

Compression A physical force that presses on an object.

Dual rope system A rope system in which access equipment is used on one rope and backup/safety equipment is attached to another.

Dual tension rope system (DTRS) A two-rope system in which both the primary and the backup are both loaded.

Fall line The path a ball would following rolling down an incline.

Force The push or pull an object experiences as a result of its interaction with another object.

Load ratio Term used to express the difference between the strength of something and the anticipated load that is expected to be placed on it.

Mass The amount of material in an object.

Rescue load Generally accepted amount of weight expected to be borne by a rescue system.

Resultant The net force exerted on an object by the combination of all force vectors acting upon it.

Rope rescue system Rope system used to effect a rescue.

Rope system Rope and equipment that is rigged together to perform a specified function.

Single rope technique (SRT) A rope system in which where all attachments are made on a single rope.

System safety factor A load ratio expressing the difference between the breaking strength of the weakest point of a rope rescue system as compared with the maximum anticipated load at that point.

Tension Term used in rope rescue to describe the stress in a system or system component due to transmitted forces that pull in equal and opposite directions.

Tension member Component parts of a system used to support force in a rope-based system (rope, webbing, and other similar equipment).

Two-tensioned rescue system (TTRS) A particular form of a dual rope system in which the primary system and the belay line are more-or-less identical.

Vector Direction and magnitude of a given force.

Weight The effect of gravity attracting two masses.

On Scene

1. In a rope rescue lowering system with a long, sloping edge at the top, what effect will the edge have on the system?

2. If the system described in #1 is converted to a raise, what effect will the edge have on the system?

3. What principles of physics should be considered when creating a rescue system?

4. What considerations should be factored in to a decision of whether to use a fixed or moving system for access to a rescue subject?

5. What are the advantages and disadvantages of a TTRS?

Chapter Opener: © Jones & Bartlett Learning. Courtesy of Loui McCurley; On Scene siren: © Bildgigant/Shutterstock.

CHAPTER 11

Operations

Anchorages

KNOWLEDGE OBJECTIVES

- Define common anchorage terms. (pp. 198–199)
- Explain how to weigh the considerations of strength, direction, and position in anchor selection or placement. (**NFPA 1006: 5.2.5, 5.2.6**, p. 199)
- Describe the considerations for positioning anchor systems. (**NFPA 1006: 5.2.5, 5.2.6**, p. 200)
- Identify the considerations for back up anchor systems. (**NFPA 1006: 5.2.5, 5.2.6**, pp. 200–201)
- Describe the principles of single point anchor systems. (**NFPA 1006: 5.2.5**, pp. 202–205)
- Describe how force is distributed among anchor points. (**NFPA 1006: 5.2.6**, pp. 207, 209)
- Explain the difference between load sharing and load distributing anchor systems. (**NFPA 1006: 5.2.6**, pp. 207, 209)
- Explain the principals of multi-point anchoring systems. (**NFPA 1006: 5.2.6**, pp. 206–211)
- Discuss the use of directionals. (**NFPA 1006: 5.2.5, 5.2.6**, pp. 212–213)
- Explain how and why to back up anchors. (**NFPA 1006: 5.2.5, 5.2.6**, pp. 214–215)
- Describe how to safety check anchor systems. (**NFPA 1006: 5.2.5, 5.2.6, 5.2.7**, p. 216)

SKILL OBJECTIVES

- Tie a tensionless hitch. (**NFPA 1006: 5.2.5**, p. 202)
- Construct fixed and focused multi-point anchor system (**NFPA 1006: 5.2.6**, p. 208)
- Construct a protected self-adjusting anchor system. (**NFPA 1006: 5.2.6**, p. 210)
- Back up an anchor using a pretensioned tie back. (**NFPA1006: 5.2.6**, p. 213)
- Complete an anchor system safety check. (**NFPA 1006: 5.2.5, 5.2.6, 5.2.7**, p. 216)

You Are the Rescuer

You are in the first team to arrive at the scene of a high-angle rescue. The incident commander (IC) informs you that the plan is to send a rescuer down to first assess and prepare the subject for packaging. Then, a rescuer with litter will be lowered so that the subject may be packaged and raised back to the top of the drop. An edge attendant will assist with the litter. Your team is assigned to establish anchorage systems to support all these tasks.

1. Name three things you will need to take into consideration as you prepare to set the anchorages.
2. How many anchor points will you need, and how strong does each need to be?
3. Should your anchor systems be fixed and focused or self-adjusting? Why?

 Access Navigate for more practice activities.

Introduction

All rope rescue systems must incorporate a secure well-constructed anchorage as a foundational element. The best anchorages are well positioned to achieve the objective, and both strong and stable enough to support the intended load. Anchoring is to the rope rescue system what a foundation is to a building. Without suitable, secure anchors, the remainder of the rope rescue system (ropes, hardware, and other gear) is in danger of failing, no matter how well established they are. Just as a solid foundation is the primary concern before construction of a building, you must have a suitable, secure anchorage before you rig the rest of the rope rescue system.

Anchor Terminology

The most confusing part of anchor systems is terminology. This is largely because terms have evolved differently in different industries, but as the world becomes smaller and regulatory language overlaps, these have begun to collide. Historically, rope users typically borrowed terminology from recreational climbing and mountaineering usage—in which the big differentiation was between "natural" (as in, a large boulder or tree that would be left intact after use) and "artificial" (as in, a chock or other climbing pro that would be removed at the end of a climb.) As time went on and climbers began to install bolts in rock, and slings were left on protrusions, the lines between natural and artificial anchors began to blur, as these are types of artificial anchors that depend largely on natural features, and are not removed at the end of the climb.

Likewise, similar concepts were evolving in the fall protection world, where the primary considerations revolve around the level of training required for a worker who is expected to use these systems. For regulatory purposes, the Occupational Safety and Health Administration (OSHA) and American National Standards Institute (ANSI) use the terms *permanent anchorage* and *temporary anchorage* to help hone in on this critical distinction: that is, installing and using temporary anchorages takes more training and experience than clipping in to an existing permanent anchorage.

For purposes of this text, we will follow the OSHA and ANSI approach, and will use the following anchorage-related terminology: anchorage (or anchor point), anchor system, and anchor connector. An **anchorage (or anchor point)** is a single, secure connection for an anchor. An anchor point is used either alone or in combination with other anchor points to create an anchor system capable of sustaining the actual or potential load on a rope rescue system. Anchor points take a number of forms, from structural beams to trees. An **anchor system** is one or more anchor points rigged to provide a structurally significant connection point for rope rescue system components. An **anchorage connector** is how rope rescue system components are secured to an anchor. For clarity and precision, rope rescuers should use the full terms, such as anchor point or anchorage connector, instead of referring any type of anchorage system or component as just an "anchor".

The two final common anchorage terms indicate the type of anchor point. **Structural anchorages** already exist in nature or in engineered structures. The most commonly used structural anchorages in the natural environment are trees and rocks, around which anchorage connectors such as webbing or rope can be wrapped or tied. Before using a tree as an anchorage, you should examine it for weakness, such as possible rot or a shallow root system (**FIGURE 11-1**).

CHAPTER 11 Anchorages

FIGURE 11-1 A. An example of a structural anchorage. **B.** An example of installed anchorage.

Even a tree with sound wood may not be a good anchorage if the root system is shallow or thin or if the tree is in wet soil. Boulders weighing tons can be pulled over by stresses applied in the wrong

TIP

An additional note about installed anchorages. In the workplace, there is a great deal of variation in industrial fall protection. Rope rescuers should avoid making assumptions especially about installed anchorages, as they may not always be what they appear to be. Most fall arrest anchor systems in industrial settings are rated to 5000 pounds (2267 kilograms), or to a strength that is at least twice the maximum anticipated load during a fall event. However, some workers using rope systems for access may operate with anchor system strengths as low as 2700 pounds (1225 kilograms). Restraint anchor systems, which may look a lot like full-strength anchor systems but are intended only to prevent a worker from reaching an edge, may be rated to only 1000 pounds (453 kilograms).

direction. In industrial and urban environments, structural anchorages may include building elements and supports.

Installed anchorages are anchor points that have to be created. In industrial and urban environments, installed anchorages may consist of davit arms and D-rings.

Principles of Anchoring

Anchorage Strength

Anchorages must be able to sustain the maximum anticipated load that is likely to be imparted on them, plus an appropriate safety factor. Anchor points deemed to be of a strength that far exceeds any force the rope rescue system could possibly deliver to them are said to be **bombproof**. Remember, there is no magic number that is necessary for the rescuer to achieve in anchoring. This is a judgment call that only a competent person should make, but should be of a strength that is sufficient to withstand the maximum anticipated force times a relevant safety factor. See Chapter 10, *Principles of Rigging* for a discussion on safety factors.

The ability of an anchorage to withstand the forces that may potentially be imparted upon it depends on a number of factors, including the following:

- The condition of the anchorage. A live tree, for example, usually can withstand greater forces than a dead one.
- The structural nature of the anchorage. A load-bearing structural column in a building generally can withstand greater forces than a handrail.
- The location of force on the anchorage. A tree with the force pulling on it near the ground generally can withstand greater force than a tree on which the stress is located higher up.

Direction of Pull

Anchorage strength may be greatly influenced by the **direction of pull**. Direction of pull refers simply to the specific direction in which a vector is applied. An anchor system that is strong in one direction may not be so strong in another. For this reason, try to set anchor points that are in line with the direction of pull. Also consider the effects if the direction of pull changes—for example, as a result of variance in the fall line, or due to the difference between lowering and raising. With some anchor point placements, if the direction of pull changes, the anchor system could weaken or fail.

Positioning of Anchorage Systems

An anchor system that is in close proximity to and directly above the subject to be rescued will provide the most stability and will best facilitate edge transitions and loading of the subject. However, in some circumstances it may be preferable to have the anchor system off to the side, such as the following situations:

- Rocks or other dangerous objects may fall on the rescue subject or rescuers.
- Conditions exist between the anchor point and the rescue subject that could endanger rescuers or damage rope or other equipment.
- A flashover and fire erupt from a window.
- A hostile or deranged person is in the area.
- No suitable anchorages are available directly above.

In some rescue systems, it may be preferable to have additional systems pre-set off to the side to provide adequate working space and to avoid entanglement, such as in the case of a separate haul system or a nonloaded belay line. In any case, care should be taken to ensure that the angle between systems is not likely to create additional hazard by dragging debris over the edge, or by creating unnecessary swing-fall (or pendulum) potential. However, anchoring techniques such as directional deviations or rebelays, discussed later in this chapter, may be worthy of consideration in situations such as these.

Redundancy

Most rescue system anchorages are made of two subsystems: a **working system** for moving the load, and a **backup system** (or "belay") to protect against potential failure. Each of these systems should be anchored independently from the other, although it is not uncommon to make a connection from the working system anchor to the backup anchor system, and vice-versa, just for added security.

Anchor systems present a number of opportunities for failure, including the following:

- Uncertain strength of anchor points. Anchor point failure is one of the most common causes of anchor system failure.
- Failures in human judgment and experience. A poor anchor may be chosen, knots may be tied incorrectly, or carabiner gates may be left unlocked.
- Equipment failures. Abrasion and cutting of rope and webbing and stressing of both hardware and software can occur.

Because of the potential for anchor system failure and because the rest of the rope rescue system depends on the anchor system, it is good practice to back them up. Backing up is the implementation of redundant anchorages for safety. There are two primary methods of backing up an anchor system.

1. Backing up to the same anchorage (**FIGURE 11-2A**). Do this only if you are absolutely certain the one anchorage can sustain any forces to which it may be subjected by the rope rescue system (i.e., it is bombproof). This type of backup is used when the possibility of failure exists in other portions of the anchor system (e.g., carabiners, knots, and slings).
2. Backing up to a separate anchorage (**FIGURE 11-2B**). This requires a system that incorporates multiple anchorages. If the direction of loading will be shifting from side to side, you should make the multiple anchorages load sharing. Load-sharing systems are explained later in this chapter.

The specific method of backing up the anchor system and the number of anchor points you need depend on a number of variables, including the following:

1. The condition of the anchor points. If the potential exists for failure of one of them or of the equipment attached to it, more than one anchor point is needed.
2. The nature of the rope rescue operation. If both a main line and a belay line are used, for example, you should use separate anchor systems for each. If the main line and the belay line originate from the same anchorage, the danger exists of line tangles or heat-fusion damage to the rope if the ropes cross. The belay system would also have to be substantial enough to catch a fall, which means you would need a substantial second anchorage. See Chapter 9, *Ropes, Knots, Bends, and Hitches* for information on avoiding heat-fusion damage from ropes crossing.
3. The loads and stresses involved in the system. These vary in intensity depending on the ways in which the anchorages are used, such as for the following:
   - Supporting only equipment
   - Supporting the weight of only one person
   - Creating directionals that generate forces greater than the load (load amplifiers) (**FIGURE 11-3**)
   - Potential shock loads resulting from a fall or a partial system failure

CHAPTER 11 Anchorages 201

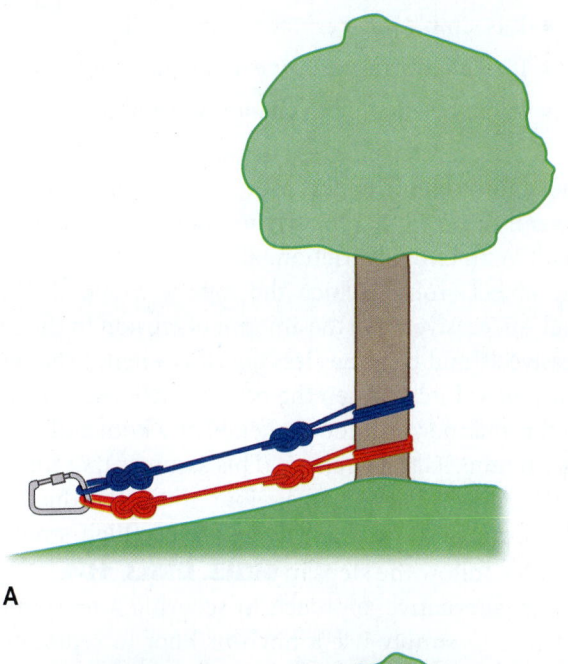

A

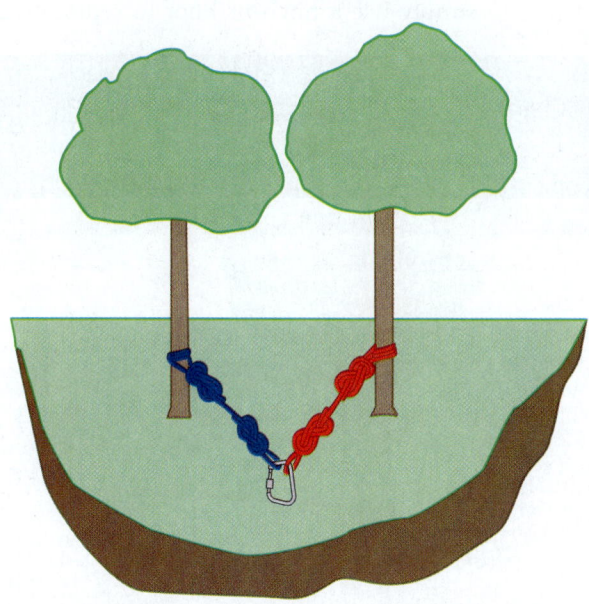

B

FIGURE 11-2 Backing up anchorages.

- Rescue lowering operations
- Hauling systems
- Highlines (a system of using a rope suspended between two points to move people or equipment)

TIP

Rescuers should monitor anchor systems continuously during an operation. The shifting of loads and other actions can cause anchorages to become less secure than when they were first rigged.

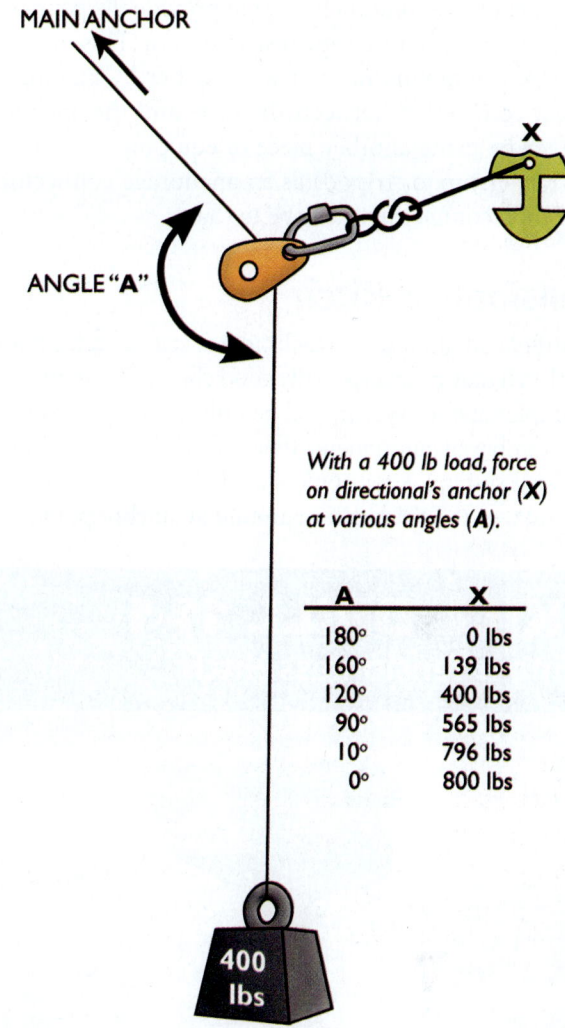

Force on a direction pulley's anchor changes with the angle. Maximum load amplification is two times the applied load.

With a 400 lb load, force on directional's anchor (**X**) at various angles (**A**).

A	X
180°	0 lbs
160°	139 lbs
120°	400 lbs
90°	565 lbs
10°	796 lbs
0°	800 lbs

FIGURE 11-3 Forces on a directional anchorage.

Connecting to an Anchorage

There is a difference between an anchorage and an anchorage connector. While, for example, a stairwell beam may be considered a good potential anchorage, it only becomes viable if you can attach to it effectively. Connecting to an anchorage may be done with a strap, a carabiner, or a special device such as portable beam clamps and removable bolts. In this text, we will refer to artificial devices specifically designed for creating temporary anchor points or protection in places where no natural anchors are available as anchorage connectors.

Many artificial anchorage connectors are inserted into rocks or spaces in rock. These types of hardware include bolts, nuts, chocks, and cams. If such devices are to serve as safe, effective anchor points, they must be placed by a person skilled and practiced in their use.

Single Point Anchorages

A single point anchorage is a type of anchorage system that utilizes one anchorage to provide the primary support for the entire rope rescue system. This may be achieved in any number of ways, either by making a direct connection between the rope and the anchorage, or by using another piece of equipment (such as a strap, clamp, or tripod) as an anchorage connector, and then connecting the rope to that.

Tensionless Hitch

In urgent situations, in which time is critical, a tensionless hitch may be an especially good choice. It also means a simpler anchor system will be utilized, which reduces the number of components that may fail in a more complicated system. A tensionless hitch offers the following advantages as a termination around an anchor point:

- It is simple.
- It reduces stress on rope and equipment.
- It can be adapted to changing conditions.

To create a tensionless anchor system, simply wrap the standing end of the rope around the anchorage several times so that the friction of the wrap takes the load. With enough friction, all the force is applied to the object around which the rope is wrapped. With each successive wrap the amount of tension in the successive strand becomes less and less—hence the term *tensionless* hitch. When the rope is anchored correctly in this manner, it is not weakened by a knot and therefore retains its full strength. This assumes that the size of the object is more than four times the diameter of the rope. To tie a tensionless hitch (high-strength tie-off), follow the steps in **SKILL DRILL 11-1**.

An alternative approach to securing a tensionless hitch is to simply use a hitch or knot to replace the

SKILL DRILL 11-1
Tying a Tensionless Hitch

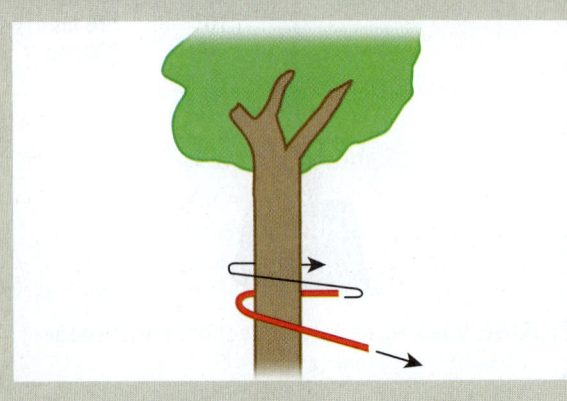

1 Take at least three wraps with the rope around the object to be used as the anchor point.

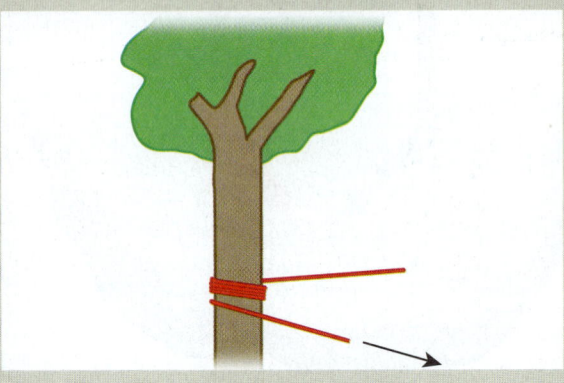

2 Make sure there is no rope cross in the turns.

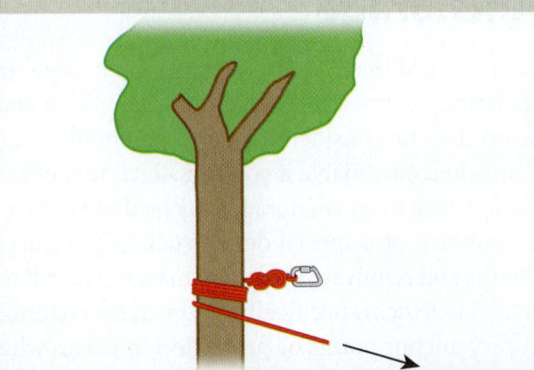

3 Tie a figure 8 on a bight in the running end of the rope and clip a locking carabiner into the figure 8 knot.

4 Clip the carabiner across the standing end of the rope at the bottom of the spiral.

figure 8 knot and carabiner. Tie off the end in a way that prevents the rope from unwinding around the object and without placing any bends in the loaded section of the rope.

Knots

Another type of direct attachment occurs when a line is simply wrapped around an anchorage and then terminated with a knot. This is often a good choice for creating an anchor point where there is low likelihood of needing to move or adjust the anchor system quickly.

If you are using a vertical member as an anchor point that is short enough for you to get a loop of rope over it easily, you can tie a figure 8 on a bight in the end of the rope and place the loop over the anchor. Otherwise, a loop of rope may be tied around the structure with a knot such as a figure 8 retrace or a bowline.

Do not forget to factor in the strength of the knot when analyzing the strength of the anchor system and take care to keep the inside angles of the loop narrow enough to maintain the shape and structure of the knot and to ensure good management of forces.

A disadvantage of rigging the mainline rope directly to the anchorage is that it limits the ability to modify the anchor system, and modifications could become necessary because of the changing conditions that can occur in rescue situations. A possible solution is to use a separate piece of rope or webbing for the anchor system that is as strong as or stronger than the main line and to attach the main line to it.

Soft Slings as Anchorage Connectors

Webbing is a convenient material for connecting to an anchorage. It also is convenient for making continuous loops known as runners or slings. The advantages of using manufactured webbing loops instead of ropes for anchor points are that it is less expensive and there are fewer knots to learn. However, webbing cannot be tied into as many different knots as rope can and it does not absorb shock loading as well as most ropes.

You can create a sling from a piece of webbing by tying it into a loop using a ring bend knot (also known as a water knot, overhand bend, or tape knot).

Presewn Slings

Properly sewn slings are an easy alternative to tied slings. They may be used as a quick, convenient means

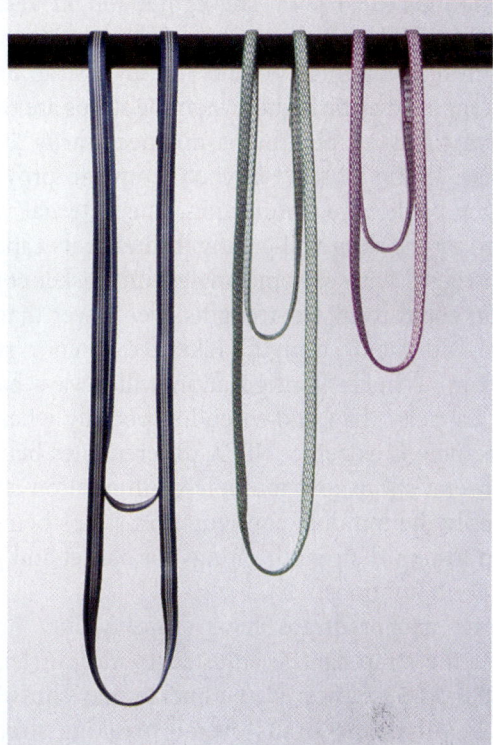

FIGURE 11-4 Presewn slings are available in various widths and lengths.

of connecting to an anchorage. Two advantages of presewn slings are that they are quicker to use and there is less chance of tying the wrong knot when making a continuous loop. Several brands of presewn slings that can be used for anchoring are available.

Presewn slings are available in a variety of webbing widths and lengths (**FIGURE 11-4**). When buying presewn slings, make sure they have adequate tensile strength for the safety factors you are likely to need. Also, make sure you know how the manufacturer of the presewn slings tests them for strength. Anchor slings can have strength ratings based on being pulled end to end, pulled when rigged using a basket technique (wrapped around the anchorage), or pulled when girth hitched around an object. Before and after you use presewn webbing, always inspect it for wear on the stitching and the web material.

Anchor Straps

Anchor straps are webbing lengths with hard rings sewn into each end where a carabiner can be clipped. These straps can be a quick way of setting reliable anchor points. Anchor straps are available with different strength ratings, as outlined in NFPA 2500 (1983). The heavy-duty ones (with the NFPA G designation) typically have an end-to-end breaking strength of about

8000 pounds (35.6 kN). The lighter-weight versions (NFPA rated T) typically have an end-to-end breaking strength of about 4945 pounds (22 kN). Strap breaking strengths may be higher when the straps are rigged to form a basket, but this is not necessarily always the case. While a basket-hitched strap can provide a stronger anchorage connection, the internal angle formed by the sling will greatly influence its capacity. When rigged with extreme angles, the basket configuration could result in strengths even lower than the end-to-end rated strength. Likewise, anchor straps rigged in a choker configuration will always have a lower capacity than end-to-end, especially when the choke angle is extreme. NFPA differentiates between end-to-end straps and multiple-configuration straps, with only the multiple-configuration straps being required to report strength ratings for basket and girth hitch configurations.

Some anchor straps have a heavy-duty buckle so that the strap can be adjusted to various lengths (**FIGURE 11-5**). However, the buckle may slip with a force less than the strap's overall breaking strength. In addition to anchoring, some of these straps can be used for such purposes as litter bridles or as a quick anchorage connector for an edge attendant. As with other presewn slings, make sure the anchor straps you use have adequate tensile strength for your system safety factor. In addition, make sure the manufacturer can provide you with specifications on how much loading causes the adjustable buckle to slip.

The problem with a simple loop of webbing around a vertical anchorage is that it tends to slip down on the anchorage. There are several techniques for holding webbing in place on a vertical anchor point. One is to tie the webbing around the anchor point in a girth hitch. The drawback to this technique is the temptation to cinch the webbing back on itself. This should not be done because it puts potentially dangerous stress on the webbing.

A stronger alternative to the girth hitch is the wrap 3, pull 2 (W3P2) system (**FIGURE 11-6**). In this system, there are three wraps around the anchorage with two loops pulled out. One advantage of the W3P2 over a girth hitch is that it tends to stay in place better once snugged up and does not slip down the vertical anchorage like a girth hitch often does. Note that for maximum strength, the webbing knot is on the interior loop, where it will incur the least stress. Again, however, specific rigging methods—including internal angle of the anchorage connector material as it protrudes from the anchorage, and how layers of the anchorage connector material interact with one another—will have a strong influence on overall performance. Chapter 10, *Principles of Rigging* covers these concepts in further detail.

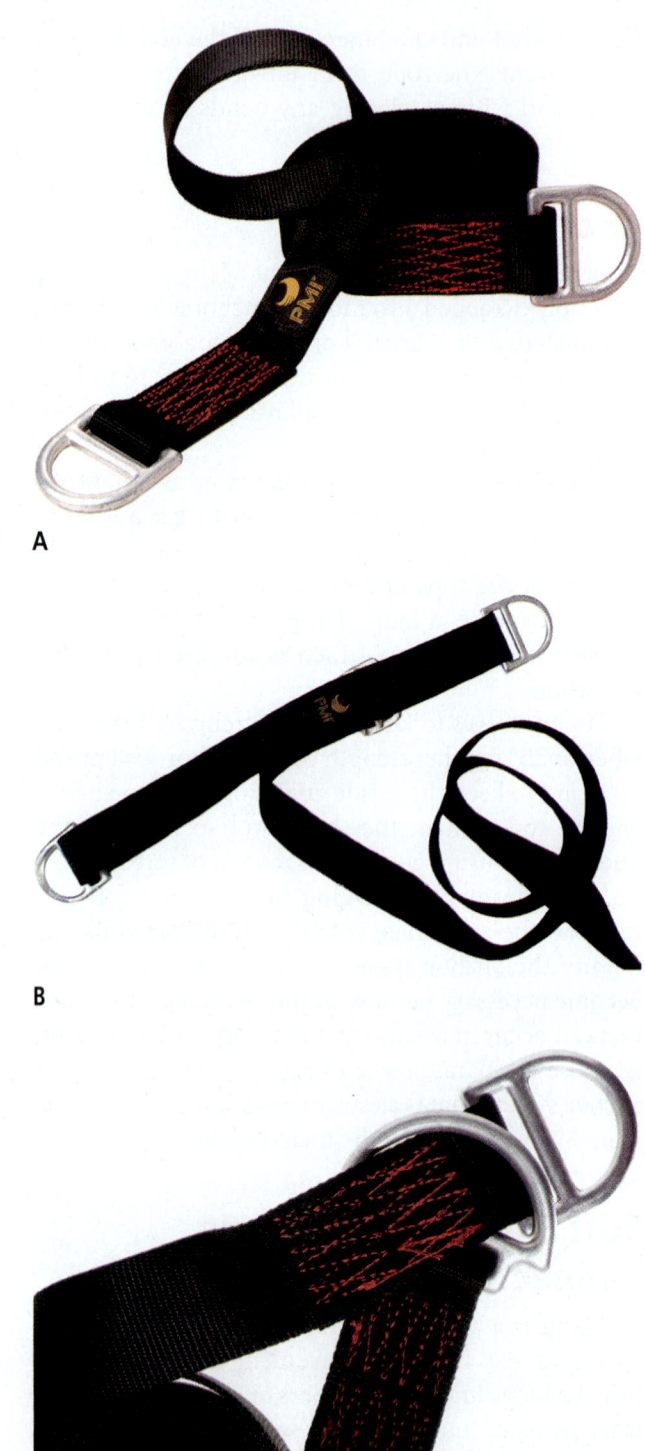

FIGURE 11-5 Anchor straps with heavy-duty buckles allow adjusting to various lengths.
A–C: Courtesy of Pigeon Mountain Industries.

A secure method of placing webbing around an anchorage is to tie it in a loop around the structure using a ring bend knot (**FIGURE 11-7**). However, if the object used as the anchorage is very large, this procedure can be awkward and time-consuming for one person. An alternative method is first to tie

CHAPTER 11 Anchorages

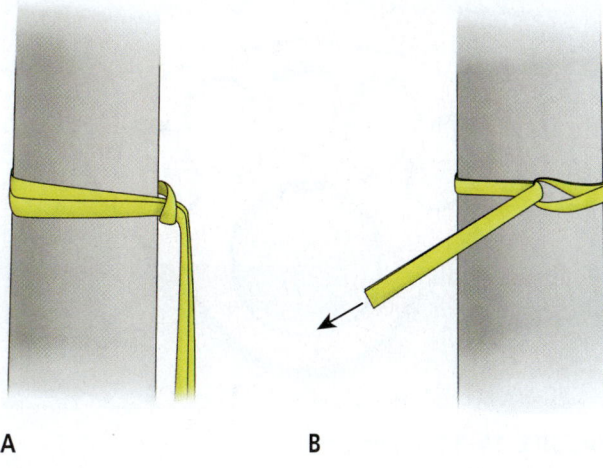

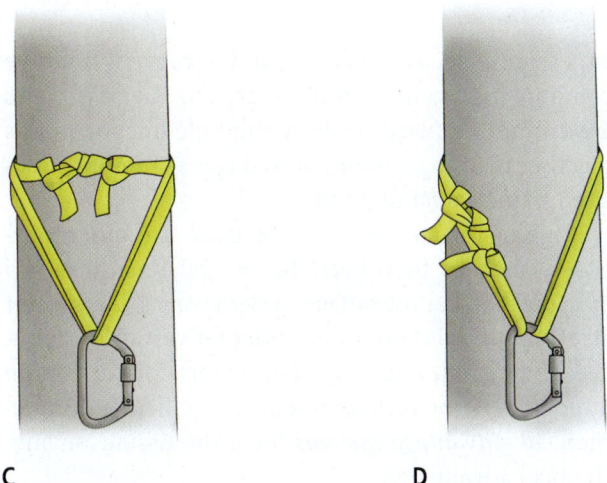

FIGURE 11-6 A. and **B.** Girth hitch. **C.** Wrap 3 pull 2 is preferable to the girth hitch. **D.** Basket hitch.

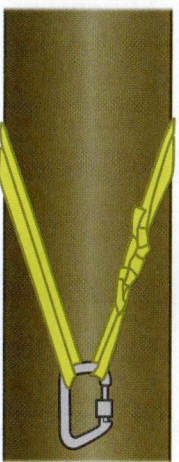

FIGURE 11-8 Wrapping webbing around an anchorage to create an anchor point.

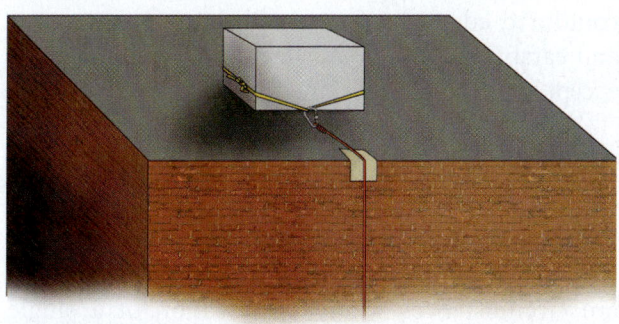

FIGURE 11-7 Tying webbing around an anchorage.

the runner into a loop, then wrap it around the anchorage, and then clip the two ends together with a carabiner (**FIGURE 11-8**). Such tied runners, however, require double the length of an untied length of webbing.

TIP

If you use tied webbing for slings, do not leave knots tied in the webbing unless each runner is carefully inspected before the webbing is returned to the equipment cache and then inspected again before the next use. You should be aware of the following problems:

- You may not know who tied the knot and what kind of knot it is.
- Knots in webbing often work their way out.
- The knots may have been spot welded in webbing that has been shock loaded.

Portable Anchorages

Portable anchorages are prefabricated anchorages that can be moved from place to place. They may serve as an actual anchor point, or as a directional aid to help route a rope for more optimum rigging. A common example of a portable anchorage is the tripod. A tripod may be used to create a standalone anchor system over a space such as a maintenance hole cover, or it may be used as an artificial high directional to assist with load management. Monopods, bi-pods, and quad-pods are variations on the tripod; these, too, are often used in rope rescue systems.

Another example of a portable anchorage is the beam clamp (**FIGURE 11-9**). Beam clamps can be quickly attached to overhead structures such as steel beams. Some meet the anchor system requirements for one-person loads set by the ANSI, the Canadian Standards Association (CSA), and the U.S. Occupational Safety and Health Administration (OSHA):

FIGURE 11-9 Beam clamp.
Courtesy of Pigeon Mountain Industries.

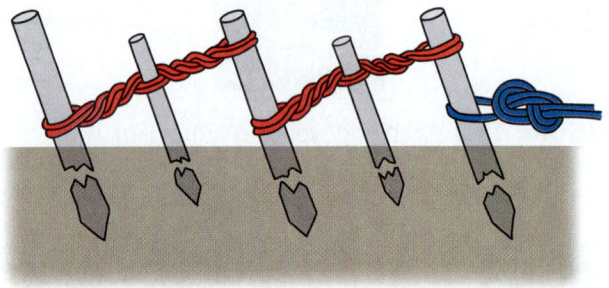

FIGURE 11-10 A picket system.

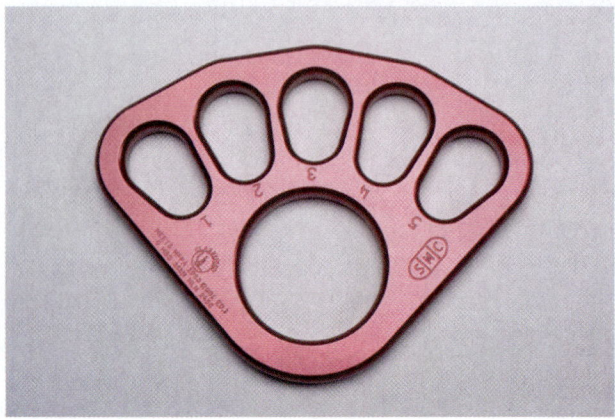

FIGURE 11-11 Rigging plate.
Courtesy of Pigeon Mountain Industries.

5400 lbf (24 kN). For large or heavy loads, two or more beam clamps may be used together as multi-point anchorages as needed to ensure appropriate system safety factors.

Removable bolts are another type of portable anchorage that is useful in certain situations. For these, holes of adequate depth and diameter must be pre-drilled. When placing removable bolts, take care to ensure that the version you are using is the correct size and strength for the application in which you are using it, and that the interface is sufficient to ensure adequate holding power.

A picket system is an alternative in a natural area where there is a shortage of natural anchorages (**FIGURE 11-10**). Different types of pickets are available for soil versus snow. Although a picket system can work very well when correctly rigged, establishing it properly requires proper training, and setting a system up can take a great deal of time for an inexperienced user. In addition, not all soil or snow types can hold pickets securely. Loose, sandy, or muddy soil, or snow, may not hold well regardless of the number of pickets used. Most picket anchorage systems consist of several rows of pickets.

Rigging Plates

Rigging plates may be used as a part of an anchor system to help organize rigging (**FIGURE 11-11**). They can help spread out the various components so that they are easier to see and manage, and to prevent multiple components from jamming together. Rigging plates are commonly used to draw multiple anchor points together to a single point, or to disperse multiple lines from a single anchor point.

Rigging plates also can be used for other purposes, such as to collect the several lines involved in a mechanical advantage system, or as the **master attachment point** in a litter spider for vertical systems. See Chapter 14, *Lowering Systems* for a discussion on litter spiders for vertical rescue, and Chapter 15, *Mechanical Advantage Systems* for a discussion on mechanical advantage.)

Because the very purpose of rigging plates is to collect multiple points, special care must be taken when using these to ensure that the anchorage and anchor system are not being overloaded. Look for rigging plates that are strong (NFPA rated G) and that have contoured edges that are less likely to damage rope and carabiners. The holes should be large enough to accept large locking carabiners easily.

Multi-Point Anchor Systems

Up to this point, the discussion has focused on single anchor points and anchorages. However, there are circumstances where the location of a single anchor point is not suitable for the task, where the strength of a single anchor point may not be sufficient to provide desired levels of safety, or where the anchorage is not in line with the intended direction of travel.

Both of these situations may be remedied by collecting multiple anchor points together to form a single connection point. This type of anchor system is commonly known as a **multi-point anchor system**. When the load is applied at the point of collection, the force is distributed and shared by the all the anchor points. How these forces are divided among the anchorages will depend on the method of connection and the relative angle between the various points.

While the goal is to build a robust multi-point anchor system, anticipating—and mitigating—the probability and consequences of a potential failure is a necessary part of the rigging.

Multi-point anchor systems work best when anchorages are positioned somewhat adjacent to one another and not too far apart. Properly positioned anchorages lend to well-placed anchor points, which help to reduce probability of failure. When arranged such that the load-connection point is **fixed and focused**, so that it does not move under load, the force on individual anchor points may vary as the direction of pull changes. This type of anchor system is sometimes called a **load-sharing anchor system**. When using a fixed and focused multi-point anchor system, care must be taken to keep the forces distributed as evenly as possible across the different anchor points. Alternatively, the legs of a multi-point anchor system can be rigged to float, so that the load-connection point is self-adjusting as the direction of pull changes.

Regardless of which type of multi-point system is used, it is determined to be one anchorage. Multiple anchor points rigged into one fixed and focused anchorage system equals one anchorage, just as multiple anchor points rigged into one self-adjusting anchorage system equals one anchorage.

The simplest way to create a load-sharing anchor system is to use two anchor ropes or slings, and run one from each of the two anchor points, clipping them together into a single point using one or two large locking carabiners (**FIGURE 11-12**). The place at which you clip the two lines together with the carabiners is known as a **focal point**. A similar effect may be created by placing a single sling between the two anchor points and tying a knot between the two where you wish to fix the focal point (**FIGURE 11-13**). Using this method, the location of the knotted connecting point can be adjusted to accommodate a variety of fall-lines.

To utilize fixed and focused multi-point anchor systems, follow the steps in **SKILL DRILL 11-2**.

The best application of a fixed and focused multi-point anchor system is where the direction from which the load is applied does not change significantly

FIGURE 11-12 Two slings back up one another on a single anchorage.

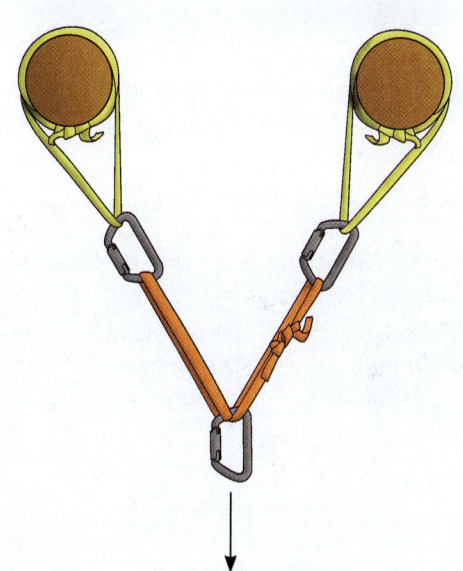

FIGURE 11-13 A single sling between two anchor points.

TIP

When rigging a multi-point anchor system, care must be taken to avoid creating too wide an angle between the anchor points. Ideally, this angle should not exceed 90 degrees, because beyond this angle the forces on each anchor point begin to multiply drastically. Beyond 120 degrees, the forces on each anchorage would actually begin to exceed that of the supported load. **FIGURE 11-14** illustrates how angles affect the forces on anchor points and other elements of the system.

SKILL DRILL 11-2
Creating a Fixed and Focused Multi-Point Anchor System

1 Identify two anchor points relatively adjacent to one another. Clip a separate sling into each of the two anchor points.

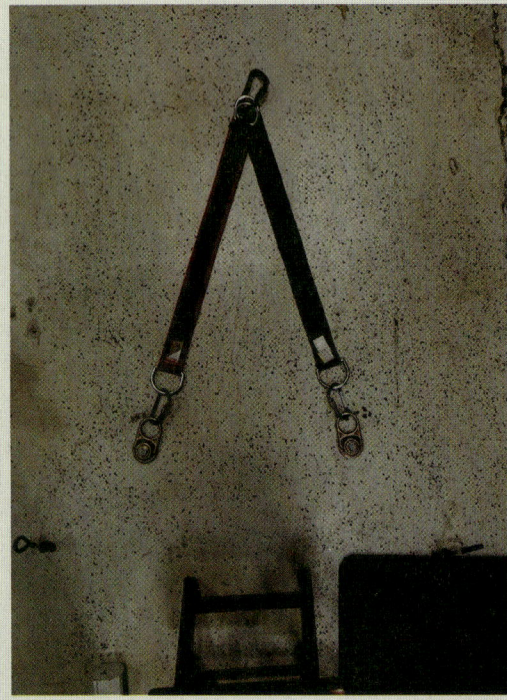

2 Bring the loose ends of the slings together and clip with a connector, ensuring that the angles made by the slings do not exceed 120 degrees.

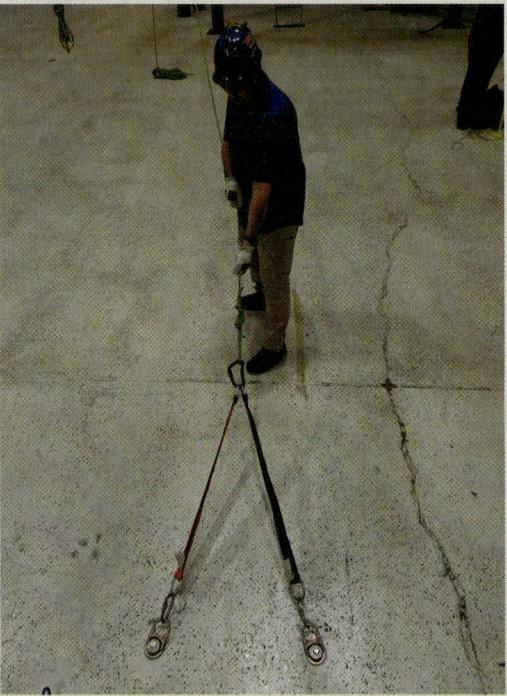

3 Test the system to ensure that the focal point is in line with the anticipated direction of pull so that the force is shared equally between the two anchor points.

CHAPTER 11 Anchorages 209

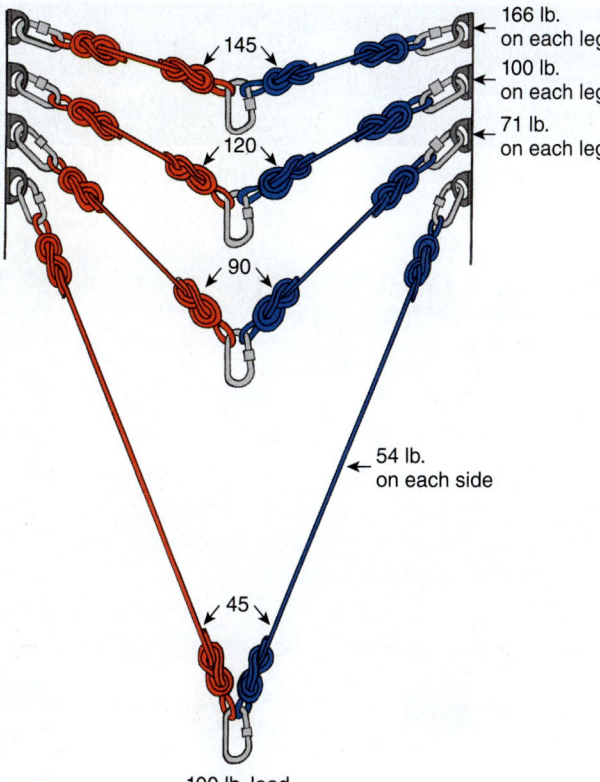

FIGURE 11-14 As the angle between legs of an anchor increases, so does the amount of force applied to each of the anchor points.

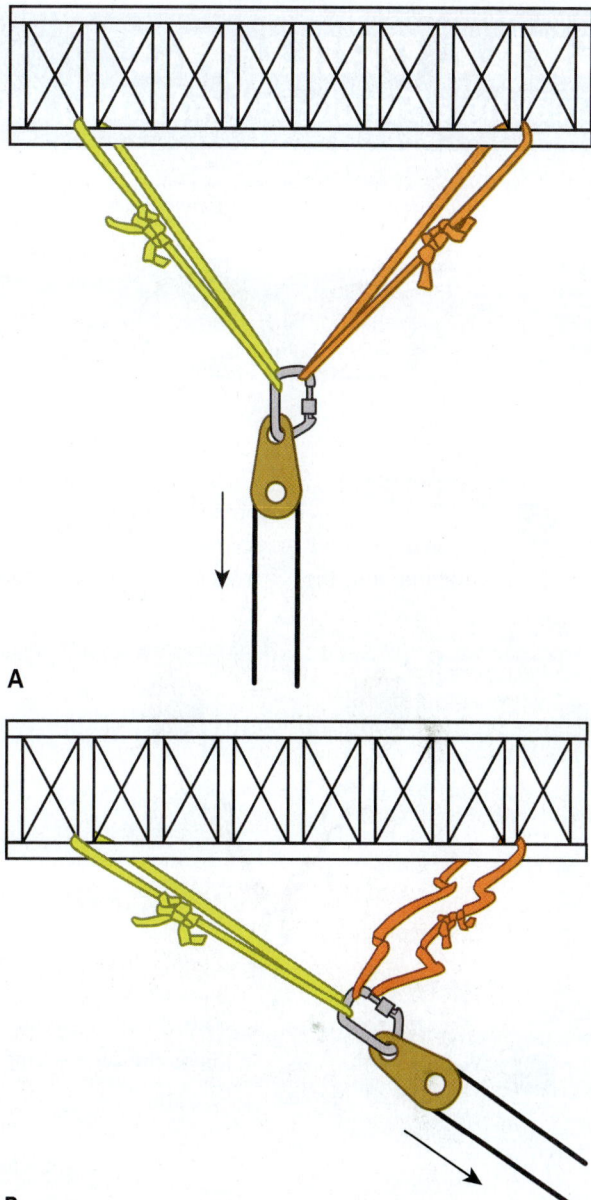

FIGURE 11-15 Forces on fixed and focused anchor systems. **A.** Stress on each anchor point is equal only when force pulls in one direction. **B.** Side-to-side forces increase stress on the fixed and focused anchor system.

during use. When force from the load pulls directly on the center of the angle, the stress is applied relatively equally to the different anchor points (**FIGURE 11-15**).

In a fixed and focused system where the direction of pull changes during the course of the operation, the entire rescue load could be supported by just one of the anchor points in the system. Presuming that the multi-point anchor system was created for a reason, this would seem to be an undesirable outcome and should be avoided.

One way to achieve more equitable distribution of the load in a multi-point anchor system is to arrange it as a **self-adjusting anchor** (or **load distributing anchor system**). This type of multi-point system involves two anchor points and uses a sling to form the letter X and a carabiner to lock it in place. Clipping the carabiner across the X in the line helps to ensure that should one anchor point fail, the webbing will reset itself to pull on the other anchor point. Even with this built-in redundancy, if there is any significant distance between the two anchor points, the failure of one could impose a significant shock load on the other. Self-adjusting anchor systems should only be considered when the probability and/or the consequence of associated risks can be mitigated. To create a simple self-adjusting anchor system, follow the steps in **SKILL DRILL 11-3**.

TIP
While redundancy can be a good thing, when redundancy begins to diminish efficiency, it can actually compromise safety.

There are a couple of ways to mitigate potential shock load. One is by creating a protected self-adjusting anchor system with built-in redundancy. Redundancy simply describes the concept of incorporating parts that protect against failure of other components. A rescue system is truly redundant when the failure

SKILL DRILL 11-3
Creating a Simple Self-Adjusting Anchor System

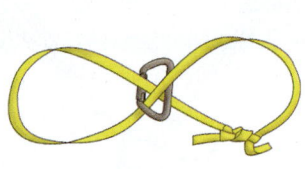

1 Configure a loop of webbing or rope in the shape of an 8 and clip a large locking carabiner into the X across the inside loop.

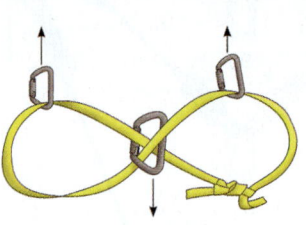

2 Take each end of an outside loop and clip it into an anchorage. Clip the carabiner on the inside loop into the main line.

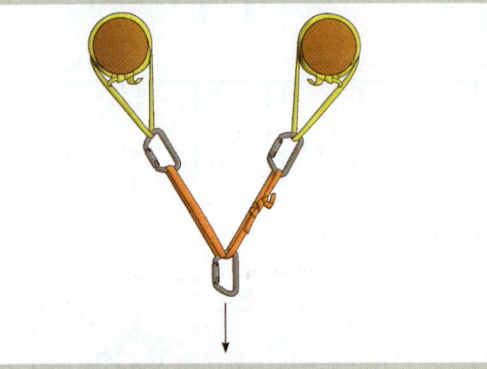

3 Make sure the angles made by the sling do not exceed 120 degrees.

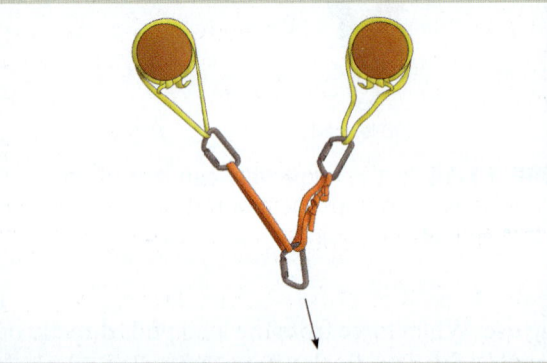

4 Test the system by hand. Simulate failure of each one of the anchor points to make sure the system catches the load. Whatever the direction of the pull, the central carabiner should slip along the sling to equalize the forces.

of any one point will not result in the failure of the entire system.

Another way to mitigate the potential shock load is to reduce potential fall distance by extending the anchor points and reducing the length of the adjusting component in the multi-point anchor system as shown in **SKILL DRILL 11-4**.

1 Create a large loop and tie the two ends together using a figure 8 bend.
2 Place the knot at about 3 o'clock or 9 o'clock on the circle so that it remains out of the way, and take a large bight of rope and flip it back inside the circle about two-thirds of the way up.
3 Gather these four strands of rope together and tie a figure 8 knot with them.
4 At the top of the circle, there is now a large loop with a smaller loop inside. At the bottom of the circle is a much smaller loop created by tying the figure 8 from the four strands.
5 Take the larger loop at the top of the circle and clip it into all the anchor points using a locking carabiner at each point. Take the smaller loop at the bottom of the large loop and clip it to the larger loop between each anchor point. Take the small loop below the knot created from the four strands and clip it into the main line.

To simply reduce the length of the adjusting component in a multi-point anchor system, simply use a sling to extend each anchor point to a smaller self-adjusting focal point. In this case, a small sling made of slick material such as HMPE provides a smooth self-adjusting function. Using a cordelette or nylon sling to extend the anchor points will also help to mitigate the low elongation of HMPE. This type of system, when correctly constructed and when conditions are right, can have some important advantages:

1. The forces on anchor points should remain distributed and shared by all anchor points (albeit not equally), whatever the direction of pull.
2. If any anchor point fails, the system should readjust to help redistribute loading on the remaining anchor point or points—albeit not equally, but without imparting excessive impact forces.

Self-adjusting anchor systems are sometimes called self-equalizing anchor systems. However, it is important to note that no anchor system can be made completely "self-equalizing" because of the following:

- With the angles involved in rigging any self-adjusting anchor system, the elements of the system are subject to varying forces, and the system therefore can never be completely equalized.
- In a shock-loading situation, redistribution of forces does not occur instantaneously. During this transition to redistribution, some elements of the self-adjusting anchor system receive greater loads than others.

Self-adjusting anchor systems have fallen out of vogue due to hazards associated with misuse, but they should not be completely discounted as they are a useful option when implemented by appropriately trained and skilled rope rescuers. The following guidelines can help ensure the best distribution of forces and adaptation to shock loading with failure of an anchor point:

- Keep the angles small, both to reduce magnification of forces on anchors and to help the system readjust to the new loading.
- Design the systems so that as little drop as possible would occur should any anchor point fail. One way to do this is to keep anchor point legs as short as possible.
- Choose less bulky rope or webbing for the self-adjusting mechanism, and adjust the system so that knots are less likely to run through carabiners when the system readjusts.
- Use Kevlar or Spectra in an anchor system only if you include shock-absorbing materials somewhere in the overall rescue system to help reduce impact force in the event of uncontrolled loading; rope and webbing made of materials such as Kevlar and Spectra do not have the shock-absorbing qualities of materials such as nylon.
- Make all the anchor points in a self-adjusting system as bombproof as possible, given the constraints of time and efficiency.

Multi-point anchor systems, whether fixed or self-adjusting, are not for casual use in rescue. While the probability of overloading a single point may be higher in a fixed and focused multi-point anchor system, the consequences of doing so may be higher in a self-adjusting anchor system. Construction and use of these systems requires a knowledge of how the orientation of the anchorages and their method of connection affect the ability to support the intended load. Failure to recognize how these forces are distributed in a multi-point anchor system can lead to overloading of the system or one of its elements, possibly resulting in catastrophic failure. For these reasons they should be used with caution and only when the benefits outweigh the risks.

> **TIP**
>
> Self-adjusting anchor systems have the potential for dangerous shock loading as one anchor point pulls out and others take the load. Reduce potential shock loading by keeping angles small and slack to a minimum. Also:
>
> - Try to keep anchor points close to each other. If this is not possible, it may be better to extend faraway anchor points with static rope to keep the load-sharing anchor system's loop as small as possible.
> - Keep the outside angle to less than 90 degrees. Even better, limit the angle to 60 degrees. This limits the forces on the remaining anchor points if one of the inside anchors fails in a self-adjusting anchor system with three or more points. The outside angle is measured from the two outermost anchor points down to the focal point.
> - Rig self-adjusting anchor systems with a minimum of slack in the system. Keep rope or webbing length to a minimum (approximately 8 feet [2.4 m] or less) in the tied loop of a three-point system. Rig for a maximum 1-foot (0.3-m) drop with the failure of an anchor point.

Directionals

A **directional** is a technique for bringing a rope into a more favorable position or angle. A directional (in some cases, known as a *deviation*) can be created in many ways, and each method must be judged on its advantages and disadvantages specific to a given situation. **FIGURE 11-16** shows two trees, wide apart, that could serve as strong anchors at the top of a cliff. However, say, for example, that you want to use the rope to reach a spot that is between the trees:

1. If you anchor the rope to either of the two trees (as has been done to the tree on the right in **FIGURE 11-16A**), it would be difficult for you to reach the desired location without a significant pendulum.

A

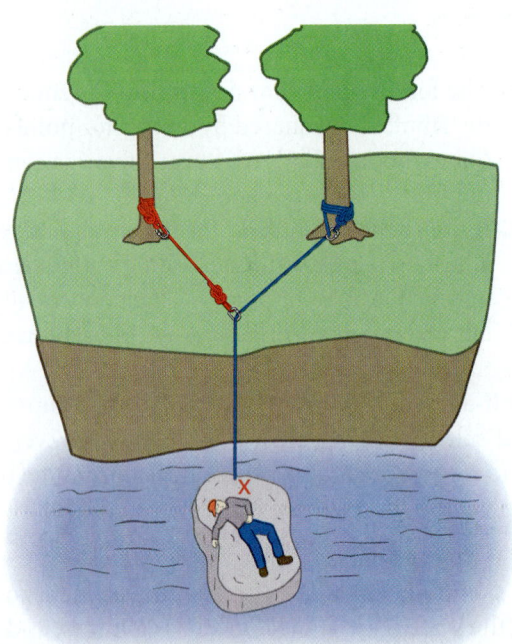

B

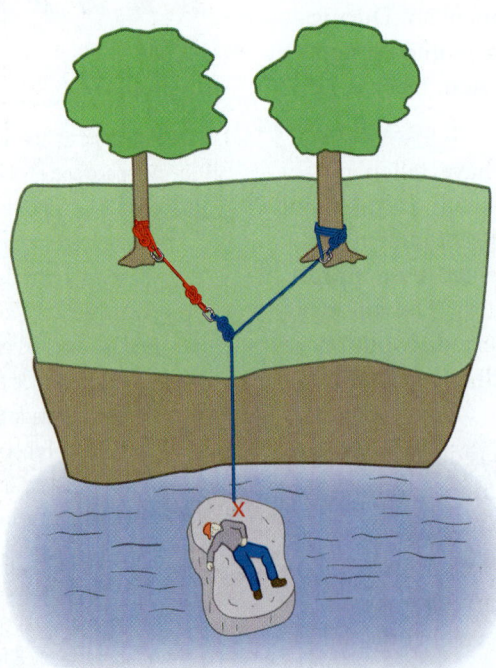

C

FIGURE 11-16 Setting a directional.

2. Now, say you add a secondary anchor, a directional, to the smaller tree on the left (**FIGURE 11-16B**). In the end of this second rope, you tie a figure 8 on a bight knot and clip a locking carabiner into the figure 8 knot. Finally, you clip this locking carabiner across the main anchor rope. Thus, the main rope is now at a better angle for you to reach your target. One additional advantage with this approach is that the main rope runs freely through the carabiner so that as the exact position of the activity moves back and forth, the angle can change slightly.

3. A less-desirable alternative would be to tie a figure 8 overhand knot in the main line and clip it directly into the carabiner on the directional (**FIGURE 11-16C**). This would prevent the main line from sliding freely through the carabiner. It would also create a multi-point anchor, but without the advantages of the example in Figure 11-16B. When the rope moved to one side or the other, it would impose a great deal of loading on one anchor but very little on the other.

TIP

When rigging anchorages, it is important to always be aware of the need for protecting rigging materials, such as rope and webbing, from damage from sharp edges. Loaded rope and webbing can easily fail when damaged by unprotected edges. (See Chapter 10, *Principles of Rigging* for additional information on edge protection.)

Location of Directionals

When you establish anchorages for primary systems and directionals, you must keep in mind how safe and accessible they are for the rope rescuers who work with them. Keeping anchor points, such as the trees, far enough back from the edge of a drop, offers several advantages:

1. Those rigging the anchor points and the directional would be in less danger of falling.
2. The rigging would be more accessible and therefore more under the control of the rescuers.
3. With the rope running over the edge, not all the weight would be directly on the anchor system. Part of it would be taken by the edge of the drop. (The drawback would be possible abrasion on the rope.)

Back-Tie

A back-tie is a good method of providing additional security to an anchor that is in the right position but not strong enough alone for a rescue load. In **FIGURE 11-17**, the anchor point on the right is near the edge and in a good position, but it does not have the strength to take the load. You may consider this a primary anchor. Behind the primary anchor and directly in line with it is a second anchor. The strength of this secondary anchor, combined with that of the primary anchor, will be enough to sustain the rescue load.

To combine the two anchors, a rope or sling is run between them; this connector is known as a **back-tie**. Two things must be kept in mind when using a secondary anchorage as a backtie:

1. The back-tie anchorage must be as strong as or stronger than the primary anchorage.
2. The back-tie must have no slack that would allow shock loading, which could cause both anchorages to fail.

Pretensioned Back-Ties

Inherent slack is sometimes caused in anchor systems by rope or webbing stretch and by flexing of anchor points. If an untensioned anchor system receives shock loading, elements could fail. One way to reduce the chance of shock loading is to pretension the system with a pretensioned back-tie. A pretensioned back-tie is a method of removing slack in a back-tied anchor system before it is loaded.

One technique for creating a pretensioned back-tie is to connect the back-tie to the anchorage with a simple mechanical advantage system. The following steps are used to construct the system shown in **FIGURE 11-18**:

1. Place wrap 3, pull 2 webbing attachments on each anchorage facing one another.

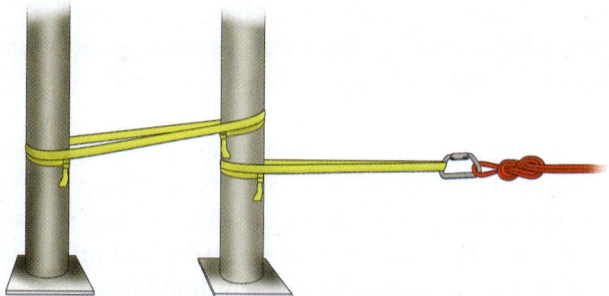

FIGURE 11-17 A back-tie uses a means of connection to combine the two anchorages.

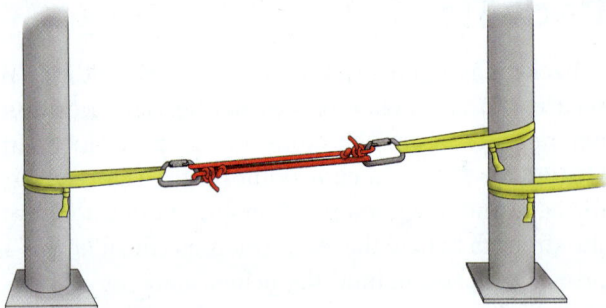

FIGURE 11-18 Pretensioned back-tie.

2. Place a large locking carabiner in each of the web attachments.
3. Use a length of rope to create a simple 3:1 mechanical advantage system using the carabiners.
4. Start by attaching the rope to the carabiner at the rear anchorage.
5. Run the rope through the forward carabiner and back through the rear carabiner, and pull it forward until it almost reaches the front carabiner.
6. Tension the simple hauling system using two people. After the initial tensioning, vector the system by pulling it back and forth sideways to remove additional slack.
7. While maintaining the tension, tie off the system near the front carabiner using half hitches.

Note that the pretensioned back-ties should be close to in line with the direction of pull of the lowering or hauling system or highline. They should never be more than 15 degrees off to either side.

Other methods, such as a trucker's hitch, can also be used to pretension between two points.

Structural Anchorages

As a rope rescuer, you should become adept at finding and analyzing structural anchorages in any and all environments in which you are likely to work, as these are what rescuers use most often. Experience and training are necessary to have when analyzing structural anchorages for rescue, because rescue often occurs in places where preplanned anchorages do not necessarily exist.

The most common structural anchorages in natural environments are boulders and large trees. However, these types of anchorages are less prevalent in urban environments, especially on buildings. When you rig anchor systems on buildings, choose anchorages that are inherently part of the building's structure or specifically constructed to support high loads. Some examples of anchorages in urban environments most likely to be acceptable would be the following:

- Structural columns
- Projections of structural beams
- Supports for large machinery
- Stairwell support beams
- Brickwork with large bulk (e.g., corner walls)
- Anchorages for window-cleaning equipment

Some examples of structures that may be deteriorated and inappropriate for use as an anchorage are as follows:

- Corroded metals
- Weathered stonework
- Deteriorated mortar in brickwork
- Vents constructed of sheet metal
- Gutters and downspouts
- Brickwork without bulk (e.g., small chimneys)
- Fire hydrants/standpipes

Less-Obvious Anchorages

Some manmade structures and buildings at first may appear to have no anchorages. However, after some practice in working this type of problem, riggers may find some unexpected but good anchorages.

Elevator and Machine Housings

Elevator and machine housings often are larger and more substantial than what is expected for an anchorage (FIGURE 11-19). However, by taking a length of rope, running it around the housing several times, and tying the ends together with a figure 8 bend, you may be able to create a secure anchor point for several lines.

Scuppers (Roof Drain Holes)

Many buildings have low parapets with drain holes set in them at roof level. In a pinch, you can create an anchor point by running rope or webbing through the drain hole and back over the top of the parapet.

If possible, you should use the scupper on the side of the building opposite the one over which you will run the main line. This allows space on top of the building for rappellers to rig into the rope and to set other rigging, such as lowering and raising systems.

The more substantial parapets are those constructed of reinforced concrete. If the parapet is

FIGURE 11-19 Anchor around elevator or machine housing.

constructed of brick or block, riggers should strongly consider other options. If this truly is deemed the best option, at least make sure that several brick or block courses are involved and that the mortar is in good condition. Even under the best of conditions for a brick or block parapet, it would be wise to rig with at least two anchor points. Be sure to pad all sharp edges.

Wall Sections Between Windows and/or Doors

If windows and/or doors are close enough together, you can use them to create a substantial anchor point. Pass the anchor rope or webbing through an open window or door, around the intervening wall, and back through an adjacent window or door to tie off the rope or webbing. The anchorage wall should be on the side opposite where the main line will run out of the building. This provides more safe space for rappellers to rig into the rope or to set rigging, such as lowering and raising systems.

Stairwell Beams

If you anchor to stairwell beams, make sure they are the structural members; these are the open steel beams to which the stair risers are secured.

Installed Anchorages

Installed anchorages are placed expressly for the purpose of forming an anchor for a safety or rescue system, either permanently or temporarily. Placing bolts takes time, and a great deal of practice and training are required to learn to set them correctly. They also do permanent damage to the material or location in which they are installed, which may be frowned upon in managed areas. Preplanned permanent anchorages such as bolts (or pre-drilled holes for removable bolts) should be installed in locations where work at height or rescue is frequent. Many newer tall buildings are likely have permanently installed anchor points on the roof for window-cleaning equipment. Documentation will note if the installed anchorage was professionally installed and certified by an engineer. Industrial sites often will have pre-engineered fall protection anchorages over the entrances to confined spaces such as tanks and vaults.

> **TIP**
>
> Do not confuse window-cleaner eyebolts with guy-wire hooks. Window-cleaner eyebolts are substantial (usually ¾-inch [19.1-mm]), closed eyebolts in structural concrete. Guy-wire hooks are used to stabilize items such as signs and antennae; they are not designed for life support.
>
> Many window-cleaner eyebolts are designed to be pulled vertically with cantilevered rigging; these may fail if pulled sideways. Always back up window-cleaner eyebolts with other anchorages.

Adapting to a Lack of Anchors

Extending Anchor Points

When anchorages are not found nearby, you may be able to establish them by running lengths of rope, sometimes for a few hundred feet, to suitable objects. This should be done only with static rope. Polyester core static ropes typically have even less stretch than nylon static ropes of similar construction, making them a good choice in this application of extending long anchor ropes. Attempting to extend anchor points in such a manner with dynamic rope could create a dangerous situation because of the large amount of stretch. Even with static rope, undesirable stretch may occur if only one line is used. Depending on the load, the lines can be doubled, tripled, or quadrupled to reduce the stretch.

An example of a situation in which extended anchor points may work would be the roof of a building where (absolutely, positively) no anchorages exist. Often, static rope can be run through a stairwell or the top-floor windows to lower floors where anchorages exist.

Using Vehicles for Anchorages

One sort of portable anchorage that usually is available is an emergency vehicle. As these are not generally specifically designed to serve as anchorages, and performance can vary so widely based on conditions, these should be considered and used only under the direction of a competent person. The following safety guidelines should be observed when using a vehicle for an anchorage:

1. Park the vehicle on a solid surface. High load forces can drag a vehicle across loose material, such as sand or gravel.
2. Set the parking brake.
3. If the vehicle has an automatic transmission, set it in *Park*. If it has a manual transmission, set it in a gear in opposition to the pull (e.g., *Reverse* if the pull is from the front).
4. Chock the wheels. Forces created in a high-angle system can move a vehicle with its brakes set. If no chocks are available, use spare tires.
5. "Error-proof" your portable anchorage by removing the ignition key and perhaps even applying caution tape to the steering wheel or driver-side door.
6. Avoid contact between the rope and hot exhaust, sharp edges, and substances which may cause damage.

Potential anchor points in a vehicle include structural parts, such as axles and cross members. However, be sure to protect rope and webbing from oil and grease. Furthermore, rope and webbing also must be protected from destructive substances such as battery acids, which often are found around vehicles.

Some parts of a vehicle such as tow eyes and bumpers may appear to be both convenient and substantial but may not make suitable anchorages because they are not actually connected to a structural element of the vehicle or may be weakened by damage. Always perform an inspection of the connections and condition of the anchorage connections on vehicles to ensure that they are adequately connected to the vehicle and in good repair.

Placement of Anchor Systems

Regardless of the type of anchorage or anchor system, placement of secure anchors depends very much on good judgment, which is developed through experience and practice. Although their specifics may vary from place to place, all anchor systems share certain characteristics:

- Aside from the physical aspects of the anchorage itself, it is important also to consider how safe and accessible they are for the rope rescuers who work with them.
- It generally is good practice to set an anchorage connector as close as possible to an intersecting structural connection point to minimize the torque and leverage applied by the rope system. For vertical elements such as posts and trees, this is typically closer to the base. However, more robust components can be used as an anchor mid span or higher up on a vertical element when the position better suits the needs of the operation, such as:
  - Creating a better angle for an edge transition.
  - Reducing abrasion on an edge.
  - Managing a loaded litter
  - Reducing friction on mechanical advantage systems.

Evaluating an Anchor System

Before any anchor system is subjected to a live load, a safety check should be made to ensure the following:

- The anchor system components are physically checked to ensure that they are correctly oriented, knots are properly tied, and all mechanical components such as carabiners and links are closed and locked.
- The overall strength of each individual anchorage is sufficient to support its intended load and offers an adequate margin of safety.
- The configuration and alignment of the anchor system is checked by applying a nominal force in the direction of the intended load and observing the angle and tension on the individual elements to ensure they react as intended.
- The anchorage and anchorage connector(s) are in good condition.

After-Action REVIEW

IN SUMMARY

- Utilizing standardized terminology is key in maintaining safety during a rope rescue. Because the terminology of anchor systems evolved differently in different industries, ensuring that all team members utilize a common terminology is essential.
- Anchorages must be able to sustain the maximum anticipated load that is likely to be imparted on them, plus an appropriate safety factor.
- The ability of an anchorage to withstand the forces that may potentially be imparted upon it depends on a number of factors, including the condition of the anchorage, the structural nature of the anchorage, and the location of force on the anchorage.
- An anchor system that is in close proximity to and directly above the load will provide best stability and usability.
- Because of the potential for anchor system failure and because the rest of the rope rescue system depends on the anchor system, it is good practice to back them up. Anchor systems may be backed up by backing up to the same anchorage or backing up to a separate anchorage.
- An anchorage is what a system connects to, and an anchorage connector is how a system connects to an anchorage.
- A single point anchorage is a type of anchorage system that utilizes one anchorage to provide the primary support for the entire rope rescue system. This may be achieved in any number of ways, including a tensionless hitch, knots, soft slings, presewn slings, and anchor straps.
- Portable anchorages are prefabricated anchorages that can be moved from place to place. They may serve as an actual anchor point, or as a directional aid to help route a rope for more optimum rigging. They include beam clamps and picket systems.
- Rigging plates may be used as a part of an anchor system to help organize rigging. They can help spread out the various components so that they are easier to see and manage, and to prevent multiple components from jamming together.
- Rigging plates are commonly used to draw multiple anchor points together to a single point, or to disperse multiple lines from a single anchor point.
- Multi-point anchor systems collect multiple anchor points together to form a single connection point so that force is distributed and shared by the all the anchor points. Multi-point anchor systems may be fixed and focused or self-adjusting.
- A directional anchorage may be placed midline to change the direction of travel of the load.
- Backing up is essential when you have an anchor that is in the right position but not strong enough alone for a rescue load. This may be achieved with pretensioned back-ties.
- Structural anchorages are both natural (boulders) and human-made (structural columns).
- If no anchors are available, extended anchor points or vehicles may be used.
- Before any anchor system is subjected to a live load, a safety check should be made.

KEY TERMS

Anchorage connector How rope rescue components are secured to an anchor.

Anchorage (or anchor point) A single, structural component used either alone or in combination with other components to create an anchor system capable of sustaining the actual and potential load on the rope rescue system.

Anchor system One or more anchor points rigged to provide a structurally significant connection point for rope rescue system components.

Backup system Secondary rope system, or "belay" used to provide protection against potential failure of a primary system.

Back-tie A connector from a primary anchor to a second, backup anchor.

Bombproof Jargon for an anchor or anchor system believed to be very secure.

Directional A technique for bringing a rope into a more favorable position or angle.

Direction of pull The specific direction in which a vector is applied

Fixed and focused A multi-point anchor system in which the anchor points are rigged such that the load-connection point does not move even as the direction of pull might change.

Focal point The point where rigging comes together for maximum strength.

Installed anchorages Anchor points that need to be created and that do not exist in nature or engineered structures.

Load-distributing anchor system See *self-adjusting anchor system*.

Load-sharing anchor system See *fixed and focused anchor system*.

Master attachment point The main point at which a load (litter, spider, rescuer) is attached to a rescue system.

Multi-point anchor system An anchor system in which multiple anchor points are connected together to form a single connection point.

Protected self-adjusting anchor system A self-adjusting anchor with built-in redundancy.

Self-adjusting anchor A multi-point anchor system that is rigged so that the load-connection point floats as the direction of pull changes.

Structural anchorages Anchor points that already exist in nature or in engineered structures.

Working system Primary rope system used to move a load; may be a raising system, lowering system, horizontal system, etc.

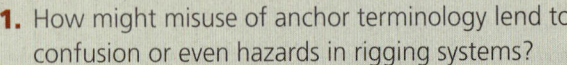

1. How might misuse of anchor terminology lend to confusion or even hazards in rigging systems?
2. What factors should be considered when choosing to rig a single-point anchor system as compared with a multi-point anchor system?
3. When might a fixed and focused multi-point anchor system be preferred over a self-adjusting anchor system, and vice-versa?
4. What purpose might a directional change anchor system serve in a lowering system?
5. How does redundancy play into anchor rigging?
6. What considerations are involved with inspecting an anchor system for use?

CHAPTER 12

Operations

Fall Protection and Belay Operations

KNOWLEDGE OBJECTIVES

After studying this chapter, you should be able to:
- Identify conditions where a belay should be used. (pp. 220–221)
- Explain the purpose of a belay. (**NFPA 1006: 5.2.9, 5.2.10, 5.2.11**, pp. 220–223)
- Define active and self-belay. (**NFPA 1006: 5.2.9, 5.2.10, 5.2.11**, p. 221)
- Describe commonly used active fall protection equipment and systems. (**NFPA 1006: 5.2.9, 5.2.10, 5.2.11**, p. 223)
- Describe the components of active belay systems. (**NFPA 1006: 5.2.9, 5.2.10, 5.2.11**, pp. 223–224)
- Describe the controlling actuation of a belay system. (**NFPA 1006: 5.2.10, 5.2.11**, pp. 224–237)
- Explain the purpose of assisted-braking devices. (**NFPA 1006: 5.2.9, 5.2.10, 5.2.11**, pp. 224–237)
- Explain the considerations for belaying a rescue load. (**NFPA 1006: 5.2.9, 5.2.10, 5.2.11**, pp. 224–237)
- Explain the process of anchor selection and use in a belay system. (**NFPA 1006: 5.2.9, 5.2.10, 5.2.11**, pp. 224–237)
- Describe the considerations for belaying a load during a raising or lowering operation. (**NFPA 1006: 5.2.10, 5.2.11**, pp. 224–237)
- Describe the steps of a preoperational safety check of a belay system. (**NFPA 1006: 5.2.9, 5.2.10, 5.2.11**, p. 229)
- Describe the safety considerations when operating a belay. (**NFPA 1006: 5.2.10, 5.2.11**, pp. 238–239)

SKILL OBJECTIVES

After studying this chapter, you should be able to:
- Practice utilizing a light load belay system. (**NFPA 1006: 5.2.10**, pp. 226–228)
- Conduct a preoperational safety check on a rigged belay system. (**NFPA 1006: 5.2.9, 5.2.10, 5.2.11**, p. 229)
- Communicate belay status effectively. (**NFPA 1006: 5.2.9, 5.2.10, 5.2.11**, pp. 228–229)
- Practice rescue load belays. (**NFPA 1006: 4.2.11**, pp. 230–231)
- Construct a radium release hitch. (**NFPA 1006: 5.2.9**, pp. 232–233)
- Construct the tandem Prusik belay system. (**NFPA 1006: 5.2.9**, p. 234)
- Operate (tend) the tandem Prusik belay. (**NFPA 1006: 5.2.10**, p. 235)
- Release the radium-release hitch. (**NFPA 1006: 5.2.10**, p. 236)
- Rig the Prusik minding pulley. (**NFPA 1006: 5.2.9**, p. 237)
- Operate Prusik minding pulley. (**NFPA 1006: 5.2.10**, p. 237)

You Are the Rescuer

You are among the first responders arriving at a rescue site where there are injured persons on an exposed platform without handrails. The area is accessible by foot, but a fall could have catastrophic results. You have been assigned to set up and then belay medical personnel who need to access the injured subjects via an area with fall potential.

1. How will you rig for a belay with a minimum of equipment?
2. How do you keep your body from being a link in the belay system?
3. What are the correct voice communications with the person you are belaying?

 *Access Navigate for more practice activities.*

Introduction

Belay is a term that is most often heard in recreational climbing circles. It means applying the use of equipment and methods to stop a load from falling. In the rope rescue environment, the principle of belaying is used to keep a person or load from falling. The ability to belay is a critical skill for anyone operating in the rope rescue environment. If you accept the assignment as a belayer, you have made a very serious commitment. It means that the well-being, perhaps even the life, of the person(s) at the end of the rope is in your hands. Saying that you can belay when you cannot or allowing your attention to lapse from the job of belaying could result in severe injury or death for the person at the end of the rope.

> **TIP**
>
> **Prerequisites for Practicing this Chapter's Skills**
> Before attempting the activities described in this chapter, you must have demonstrated that you can properly do the following:
> - Use and care for rope.
> - Use and care for other equipment needed in the high-angle environment.
> - Apply the principles of anchoring and rig a safe and secure anchor.
> - Apply the principles of belaying and safely and confidently belay another person using either a Münter hitch or a personal belay device.
> - Tie correctly and without hesitation the knots described in Chapter 9, *Ropes, Knots, Bends, and Hitches*.

Situations Requiring a Belay

A belay may be called for anytime exposure to the danger of falling arises. A belay would be required in the following situations:

- A rescue situation involves the danger of falling. For example, it would be appropriate to attach a belay line to a rescuer attempting to aid a person threatening to jump from a bridge.
- A person is crossing an area not generally considered to be dangerous, but that includes a small area of exposure to falling.
- A person is unsure in attempting a new skill, such as rappelling for the first time.
- A person's physical or mental capabilities are diminished, for example, because of injury, vertigo, exhaustion, or hypothermia.
- Environmental factors, such as possible rock falls or areas slick with ice, increase the danger of falling.
- A person is rock climbing or mountaineering in hazardous terrain. Should the climber slip, a proper belay can hold the individual.
- A secondary, or backup, means of safety is desired for one or more people who are being lowered by rope, such as in a rescue.
- A secondary, or backup, means of safety is desired for one or more people who are being raised by rope, such as in a rescue.

Decisions on When to Belay

In some cases the need for a belay is not completely clear-cut, such as in the following situations:

- A belay may cause a greater problem than not having one. For example, several rope lines may already be involved, and an additional line from a belay could cause entanglement.
- A free drop is involved, in which the load may spin. A belay line could entangle the main line and stop everything from moving.

- The system is moving through flowing water or other medium where accidental jamming of the belay could have catastrophic results.
- The need for conservation of resources or time is greater than the potential risk posed by not having a belay.

Judgments about these situations must be made locally by well-trained and experienced individuals. At a minimum, belays should be used anytime the combined potential for and consequences of a fall are unacceptable to the situation. As a general rule, if you are not sure, it is probably best to belay.

Belays as Safety Backups

If the failure of a rope rescue system would lead to the injury or death of a person, then a backup system should be introduced into the system for safety. There are two general approaches to this. The first, called **self-belay**, is a form of fall protection that a person manages themselves, such as a rope grab. The other, called an **active belay**, generally refers to a rope rescue system that consists of a secondary brake device operated by a belayer.

> **TIP**
>
> The term *belay* refers to the act of securing or fastening something. With both Old English and Dutch roots, the term was used as the sailing world early as the 1500s to refer to the act of wrapping loops of rope around a cleat or pin to secure it. As a slang term, it was used to simply mean *stop*. From this etymology, the word also became used among mountain climbers to refer to the system used to catch a falling load—which is precisely the sense in which it is used in rope rescue applications.

As previously discussed, the determination of whether or not a belay is needed is a judgment call, but if the failure of the rope rescue system would result in injury or death, then a backup should be considered. Some means of secondary safety should be considered for all rescue operations, with belay considered compulsory the following instances:

- There is a high likelihood of failure of the main system.
- The consequence of failure of the main system would be severe.
- The protection provided by the belay outweighs the potential disadvantages or hazards that the belay might create.

Sometimes, rope rescuers may determine that a belay is not necessary—for example, where the probability of failure is so low that the additional equipment, personnel, and complexity of a belay system is simply not warranted. As an example, rope rescuers responding to a car-over-the-edge incident may use a low-angle scree evac–type system to raise subjects up a slope and back to the road. Exposed only to low loads, rescue equipment and rigging are unlikely to become overloaded or to fail in such an operation. In many cases, such a failure would be unlikely to result in injury to either the victim or rescuers. Under these conditions, the additional equipment, personnel, and complexity of a belay system may not be warranted. Now consider an incident that involves a school bus with several children inside. Applying the equipment and personnel that would otherwise be committed to a belay could potentially double the speed at which the subjects could be brought to safety.

Other situations where a belay might be contraindicated for rescue systems include those where the function of a belay is likely to create greater risks for personnel, such as where a high probability of entanglement exists, where inadvertent activation of a belay is likely to leave the rescuer(s) and/or subject(s) in a precarious position, or where a belay would not protect against any real hazard.

Self-Belay

Self-belay is a method used by recreational climbers who wish to climb solo, without the need for a partner to belay them, while still maintaining a backup rope for safety. There are a number of specific techniques for this, but generally speaking, most involve setting a fixed rope to which a device is attached that will slide along the slack rope as the climber ascends. If a fall occurs, the device is supposed to catch the climber. Self-belayed climbing is also sometimes called roped-solo climbing and, in some circles, passive belay.

These methods are not completely unlike methods used for **fall arrest** in workplace environments. Industrial safety professionals do not use the term *self-belay*, but instead refer to these systems as **active fall protection**. Some active fall arrest systems used for workplace safety include the following:

- **Lanyard with force absorber**
- **Vertical lifeline** fall arrest system
- **Self-retracting lifeline**
- **Rope access backup**

While the mechanisms used by these systems differ, the premise is the same: a person, wearing a harness, is connected to some secure point by means of a safety system, which activates to catch them in the event of a fall. The inverse of active fall protection systems are known as **passive fall protection**, a term that refers to stationary systems that do not move, do not require the use of personal protective equipment (PPE), and require no active participation from the worker. This might include such methods as guardrails or netting systems.

Rope rescuers should become familiar with the types of equipment used for active fall protection for two reasons:

1. They may have occasion to rescue workers using this type of equipment during responses in their area.
2. They may have occasion to use this type of equipment to protect themselves while performing a rescue.

While a lanyard with force absorber is a very common type of fall protection system used by workers, it is also the least versatile (**FIGURE 12-1**). These systems function best in an environment where a worker is able to stand on secure footing, and is likely to remain in one place for an extended period of time. It requires the use of a fixed-point anchorage above the worker, to which a lanyard (generally not exceeding 6 feet [1.8 m]) with a force absorber is attached. If the worker falls, the built-in force absorber dissipates some of the energy of the fall to reduce the force on the worker's body and on the anchor. Fixed systems like this are of limited use, which drives many employers to seek more dynamic systems that protect the worker even while they are moving.

A self-retracting lifeline (SRL) is one type of such a system (**FIGURE 12-2**). In this system, a cable or fiber line is wound around a tensioned reel (or block) inside an enclosed case. A centripetal brake in the block mechanism locks off in the event of sudden impact. Some SRLs feature a block fixed directly to the dorsal D ring, with the line anchored above, while others are designed for the block to be attached to an anchorage above the user. As the user moves, the tensioned reel allows the line to extend and retract as the user moves. If there is a sudden increase in velocity, such as in a fall, the brake will engage to stop the fall. The brake will not release until weight is removed.

Similarly, a vertical lifeline fall arrest system is another type of fall protection system that moves with the user. It is most often used by workers who are working on an elevated platform or climbing a ladder. In these systems, a rope grab travels along a rope or cable that is anchored from above. The user is connected to the rope grab with a lanyard up to 6 feet (1.8 m) in length between it and their harness, typically at the dorsal attachment point. If there is a fall, the rope grab activates. Again, once it is locked, weight must be removed to release the brake. Rope grabs are discussed in Chapter 8, *Rescue Equipment*.

A rope access backup is perhaps the most versatile of the active fall protection systems described here (**FIGURE 12-3**). Similar to the vertical lifeline system, the user relies on a rope grab that travels up and down along a rope that is anchored from above, but in this case, the lanyard between the rope grab and the user is much shorter in length—typically 3 feet (0.9 m) or less—and connects to the sternal harness attachment point rather than the dorsal attachment point.

FIGURE 12-1 A worker using a fall arrest lanyard with force absorber.
© Don Mason/Getty Images.

FIGURE 12-2 A worker using a self-retracting lifeline for fall protection.
© King Ropes Access/Shutterstock.

FIGURE 12-3 A bridge engineer using rope access for safety during an inspection.
© King Ropes Access/Shutterstock.

FIGURE 12-4 A tubular-type belay device in use.
© Salienko Evgenii/Shutterstock.

Rescuers who work in urban environments and who may respond to incidents involving construction, workplace, or industrial fall protection should be familiar with these types of systems, both as an option for their own use and in case they might need to rescue someone who has fallen while using such a system.

The main advantages of using personal fall arrest systems for rescuer safety are that the systems require no extra personnel, and that a trained rope rescuer ascending a rope or climbing a structure with an active fall protection system already anchored above can operate the system themselves.

However, there are limitations to personal fall arrest systems. One is that the system must be able to be anchored above the work to be performed, in advance of their use. Another is that the user must be capable of monitoring and ensuring the proper function of the system at all times while working. Finally, passive belay systems are generally suited only for single-person loads, not for rescue loads.

Active Belay

An active belay is one that requires a trained belayer to operate. Variations on this method may be used for belaying one person, or for a larger rescue load. Active belay methods used by climbers often involve a tubular-type belay device that permits the rope, managed by the belayer, to run freely until/unless the belayer applies hand tension to lock it off (**FIGURE 12-4**). While the human aspect of an active belay is an oft-debated topic due to the possibility of human error, this method does offer more flexibility in use, particularly in complicated environments, and permits a modicum of control by the belayer. These factors may, in fact, result in an active belay being a "safer" choice for a given operation.

The Belay System

The following elements constitute a one-person belay system (**FIGURE 12-5**):

- A load (in this case a person) that is at risk of falling (this may be a person rappelling, ascending, or climbing).
- A harness, worn by the belayed person. The harness should be fit to the climber so that it is able to safely hold them without causing injury when caught by the rope.
- A rope, connected to the harness by means of a carabiner or a knot, and extending toward an anchorage system.
- An anchorage system that is capable of withstanding the highest potential shock load that could result from the fall of the person being belayed.
- A **belay device**. A belay device is a type of braking mechanism through which rope can be easily fed in either direction, but that is easy to lock off in the event of uncontrolled loading. A rope, called the **belay line**, is rigged into the belay device according to manufacturer's instructions.
- The **belayer**. This is a properly trained individual who controls the rope as it passes through the belay device so that, in the event of an uncontrolled descent or fall by the load, they are able to exert a slowing or stopping action on the line. The belayer must manage the rope so that excessive slack does not increase the shock load to the rope if the load should fall.

FIGURE 12-5 Elements of a belay system.

A good compromise belay option is a hybrid system wherein the device operated by the belayer offers an assisted catch by self-activating when the rope or line is pulled rapidly, such as in a fall. With such a device, the belayer can quickly feed rope in either direction, maintaining versatility, but if the working end of the rope is pulled rapidly, a secure locking mechanism will engage. Sometimes referred to as an autolocking or autoblock function, manually operated belay devices with this feature generally also offer a mechanism—such as a handle—that permits the belayer to release tension from the locked-off device without having to lift the load. Some descenders also function well as a belay device, but the methods used in operating the device will differ depending on whether the operator is lowering or belaying. See Chapter 18, *Personal Vertical Skills* for a detailed discussion on descenders.

Light Load Versus Rescue Load Belays

Belaying of light loads (single-person or equivalent) differs from belaying heavier loads. In this resource, we will refer to loads ranging from 50 to 300 pounds (23 to 136 kg) as light loads and loads of 300 to 600 pounds (136 to 272 kg) as rescue loads. This is not terminology from the National Fire Protection Association (NFPA). However, we use it in this resource to provide perspective related to the differences associated with handling loads of different weights. Some devices that may work well for belaying light loads create insufficient friction for catching heavier loads—and some devices that may work well for belaying heavy loads are not easy to feed rope through when belaying light loads—so understanding the capabilities and limitations of the system(s) used in a rescue is imperative.

Active Belay Systems: Light Loads

The most frequently belayed light load that a rescuer will handle is that of another rescuer. There are a wide variety of belay systems that are appropriate for belaying light loads. They all work essentially the same way: they create a braking action on the rope to prevent the person at the end of the rope from falling far enough to be injured. In this chapter, some of these light load belay techniques are examined in detail. These techniques are among those considered to be simple to learn and easy to use because they employ minimal equipment and straightforward concepts. Light loads and short fall distances pose less of a danger to the belayer and the person on the rope.

The Münter Hitch

The **Münter hitch** is the subject of some controversy because it may not offer sufficient friction to control heavy rescue loads. However, it does have one distinct advantage: it is a versatile and useful tool that is always in a rope rescuer's toolkit as long as the rescuer has a rope and a carabiner. For this reason alone, it is worth knowing.

The ideal carabiner for use with a Münter hitch is one specifically designed for this purpose. Such carabiners are called HMS (an acronym for the German word *Halbmastwurfsicherung*) or Münter hitch carabiners. The HMS carabiner has a pear-shaped design with gentle curves at the base that allow the Münter hitch to move freely back and forth through the carabiner. Angulated carabiners such as a D design may be used, but are prone to causing a jamming condition.

Factors to consider when using a Münter hitch are (1) where the force will ride on the carabiner in the event of a failure and (2) how the rigging can be created to reduce the probability of the moving rope opening the carabiner gate. Ideally, the part of the line that carries the load should extend from the spine-side of the carabiner, as this results in optimum strength and alignment for the system. This, however, places the working part of the rope controlled by the belayer's hand nearest the carabiner gate. Therefore, the belayer must take care to avoid running the moving rope across the carabiner gate, as this could cause it to unlock or even open.

The Münter hitch is particularly useful for belaying light loads, and in testing has even proven successful with heavier loads as long as there is additional friction (such as an edge) in the system. For maximum control when using the Münter hitch, keep the angle between the braking side of the rope and the rope going to the load as narrow as practicable to achieve as much friction as possible.

TIP
An advantage of using the Münter hitch is that it is easy to reverse rope direction without unclipping the rope.

Using a Free-Running Belay Device

Another way to belay a light load is to use a tubular or slot-type, free-running device. There are numerous designs of light load belay devices, many of which use variations on the old sticht plate design (c. 1970s). Originally designed as a flat plate, modern tubular versions are less prone to jamming. These typically feature elongated openings, through which a bight of rope is pushed and then clipped with a locking carabiner to the anchor. This simple bend in the rope is easy to feed in either direction, but should the rescuer fall, the resulting friction against the rope as it is pressed between the carabiner and the belay device provides sufficient braking assistance to stop a fall. While these devices typically have two slots (to accommodate a dual-rope rappel or double-rope climbing techniques), only one slot is used for belaying in a single-rope system.

TIP
Maintaining proper hand position on the rope as it is moving is one of the most critical operations in belaying. The person at the end of the rope could fall at any time. If you do not have your brake hand on the rope at all times, ready to grasp it in an instant, you may drop the person and cause severe injury or death.

TIP
Some figure 8 descenders are designed with a belay plate either in the small ring or between the two rings. However, you should use figure 8 belay plates with caution for the following reasons:
- The slot in the figure 8 may not be the correct size for the rope you are using.
- Some figure 8s have slots that are not well designed for use as a belay plate.
- A figure 8 is not as well balanced as some devices and may not be as easy to use.

Assisted-Braking Belay Devices

Some personal belay devices are designed to provide an assisted catch. That is, in the event of sudden loading, a mechanical action helps brake the rope. This action is also sometimes called autolocking or autoblock. Popular devices include the GriGri 2 (Petzl), the Birdie (Beal), and the Alpine Up (Climbing Technology).

Assisted-catch devices are designed to stop a falling climber more quickly than free-running devices do. They should always be used according to manufacturers' directions. Always belay with these devices in such a way that you are not relying on the assisted catch and always consider your brake hand to be the primary source of belay activation and arrest.

> **TIP**
>
> Practice with the belay device you will be using. The devices are different and require skill to operate smoothly and correctly. The only way to gain that skill is actual practice with the device you will be using.

Practicing a Light Load Belay System

The practice system for a light load will consist of the following elements:

- A belay system, consisting of anchorage, rope, belay device, and enough connectors to hold it all together
- A weight or a dummy of sufficient mass to simulate the weight of a falling person
- A method of raising the weight on a separate system, such as a winch or mechanical advantage hauling system
- A quick-release device capable of releasing the load while under tension
- A belayer
- An instructor
- Additional personnel as required for raising the load or assisting with the training

To practice utilizing a light load belay system, follow the steps in **SKILL DRILL 12-1**.

> **TIP**
>
> When you are belaying, it is essential that you use standard voice signals (also called commands or calls) that will be understood by teammates. Otherwise, even momentary confusion could cause an accident.

SKILL DRILL 12-1
Practicing Using a Light Load Belay Practice System

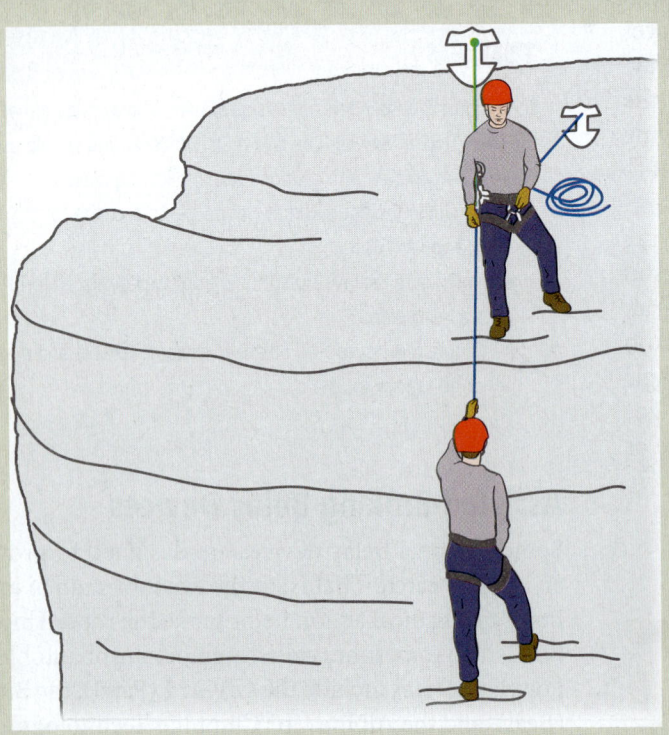

1 First, allow the belayer to practice operation of the device on level ground or low-angle terrain until they are comfortable with it. Then, set up a belay system as you would for a typical personal belay system, in accordance with manufacturer's guidelines for the device you are using. The belayer should position themselves as they would for operating a typical personal belay system. The belay system is attached to a light load as it would normally be for protecting another rescuer. A raising system is also attached to the load, with a quick release mechanism at the load.

SKILL DRILL 12-1 (continued)
Practicing Using a Light Load Belay Practice System

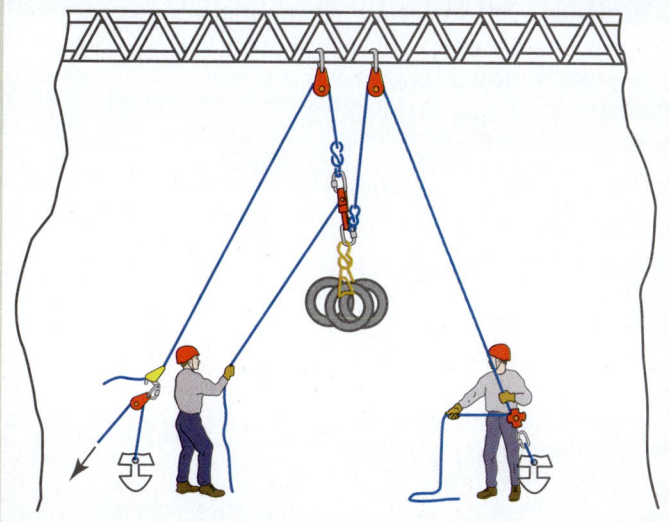

2 Raise the weights to the desired height, keeping as much tension as possible in the belay line. When sufficient ground clearance is achieved, gently lower the weights onto the belay device, so that the student belayer takes the full load. The belayer lowers the load to ground (or as far as necessary to feel comfortable with operation of the device).

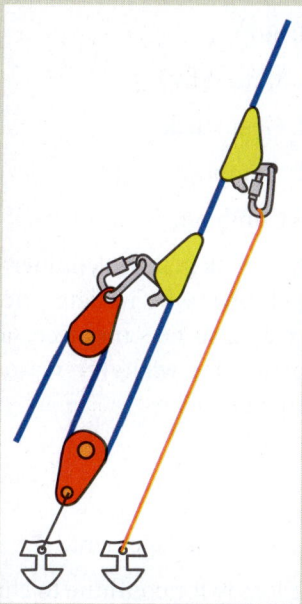

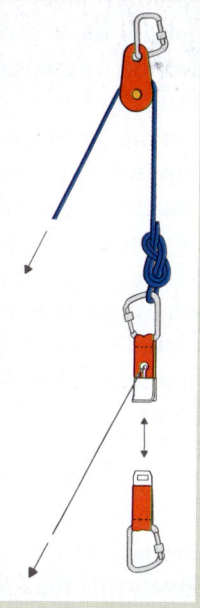

3 Reconnect the haul system to the weights and once again raise the weights to the desired height, this time leaving slack in the belay line (approximately 1 foot [30 cm] is good to start) plus sufficient clearance above ground to account for stretch in the line and reaction time (at least four times the amount of slack in the system is a good rule of thumb).

4 Without warning the belayer, release the weights and allow the belayer to attempt to arrest the fall of the weight. Repeat the exercise, increasing the fall distance as appropriate.

(continues)

SKILL DRILL 12-1 (continued)
Practicing Using a Light Load Belay Practice System

5 When the belayer is comfortable with catching a fall on a stationary system, reset the exercise as follows (Steps 5–8). Set up a method for simulating movement of the load. A winch may be used, or a mainline (raising or lowering) and belay system as would be used for a typical rescue operation, using weights as the load.

6 Have the belayer position themselves as they would for belaying another rescuer. Attach the belay system to the load as it would normally be. The mainline is also attached, with a quick-release mechanism at the load.

7 Engage the simulated load movement system (raising or lowering) with the belay system maintaining pace as it would in a normal rescue.

8 Without warning the belayer, and with the system still in motion, release the weights and allow the belayer to attempt to arrest the fall of the weight.

Communications

Using appropriate commands during belay operations is essential. Communications guidelines for individuals forms a foundation for those used in rescue operations. This topic will be covered in further detail relative to rescue systems in Chapters 14, *Lowering Systems* and 15, *Mechanical Advantage Systems*.

Any life safety communications procedures should be grounded in the concept of closed loop communications. Closed loop communications are those in which persons on both the initiating side and the receiving side of a message confirm the message. In the case of a belay operation, this would involve the person who is on rope and ready to be belayed asking a relevant question, such as "On belay?" and the person providing the service of belay to respond in kind with a statement such as "Belay on." Establishing a system in which the cadence, rhythm, number of syllables, and order of usage are unlikely to be confused will help to maintain safety during the operation. Allowing the person who is in the position of risk to initiate the communications is generally preferred. For example:

RESCUER: On Belay?

BELAYER: Belay on.

RESCUER: Climbing.

BELAYER: Climb on.

At times, it may become necessary for the rescuer to call a stop, whether to tend to a subject or simply to negotiate an obstacle. This is easily achieved.

RESCUER: Stop.

BELAYER: Stop. Why?

RESCUER: Gear stuck.

RESCUER: Climbing.

BELAYER: Climb on.

Note, in the event that the climber calls a stop, it should be the climber who initiates movement again. Of course, it may also become necessary for the belayer to call a stop for whatever reason. That might look something like this:

BELAYER: Stop.

RESCUER: Stop. Why?

BELAYER: Rope management.

BELAYER: Ready for continue to climb.

RESCUER: Climbing.

BELAYER: Climb on.

The key points are as follows:
- The person on the sharp end of the rope should always be the one to initiate movement
- Anyone can call a stop
- Communications should be closed loop

Refer to Chapter 3, *Hazards Associated with Rope Rescue* for more on closed loop communication.

> **TIP**
>
> When anchoring yourself as a belayer, never place yourself into the system. That is, do not belay with a device clipped to the front of your harness and the anchor clipped to the back of your harness. Doing so essentially makes you a link in the anchor system of the belay. Should the climber fall and the system shock load, you could be severely injured, possibly suffering a fractured pelvis. You could be disabled or incur a life-threatening injury. Even if you are not injured, making yourself part of the anchor system for a belay limits your ability to correct problems like snagged ropes. Belay devices should go directly to an anchor system. If you prefer to be clipped in, clip into the anchor in a manner that allows you to tie off the belay easily and move away to resolve any problem that may occur.

> **TIP**
>
> It is not enough to have intellectually learned belaying. That is, it is not enough to have read about belaying, to have watched it, or even to have practiced the hand positions. You must have combined both the mental and physical experience of the actual belay situation before an emergency.

Belaying a Rescue Load

Rescuers must use critical thinking in analyzing their rescue system's potential for failure. Simply put, this just means that rescuers should think about every part of the rescue system to try to identify the weakest points, the parts of the system most likely to fail, and the greatest risks to the system. This can be accomplished by examining every element of a rope rescue system and applying the **critical point test**. A critical point is one that, if it fails, will result in failure of the entire system. Examples could include anchor points or rope. To perform a critical point test:

1. Examine the system.
2. Identify any point or points at which failure would result in catastrophic failure (that is, complete failure of the entire system).
3. Evaluate methods by which protection can be provided in the rope rescue system to prevent failure of the critical point(s) from resulting in catastrophic failure.
4. Determine whether the methods of protection improve safety enough to warrant the complexity and resources that they would add to the rope rescue system.

The critical point test can be applied both to equipment and to team members. When there is high risk, characterized by an unacceptably high probability and an unacceptably high consequence of failure, a belay may be used to build redundancy into the system to ensure that the failure of any one point does not result in catastrophic failure.

As previously stated, redundancy must be balanced with complexity. More redundancy does not necessarily mean more safety. As a system becomes more complex, it can become more difficult to manage, which in turn can increase potential for failure. For example, in a situation where adding a belay would require placing extra people at risk of a fall, it may not be warranted. A savvy rescuer strives to optimize the level of redundancy with efficiency because when redundancy begins to reduce efficiency, safety is compromised.

When considering the options for belaying during rescue, take into account the effects of the heavier load, the fall line, the equipment being used, and the consequence of a potential failure. There is no single perfect solution for all circumstances. The unique aspect of belaying a rescue load lies in the fact that you are usually protecting a relatively heavy load (perhaps even as much as 600 pounds [272 kg]) with relatively short fall potential on a static or low-stretch rope, whereas the single person belay methods described earlier in this chapter are generally intended to catch a lighter load (less than 300 pounds [136 kg]) with potentially larger fall distances, and often on a dynamic rope.

Manually Operated Assisted Belay for Rescue

The same basic premise that applies for assisted-catch or autolocking belay devices with light loads may also be applied to belaying rescue loads, with the caveat that the devices used for belaying rescue loads are typically more robust than those used for light or personal loads.

Often, devices that are sufficient for belaying a rescue load have as their primary function the role of braking, or lowering. Of primary concern when catching a rescue load is to ensure that the force will not cause damage to the device or to the rope, and that it is able to be released readily after the catch. Some of the devices that offer an autolocking feature conducive to belaying rescue loads include the ISC D4, the CMC MPD, and the Petzl I'D (**FIGURE 12-6**).

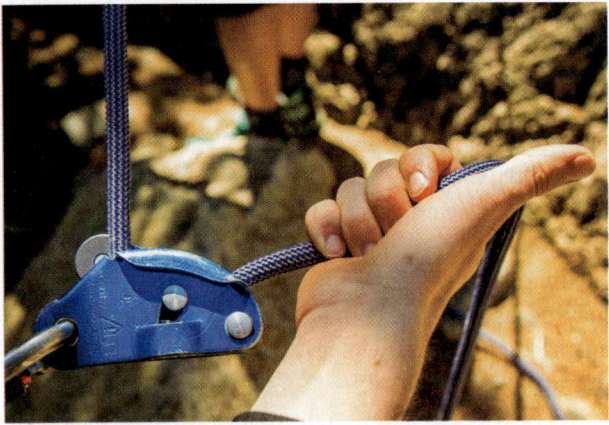

FIGURE 12-6 Manually operated assisted belay for rescue.
© Georg Berg/Alamy Stock Photo.

> **TIP**
>
> **One-Person and Two-Person Loads**
> Previous versions of some NFPA standards differentiated between one-person and two-person rescue loads, but this terminology has since been eliminated. NFPA now refers to technical-use and general-use equipment, each of which has different equipment performance requirements, but neither of which are dictated by the number of people on the end of the rope. Decisions about how to belay and which equipment to use should be based on an analysis of circumstances, including weight of the load, maximum potential fall distance, desired stopping distance, preferred maximum arrest force, and other relevant system factors.

> **TIP**
>
> Practice with the belay device you will be using. Each device is different and require skill to operate smoothly and correctly. The only way to gain that skill is actual practice with the device you will be using.

Each of these devices is unique in its specific function, and should always be used according to manufacturer's directions. Always belay with these devices in such a way that you are not relying on the assisted catch. You should always consider your brake hand to be the primary source of belay activation and arrest.

Rescue Load Belay Practice

The practice system for rescue load belay practice consists of the following elements:

- A belay system, consisting of anchorage, rope, belay device, and enough connectors to hold it all together
- A weight or a dummy weighing enough to simulate the weight of your typical rescue load
- A method of raising the weight on a separate system, such as a winch or mechanical advantage hauling system.
- A quick-release device capable of releasing the load while under tension
- A belayer
- An instructor
- Additional personnel as required for raising the load or assisting with the training

Remember, when you are belaying, it is essential that you use standard voice signals (also called commands or calls) that will be understood by teammates. Otherwise, even momentary confusion could cause an accident. To practice rescue load belays, follow the steps in **SKILL DRILL 12-2**:

1. First, allow the belayer to practice operation of the device on level ground or low-angle terrain until they are comfortable with it. Then, set up a belay system as you would for a typical rescue belay system, in accordance with local protocol and manufacturer's guidelines for the device you are using.
2. The belayer should position themselves as they would for operating a typical rescue belay system. The belay system is attached to a rescue load as it would normally be during a raising or lowering operation. The haul system is also attached to the load, with a quick release mechanism at the load.
3. Raise the weights to the desired height, keeping as much tension as possible in the belay line. When sufficient ground clearance is achieved, gently lower the weights onto the belay device, so that the belayer takes the full load. The belayer lowers the load to ground (or as far as necessary to feel comfortable with operation of the device).
4. Reconnect the haul system to the weights and once again raise the weights to the desired height—this time leaving slack in the belay line (approximately 1 foot [30 cm] is good to start) plus sufficient clearance above ground to account for stretch in the line and reaction time (at least four times the amount of slack in the system is a good rule of thumb).
5. Without warning the belayer, release the weights and allow the belayer to arrest the fall of the

weight. Repeat the exercise, increasing the fall distance (and clearance distance) as appropriate.

6. When the belayer is comfortable with catching a fall on a stationary system, reset the exercise as follows (Steps 6–7). Set up a mainline (raising or lowering) and belay system as you would for a typical rescue operation, using weights as the load. The belayer should position themselves as they would for a typical rescue. The belay system is attached to the load as it would normally be in a rescue operation. The mainline is also attached, with a quick release mechanism at the load.

7. Engage the rescue system (raising or lowering) as you would in a normal rescue, with the belay system maintaining pace as it would in a normal rescue. Without warning the belayer, and with the system still in motion, release the weights as the belayer attempts to arrest the fall of the weight.

Tandem Prusik Belay System

The tandem Prusik belay system has been, for many years, the de-facto standard in rescue. While it may be used with any rescue system, it was devised specifically to provide hands-free safety to free-running lowering devices such as the brake rack and brake tube. See Chapter 14, *Lowering Systems* for more on lowering systems.

A tandem Prusik belay is composed of two triple-wrapped Prusiks placed in line about a hands-width apart on the belay rope and anchored securely. If rigged and managed correctly, in the event of a failure the Prusiks exert a clutching action, working together to grab the rope amid just enough slip to prevent an abrupt shock load from occurring.

There are limitations with tandem Prusiks belays in that under certain conditions, they may not catch as anticipated. For example, if the equipment is icy or muddy, if Prusik cord of the wrong diameter or material is used, or if the hitches are not tied tightly enough, the belay rope may slip through the Prusiks. Tandem Prusiks can fail under other conditions, too. In high-impact situations, for example, tandem Prusiks can cause the belay line to fail by pinching it until the rope separates. Also, the Prusik material itself can fail, particularly if the wrong size or type of Prusik cord is used to build the tandem three-wrap Prusik belay system.

The success of a tandem Prusik belay depends on the characteristics of materials used, and on the interaction of the Prusik cord material with the rope material. While there are no specific rules for this, there are

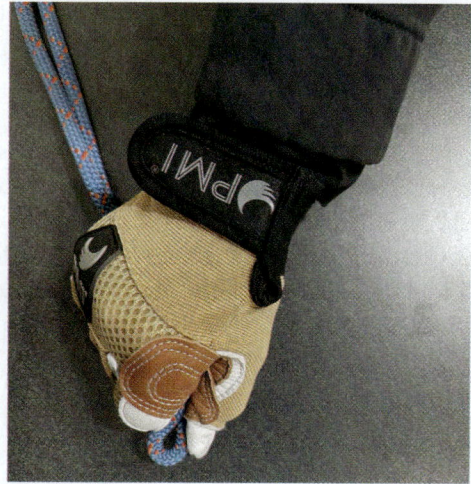

FIGURE 12-7 The Prusik pinch test.

a few rules of thumb that have emerged from various testing performed over the years:

- The first, known as the pinch test, is derived from testing and subsequent recommendations from Arnor Larsen, British Colombia Council of Technical Rescue. To perform the pinch test, simply pinch a bight of Prusik cord firmly between your thumb and forefinger. The Prusik should be stiff enough that it does not pinch flat against itself, but leaves a small eye that is just a bit smaller than the diameter of the Prusik cord (**FIGURE 12-7**). If the cord is too stiff, it will not grab, but if it is too floppy, then it will grab too aggressively, resulting in failure of the system.

- Prusiks used for a tandem Prusik system should be about 4 mm smaller in diameter than the host rope. The bigger the difference between diameters, the more readily the Prusiks will grab and hold. Prusiks with less than 2.5 mm of difference may not grab consistently, and Prusiks with greater than 5 mm of difference may be too grabby.

- Prusiks of two different lengths should be chosen such that when applied to the host rope there is about 4 inches (10 cm), or a hand's width, of distance between them.

- While Prusik loops may be tied, commercially produced Prusiks with sewn terminations provide more consistent length and performance, and are often considered easier to use.

Load-Releasing Hitches

A properly constructed Prusik belay system requires some means of releasing the Prusiks should they become engaged. There are several ways to release the load, and these techniques usually involve a **load-releasing (LR) hitch**, also known as an LRH. The type of LR hitch described in this resource is the radium-release hitch. The radium release hitch is simply one particular method for tying an LR hitch; it is preferred for its ability to withstand high shock loads and to lower the arrested load under control. Note that not all LR hitches provide shock-absorbing capacity.

A radium release hitch serves two primary purposes:

1. If the belay line becomes loaded accidentally, the LR hitch can be used to shift the load back to the main line.
2. The radium release hitch has some shock-absorbing capacity.

In addition, the LR hitch can be used for some purposes other than belaying, such as changing over from a raising system to a lowering system or from a lowering system to a raising system. To construct a radium-release hitch, follow the steps in **SKILL DRILL 12-3**.

Constructing the Tandem Prusik Belay System

Based on a compendium of research that has been compiled over the years, proper function of the tandem Prusik belay system is highly dependent on precision and detail in rigging. Before relying on a tandem Prusik belay, always test the exact setup you plan to use to affirm its catching ability and holding power. To construct a tandem Prusik belay system, follow the steps in **SKILL DRILL 12-4**.

TIP
Tandem Prusik belays do not work in all rescue conditions. Prusiks can slip on ropes that are muddy or icy. A Prusik-type belay may not be appropriate in a hazardous environment in which a hang-up or a failure to catch could cause severe injury or death.

SKILL DRILL 12-3
Construct a Radium-Release Hitch

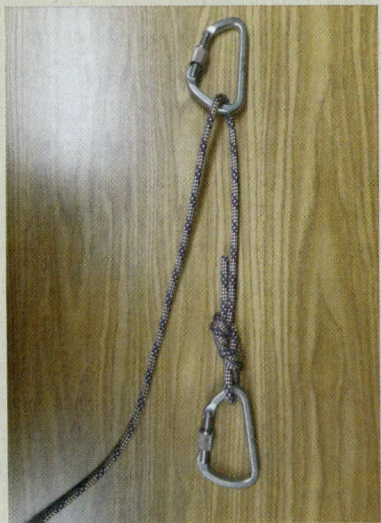

1 Collect the materials needed for a radium-release hitch: two locking carabiners of suitable strength for rescue loads, spaced about 4 inches (10 cm) apart, and 33 feet (10 m) of 8-mm nylon static kernmantle rope or 8-mm nylon accessory cord.

2 Tie a compact figure 8 on a bight in one end of the cord and clip the loop (bight) of the knot into the load-side carabiner with the loop close to the carabiner's spine. This will become the load-end carabiner where the tandem Prusiks are attached to the LR hitch.

CHAPTER 12 Fall Protection and Belay Operations 233

SKILL DRILL 12-3 (continued)
Construct a Radium-Release Hitch

3 Pass the standing part of the cord through the other carabiner, leaving about 4 inches (10 cm) between the two. This will become the anchor-end carabiner.

4 Now reeve the standing part of the cord back down and through the load-end carabiner next to the previously tied figure 8 on a bight.

5 Bring the cord back up to the anchor carabiner and tie a Münter hitch onto the anchor carabiner next to the previous wraps on the gate side of the end.

6 Secure the hitch by taking a bight in the standing end of the rope and tying a half hitch around the three parallel cords just below the Münter hitch. Back it up with an overhand-on-a-bight knot around the bundle. Tie a figure 8 on a bight in the standing end of the cord and clip it into an anchor as a backup.

SKILL DRILL 12-4
Constructing the Tandem Prusik Belay System

1. Attach the radium-release hitch built in Skill Drill 12-3 to a secure anchor.

2. Select two sewn or tied Prusik loops with about 6–7 inches (15–18 cm) difference in length. The cord should be about 4 mm smaller in diameter than the host rope. For ½-inch rope, the authors recommend 8-mm Prusik cord. For 7/16-inch rope, 7-mm Prusik cord works well.

3. Place the longer loop across the rope with approximately one-third of the loop on one side of the rope and two-thirds on the other side. Bring the longer section of the loop around the rope and through the bight formed on the other side of the rope by the one-third portion. Repeat this two more times, forming a Prusik with three wraps on the rope.

4. Using the shorter loop, tie a second three-wrap Prusik as described in step 2. Position this second Prusik between the first Prusik and the anchor. Be sure to wrap the two Prusiks in the same direction on the rope. Position them so that the double fisherman's knots are between the Prusiks and the anchor carabiner. Dress the Prusik hitches. There should be about 4 inches (10 cm) of space between them when both are fully extended from their anchor.

5. Clip both Prusiks into the load-end carabiner of the radium release hitch. Clip the long Prusik in first, then the short one.

TIP
As with all belay systems, consider how tandem Prusiks will align with the other parts of the system in case of shock loading. Will they, for example, make contact with rescuers or other equipment, or cause anchor failure?

TIP
A tandem Prusik belay must be tended by a person who is knowledgeable in its operation and who remains alert at all times during its use.

Operating the Tandem Prusik Belay

The basic technique for Prusik tending is to keep one hand cupped on the Prusiks as the rope is pulled in or let out. Use your other hand to take up or pull out the belay slack and to feel the tension so that you can decide if more or less rope is needed. To operate a tandem Prusik belay, follow the steps in **SKILL DRILL 12-5**.

TIP
Tandem Prusiks should be tended by a belayer wearing gloves. The belayer should keep the following points in mind:
- Before use, the belayer must inspect the Prusiks to make sure they have been tied correctly, they are neat and dressed, and they are the appropriate distance apart.
- The belayer must ensure the Prusiks remain snug on the rope at all times, but not so tight as to catch.
- When Prusiks are appropriately snug, they will make a soft swishing sound on the rope.
- Should they become too loose, system operations should be stopped until rectified.

SKILL DRILL 12-5
Operating (Tending) the Tandem Prusik Belay

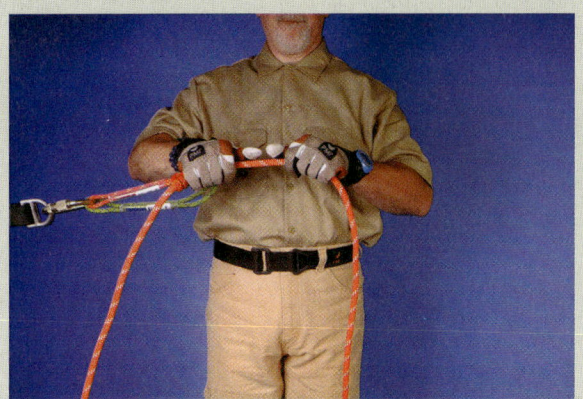

1 To give slack or feed out the belay in a lowering operation, start with your hands together. Keep one hand cupped on the Prusiks to hold them in place and to prevent them from be coming tight when you do not want them to do so.

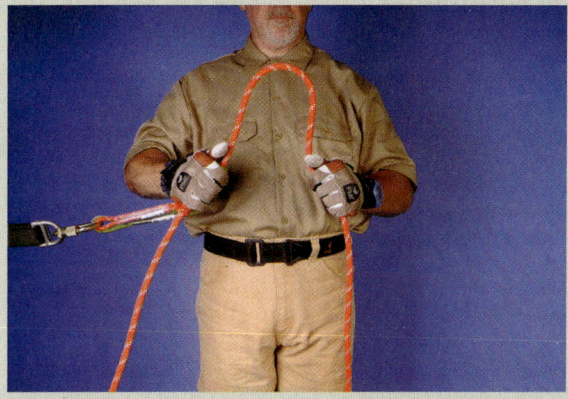

2 With a twist of the wrist of the other hand (the feeling hand), pull out about 18–24 inches (46–61 cm) of the rope through the Prusiks.

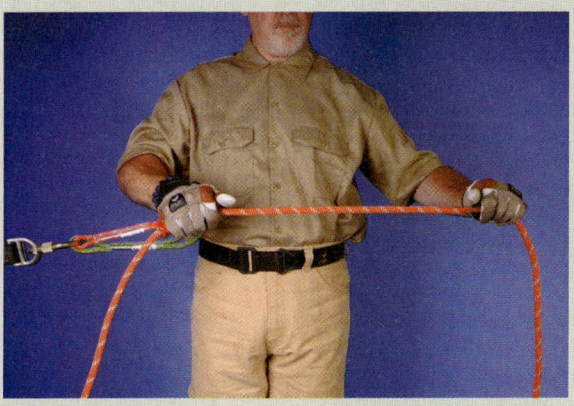

3 As the belay begins to come tight, the S curve will flatten out, thus straightening out the wrist of your feeling hand.

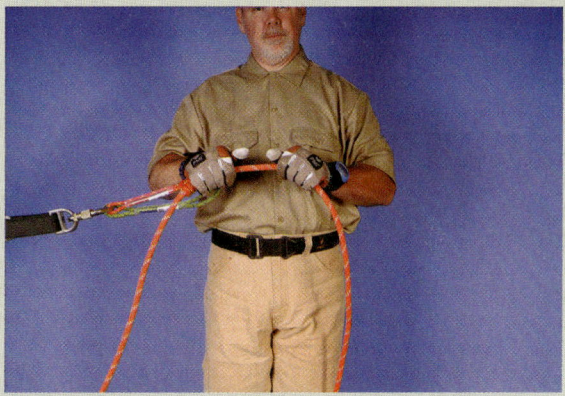

4 This is the signal to slide the feeling hand back up against the Prusik hand and again pull out another 1–2 feet (0.3–0.6 m) of slack, forming the S again in your feeling hand.

5 To take in slack while belaying in a raising operation, pull any available slack through the Prusiks by applying tension to the belay line with the feeling hand.

6 As slack develops, remove it by sliding the Prusiks forward toward the load with the Prusik hand and keeping tension behind the hitch with the feeling hand.

7 Take care not to run the Prusiks too tight, because the load may suddenly change direction, and the Prusiks would quickly jam.

Courtesy of Steve Hudson and Kris Green.

Releasing the Radium-Release Hitch

If the tandem Prusiks become loaded for any reason, follow the steps in **SKILL DRILL 12-6** to release them.

Prusik Minding Pulley

A Prusik minding pulley (PMP) may be used in raising operations to help operate the Prusiks. The tandem Prusiks, when rigged correctly, catch on the edge of the pulley side plates as the rope enters the pulley. The side plates of the PMP are designed to keep the Prusik knots sliding on the rope and to prevent them from binding in the pulley. The PMP is designed such

> **TIP**
>
> When releasing an LR hitch:
> - Never release any LR hitch until you are certain the load can be successfully transferred to some other system or will reach the ground before all the cord in the LR hitch is let out.
> - Make sure you will be able to control the load as you release it. In extreme load situations, the job may require two or more people.
> - Prevent your fingers, hair, and clothing from getting caught in the Münter hitch.

SKILL DRILL 12-6
Releasing the Radium-Release Hitch

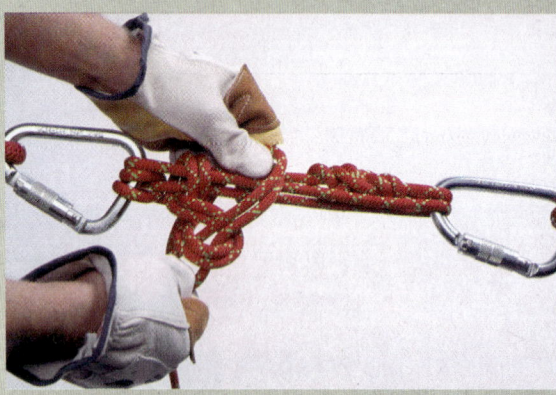

1 Untie the overhand-on-a-bight knot that is secured around the LR hitch. Leave the figure 8 on a bight in the loose end of the hitch attached to the anchor.

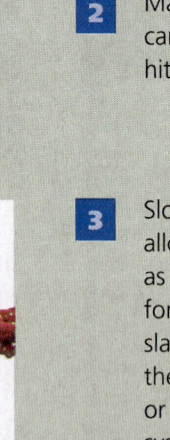

2 Maintaining tension on the Münter hitch, carefully untie the half hitch below the Münter hitch.

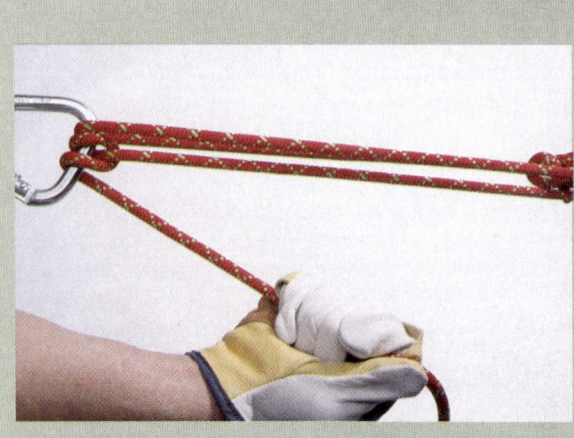

3 Slowly ease the tension on the Münter hitch, allowing the radium-release hitch to lengthen as the Münter hitch controls the load. If the force applied by the load is insufficient to pull slack through the hitch, slowly feed cord into the Münter hitch. Resecure the Münter hitch or rerig the system as needed. Extend the system just far enough to take the load back onto the main system.

Courtesy of Steve Hudson.

CHAPTER 12 Fall Protection and Belay Operations   237

SKILL DRILL 12-7
Rigging the Prusik Minding Pulley

1 Rig the PMP to an LR hitch on a suitable anchor.

2 Run one end of the belay rope through the pulley and attach the Prusiks to the belay rope on what will be the load side of the pulley.

3 Secure the Prusiks in the load-end carabiner of the LR hitch with the PMP next to them.

> **TIP**
> Belayers must be alert at all times. Failures often occur with no warning. While belaying with tandem Prusiks, do not allow the rope to contact any part of your body other than your hands. You could be injured if the belay activates.

that should a failure occur, the tandem Prusiks will grasp the rope and catch the load. To rig the PMP, follow the steps in **SKILL DRILL 12-7**. To monitor the PMP, follow the steps in **SKILL DRILL 12-8**:

1 As the load is hauled, pull the belay rope through the PMP. When the hauling operation is at rest and the slack is out of the belay, the hand grasping the rope on the load side should move toward the Prusiks and snug them back toward the load to limit excess slack should the belay be suddenly loaded.

2 Whenever the opportunity arises (e.g., during pauses in hauling), adjust the Prusiks as needed. For the best efficiency, keep the angle between the ropes going in and out of the PMP as close to 0 degrees as possible.

As with any Prusik safety system, the Prusiks must be tight enough to grasp the rope automatically if the rope should start to slip through them. The rescuer who tends the PMP must make sure the Prusiks remain tight and properly dressed.

Additional Words of Caution for Belayers

Arranging the Belay Direction

The main elements of the belay system—the anchor, the belay device, and the load—must be in as direct a line as possible so that the instant the load falls, the force comes directly onto the belay device and the anchor. If these elements are not in a direct line, then any or all of the following could happen:

- The belay device could fail to work properly.
- The belayer could be thrown off position.
- The anchor could fail.
- The system could be shock loaded.

Also, the belay rope must not be around or against the belayer or any other person, because that individual could be injured by the rope suddenly coming taut.

Maintaining Proper Slack in the Belay Rope

The appropriate amount of slack in a belay rope varies with the device type and the rescue system you are using. Pay attention to the amount of slack that you are maintaining in the belay rope during practice. If you are belaying an individual to protect their progress on another rope or along a surface, a rope that is too taut could interfere with their movement. If the belay is a safety for a separate lowering system, a too-taut rope could interfere with the brakeman's actions. With practice, you will develop good judgment regarding the amount of slack to allow in the belay rope. Unless you are operating a two-tensioned rope system, the rope should have at least some visible slack, but not so much that intense shock loading of the rope would occur during a fall. See Chapter 14, *Lowering Systems* for more on two-tensioned rope systems.

Securing the Belayer

If the belayer is near a place where they could fall, they should be secured to an anchor by a safety line. If you are the belayer, connect your safety line to an anchor separate from the belay if possible. This helps ensure that you, as the belayer, are not endangered by a falling load and that whatever happens, you will remain secure and capable of continuing the belay or otherwise assisting with the operation.

Bottom Belay (for Rappelling Only)

A bottom belay is a pull on the rappel rope from a position below the rappeller (**FIGURE 12-8**). In essence, it

FIGURE 12-8 Bottom belay.
Courtesy of Pigeon Mountain Industries.

is a substitute for the rappeller's control hand. A bottom belay can be performed by an individual at the bottom of the belay or by a second rappeller on a separate line who is lower than the belayed rappeller. Pulling on the belayed rappeller's line increases friction on the rappeller's descender. This method may be used to assist a rappeller who is in danger of losing control.

A bottom belay is not a substitute for the belays previously described. It should be used only in an emergency and when a top belay is not available. Bottom belays have significant drawbacks, including the following:

- The rope can easily slip out of the grip of the person at the bottom.
- The belayer can exert only as much pressure as their body weight. This often is sufficient if the belayer is directly below the rappeller and the rappeller has not gained too much momentum out of control. However, it frequently is not effective if applied from an angle or from so far beneath the rappeller that the pull is rendered ineffective by stretch in the rope.
- A bottom belay is not effective with all rappel devices.
- A person doing a bottom belay is in danger of being hit by objects such as rocks dislodged by the person or rope above.

- A bottom belay does not provide backup for failure of the mainline rope, anchor, or rappel device, but only for an out-of-control rappel.
- The effectiveness of a bottom belay may depend on the length of the drop. A longer drop, involving greater lengths of rope, may not be as effective as a short drop involving shorter lengths of rope.

Body Belays

An additional technique for belaying is known as the body belay. With this technique, the belayer creates friction by running the rope around their body, usually around the waist. Except in emergencies, this technique is not recommended, for the following reasons:

- It is not as easy to stop a fall as with a belay device.
- It can injure the belayer.
- The belayer's ability to hold a fall is only as strong as their pain threshold.
- If the belayer is at the top of a drop, they could be pulled over.
- The belayer can become entangled in the rope.
- If the climber falls, the belayer may be entrapped in a position from which they cannot assist the climber.

Belay Failure: The Human Element

When a belay fails, it is typically due to the weak link in the system: failure of the belayer. Such accidents are most often caused by one factor or a combination of two factors:

1. The belayer's attention drifts momentarily just as the load falls.
2. The belayer does not perform the correct actions due to either lack of training or complacency.

Because of the potential for belay accidents, some organizations adopt a philosophy that a belay should catch when the belayer is "hands free." However, this is not effective in every case, because it does not always provide the most efficient or maneuverable system. It can also be counterproductive, because a human's natural reaction at the moment of truth may not necessarily be to let go.

One basic principle about human nature is this: in a sudden emergency, humans respond with what is instinctive. Such emergency situations could relate to driving experience under the threat of a vehicular accident or to weapons training in a law enforcement confrontation. Belaying is a similar activity in that it must be thoroughly learned under realistic conditions, followed by constant practice, until it becomes instinctive. Otherwise, the belayer may fail to take the correct action when the emergency suddenly occurs. Such failure could result in severe injury or death.

Dual-Tension Rope Systems as Backup Systems

Dual-tension rope systems (DTRS) can be any one of a number of lowering systems in which both the primary and the backup system are loaded. The two-tensioned rescue system is perhaps the most commonly used of these, and are an operations-level skill as they may be utilized with or without a rescuer being suspended from the system(s). Another approach to DTRS is the traveling brake system (TBS). The TBS is a technician skill, as the second system is operated by a rescuer who is on rope. Dual-tension rope systems are discussed in Chapter 10, *Principles of Rigging*. The traveling brake system is covered in Chapter 19, *Pick-off and Litter Management*.

Two-Tensioned Rescue System (TTRS)

The two-tensioned rescue system (TTRS) is a type of moving DTRS in which both the primary system and the belay line are essentially identical, often a mirror image of one another, both operated from secure (but usually separate) anchor points.

Some TTRS setups involve two ropes running through one braking device, such as a brake tube or multipurpose device. Other systems use two brakes, each controlling a rope attached to the rescue load. In two-brake systems, each lowering device backs up the other.

TTRS setups offer several advantages in certain situations. For example, they do not require special equipment solely for the belay, and the roles of brakeman and belayer require the same skills and can be interchanged easily. Also, TTRSs usually do not accidentally engage and hang up, they do not require LR hitches, they provide a soft catch, and they are predictable in the way they will perform.

However, in such a system, the individuals operating the brakes must be well coordinated and alert for failure of the other system. With certain devices, there may also be difficulty with changing direction suddenly from lowering to hauling. As with all rescue operations, equipment and methods should be selected based on circumstances and need.

The TTRS may be operated using any descent device that is also capable of being used as a belay. In remote rescue operations, even tube-style sticht plate-type devices are sometimes used for this purpose. In general, considerations for evaluating a device to perform these dual functions include the following:

- Load range/maximum rated load
- Ease of use
- Smooth transition during initial release of heavy load
- Ease of transition from lower to raise (and vice-versa)
- Whether additional friction is required for heavy loads
- Effective function as a progress capture

See Chapter 8, *Rescue Equipment* for a discussion of specific equipment.

Operating the TTRS

Operation of the TTRS with a braking/belay device is addressed in Chapter 14, *Lowering Systems* with a focus on lowering. Here, we will focus on the belay aspects of the TTRS.

Use of this system requires approximately the same number and distribution of personnel as any other vertical system—the main difference being that instead of a belayer, there is simply an extra brakeman. That said, it should be established between the two brakemen at the outset which will serve as "Brake" and which as "Belay," as responsibility for system control will rest primarily with the "Brake" (in accord with the attendant).

The braking/belay device used for both ropes should be capable of controlling the full load, if necessary. The two devices need not be identical, but they should be essentially interchangeable; that is, each individually capable of both braking and belaying. The two should be able to see and hear one another, and both should be aligned with the fall line. Because these systems are typically rigged very similar to one another, this is often referred to as a "mirrored system."

With both ropes attached to the load in an approximately equal manner, and such that the failure of any one point will not result in catastrophic failure, the two ropes should be passed through the braking/belay devices. As the attendant prepares to back over the edge, both brake and belay should feed rope through the system such that the brake is taking about 75 percent of the load and the belay has just light tension on it. If there is too much friction for the attendant to pull both lines through both devices smoothly, the primary brake should serve as primary while the belay actively feeds line through the secondary device, keeping it just hand tight. This takes practice, but with experience becomes a smooth operation. A high directional can be helpful during this transition, allowing the attendant to pull a bit harder.

If, at any time, there is reason for the brake and belay to switch roles, this should be communicated directly between the two. Ideally, such changes are initiated by the brake, with language such as "Transition to your load." Once majority of the load has been transitioned to the belayer, they should respond by saying "My load, I am brake," to which the original brakeman (now belayer) should respond "Your load, I am belay."

The beauty of a TTRS that incorporates autolocking bollard-type devices is that it can be easily transitioned to become a raising system. This is best achieved by simply attaching a pulley (with a rope grab) to the mainline of each system to create a 3:1 mechanical advantage on each of the lines. If additional mechanical advantage is needed, this can easily be adapted to an in-line 5:1 in the usual manner. With coordination, approximately equal tension can be maintained on the two lines by hauling on both at the same time. The goal is to maintain roughly the same amount of tension on both lines so that they have approximately the same amount of stretch. If resources are limited, it is possible to rig the haul system on just one line and use the second line as belay only, but the efficiency of this system will be somewhat reduced, and in the event of a failure, the load will travel further. Mechanical advantage is described in Chapter 15, *Mechanical Advantage Systems*.

After-Action REVIEW

IN SUMMARY

- A belay may be called for any time the risk of falling arises. At a minimum, belays should be used any time the combined potential for and consequences of a fall are unacceptable to the situation.
- There are two approaches to belay. Self-belay is a form of fall protection that a person manages themselves. Active belay refers to a rope rescue system that consists of a secondary brake device operated by a belayer.

- Some industries use the term active fall protection instead of self-belay. Some active fall arrest systems used for workplace safety include the following:
  - Lanyard with force absorber
  - Vertical lifeline fall arrest system
  - Self-retracting lifeline
  - Rope access backup
- Passive fall protection refers to stationary systems that do not move, do not require the use of personal protective equipment, and require no active participation from the worker. This might include such methods as guardrails or netting systems.
- A belay system includes a load, a harness, a rope, an anchorage system, a belay system, a belayer.
- How a load is belayed will depend upon its weight.
- Rescuers must think about every part of the rescue system to try to identify the weakest points, the parts of the system most likely to fail, and the greatest risks to the system. This can be accomplished by examining every element of a rope rescue system and applying the critical point test.
- Dual-tension rope systems (DTRS) can be any one of a number of lowering systems in which both the primary and the backup system are loaded. The two-tensioned rescue system is perhaps the most commonly used of these, and are an operations-level skill as they may be utilized with or without a rescuer being suspended from the system(s).

KEY TERMS

Active belay A means of protecting another person from a fall by means of a system that requires active participation on the part of a belayer, as in by exerting tension on a trailing rope.

Active fall protection A form of protecting a person from a fall through the use of harness-based personal protective equipment, and involving active participation of the user.

Belay To protect against falling by managing an unloaded rope (the belay rope) in a way that secures one or more individuals in case the mainline rope or support fails.

Belay device A braking mechanism through which a secondary line, also called the belay line, is rigged. The device must allow free run of the belay rope through it when the system is operating correctly; it must exert a slowing or stopping action on the belay line if an uncontrolled descent or fall occurs on the mainline.

Belay line The line attached to one or more individuals that provides protection against a fall or system failure.

Belayer Person who operates the belay system.

Critical point test An assessment that determines if there is a point in the system that, if it fails, will result in failure of the entire system.

Fall arrest The form of fall protection that stops a load (or person) that is falling.

Lanyard with force absorber A length of rope or webbing with connectors at each end that is designed with an integrated means of mitigating arresting force on the body in the event of a fall.

Münter hitch A simple method of configuring a rope around a carabiner in a way that applies friction for the purpose of controlling a load, as for belaying.

Passive fall protection A form of protecting a person from a fall through the use of methods or equipment that does not require the wearing of personal protective harness or gear, and that does not require interaction on the part of the user.

Rope access backup A vertical lifeline-like type of safety system used in rope access, wherein the backup system is completely interchangeable with the progress system.

Self-belay A form of fall protection that a person manages themselves, such as a rope grab.

Self-retracting lifeline A type of fall arrest device that contains a drum-wound line that pays out from and automatically retracts back onto an enclosed drum during use, and that automatically locks at the onset of a fall to arrest the user.

Vertical lifeline A rope, a cable, or a track to which a worker is attached via a full body harness (usually dorsal attachment), a lanyard, and a traveling rope grab.

On Scene

1. Are there ever situations in which a belay or secondary system might be contraindicated for a rescue load? If so, under what conditions might this occur?

2. In what situations might an active belay be preferred over a passive belay?

3. Under what circumstances might a free-running belay device be preferred over an autolocking device?

4. What are the limitations of a tandem Prusik belay system?

5. What are the advantages/disadvantages of a TTRS?

CHAPTER 13

Operations

Patient Evacuation

KNOWLEDGE OBJECTIVES

After studying this chapter, you should be able to:

- Identify the types of litters used in rope rescue operations. (**NFPA 1006: 5.2.20**, pp. 244–247)
- Describe the considerations for choosing a litter for a specific operation. (**NFPA 1006: 5.2.20**, pp. 244–247)
- Provide an overview of medical care priorities during a rope rescue. (**NFPA 1006: 5.2.20**, pp. 247–251)
- Discuss the impact of medical technologies on rope rescue. (**NFPA 1006: 5.2.20**, p. 250)
- Describe the medical assessments and care common in rope rescue. (**NFPA 1006: 5.2.20**, pp. 247–251)
- Identify the patient packaging concerns that must be considered. (**NFPA 1006: 5.2.20**, pp. 251–259)
- Describe the methods of protecting a subject in a litter from hazards. (**NFPA 1006: 5.2.20**, pp. 254–255)
- Describe the methods of securing a subject in a litter. (**NFPA 1006: 5.2.20**, pp. 255–259)
- Describe the process of hand carrying a litter. (**NFPA 1006: 5.2.22**, pp. 259–261)
- Describe the methods of transitioning litter attendants. (**NFPA 1006: 5.2.22**, p. 261)
- Describe the use of a lift-assist sling to carry a litter. (**NFPA 1006: 5.2.22**, pp. 261–262)
- Describe the methods of negotiating obstacles with a litter. (**NFPA 1006: 5.2.22**, pp. 263–264)
- Describe the process of transferring the subject to the care of emergency medical services (EMS). (**NFPA 1006: 5.2.20**, p. 264)

SKILL OBJECTIVES

After studying this chapter, you should be able to:

- Remove a subject's helmet. (p. 254)
- Package a subject in a litter. (**NFPA 1006: 5.2.20**, pp. 251–258)
- Perform a burrito wrap to package a subject in a litter. (**NFPA 1006: 5.2.20**, p. 256)
- Perform upper 30 tubular restraint. (**NFPA 1006: 5.2.20**, p. 257)
- Perform lower 30 tubular restraint. (**NFPA 1006: 5.2.20**, p. 258)
- Carry a litter as part of a team. (p. 260)
- Perform the offset litter lift. (p. 260)
- Perform the tap litter attendant rotation method. (p. 261)

You Are the Rescuer

You and your team have been called to a rescue at a train trestle near the edge of town. Four kids had been riding their bikes along the tracks when one went over the edge, resulting in injury. The area is rugged and will require carrying the subject up a 35-degree slope over uneven terrain.

1. What litter will you use to package and transport the subject?
2. How should you package the victim for optimum stabilization, comfort, and protection?
3. What is the best way to carry the litter to avoid becoming fatigued and dropping the litter?

 *Access Navigate for more practice activities.*

Introduction

The process of extricating the subject from their predicament and transporting them to definitive care will often require that they be packaged into a patient carrying device, known as a **litter**. Rescuers should consider whether or not a litter is actually needed during technical rescue operation, as a subject with minor injuries may be more easily evacuated simply using a harness and a rope-raising or -lowering system.

Knowing how to properly secure the subject in a litter with consideration to their injuries, the type of evacuation methods to be used, and the duration of time the subject will be confined to the litter is a foundational skill for every rescuer. Understanding how to transport and move the litter once the subject is in it is also an essential skill for the rope rescuer. This chapter reviews common types of litters used for rescue, including methods for packaging, protecting, and caring for a subject, and some general guidelines for moving a litter across uneven terrain. Attaching the litter to a rope rescue system is not detailed in this chapter, but may be found in Chapter 14, *Lowering Systems*.

Litter Function

Litters provide the following basic functions in the rescue environment:

- They serve as a means of transporting a sick or injured subject.
- They help physically stabilize the subject during transport.
- They protect the subject from physical and environmental hazards and from further injury.
- They provide a means of attaching the subject to the rescue system.
- They provide a platform for medical interventions and equipment.

Types of Litters

Rigid Litters

A common type of litter traditionally used in rope rescue operations is the **rigid basket litter** (**FIGURE 13-1**). The word *basket* suggests a stiff, hard-framed litter that has enough depth to contain the subject with at least some protection to the sides as well as beneath. When rescuers use a rigid basket litter, they are said to place the subject *in* it; this differs from the terminology used with backboards or with ambulance gurneys, where a subject is said to be placed *on* the stretcher.

Rigid basket litters for rescue are constructed of rugged materials that help protect the subject and

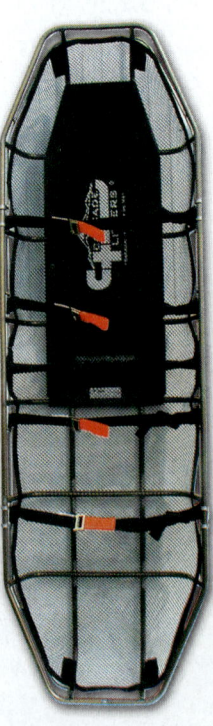

FIGURE 13-1 Basket litter.
Courtesy of Cascade Rescue.

resist damage in the rescue environment. Rigid basket litters are typically constructed of a tubular frame that is augmented with metal parts, plastic, or fiberglass for subject protection. Rigid basket litters may have a uniform rectangular shape, or they may be tapered at the foot end. A rectangular-shaped litter allows the subject to be loaded with the head at either end and permits use of wider spine boards, but the tapered style is more compact and not quite as heavy.

Rigid basket litters may be of a one-piece or two-piece design. While one-piece litters are somewhat less expensive, two-piece litters are easier to transport. Two-piece litters with a tapered foot usually nest, making them easier to store and carry (FIGURE 13-2).

Traditional, rectangular rigid basket litters may be difficult to maneuver through confined spaces; some rigid basket litters are constructed with a particularly narrow profile, allowing them to be moved more easily through confined spaces. A larger-sized subject may not fit as well in these narrow litters, and these also do not accommodate many backboards.

Metal Basket Litters

The classic tubular metal-framed basket litters with wire mesh insert are sometimes called a Stokes litter. More robust versions of these type of litters have support members constructed of tubular stainless steel; litters made of conventional mild steel tubing and metal straps are less expensive but are not as durable or long lasting as the stainless models, and they require more maintenance to prevent rusting. The advent of titanium tubing in litters is an innovation that provides an extremely lightweight and strong basket. Titanium models cost more than traditional metal basket litters.

Metal basket litters are a good choice for high-angle rope rescue, because they do not flex much even when suspended horizontally by a limited number of attachment points. They also offer good protection for the subject. Some metal basket litters have tie-in points specifically designed for rigging bridles for high-angle rescue.

Metal basket litters are typically lined with some sort of insert, attached with straps to the bottom litter rail. Inserts in older litters were often fabricated from chicken-wire, but these days various forms of mesh have replaced the wire. Pulled tight, an insert forms a flat, raised surface above the bottom of the litter. The insert adds to the subject's comfort and makes it easier to slide the subject into the litter. However, inserts should be used with care as they can raise the subject higher in the litter, changing the loading dynamics and balance.

While an open mesh litter does not provide much of a barrier to protect the subject from ground protrusions, it also does not hold water or create an airfoil, which can be an advantage in some situations.

Enclosed Basket Litters

Used by mountain rescue and ski patrollers for many years, enclosed basket litters have become increasingly popular among technical rescue teams for their lighter weight and the amount of subject protection they offer. Usually beginning with an aluminum, steel, or titanium frame, the body of the litter is then laminated with carbon fiber or fiberglass composites, much like a modern ocean racing sailboat. While carbon fiber is super lightweight, fiberglass offers the benefit of being less stiff, far less brittle, and less expensive. Some litters incorporate plastic resins for a similar effect.

This style of litter provides excellent subject protection and an excellent strength-to-weight ratio. They are available in similar size options as metal basket litters (tapered, rectangular, wide body, standard width, confined-space width) and in one- and two-piece models.

Composite or plastic enclosed basket litters can be carried, pushed, pulled, and dragged across rocks and other rugged terrain (FIGURE 13-3). The enclosed shell will protect the subject better than an open frame

FIGURE 13-2 A two-piece litter.
Courtesy of Cascade Rescue.

FIGURE 13-3 Composite basket litter.

TABLE 13-1 Advantages and Disadvantages of Enclosed Basket Litters	
Advantages	**Disadvantages**
■ Slide easily along rough surfaces and snow ■ Offer more protection for subjects	■ May retain water and snow if not properly drained ■ May be blown about in high winds and helicopter rotor wash

TABLE 13-2 Advantages and Disadvantages of Flexible Litters	
Advantages	**Disadvantages**
■ Lightweight ■ Can be stored or carried in a compact backpack ■ Adaptable ■ Drag easily over rough surfaces even when pulled only from one end ■ Reasonable cost	■ Require additional spine immobilization ■ Not rigid enough to be carried by two people at either end ■ Not as convenient as basket litters for litter teams to carry when loaded with a subject ■ Cannot be used with most litter wheels

metal basket. Most of these shells will hold up well under a severe beating, but they may look pretty bad afterward. Plastic litters are not as strong and usually do not last as long as the metal (stainless steel or titanium) basket litters.

Choosing an enclosed basket litter with a smooth bottom will help it to slide more easily over snow and rock scree. Having a clear top rail is useful for subject tie-ins and also facilitates easier carrying and rigging by rescuers **TABLE 13-1**.

Basket Litter Strength and Performance

At one time, basket litters were tested to a military specification. This is now defunct and has been withdrawn, and in its place a new, robust requirement has been included in NFPA 2500, *Standards for Operations and Training for Technical Search and Rescue Incidents and Life Safety Rope and Equipment for Emergency Services*. The NFPA standard requires that litters be subjected to a special test for strength and deformation. A similar test method may be found in ASTM Standard F2821, Test Methods for Basket Type Rescue Litters. The test method pulls on the litter with a four-point bridle from the center single attachment point of the bridle, similar to a typical method of rigging a litter for lift, and requires that it withstands a minimum load of 11 kN (2473 lbf) without failure. The litter is also not allowed to deform more than 2 inches (50 mm). In addition, if the litter is designed to be lifted only from the head end of the litter for a vertical orientation of the litter, the attachment points for that rigging must pass the test, as well as be NFPA 2500 (1983) certified. Unless a manufacturer provides verification from a third-party that their litter has passed this test, the user has little to go on to confirm strength and performance of the device.

Flexible Litters

Unlike a rigid basket litter, a **flexible litter** does not have an inherent rigid structure; rather, it wraps closely around the subject. Usually constructed of flexible plastic or other durable material, their flexibility makes them a good choice especially for confined spaces. These types of litters are also more likely to fit in smaller aeromedical transport helicopters. Advantages and disadvantages of flexible litters are listed in **TABLE 13-2**.

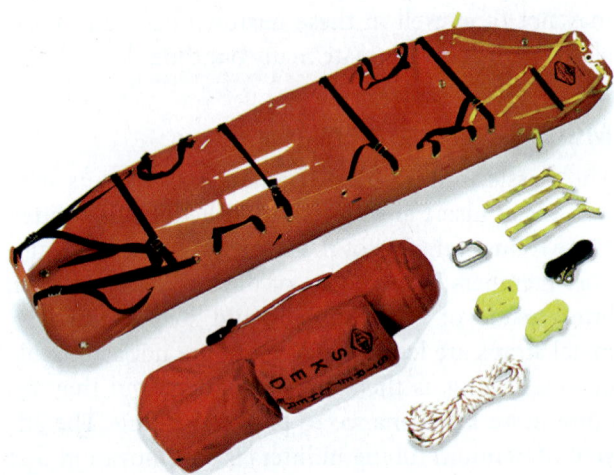

FIGURE 13-4 A Sked litter.
Courtesy of Steve Hudson and Kris Green.

Sked Litter

The Sked litter is a commonly used flexible litter that consists of a dense sheet of polyethylene plastic (**FIGURE 13-4**). When the Sked is conformed around the subject like a cocoon, the litter becomes more rigid. The litter has built-in straps and buckles to help secure it around the subject (**FIGURE 13-5**). Because it conforms to the subject's size, the package is sometimes easier to move through confined spaces than basket litters.

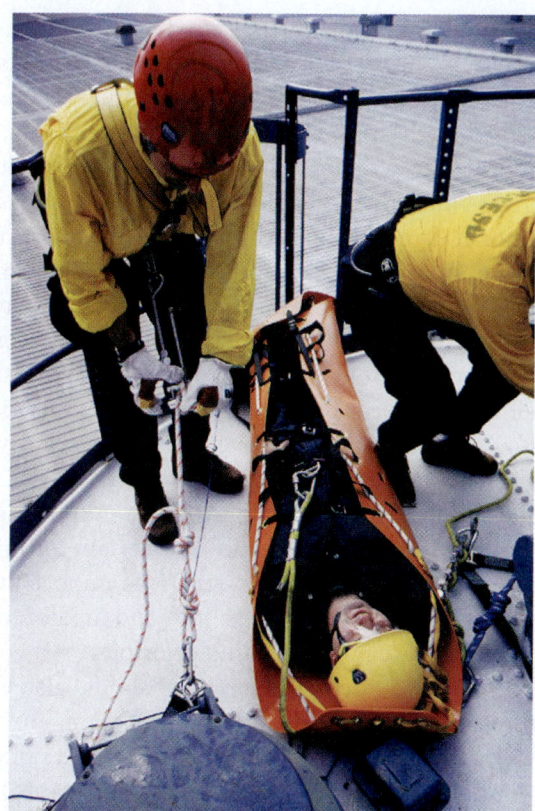

FIGURE 13-5 A Sked litter conformed around a subject.

Carrying a subject packaged in a flexible litter for long distances is an acquired skill as the number and configuration of carrying handles can be limited.

Choosing a Litter for Rescue Operations

When selecting a litter for rescue operations, consider the following:

- Whether the litter will be used for low- or high-angle rope rescue
- Strength of the litter
- Subject restraint
- Level of subject protection
- Presence and nature of injury to the subject
- Rigging attachments and tie-in points
- Number and location of handholds for carrying

Medical Care

The delivery of medical care during rescue is made even more challenging by hazardous settings such as electrical towers, structure fires, rockfall, avalanche danger, and lightning strikes. Under these conditions, a balance must be achieved between the delivery of on-scene medical care and expedient evacuation/extrication from dangerous environments. This example is only one of the high-stakes decisions that must be made during technical rescue. The wide range of conditions common to technical rescue requires that rescuers view each case independently and resist adopting a "one-size-fits-all" protocol-driven approach to medical care, packaging of the subject, and rescue rigging.

In recent years, there have been important advances in reducing the division of rescue team members into riggers and rescuers. This breakdown of preexisting walls between technical and medical rescuers has been achieved through cross-training in all disciplines critical to integrated technical rescue. A well-rounded rope rescuer should be sufficiently competent with both technical systems and medical care. As always, follow local standard operating procedures and the medical directives for your level of emergency medical training.

Medical Technologies

Until recently, medical technologies employed in technical rescue were inevitably adopted from other emergency medical care settings, in particular, prehospital emergency medical services (EMS). In recent years, however, the rising demand for small, portable, robust

Sked litters are designed with a configurable harness system that accommodates connection of a technical lowering or raising system to move the packaged subject in either a horizontal or a vertical alignment. For added rigidity and protection, the subject can be secured into an Oregon Spine Splint II (OSS II) before being packaged into the flexible litter. The OSS adds structural stability and an additional barrier for the subject, including a spreader at the shoulder to reduce the feeling of shoulder squeeze for the subject.

Some of the newer versions of flexible litters offer built-in subject restraint systems, somewhat like a harness. Often they are encased in a durable, abrasion-resistant fabric. There are pros and cons to these designs. While such adjuncts increase comfort and security, anything added to the system increases weight, and any coverings over the bare plastic reduce the ability of the litter to slide as well across rugged surfaces.

One advantage of this type of litter is that it can be rolled into a compact shape that is stored or transported in its own backpack. At less than 9 inches (22.9 cm) in diameter and under 37 inches (91.4 cm) long, this litter weighs only 17 pounds (7.7 kg) (inclusive of straps), and is relatively easy to carry into a confined space or transport long distances.

medical monitoring devices has prompted the development of a number of rescue-viable technologies, which are beginning to appear in rescue teams. Examples of newly developed technologies suitable for technical rescue include new wireless systems that allow packaged, hypothermic subjects to be monitored without unpacking them; smartphone applications; miniaturized ventilators; and compact heat systems for the treatment of hypothermia. Some simple preexisting technologies can also be adapted to work nicely in the vertical world. For example, applying a heart rate monitor to the subject's chest that transmits their heart rate to your watch can allow you to continuously monitor the subject even while several feet away.

Key attributes of technologies found to be helpful for subject care during rescue include the following:

- Simple, adaptable, and durable
- Small, light, and easy to handle
- Durable extended power source
- Valuable data and quality display or alarm settings
- Reasonable cost with high return on investment

Whether equipment is derived from standard EMS settings or is intended for the rescue environment, the team must consider how and how often the equipment will be used for subject care. A key variable is the durability of the equipment: technical rescue exposes equipment to inclement weather, extremes of temperature, and physical damage caused by impact against elevated platforms or rock faces. Teams should anticipate that destruction of valuable equipment is likely. The use of expensive equipment should be carefully reviewed to determine the risk–benefit ratio for each rescue.

Many subject monitoring devices designed for urban prehospital EMS are too large, heavy, or complex for use in helicopter short-haul and high-angle rope rescue. Dependence on fragile, expensive, and potentially unreliable equipment may ultimately be a disservice to subjects. For example, medical equipment designed to clamp onto ambulance stretcher railings may interfere with the rope rescue system if it is attached to the basket litter rail, where it may interfere with the litter bridle.

In some cases, minimal changes in standard technique may be all that is required to accomplish a medical treatment objective. One example of this involves the controversy over using rigid femur traction splints within litters. There are multiple examples from the field of the splint becoming entangled with litter bridles or becoming hung up on obstructions during raising or lowering operations. In the case of femur fractures, foregoing stabilization of a fractured femur is undesirable because the severity of damage to adjacent muscle and neurovascular structures can potentially increase with the length of time the femur remains unaligned, in addition to a potential for increased blood loss. There are a number of small, low-cost alternatives to traditional femur splints that are more appropriate for technical rescue. Examples include the Slishman Traction Splint traction device or equivalent equipment. The rescuer should keep in mind that even a traction splint could have negative consequences by causing pressure necrosis from prolonged application. Virtually all rescue teams have found it necessary to alter the rescue system to reduce interference with the care of the subject, or to change the manner in which medical care is provided to allow for certain technical rescue configurations. Either way, it is important to remember that rescue of the subject and medical care of the subject are not conflicting priorities.

The choice of medical equipment should also vary with resources, the needs of the rescue team, and the needs of the subject population served. For example, metropolitan fire departments and rescue teams that perform technical rescue at amusement parks and in metropolitan areas may expect to encounter a higher percentage of medical problems than a mountain rescue team that mainly responds to climbing accidents.

Benefit to Subjects

Technical rescue is a treatment environment that imposes inherent limitations on the ability to deliver medical care. Some teams, therefore, may find it impossible to provide certain levels of care to subjects. Consequently, the rescue team should critically review each piece of medical equipment to judge its benefit to the subject. The advantages of specific medical devices should be weighed against alternative methods of performing the task required of the device. For example, patient monitors, which record cardiac rhythm, capnography (exhaled carbon dioxide), oximetry (oxygen saturation), and blood pressure, are usually too expensive, cumbersome, and heavy for technical rescue. More important, the data provided by these devices can be acquired through a combination of skilled manual assessment of vital signs and the use of small, cost-effective devices such as finger-tip oxygen saturation monitors.

Tragedies in patient care can be avoided if rescuers resist the loss of skill that occurs when care providers become dependent on medical equipment. Specifically, it is unwise to rely on systems that lack redundancy or have functions that can be duplicated with manual skills. An example of proliferating equipment that poses a risk to agile, responsive operations is reliance on communication gear, global positioning system (GPS) units, or medical telemetry equipment. Eventually, such equipment will be unavailable or inoperable, endangering mission readiness and subject care. For this reason, teams should depend primarily on the assessment skills of the medical providers and traditional field craft such as map and compass skills rather than on the electronic readout of subject monitoring devices or miniaturized GPS. Despite these concerns, electronic monitors can be useful to teams called to rescue subjects with a history of cardiac disturbances or complex medical illnesses. The final decision regarding these technologies can be made only through experience and the opportunity costs resulting from purchase, training, and additional load to the team.

Technical Rescue Medical Kit

A technical rescue medical kit (TMK) is designed to work within the confines of a technical rescue system and should be used by the primary medical provider and/or **medical attendant**. The TMK is designed to be light, portable, and easily accessible to provide medical resources for the duration of a technical operation. Some providers will choose to have a TMK as part of their litter setup while others will choose to integrate this kit into the overall larger field medical kit. Under field conditions, the best TMK has the following characteristics:

- It is small and durable.
- It is constructed of waterproof, durable fabric. For caving or canyoneering and similar rescue environments, a hard shell may be preferable.
- It has a load-bearing connection point so that it may be suspended or secured.
- It is built so that items do not fall out easily.
- It has sufficient chambers for organizing the equipment to be carried.
- It has internal fasteners for securing equipment (e.g., intravenous fluid bags).
- It has contents appropriate for treating high-probability injuries.
- It is appropriate to the provider's level of certification.

The TMK design and contents require careful preplanning, preparation, and evaluation. Adapting the TMK to the skills and needs of the teams is an important first step. TMKs for a basic life support (BLS) provider likely will contain basic trauma dressings, splints, oropharyngeal airways, and suction devices. At the advanced life support (ALS) level, the TMK likely will contain a few appropriate injectable medications; an IV/IO (intravenous/intraosseous) start kit, IV fluids; and ultraportable monitoring devices, such as fingertip oxygen monitors (e.g., Nelcor). The medical attendant must also be competent in managing the technology or intervention and know what to do in the event of an emergency. This skill is particularly critical with intubated subjects transferred to medical attendants for lowering or raising operations. In this situation, the tender must know what to do if the bag mask device becomes disconnected, the subject expels the endotracheal (ET) tube, or the artificial airway becomes compromised or dislodged.

Assessment and Interventions

The following sections review the medical needs that most often arise in a technical rescue.

Triage

Technical rescue scenes can involve multiple injured subjects. It is common to encounter multiple stranded and injured climbers, multiple subjects on a tower in an industrial environment, or multiple subjects in urban settings when mass disasters strike. When you are faced with multiple subjects, it is advised to use a standardized triage protocol to aid in medical rescue decision making. Assigning a triage priority to a subject will involve the following assessment questions:

- Is this a rescue or recovery?
- What are the available resource capabilities?
- What are the resource priorities?
- What are the severities of subject injuries?

Several triage assessment tools have been created in emergency medicine to aid the provider in making rapid decisions when faced with a multiple-casualty emergency. One such tool named Simple Triage and Rapid Treatment (START) triage is typically taught in EMS courses and can be adapted to the technical rescue environment. While specific guidance for assessing the different levels is at the discretion of the agency's local medical authority, START triage separates the injured into four groups:

- The expectant, who are unlikely to survive given the severity of injuries. These subjects are given a BLACK triage tag.
- The immediate, who can be helped by immediate interventions or transport. These subjects are given a RED triage tag.
- The delayed, whose injuries are serious and life threatening but who are not expected to deteriorate significantly over the next

several hours. These subjects are given a YELLOW triage tag.
- The minor, who have minor injuries and are unlikely to deteriorate over the next few days. These subjects are given a GREEN triage tag.

Assigning a triage tag will also set priorities for evacuation and transport. The expectant (black) subjects are left where they are found and can be evacuated after all other triaged subjects have been evacuated. The immediate (red) subjects should be evacuated by the fastest means possible (e.g., helicopter medevac if available). The delayed (yellow) subject can have their evacuation delayed until all immediate subjects have been transported. The minor (green) subjects are not evacuated until all immediate and delayed subjects have been transported. It is important to continually reassess all triaged subjects to ensure that subjects do not need to be reclassified into a different triage category.

Severe Bleeding

Severe bleeding from wounds to the arms and legs can result in shock and death. If bleeding cannot be stopped with direct pressure, a tourniquet should be applied above the wound. Follow all local medical protocols and your level of training.

Airway and Breathing

Airway and breathing are primary concerns in all emergency medical cases. In a technical rescue, assessment and management of airway and breathing are more complex than in other types of rescues. Airway emergencies that occur during rescue operations, such as an airway becoming blocked by blood, bone fragments, or vomit, require immediate action under difficult circumstances.

If the subject is unconscious, the airway must be under constant supervision to prevent occlusion that can occur secondary to improper positioning, deterioration of the subject's condition, or accumulation of body fluids. It is advisable that a medical attendant be with any subject at risk for airway or breathing difficulties, including unconscious subjects and any subject with altered mental status. In addition, consideration should be given to ensuring that any subject with spinal motion restriction have a medical attendant available to intervene in case of airway or breathing compromise. Ultimately, these criteria indicate that the majority of technical rescue subjects will require a medical attendant.

The following questions about the airway and breathing should be addressed before packaging and moving the subject:

- Does the subject have a clear airway and normal mental status?
- Could the injuries or illness result in the onset of airway difficulties during technical rescue?
- Is the mouth free of tooth fragments, blood clots, gum, tobacco, and dentures?
- Is the subject at risk of vomiting (e.g., faint, nauseated, hypotensive)? (Note: Vomiting often occurs suddenly and sometimes regardless of the individual's medical or trauma status.)
- How will the medical attendant clear the airway during lowering or raising operations?
- Can respiration be monitored while the medical attendant directs the litter over obstacles?
- How will the medical attendant perform rescue breathing if the subject develops respiratory failure?

Assessment and management of the airway and breathing in a subject may require BLS measures and devices, as well as ALS techniques. Although basic and in-depth medical interventions are discussed here, it is important to ensure medical rescue providers work with their respective levels of certification/licensure and within the parameters set forth by the rescue service medical director.

Head and Spine Considerations

Spinal motion restriction (or spinal immobilization) in austere and technical terrain can have significant ramifications on the chosen path and equipment needed to extricate a subject. Prehospital spinal motion restriction has been the topic of recent debate in the medical community, with several key organizations, most notably the Wilderness Medical Society and the National Association of Emergency Medical Services Physicians, now recommending through published guidelines a more liberal approach to cervical spine precautions. Asking your current rescue service medical director to review these guidelines can be a way of eliciting healthy conversation on such a controversial topic. It is beneficial in technical rescue operations to have a preexisting protocol to rule out cervical spinal trauma.

Indications of spinal injury can include blunt trauma with a significant mechanism suspicious of spine trauma; a severely injured subject; altered mental status (Glasgow Coma Scale [GCS] <15); or evidence of intoxication, neurologic deficits, thoracic or other significant distracting injury, and significant spine pain or tenderness (>7/10). Generally speaking, if any of these indicators are present, most medical protocols will call for spinal motion restriction.

If cervical spine precautions must be taken, EMS providers have traditionally used rigid backboards, but these are very uncomfortable for long periods and can cause pressure ulcerations. For prolonged evacuation and evacuation in technical rescue, a conforming backboard, such as a vacuum mattress, is the international standard and is recommended by the International Commission for Alpine Rescue. Conforming backboards must be properly secured with the subject in the litter so that no lateral or longitudinal movement occurs.

Head trauma subjects are prone to nausea, vomiting, altered mental status, other critical injuries, hypertension, and bradycardia. Bleeding from head wounds may be considerable, and if not controlled may result in shock. Head trauma should be considered in subjects with alterations in consciousness; inadequate or chaotic respirations; skull fractures; bloody fluid from the ears, nares (nostrils), or mouth; facial injury; head cuts; lacerations or bruises; amnesia; ataxia (difficulty balancing); or vertigo/dizziness. In the presence of suspected head trauma, the EMS provider should assess for the presence of Cushing's reflex: increased blood pressure, irregular respirations, and a reduction in heart rate, indicating an increase in intracranial pressure.

Questions about the status of the head and spine that should be answered before packaging and moving the subject include the following:

- Is there a possibility of head or spinal injury?
- Should spinal precautions be performed before the subject is moved?
- Should the subject be secured to a spinal motion restriction device before being immobilized in a litter? (Virtually all multisystem trauma subjects are treated in this manner.)
- Can the litter tender support the subject's breathing and circulation during evacuation if needed?
- Will the risk of immobilizing the subject dramatically increase the overall rescue risk to the subject or rescuers (i.e., can the subject move under his or her own power out of a technically dangerous environment—avalanche slope, cave, etc.)?

Arguably the most beneficial treatment that can be done for a head trauma subject with suspected increased intracranial pressure is simple positioning of the subject in an elevated head-up position. In a technical rescue setting, this can be accomplished by positioning the litter head up 30 degrees. Adjustable litter bridle systems can quickly accomplish this.

Packaging the Subject in a Litter

Packaging is the term used for placing a subject in a litter and securing them for evacuation. Rope rescuers may choose to evacuate a subject by litter, regardless of whether or not they are injured or require medical care. Examples of situations where rescuers may choose to evacuate an uninjured subject by litter might include when a subject is very young or very old, when a subject is particularly frightened or distraught, or when a subject has been awaiting rescue for an extended period of time.

The moment rescuers begin to apply emergency medical care to a subject, the subject becomes a *patient* and local medical directives apply to packaging and transport. For clarity, we will continue to use the term *subject* for the person being rescued in this chapter. While the local authority having jurisdiction (AHJ) may have more to say on this subject, some of the common packaging concerns (whether for an uninjured subject or an injured subject) include the following:

- Immobilization (or spinal motion restriction) and provision for medical care, as described previously, including first aid, and BLS and ALS care at the level of the responder's expertise
- Providing for the subject's comfort and protection from physical hazards
- Securing the subject for the type of evacuation to be performed

Before packaging the subject, secure the litter from potential effects of gravity, such as falling over an edge. This can be accomplished with a safety line or belay lines, or perhaps even by changing locations. If possible, package the subject in a level area. If on a slope, normally the subject should be packaged with their head upslope unless medical conditions prevent this. Having the subject's head higher on a slope helps keep the subject more comfortable and spatially oriented. Package the subject so that they will not shift or fall out, regardless of the angle of the litter.

Spinal Motion Restriction and Medical Considerations

Airway Management

As with other forms of emergency medical care, any subject must be packaged with concern for airway management. Package the individual so that you can roll them, or turn them on their side, to allow the airway to be cleared in the event of vomiting or other airway threats. Keeping a suction device handy is always a good idea. If a mask or pharyngeal airway (ET or nasotracheal tube) is required, it should be placed before the litter is moved, and stabilized to prevent displacement during movement.

Most technical rescuers wear protective leather or synthetic gloves to protect their hands during an operation. As a practical matter, it can be difficult to clear the airway effectively—or perform other medical interventions—while wearing protective leather gloves. Medical attendants who may be exposed to subject contact should wear latex or nitrile gloves underneath the rope rescue gloves and should practice quick glove removal during training.

Immobilization (or Spinal Motion Restriction)

Philosophies surrounding prehospital care of spinal trauma subjects are constantly evolving. Once upon a time, every responder was taught as a matter of course to immediately immobilize all subjects with mechanism of or suspected spinal injury by immediately placing a cervical collar, securing the subject to a backboard, and packing material around the head to prevent any movement. In recent years, as an increasing number of studies have revealed negative consequences associated with full spinal immobilization, some agencies have adopted protocols that discourage use of full spinal immobilization, except in the case of a few specific subsets of subjects.

Rescue organizations should follow the protocols established by their medical authority, with full disclosure and dialog surrounding the differences between transporting a subject in a litter as compared with transporting a subject on a stretcher. Responders should practice their spinal motion restriction technique with the litter to determine the best system in advance.

When immobilization (or spinal motion restriction) is indicated, litters should not be assumed to be, by themselves, an adequate stabilization tool. Depending on local protocol, some subjects may require only cervical stabilization, while others may require full spinal immobilization. Whenever full spinal immobilization is required, subjects should first be fitted with a c-collars before full spinal motion restriction is performed and then, finally, placed into a litter. Whatever method of spinal motion restriction is selected, it should not interfere with technical evacuation systems or carrying methods and should allow rescuers access to the subject so that they can monitor condition during transport and offer medical intervention as needed.

A

B

FIGURE 13-6 Two examples of full rigid spine boards.
(A) © narin phapnam/Shutterstock; (B) © Josefiel Rivera/SOPA Images/ZUMA Press, Inc./Alamy Stock Photo.

Rigid backboard. A traditional means of immobilization in prehospital settings has long been the rigid backboard. Spine boards may be made of wood or plastic and are available in both long and short configurations (**FIGURE 13-6**). While rigid backboards provide very inflexible protection and can be a useful in extricating a subject from an awkward predicament or entrapment, they are also known to cause increased pain and discomfort, including pressure ulcers and aggravating injuries, especially in evacuations with long transport times.

Rescuers who plan to use a full rigid backboard in a litter should ensure that it meets the following criteria:

- Be appropriately shaped and narrow enough to allow placement inside the litter
- Be made of strong, easily cleaned material
- Have attachment points to allow attachment to the litter
- Have handles for lifting
- Allow insulation of the subject

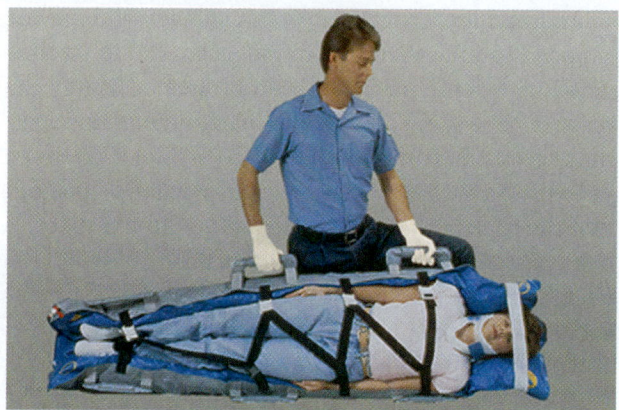

FIGURE 13-7 A full-body vacuum splint.

Full-body Vacuum Splint. A full-body vacuum splint is an excellent, albeit bulky, alternative to a rigid spine board (**FIGURE 13-7**). Sometimes referred to as a beanbag, a full-body vacuum splint is an airtight mattress-like device filled with polystyrene beads that become interlocked when air is sucked out of the container. In addition to being lighter weight and easier to transport than a rigid backboard, the full-body vacuum splint has been shown to provide the same degree of spinal motion restriction as traditional backboard methods while conforming comfortably to the subject. Because they contour to the body curves, they are less likely to cause pain and tissue damage from pressure.

Vacuum mattresses have the same overall weight as full spine boards, but they offer some advantages:

- Insulation from the cold
- Protection from water and snow
- The ability to conform to the individual subject
- Greater subject comfort in the extended transport environment (e.g., greater than 1 hour to the emergency department)

Vacuum splints also have some disadvantages, although many of these have been improved upon in recent years. The disadvantages include the following:

- Higher cost
- Risk of puncture (although the splints can be repaired in the field with a repair kit)
- Retention of water and snow
- Possible restriction of access to some areas of the subject, particularly the posterior aspects

Flexible Litters. Particular care should be taken when transporting subjects in flexible litters, such as the Sked. Even those who are not suspected of having spinal injuries may benefit from at least some spinal motion restriction, as this also helps to stiffen the flexible litter and provide additional protection to the subject. Especially in the case of flexible litters, there is always a balance between bulk and mass of the device used, versus its effectiveness. The **Oregon Spine Splint** was specifically designed to be used in conjunction with the Sked, but can also be used on its own for extrication, and/or in other types of litters (**FIGURE 13-8**).

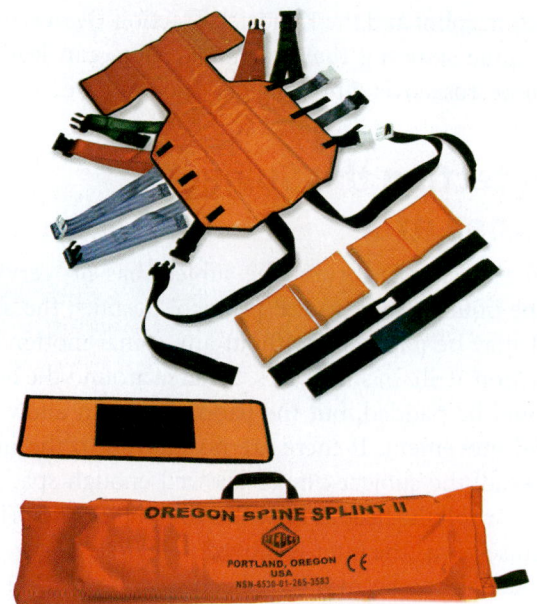

FIGURE 13-8 Oregon spine splint.
Courtesy of Skedco.

Splinting

Splinting of suspected extremity fractures is also part of the subject's packaging discussion, especially as it relates to rope rescue operations. A litter-bound subject who will be subjected to high-angle or low-angle rope rescue activities will require special consideration to optimize their protection. For starters, hard-shelled basket stretchers provide the best protection for a subject suspected of having extremity fractures, especially long-bone fractures. Those transported by flexible or open-basket litters require more precautions to prevent further damage to the injured extremity. In any case, fractures must be protected with protective splints and bandages that allow the extremity to be examined by the medical attendant to make sure the nerves and blood vessels remain intact.

Some femur traction splints commonly used in urban ambulance transport (e.g., the Hare splint) are too bulky and long to use in rescue litters. When applied to a subject, they often extend above the litter top rail and beyond the foot end of the rescue litter. This can cause additional subject discomfort, can expose the affected limb to further injury, and can complicate rescue rigging and handling. Consequently, rescuers should plan to use and practice with more compact, portable femur traction devices, such as the Sager traction splint and the Kendrick Traction Device, with the understanding that traction devices can lead to skin necrosis over a prolonged period of time.

Protecting the Subject in the Litter

If the rescuer is certain the subject has no cervical spine injuries, and if local protocol permits, the subject may be packaged without any spinal motion restriction at all. In such cases, the area around the head should be padded, but the padding should allow for head movement. If there are no injuries to the neck or head, the subject can be allowed enough space to raise or roll the head. This will allow subjects to orient themselves as the litter is moved.

A subject who may have been wearing a helmet at the time of an incident may need to be assessed for head or spinal injury, depending on mechanism. In some cases, the helmet may need to be removed so as to not interfere with proper assessment, spinal motion restriction, or the subject's airway and breathing status. If none of these hazards is present, and if the possibility of cervical spine injury has been ruled out, the helmet may be left in place. To remove the subject's helmet, follow the steps in **SKILL DRILL 13-1**:

1. *First rescuer:* Take position above the subject's head. With your palms pressed on the sides of the helmet and your fingertips curled over the lower margin, stabilize the helmet, head, and neck in as close to a neutral in-line position as the helmet allows. *Second rescuer:* Kneel at the subject's side. Open or remove the face shield if needed, and undo or cut the chin strap.
2. *Second rescuer:* Grasp the subject's mandible between your thumb and first two fingers at the angle of the mandible. Place your other hand under the subject's neck on the occiput of the skull to take control of manual stabilization. Your forearms should be resting on the floor or ground or on your thighs for additional support.
3. *First rescuer:* Pull the sides of the helmet slightly apart, away from the subject's head, and rotate the helmet with up-and-down rocking motions while pulling it off the subject's head. Move the helmet slowly and deliberately. Take care as the helmet clears the subject's nose
4. Once the helmet has been removed, place padding behind the subject's head to maintain a neutral in-line position. Maintain manual stabilization and place an appropriate-size cervical collar or other stabilization to keep the neck in neutral alignment.

Helmets are a tricky subject. A subject who may be exposed to injury from falling objects should be provided with some means of head protection before the rescue. Depending on the directives of the medical authority, this may involve some sort of shield, a helmet, or both. If a helmet is used, rescuers need to remember that the back of the helmet will tend to lift the subject's head off the stretcher, leading to forward flexion of the neck. Unintended flexion of the neck should be managed, for example by slight elevation of the shoulder and neck using a thin layer of padding behind the neck.

If no helmets are available, the head area can be packed with blankets, packs, clothing, or other soft material and the head can be taped in place with duct tape across the forehead only. Perspiration, moisture, and blood all reduce the adhesive ability of duct tape. Rescuers should not rely on duct tape alone to hold the subject's head in position. Remember that spinal immobilization straps and duct tape can accidentally slip on the subject, increasing the risk of the subject choking.

For the comfort of the subject, consider using available padding material to create a nest for the subject, particularly for lengthy carryouts. A few minutes invested in optimizing comfort at the outset will contribute to a more positive subject experience. The subject should be protected from relevant environmental conditions, including wind, temperature, and moisture. If the subject will be in the litter for more than about 20 minutes, comfort is a real concern; beyond this time, minor discomforts are amplified, and can even create safety concerns.

Subjects, especially those who are much smaller than the interior dimensions of the litter, may benefit from filler padding, constructed of blankets or other material that are rolled up and secured in place along both sides of the subject and around their head. In carryouts where the litter may be jiggled and jostled and tilted in various directions, these rolls help to prevent subject contact with the litter frame and reduce chafing. Potential chafe points which require particular attention include clavicles, shoulders, hips, and other protrusions. Beneath the subject, provide thermal insulation as well as protection from rub points in the bottom of the litter. Protecting the underside of the subject is of particular concern in a mesh or wire basket litter. This type of litter offers little protection on the bottom from protruding objects such as debris and rough surfaces. Using a beanbag-style subject immobilization system is a good way to add subject protection, immobilization, and comfort (**FIGURE 13-9**). Alternatively, the litter may be lined with material such as a closed-cell foam pad and/or blankets.

When securing a subject in a litter, pad under the hollows of the body, such as behind the knees, in the

CHAPTER 13 Patient Evacuation 255

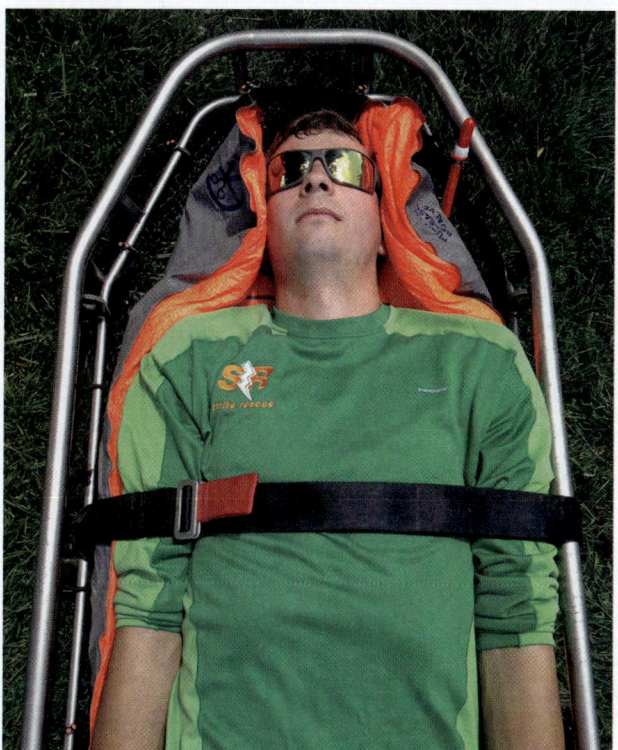

FIGURE 13-9 Using a beanbag-style subject immobilization system is a good way to add subject protection, immobilization, and comfort.
Courtesy of Jason Williams.

TIP

Considerations for Packaging a Subject in a Litter

- Do no harm. Packaging should not impede the victim's airway or circulation or cause pressure necrosis.
- Follow protocol. Directives from local medical authorities should always prevail.
- Monitor continuously. Packaging should allow rescuers to monitor level of consciousness, airway, breathing, circulation, vital signs, and other relevant medical information.
- Environmental protection. Depending on the circumstances and surroundings, this may include protection from heat, cold, wetness, debris, etc.
- Body position. Especially where duration is likely to be extended, the subject should be placed in a position of comfort that they can withstand for long periods of time.
- Bodily fluids. Whether blood, urine, or feces, rescuers should plan for the inevitable.
- Foreign objects. Check the subject's pockets and remove hard or bulky objects; ensure that the litter is free of debris and objects that could create pressure points.
- Security in the litter. The subject must be physically secured to prevent falling out of the litter in any orientation.

small of the back, and (unless cervical immobilization is a factor) behind the neck. Also pad bony parts such as the occiput (back of the skull).

For extended transport, consider adding insulation and moisture protection. While some agencies find a tarpaulin-based burrito wrap to be useful, the benefits of this should be weighed against the need to access different parts of the subject to monitor injuries or medical conditions. To wrap the subject before placing the subject in the litter, follow the steps in **SKILL DRILL 13-2**.

Where extended transport is likely, sleeping bags with zippers or Velcro closures that open all the way around the subject's torso can be useful. Remember that subjects may saturate the insulation with blood, spilled IV fluids, and waste (e.g., urine), and this increases heat loss. Wet insulation should be removed and replaced at the first opportunity.

A subject packaged securely into a litter is unlikely to be able to shield their face from light, debris, or other hazards. Head, face, and eye protection should be provided in the form of a helmet, face shield, and/or goggles. Although many subjects report experiencing claustrophobia beneath even a clear litter shield, these can provide excellent protection for the subject's face and head.

Securing the Subject in the Litter

During an evacuation, the litter will be lifted, tilted, and carried at various angles. The subject must be packaged so that they do not slip lengthwise in the litter, slide from side to side, or come out of the litter. Any tie-in system must allow access to the subject for periodic assessment and treatment, in case the subject's condition changes; it also must allow for breathing and circulation and for subject movement for comfort.

Litters often come equipped with simple strap systems that may be adequate for gentle terrain, but these often do not fasten securely enough to prevent the subject from slipping around beneath them, especially when jostled side to side or tipped lengthwise. Inclusion of a seat harness or full-body harness can be a nice touch for a subject facing a rope rescue operation, but in truth, if the litter restraint system is sufficient, a harness is not required. Avoid placement of tie-in straps directly over a subject's knees or neck and do not tighten chest straps in a manner that compromises breathing. To help prevent subject movement, straps should ideally be secured to the lower litter bar.

Equipment for litter packaging, subject protection, and medical gear, as well as other litter equipment, should be stored in their own bags and kept with the

SKILL DRILL 13-2
Performing a Burrito Wrap

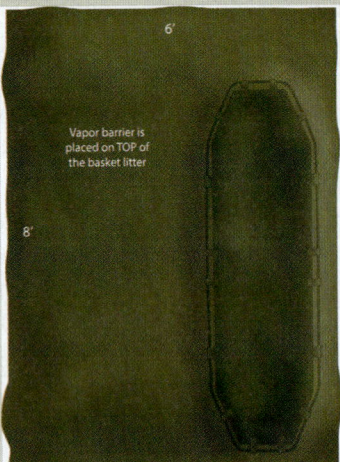

1 Lay a tarp or blanket across the litter. The blanket should be made of a nonabsorbent layer, such as polyester, to help keep the victim's skin dry.

2 Secure the subject onto a beanbag or spinal motion restriction system. Package the immobilized subject in the litter, atop the tarp. Add an insulating layer, if appropriate. Fold the waterproof tarp over the subject.

---

litter. This helps ensure that the right amount of gear is available for subject packaging and care when needed and that it is not being used in other parts of the rescue system.

TIP

Anchoring Subject Tie-Ins to the Litter
- Avoid running tie-in webbing over the top rail of the litter. Webbing over the top rail is subject to abrasion and cutting, which could weaken the subject tie-in. Also, tie-in webbing on the top rail could interfere with rescue rigging (e.g., bridles or spiders) that may need to be attached there.
- If the next lowest (second down) rail is accessible, attach the tie-in webbing to it. If the second rail is not accessible, anchor the webbing to a vertical support member, the structural tubing that runs under the litter to the opposite side. The lower the tie-in points, the more downward pull is exerted on the subject and the more secure the person will be.
- Some litters, such as the Ferno-Washington 71, have a rope laced around the lower perimeter of the basket. This is designed as an anchor for the subject tie-in.

Improvised Litter Restraint

Although manufactured restraint systems are faster and more secure, every rescuer should be capable of tying a subject into a litter using nothing more than a couple of long pieces of webbing. There are a number of different methods appropriate to accomplish this, so rescuers should choose one to commit to memory.

One such system is the **30-30 Tubular Restraint** method—so called because it utilizes two pieces of

TIP

- Do not lash webbing horizontally across the upper chest or neck. If the subject slides down in the litter, a line of webbing across this area could strangle the subject.
- If the webbing is too tight, prolonged loss of circulation could result in serious medical problems, such as compartment syndrome. In extreme cold, reduced circulation can increase the potential for frostbite or for burns from rewarming sources such as hot water bottles or heating pads. Pad the pressure points created by tie-ins. Check the subject's circulation after completing the tie-in. Recheck circulation at regular intervals.
- Litter tie-ins can work loose. Rescuers must constantly monitor the litter lashing.

SKILL DRILL 13-3
Applying Upper 30 Tubular Restraint

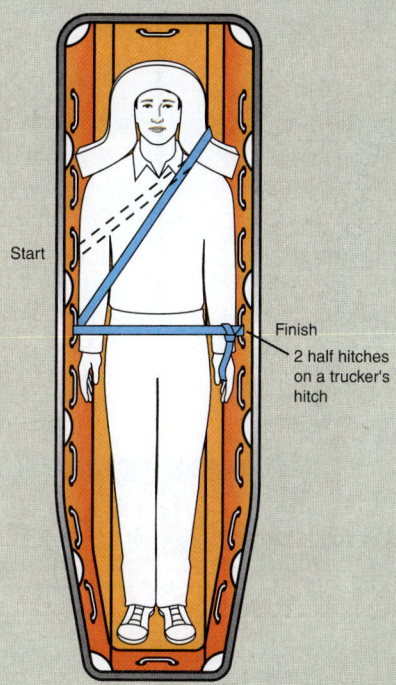

1 Using two half hitches, tie one end of each 30-foot (9.1-m) length of webbing at each side of the litter, positioned at the base of the subject's sternum. Using a horseshoe blanket roll, pad around the subject's head and shoulders. The blanket roll must cover down to the edge of the shoulders. Starting from the subject's right side, bring the webbing under the subject's right shoulder (this works best if this section of webbing is placed in the bottom of the litter before the subject is placed in it). Then bring the webbing around the subject's left shoulder, where it is padded by the blanket roll, and diagonally across the chest and right arm. Run the webbing through a litter tie-in point on the right side by the hips, bring it across the subject's hips, and tie it off at the opposite tie-in point at the subject's left side.

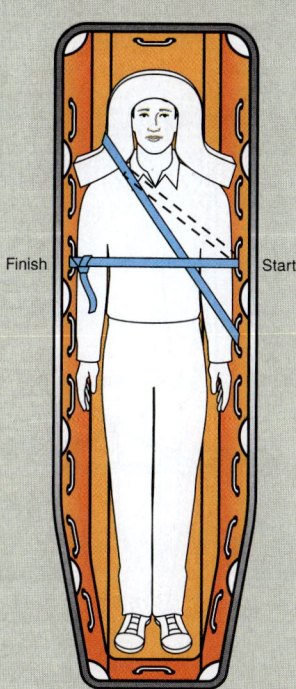

2 Now, using the webbing tied off on the subject's left side of the litter, bring the webbing under the subject's left shoulder, around the subject's right shoulder where it is padded by the blanket roll, and diagonally across the chest and left arm. Run the webbing through a litter tie-in point on the left side, bring it across the subject's chest, and tie it off at the opposite tie-in point at the subject's right side.

1-inch (25-mm) tubular webbing, each piece 30 feet (9.1 m) in length. Tubular webbing is preferred over flat webbing because it is softer and more pliable than flat webbing, making it more comfortable for the subject, and more likely to hold knots. The preference for 1 inch (25 mm) is based on this being wide enough to not exert undue pressure on the subject, while still being narrow enough for easy rigging. To stabilize a subject using the 30-30 tubular restraint, follow the steps in **SKILL DRILL 13-3**. To stabilize the subject using the lower 30 tubular restraint, follow the steps in **SKILL DRILL 13-4**.

A similar, but somewhat simpler, method of restraining a subject into a litter is the Yosemite Litter

SKILL DRILL 13-4
Applying Lower 30 Tubular Restraint

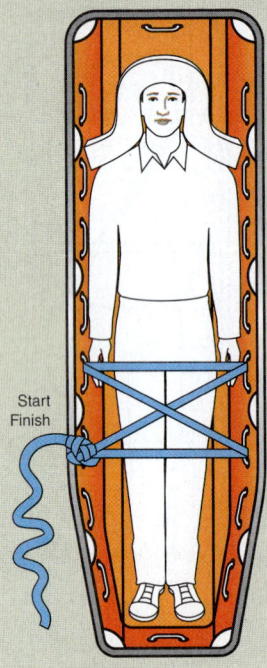

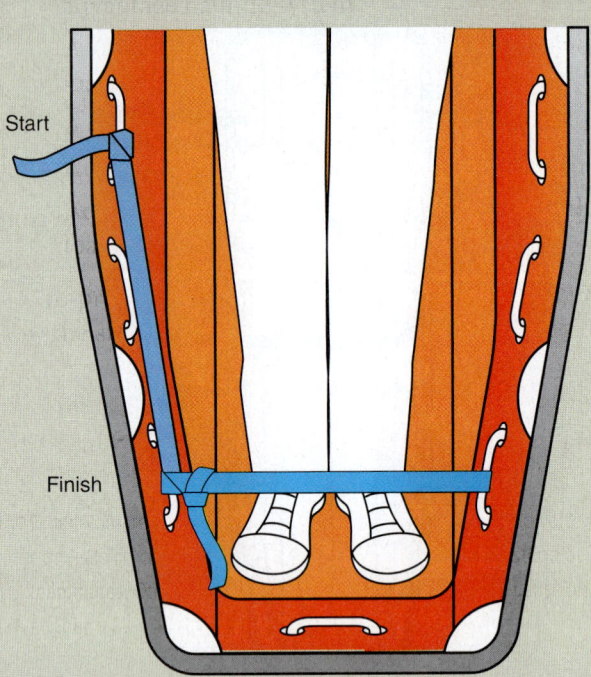

1. Find the middle of each of the remaining pieces of 30-foot (9.1-m) webbing. At the middle of the webbing piece, use an overhand-on-a-bight knot to attach the webbing to a litter tie-in point at the level of the knees at the subject's right side. Using one side of the length of webbing, run the webbing to the opposite side of the litter and run it through the next tie-in point up toward the head. Bring the webbing across the litter and run it through a tie-in point at the same level on the opposite side. Bring the webbing across the litter diagonally to a tie-in point that is the next one down toward the knees. Now bring the webbing across the litter to the tie-in point where you started and tie it off with an overhand-on-a-bight knot.

2. Take the second half of the piece of webbing, and bring it down below the feet and up to a litter tie-in point opposite the tie-in point where you started. Run the webbing through this tie-in point and back to a tie-in point at the side of the ankles. Run the webbing through this tie-in point and across the ankles and tie it off at the opposite tie-in point.

Packaging method. This works best for a subject who might not be wearing a harness and who is expected to undergo moderate jostling, as with a trail carry in uneven terrain. This method uses several nylon runners, allowing modifications as needed to accommodate for injuries or unique challenges. To secure a subject in this manner, two 18-foot (5.5-m) runners are used to perform figure 8 wraps through the groin and over the shoulders, and then several additional circumferential cross straps (approximately 12 feet (3.7 m) each in length) lock the subject down in the rescue litter.

> **TIP**
> When lacing a victim in the litter, always make certain you are not impeding breathing by lacing across the throat or compressing the chest, and that you are not causing further discomfort such as lacing across the knees.

> **TIP**
> Foot loops may go a long way toward improving the comfort of a subject who is riding in a litter, particularly when the litter must be tilted in uneven terrain or a vertical configuration.

Manufactured Litter Restraint

Manufactured, prebuilt subject tie-in systems can save rescuers time and effort when securing a subject into a litter. Some victim tie-in systems are designed to not require the subject to wear a body harness, while others are designed to require the subject to wear a body harness. Manufactured litter restraints should be preinstalled in a rigid basket litter before a rescue. Color-coded shoulder, upper torso, and leg straps assist rescuers in properly securing a subject, while wide foot stirrups keep the subject from sliding down toward the litter foot end. Strength rated buckles connect securely without knot tying. There is also a quick-secure connection for when the subject needs to be wearing a seat harness.

Carrying the Litter

A rope rescue operation rarely occurs in isolation; typically, some additional transport of the litter is required to reach a waiting ambulance, air transport, or more definitive care. Rope rescuers, and those who support these operations, should be trained and prepared to work together to transport the packaged litter across difficult or uneven terrain.

Carrying a loaded litter over any distance is hard physical work. The greater the distance, or the more rugged the terrain, the more difficult the chore becomes. Sharing the load among four to six rescuers, perhaps supplemented with some carrying accessories, can help reduce wear-and-tear on rescuers and prevent the litter being dropped. The most difficult part of carrying a loaded litter together is learning to work in unison, and not against one another.

Subject Care

Prior to moving the litter, advise the subject of the entire evacuation plan, then, continue a dialog with the subject throughout the evacuation. Being strapped into a litter and carried on your back is disconcerting at best, and spatial orientation is difficult when staring at the sky. A properly secured subject has no way of blocking the sunlight from their eyes, protecting themselves from falling debris, ducking clumsy rescuer arms, or even wiping their own nose. Anticipate the subject's needs and help them wherever possible. When rescuers take breaks for water, food, or personal relief, the subject should also be given such opportunities as well.

Litter Team Roles

A litter is typically carried with two to three rescuers, called **litter attendants** (or litter bearers) on each side, making the total cross-section (width) width of the carried litter about 6 feet (1.8 m). This being wider than most sidewalks, walking paths, trails, stairways, and other common walking surfaces, rescuers' feet will often not be on stable footing while carrying. When possible, erring toward six (rather than four) litter attendants will make for greater security for the subject and a lighter load for rescuers.

The greatest challenge with a six-person litter carry is the ability for rescuers to space far enough apart to maintain footing and keep from bumping into one another (**FIGURE 13-10**). This is particularly true when rescuers are wearing heavy clothing, equipment, or backpacks. Most people tend to be heavier through the torso than in the lower extremities, so rather than centering themselves according to litter dimensions, litter attendants should space themselves in a way that corresponds to equitable distribution of the load. This typically means erring everyone toward the head-end of the litter. A litter is only about 7 feet (2.1 m) long, so rescuers will need to adapt their spacing to ensure that they each have sufficient working space, while at the same time being sure to carry their share of the load.

FIGURE 13-10 Six rescuers carrying a liter.
Courtesy of Tom Wood, Alpine Rescue Team.

During an extended carry, litter attendants may or may not have assigned roles, such as medical attendant or **litter captain**. A medical attendant will be the litter attendant who is responsible for ensuring the medical/personal needs of the subject are attended to at any given time, at least for as long as they are in that role. This may or may not be the same person who has assumed overall medical responsibility for the subject—in fact, in most cases it is not—but for the purpose of the litter evacuation is simply the point of contact who will make a point of ensuring that medical appliances such as oxygen or splints are in place and functional, monitoring whether the subject needs a bio-break or comfort adjustment, and that the directives of the medical authority are being adhered to. There should be one medical attendant on the litter team at any given time. Sometimes this role is assigned based on position (i.e., the person nearest the head of the subject on the left-hand side of the litter) and sometimes it is identified as a specific person.

The litter captain is the litter attendant who is identified as being responsible for the forward progress of the litter and the performance of other litter attendants. Again, for the duration of time that they remain in that role, they will monitor speed and direction of the litter, stability and security of the evacuation, and any needs that the other attendants may have; they will also be responsible for issuing verbal commands to the litter team as needed. There should be one litter attendant on the litter who is identified as litter captain at any given time. Sometimes this role is assigned based on position (i.e., the person nearest the foot of the subject on the right-hand side of the litter) and sometimes it is identified as a specific person.

A litter may be carried feet-first or head-first, depending on medical needs of the subject or other parameters. When carrying a subject downhill, carrying them feet first will put their head a little higher than their feet which, as long as it is not contraindicated for shock or other reasons, can be a little more comfortable and prevent disorientation. For greatest efficiency, litter attendants should work in teams of two persons who are approximately similar in height and arm length. With one on each side of the litter, the subject will have a relatively level ride. These two attendants can switch sides periodically so that each arm has opportunity to rest. Side-sloping surfaces create an even greater challenge and require practice to negotiate effectively.

Lifting and Carrying the Litter

Numerous home and workplace injuries are attributed to lifting; as with other lifts, lifting a litter should be done with the legs, not the back. The various challenges facing rescuers, including uneven terrain, shape of the litter, weight of the subject, proximity of other rescuers, etc., can make this especially difficult. In fact, even the term *lifting* establishes a line of thinking that can lead to poor form. A better term might be to think of using offset leverage to levitate the litter. While this may seem a pedantic concept, with practice the difference will become clear. To offset the litter, follow the steps in **SKILL DRILL 13-5**:

1. Establish a litter team of six relatively size-matched litter attendants.
2. Each litter attendant should stand directly opposite their partner, each facing the litter between them.
3. Still facing their partner, attendants squat down on one knee (need not be the same knee for everyone) and grasp the top litter rail with both hands.
4. Holding the litter rail firmly with both hands, lean away from the litter with arms extended, shoulders upright, and backs straight.

The offset leverage method of lifting a litter depends on cooperation and finesse rather than brute strength and will result in the litter rising smoothly and effectively to its position of function. With practice, the same effect can be achieved with litter attendants facing forward, one hand on the litter rail.

Carrying a litter is an extension of the same concept. Litter attendants should lean slightly away from the litter, arms straight, and pulling slightly outward, against the attendant on the other side. Having even one litter attendant in the group who insists on lifting with a crooked arm as though they are curling a dumbbell will disrupt and imbalance the entire operation. Teamwork and cooperation are key here.

Finally, setting the litter down is simply the reverse operation of the lift. Head up, derriere down, litter attendants should keep their backs straight as they adjust elevation using their knees.

Rest Breaks

Rest is an important consideration for rescuers who must carry a litter long distance. Anything over about a half mile (0.8 km) can put a significant drain on resources. A litter carry can be a time- and energy-consuming endeavor, and usually requires more personnel than may seem logical. If sufficient resources are available, especially in rough terrain, switching out personnel every 200–500 yards (183–457 m) is good practice, and helps keep rescuers fresh.

Litter Team Transitions

There are a number of viable methods by which litter teams can switch rescuers on and off the litter to maintain continuity and keep personnel rested and refreshed. The easiest, and most straightforward, is simply to stop and set the litter down every so often to let litter attendants switch out, or at least switch sides. This method requires minimal coordination among team members, is simple to learn, and leaves little room for misunderstanding. However, it arguably also increases risk, offering increased opportunities for wrenched backs and pulled muscles each time the litter is set down or picked up. It also takes a few moments each time a stop occurs, requires level and secure ground on which to switch, and the repeated bending and lifting requires more energy.

The **tap litter attendant rotation method** is a more efficient method for switching out litter attendants even while the carried litter maintains forward progress, but this requires practice and experience to be effective. The tap method involves a new pair of litter attendants (left and right) inserting themselves into the litter team by grasping the trailing end of the litter on opposite sides and tapping the litter attendant in front of them who, in turn, shifts their grip forward on the litter rail and taps the litter attendant in front of them who, likewise, shifts their grip forward on the rail and taps the person in front of them. When the litter attendant at the front of the litter is tapped out, they simply let go of the litter and step up their speed to leave distance between themselves and the progressing litter behind them. Once rotated off the litter, attendants should move as quickly as possible to get in front of all of the other available reserve litter attendants, where they can catch their breath, swap sides so their next carry is with the other arm, and rehabilitate as necessary. This cycle can be repeated every few hundred yards, or at intervals dictated by terrain, availability of personnel, and duration of the evacuation. To perform the tap litter attendant rotation method, follow the steps in **SKILL DRILL 13-6**:

1. Begin a trail carry with six litter attendants, as usual, with four additional litter attendants in reserve. Reserve litter attendants should walk about 10 yards (9.1 m) in front of the moving litter.
2. When ready, the litter captain hollers "Ready to Switch!" The reserve litter attendants step off the path (one on each side) and let the litter pass.
3. Once the litter has passed, reserve attendants approach from behind to catch up to the litter so that they can grab the litter rail.
4. As soon as each reserve attendant has a firm grip on the rail, they tap the litter attendant in front of them firmly on the shoulder with a loud, "Got it," to let them know they can move forward.
5. The relieved litter attendant shifts their grip forward on the rail to the middle position, and taps the litter attendant in front of them firmly on the shoulder with a loud, "Got it," to let them know they can move forward.
6. The relieved litter attendant shifts their grip forward on the rail to the front position and taps the litter attendant in front of them firmly on the shoulder with a loud, "Got it," to let them know they can move off the litter.
7. The relieved litter attendants let go of the litter and pick up their pace to get ahead of the litter as much as practicable, switching sides with their partner for the next rotation. Repeat for several evolutions until each member has rotated on and off of the litter a couple of times.

> **TIP**
>
> **Tips for Getting the Most Out of the Tap Method**
> - Practice this method before using it on a real operation.
> - It is not a race. The litter should move swiftly and smoothly, but not at breakneck speed.
> - Litter attendants should pick partners of similar stature.
> - Litter attendants should alternate right and left arms with each rotation.
> - Establish and maintain a consistent, appropriate rotation frequency.

Some of the challenges that can disrupt the tap litter rotation method are lack of practice, lack of teamwork, and litter attendants who refuse to tap out on cycle.

Lift-Assist Sling Carry

At times, responding agencies simply do not have the resources to afford deployment of numerous rescuers to a long carryout. In such cases, a lift-assist sling can be devised from a 6- to 10-foot (1.8- to 3-m) loop of webbing. Simply girth hitch the webbing to the litter rail just behind the litter attendant's hand, run it up and over their shoulder from back to front, and then grasp the other end of the webbing with the hand furthest from the litter and press down. This will transfer the bulk of the load from the hand that grasps the litter rail to the litter attendant's shoulder. The wider the strap, the more comfortable it will be going over the shoulder.

FIGURE 13-11 Lift-assist sling.
Courtesy of Steve Hudson.

FIGURE 13-12 Cascade Trail Technician Litter Wheel.
Courtesy of Pigeon Mountain Industries.

The use of a lift-assist sling can extend the endurance of a litter attendant significantly (**FIGURE 13-11**). These can be left attached to the litter throughout an evacuation so that they can be reused by incoming litter attendants who switch into the carry. For extended carries, litter attendants who are wearing a seat harness can try clipping the end of the sling into their seat harness waist-attachment.

Litter Wheels

Even with lift-assist slings, carrying a loaded litter for any distance can be a grueling undertaking. For agencies where long carryouts are a reasonably likely event, a litter wheel is a worthwhile investment (**FIGURE 13-12**). Litter wheels are available in a variety of designs, some of which are litter dependent. For maximum versatility, use a wheel with a universal mount, to fit a wide range of litter styles and brands. Single-wheel models that incorporate a robust mountain bike or other recreational vehicle wheel offer good mobility, versatility, and optimum weight-to-performance ratio.

One example of a light, fast, and lean litter wheel is the Cascade Rescue Terra Tamer. Weighing just over 15.4 pounds (7 kg), it sports a titanium frame and fork along with a 4.25-inch (10.8-cm) fat tire bike wheel. A disk brake with an attached actuator lever helps control the load in steep terrain.

The wheel is attached to the litter by means of straps or other clamping mechanism. It can be quite difficult to load an incapacitated subject into a litter that is already mounted to a wheel, so a frame mount or quick-attachment system is especially useful. Practice using your litter and wheel combination before trying to implement them in an actual rescue.

Use of a wheel does not completely relieve litter attendants of their duties, but it can effectively reduce manpower requirements from six litter attendants to four litter attendants—or even two, if handles are used. Although it seems that adding a wheel should make transporting a litter something of a "no-brainer," in truth it is quite the opposite. Even (especially) experienced litter attendants who are accustomed to carrying litters (and especially those who rely on brute force rather than finesse) tend to meet litter wheels head-on in an engaged power struggle.

Again, it is all about finesse—not force. When using a wheel, the role of the litter attendants changes from carrying the litter to managing its stability and speed. Attendants still perform their role by grasping the litter rail and using their body to leverage the litter into place, but with the wheel already opposing the effects of gravity, lifting is no longer the issue. Instead, it becomes all about balance and rate of travel.

When transporting a litter with a wheel, attendants still work opposite a partner litter attendant at the other side of the litter, arm extended with hand grasping the top rail for control, but here the similarities end. Instead of pulling outward to levitate the litter, attendants pull in the direction they want the litter to go, with just enough offset leverage against their partner to keep the litter level. On a flat surface, they will pull forward, arm extended behind them, to drag the litter forward. On a downhill run, their arm may need to extend in front of them as they use their body weight to prevent it running too fast. When rescuers encounter a minor obstacle, such as a rock or log, the litter can be rolled up and over it just as one might lift and roll a bicycle up, over, and down the other side of an obstacle. Again, resist the urge to carry the wheeled litter.

Rescuers should practice using the wheel before using it for a rescue. Such practice can help rescuers understand the dynamics of wheeled litter handling, including how to find the center of gravity for proper balance.

Litter Handles

Another often misunderstood litter accessory are litter handles. Derived from the old Akja-style ski toboggans, special litter handles are now available to assist with the carrying of basket-style litters. Although they can add as much as 10 pounds (4.5 kg) to a trail evacuation system, with a wheel and four handles two litter attendants can effectively perform an extensive evacuation with limited person power.

The best handle systems can be removed or unobtrusively tucked into the litter when not in use, so they are easily transported into the field, and then offer adjustability in angle to allow litter attendants to easily accommodate varying terrain.

Negotiating Obstacles with a Litter

Traveling Control Lines

At times, a carryout must be performed on a low-angle surface, stairs, or other terrain that is not quite steep enough to justify a technical rescue, but a bit too sporting to leave litter attendants to deal with the situation unaided. In such cases, traveling control lines may be created using short lengths of rope (30–50 inches [76.2–127 cm]) attached to the head-end and foot-end of the litter; manned by rescuers, can help control speed on downhill runs and provide assistance uphill.

A **low-angle litter bridle** may be devised to mount traveling control lines to a litter by wrapping a short loop of webbing several times around the end of the litter and clipping it with a carabiner (**FIGURE 13-13**). Traveling control lines may be clipped into the carabiner on the bridle when needed, and the entire system simply draped inside the litter when not in use.

FIGURE 13-13 Travelling control lines.

With a traveling control line system at each end of the litter, additional personnel may be deployed as needed to assist with moving the litter. When needed to help pull the loaded litter uphill, the rescuer on the uphill line serves to assist in hauling and when the litter gets going a little too fast downhill the same rescuer can serve as slowing the litter down. Rarely will a rescue team encounter an evacuation route that is all uphill or all downhill, so it is advantageous to make these traveling control lines capable of both hauling and slowing progress.

Hand Pass

Often a litter team will encounter a short segment of particularly dicey terrain or an obstacle that is particularly unstable. It may be something as simple as a fence, a short garden wall, a piece of machinery, or a steep and rocky trail section. Whatever the obstacle, it can often be handled by using a hand-pass technique.

To implement an effective hand pass, the litter team simply stops at the point where the obstacle becomes impassable, while available reserve litter attendants pass them by and position themselves on the other side of (or even, if necessary, in the middle of) the obstacle.

The most difficult part of a hand pass is for litter attendants to commit themselves to not trying to climb up, down, or over the obstacle. With feet firmly planted, all of the litter attendants shift the litter forward, reaching toward and even over the obstacle as far as possible without letting go and without compromising the security of the load. Meanwhile, the reserve litter attendants nearest the litter on the other side of the obstacle reach out to receive the litter, with the new litter-captain-to-be informing the original team of litter attendants, "Got it!" as soon as they have a firm hold.

Without stepping forward or moving their feet, the original team of litter attendants simply begin passing the litter forward, toward the new team of litter attendants on the other side of the obstacle. As soon as any of the litter attendants from the original litter team are no longer supporting the litter, they let go and reposition themselves as needed on the other side of the obstacle. This process simply repeats until the obstacle is passed.

On particularly challenging terrain, such as in a cave, this method is sometimes used with rescuers assuming a seated position before receiving the litter, to effect what is also known as a lap pass.

Transferring the Subject to EMS

Patient Transfer Report

At some point in the operation, and perhaps at several points throughout the operation, medical care and oversight of the subject will be transferred from one medical provider to another. This might occur at various points throughout the evacuation, such as when the subject is moved from a rope rescue environment to a trail carry or ambulance, or it might occur when a medical provider of greater authority arrives on scene.

While local patient transfer forms should be used during the patient hand-off, all patient transfer reports cover a basic complement of information, including the following:

1. Mechanism of injury/illness
2. Chief complaint
3. Additional complaints
4. Vital signs
5. Treatments

Verbal Communication

Most patient handoff reports during an evacuation will be simple and verbal. To help with continuity of care, it is useful to use visual markers along the way. For example, a long strip of medical tape placed on the patient's arm next to a blood pressure cuff can be used to mark down the subject's vitals (pulse, respiration rate, blood pressure) and the times the vital signs were taken during the evacuation. A bright piece of flagging tied to the litter or to an injured extremity can help ensure rescuers avoid bumping it. A large "T" written on the subject's forehead can be used to indicate presence of a tourniquet.

The span of time available for a handoff report is typically quite limited. Rescuers should practice delivering (and receiving) information in a succinct but thorough manner so that it is easily remembered. The MIST mnemonic is part of many prehospital care report protocols. An example is shown in **FIGURE 13-14**.

Written Communication

Written reports are typically used when one provider agency or organization transfers a subject to another. Most agencies have a standardized report form specifically for this purpose. The written report is essential in that it provides the long-term memory for what happened with the subject. All who contributed to the care of the subject should also contribute to the report. Reports should be written immediately at the conclusion of an incident, so that memories are fresh and information more likely to be accurate. The written report should tell a relatively complete story of what happened, including facts about the incident, times, patient care, and other objective observations. Opinions and subjective information are not relevant here.

M	**Mechanism** or **Medical** complaint	Name, Age, Sex **Mechanism:** Speed, Mass, Height, Restraints, Number and type of collisions, Helmet use and damage, Weapon type **Medical:** Onset, Duration, History
I	**Injuries** or **Illness Identified**	Head to Toe Pain, Deformity, Injury patterns STEMI–12-lead/Stroke–Cincinnati
S	**Signs** and **Symptoms**	Symptoms and Vitals Initial, Current, Lowest confirmed BP HR, BP, SpO$_2$, RR, ETCO$_2$, BG GCS: Eyes_____ Verbal_____ Motor_____
T	**Treatments**	Tubes, Lines (location and size), Fluids, Medications and response, Dressings, Splints Defibrillation/Pacing

FIGURE 13-14 An EMS Time Out Report.
Monroe-Livingston Regional EMS Council.

A history of what happened, when and how rescue resources were activated, arrival details, and relevant care and packaging should all be carefully recorded.

Delivering the Subject to EMS

Sometimes the agency performing the technical rescue is the same as the transporting agency, sometimes not. Likewise, sometimes a rescuer will accompany the subject to the hospital, sometimes not. Depending on the situation, the transfer from technical rescue to prehospital transport may involve multiple agencies, or it may be the same agency all the way through.

Most prehospital transport resources (ground ambulance, air ambulance) use stretchers that can be fixed into place and secured during the ride. Rescue litters do not fit well into these, and so typically the subject must be removed from the litter and placed on the stretcher of the receiving agency. Medical authority is typically transferred to the receiving agency only after the subject has been placed on or in the receiving resource.

Upon arrival at the ambulance (or other receiving agency), the rescuer who has been caring for the subject up to that point should confer with the EMS provider who will take primary responsibility for the subject from that point forward. Together they will agree on a plan of action, usually utilizing personnel from both agencies to effect the transfer. Rescuers who have been involved in subject transport to that point should not leave the subject or the litter until the medical authority has confirmed that they are no longer needed.

After-Action REVIEW

IN SUMMARY

- Transportation of a subject from the rescue environment to the next stage of care often requires a litter.
- Rescuers may choose between flexible litters and rigid basket litters, depending on need.
- Enclosed basket litters offer additional protection to the subject.
- Flexible litters wrap closely around the subject and may be a good choice for confined spaces.
- Under the challenging conditions of rope rescue, a balance must be achieved between the delivery of on-scene medical care and expedient evacuation/extrication from dangerous environments. Follow local protocols and your level of training.
- Medical considerations during rope rescue include triage, spinal motion restriction, and patient packaging.
- Key packaging concerns include whether the subject should be immobilized, providing for the subject's comfort during evacuation, and securing the subject to the litter.
- Litters for high-angle rescue must be strong enough to allow rigging for rescue and to support the weight of the subject and medical attendant.
- Subjects suspected of having spinal injuries who are transported in flexible litters, such as the Sked, also require spinal immobilization.
- Devices used for spine immobilization include commercial devices such as spine boards, half boards, and vacuum mattresses.
- Basket stretchers with hard shells provide the best protection for a subject suspected of having long bone fractures.
- Before packaging the subject, secure the litter from falling downslope or over an edge.
- Any subject restraint system must allow access to the subject for periodic assessment and treatment, in case the subject's condition changes; it also must allow for breathing and circulation and for subject movement for comfort.
- Carrying a loaded litter over any distance is hard physical work, so the load should be shared between multiple rescuers and perhaps supplemented with some carrying accessories.
- A litter sling can help relieve the load of carrying a litter with one hand on the litter rail by spreading weight to the shoulder.
- Hand pass methods may be used to transfer a loaded litter a short distance across an obstacle.

- When using a litter wheel, rescuers walk beside or at the ends of the litter (or in both places), holding and balancing the litter, because the wheel supports most of the weight.
- The transfer of the patient to medical care may occur at various points throughout the evacuation, such as when the subject is moved from a rope rescue environment to a trail carry or ambulance, or it might occur when a medical provider of greater authority arrives on scene.
- Local patient transfer forms should be used during the patient hand-off, but most patient handoff reports during an evacuation will be simple and verbal.

KEY TERMS

30-30 Tubular Restraint A method of securing a subject into a litter using two 30-foot (9.1-m) lengths of tubular webbing, one from shoulder to waist, and the other from waist to feet.

Flexible litter A litter that does not have an inherent rigid structure, but rather is made of a pliable material (usually plastic) that can be wrapped closely around the subject.

Litter A transfer device designed to support and protect a subject during movement.

Litter attendants Persons assigned to participate in the movement of a litter.

Litter captain A litter attendant appointed to be responsible for the actions taken by a litter team during the time they are on the litter.

Medical attendant A member of the rescue team who provides emergency medical care to the subject during rescue and evacuation.

Oregon Spine Splint A vest-like torso splint that wraps around and can be secured to a patient's upper body to help remove them from injury sites without doing further damage to the spine.

Packaging Process and materials used to protect and keep a subject from shifting in a litter during evacuation and transport.

Rigid basket litter A patient transport device with a protective shell and depth of several inches used to carry and protect a subject during technical rescue operations.

Tap litter attendant rotation method A method used by litter transport teams to switch out litter attendants without stopping the litter.

Yosemite Litter Packaging A method of securing a subject into a litter using two 18-foot (5.5-m) runners and several additional circumferential cross straps (approximately 12 feet (3.7 m) each in length).

On Scene

1. Under what conditions might a rigid basket litter be preferred over a pram, or a flexible litter?

2. When packaging a scaffold worker who has sustained a fall and been caught by a fall arrest system, what protective measures should be taken?

3. While carrying a child out of a remote park area, your team encounters a steep, rocky section of trail about 30 feet (9.1 m) in length. How will you manage the litter down this section to reduce risk to subject and rescuers?

4. Under what circumstances might you choose to use lift-assist slings? How should these be attached to the litter?

5. What are the potential advantages of a beanbag-type backboard? Limitations?

Chapter Opener: © Jones & Bartlett Learning. Courtesy of Loui McCurley; On Scene siren: © Bildgigant/Shutterstock.

CHAPTER 14

Operations

Lowering Systems

KNOWLEDGE OBJECTIVES

After studying this chapter, you should be able to:

- Differentiate between a low-angle and a high-angle environment. (pp. 268–269)
- Describe the elements of low-angle operations. (pp. 269–270)
- Identify the components of a low-angle lowering system. (**NFPA 1006: 5.2.12, 5.2.13**, p. 270)
- Identify the minimum positions to be staffed for a rope rescue operation (**NFPA 1006: 5.2.21**, pp. 270–271)
- Identify the duties of each staffing position in a rope rescue team. (**NFPA 1006: 5.2.21**, pp. 270–271)
- Discuss the roles of litters in low-angle rescue operations.
- Describe the types of litter lowering systems in low-angle rescue operations. (**NFPA 1006: 5.2.21**, p. 275)
- Describe the methods of attaching a litter into a low-angle lowering system. (**NFPA 1006: 5.2.21**, pp. 275–277)
- Describe the methods of fall protection for litter attendants. (**NFPA 1006: 5.2.21**, p. 277)
- Explain the principles of effective communication during a lowering operation. (**NFPA 1006: 5.2.21**, p. 280)
- Describe the safety considerations when operating a low-angle lowering system. (**NFPA 1006: 5.2.21**, pp. 283–285)
- Describe the elements of high-angle operations for operations-level rescuers. (**NFPA 1006: 5.2.14**, pp. 285–295)
- Identify the components of a high-angle lowering system. (**NFPA 1006: 5.2.12, 5.2.23**, pp. 285–288)
- Describe the methods of attaching a litter into a vertical lowering system. (**NFPA 1006: 5.2.23**, pp. 291–295)

SKILL OBJECTIVES

After studying this chapter, you should be able to:

- Use webbing to create a harness. (**NFPA 1006: 5.2.13**, p. 273)
- Lower an ambulatory person. (**NFPA 1006: 5.2.13**, p. 274)
- Rig a lowering system. (**NFPA 1006: 5.2.12, 5.2.13**, pp. 274–275)
- Rig a litter for low-angle evacuation by tying the mainline rope directly to the head of the litter. (**NFPA 1006: 5.2.21**, p. 276)
- Rig a litter for low-angle evacuation by tying a closed loop directly to the head of the litter. (**NFPA 1006: 5.2.21**, p. 277)
- Secure rescuers into a low-angle lowering system, including using an adjustable tie-in. (**NFPA 1006: 5.2.22**, pp. 278–279)
- Direct a lowering operation in a low-angle environment. (**NFPA 1006: 5.2.21**, pp. 280–283)
- Communicate effectively during a low-angle evacuation. (**NFPA 1006: 5.2.21**, p. 281)

- Perform a knot pass. (pp. 282–283)
- Lower a rescuer on a dual tension lowering system. (NFPA 1006: 5.2.23, p. 289)
- Lower a rescuers litter on a mainline belay system. (NFPA 1006: 5.2.23, pp. 289–290)
- Attach tag-lines to a litter. (NFPA 1006: 5.2.23, p. 293)
- Manage a litter during a vertical lowering operation. (NFPA 1006: 5.2.14, 5.2.23, pp. 293–295)
- Prepare a two-tensioned rescue system. (NFPA 1006: 5.2.23, p. 293)
- Direct a lowering operation with two-tensioned rescue system in a high-angle environment. (NFPA 1006: 5.2.14, 5.2.23, p. 294)
- Direct a lowering operation with a mainline and belay in a high-angle environment. (NFPA 1006: 5.2.14, 5.2.23, p. 294)
- Direct a lowering operation with an aerial ladder slide in a high-angle environment. (NFPA 1006: 5.2.23, p. 295)
- Conduct a preoperational system safety check on a lowering system. (NFPA 1006: 5.2.12, 5.2.13, 5.2.21, 5.2.22, 5.2.23, pp. 271, 281–282)

You Are the Rescuer

You arrive at the scene of a partial building collapse in a major metropolitan area. There are a few subjects visible. Two of these are atop a pile of debris, obviously injured. Because of the nature of the debris, your team will not be able to hand carry a litter safely and comfortably the complete distance. Instead, they will have to attach the litter to a rope and lower it down a 25-foot (7.6-m) sloping rubble pile.

Another subject is visible three floors up in a still-standing portion of the building; the wall is gone on the collapsed side of the building, but three outer walls remain. There are no stairs left intact. The subject will have to be lowered down the vertical side of the building by rope.

1. How do you rig for a low-angle evacuation? How do you choose your route?
2. How do you rig for the high-angle evacuation? How do you choose your route?
3. What are the personnel requirements for each type of system?

 Access Navigate for more practice activities.

Introduction

Rescues should be conducted with the least possible amount of risk to rescuers. If the subject is ambulatory and capable of self-extrication, consider this option first. If a litter evacuation is required, taking the path of least resistance, and the least risk, is generally the preferred approach. The smoother and flatter the better. However, for a nonambulatory subject in steep, sloping, or broken terrain, rope rescue may become necessary.

Generally speaking, the steeper the angle of the evacuation, the greater the potential hazard to rescuers and subject. **Low-angle evacuation** is a term commonly used to describe rope-based systems used on steep but not vertical terrain. The term **scree evac** is somewhat synonymous with the term low-angle evacuation, but is generally more applicable in describing wilderness terrain than urban. In mountaineering, the word scree refers to loose and rocky slopes.

The corresponding term used to describe very steep or vertical rope-based rescue systems is **vertical evacuation**, which may also be referred to as high-angle evacuation.

The point where low-angle evacuation ends, and high-angle evacuation begins, is not always easily defined. Some organizations suggest a slope angle of 35 to 45 degrees as the differentiator, but other factors, including terrain, environment, footing, and even resources, can have an impact on this **TABLE 14-1**. The essential differences really have more to do with the way the litter and the rescue personnel are used than a specific slope angle. The National Fire Protection Association (NFPA) does not pre-define or specify a slope angle at which this differentiation is made.

According to NFPA 1006, *Standard for Technical Rescuer Professional Qualifications* and NFPA 2500, *Standards for Operations and Training for Technical Search and Rescue Incidents and Life Safety Rope*

CHAPTER 14 Lowering Systems

TABLE 14-1 Differences Between Low-Angle and High-Angle Evacuation

Low-Angle Evacuation	High-Angle Evacuation
■ The litter is supported by a combination of litter attendants and a rope rescue system.	■ The weight of the litter and the subject is supported by a rope rescue system.
■ Most of the weight of the litter attendants is supported by the ground.	■ The litter attendants' weight is supported by the rope rescue system.
■ Three or more litter attendants can be used.	■ At most, two litter attendants can be used.

and *Equipment for Emergency Services* (previously NFPA 1670, *Standard on Operations and Training for Technical Search and Rescue Incidents*), the primary difference between an operations-level rope rescuer and a technician-level rope rescuer is that the operations-level rescuer is expected to be capable of moving a subject from one stable environment to another stable environment without performing rescue skills from within an unstable environment, while the technician-level responder is expected to be capable of moving a subject from an unstable (mid-air, etc.) environment to a stable one using methods up to and possibly including exposing themselves to, and functioning in, the unstable environment.

We will learn to use lowering systems first, because here gravity is (mostly) our friend. Lowering systems teach us to work with gravity; to read the terrain, slope angles, and fall line; and to juxtapose gravity and friction to achieve an end. By the time we move on to methods for raising a load, you should expect to have a firm grasp on various ways of rigging a system.

Low-Angle Rope Rescue Overview

The term low-angle evacuation may not evoke the sense of excitement or challenge that comes with vertical, or high-angle, rope rescue. Even so, low-angle evacuation can pose significant challenges for rescuers, and these techniques require great attention to detail to perform safely. In some jurisdictions, where terrain is flat and where structures have not been built high, opportunities for low-angle rescues are even more abundant than those for high-angle rope evacuation.

Low-angle evacuation involves any inclined or rugged area over which a litter must be carried and where it is difficult or dangerous to do so without the assistance of a rope. Under some conditions, low-angle evacuation is also called broken-ground evacuation. Examples of low-angle evacuation sites include the following:

- Highway cuts and fills
- Embankments
- Construction sites
- Landfills
- Urban stairs
- Industrial environments
- Natural sloping hills
- Snow and icy slopes
- Rugged, broken terrain

Prior to evacuation, a subject should be carefully packaged in an appropriate litter using some means to ensure their security and comfort.

TIP

Prerequisites for Performing the Skills in This Chapter

Before attempting the activities described in this chapter, you must have demonstrated that you can properly do the following:
- Use and care for rope
- Use and care for other equipment needed in the high-angle environment
- Tie correctly and without hesitation the knots necessary for safe, effective work in the vertical environment (see Chapter 9, *Ropes, Knots, Bends, and Hitches*)
- Apply the principles of anchoring and rig a safe and secure anchor
- Apply the principles of rescue belaying and safely belay a rescue load (Chapter 12, *Belay Operations*)

In keeping with NFPA 1006 and 2500 (1670), a rescuer performing at the operations level may be required to access, enter, and perform rope rescue skills in a low-angle environment where the surface would otherwise be stable enough to allow them to maintain their position without a rope system but for the additional difficulties presented by the work they are doing. In such cases, rescuers may choose to use a rope rescue system to help protect themselves and/or the subject from falling or rolling down the slope.

FIGURE 14-1 Elements of a low-angle evacuation.

In short, an operations-level responder is expected to be capable of moving a subject from one stable location to another but is not expected to perform rescue skills while they themselves are in a vertical or equivalent unstable environment.

Elements of a Low-Angle Evacuation System

A low-angle rope rescue evacuation is usually composed of an anchor system, a rope, and (usually) a braking device. If the subject is not ambulatory, a litter may also be required (**FIGURE 14-1**). The number of personnel required to perform the necessary tasks will vary depending on several factors, including the weight of the subject, whether or not they are injured, slope angle, distance to be traveled, and whether or not a litter wheel is used.

Roles and Responsibilities in a Low-Angle Lowering Operation

A low-angle lowering operation can be achieved with just a couple of rescuers to assist a capable and ambulatory subject in a secure, low-risk environment, but as complexity increases, so do requirements for personnel. At the very least, any rope rescue operation requires the following:

- Someone to be in charge (operations leader)
- Someone to assist the subject (rope rescuer)
- Someone to rig the system (rigger)
- Someone to control the brake (brakeman)
- Someone to monitor safety (safety officer)

By sharing roles, as few as two rescuers might be able to carry out a rescue. For example, the operations leader might also serve as brakeman, with the rescuer also performing the duties of rigger, and the role of safety officer shared or assigned to one of the two. There are, however, risks associated with shared roles, so the goal should always be to assign separate individuals to be responsible for specific tasks.

Operations Leader

The person in charge of the rescue operations is **operations leader**. Under the Incident Command System (ICS), site operations may be the direct responsibility of the operations chief (in a small operation) or their designee (in a larger structure). For this reason, we will refer to them here simply as operations leader.

This being a weighty responsibility, it should ideally be the only role assigned to this person. The operations leader is responsible for the execution and supervision of all operational activities at a rescue site, including the method and route for evacuation, personnel assignments, and effective fulfillment of strategic goals. Depending on the size of the operation, they may have branches, divisions, groups, strike teams, task forces, or single resources working within a tiered structure or directly for them.

Rope Rescuer

One or more rescuers will be assigned roles consistent with extracting the subject. The focal point of any rescue being the subject, in a rope rescue operation it is the **rope rescuer**, the person primarily tasked with extrication and removal of the subject from the hazard, who is first assigned. When a rescuer is being lowered into place by means of a rope, that rescuer effectively takes charge of the operation for as long as the system is in play. It is at their command that the load is lowered, stopped, raised, or otherwise adjusted.

Medical intervention is not always a requirement in a rope rescue operation; sometimes the only need is to remove the subject from harm. A rope rescuer may be responsible for medical care at the basic life support (BLS) level, but need not necessarily possess advanced medical skills (unless otherwise mandated by the AHJ [authority having jurisdiction] or by circumstances). There is only so much medical intervention that anyone can do during a rope rescue operation, so advanced medical skills are generally superfluous at this point. In the event that significant medical intervention is required, it is not advisable to divide the attention of the rope rescuer between that and the rope safety system. Most importantly, the rope rescuer must be comfortable on the sharp end of a rope, and they must have sufficient technical rigging skill to adequately care for themselves and the subject in the terrain and environment where the rescue is taking place. They should also be adept in managing a distraught subject, including providing a sense of security and calm.

Rigger

Arguably the most critical position on any rope rescue operation, the rope rescue **rigger** must be capable of analyzing a rescue system in context of a rescue site, and making good decisions accordingly. They must understand equipment, systems, forces, terrain, fall lines, and evacuation routes so that they can effectively predict not just what will happen, but what will happen after that, and prepare the anchorage(s), rescue system(s), and backup system(s) (if applicable) in light of this information. Sometimes it becomes necessary for a rigger to modify systems after the rescue has commenced, due to changing circumstances.

Brakeman

When the rope runs through a braking system to create friction, the brake device must be controlled by a **brakeman**. The person in this role is responsible for the rate at which the load travels down the slope, and quite literally holds the safety of the subject and rescuer(s) in their hands. Good communication between the rope rescuer and the brakeman is essential, as it is from this individual that the brakeman takes their cues.

A brakeman should practice lowering through the device(s) used by the AHJ with different-sized loads and on different slopes so that they become familiar with what is—and what is not—a comfortable and smooth method and rate of lowering. Having some experience as a rope rescuer is also helpful, as it gives the brakeman more understanding of what is happening on the other end of the rope.

Radios are often useful for communicating between the brakeman and the load. The rope rescuer can generally add this duty to the others they are performing, but it is generally preferred that someone other than the brakeman operate the radio at the top.

Safety Officer

A safety officer, whose job it is to do nothing other than constantly monitor the entire operation to look for risks and hazards, is an important function on any rope rescue site. This individual should be an experienced person who thoroughly understands equipment, rigging, and rescue techniques so that they can survey a situation and know instantly if something is incorrect. If they do see a threat to safety, they should immediately call a stop to the operation until the matter is corrected.

The safety officer must be able to stand back with an objective viewpoint and must never be sucked into the operation itself. Safety officers check for the following:

- Hazards, including hazardous atmospheres and falling debris
- Personal safety equipment, including helmets, gloves, and breathing apparatus
- Rescue rigging, including belays, anchors, rope padding, and unlocked carabiners
- Personal rigging, including seat harnesses, locked carabiners, and knots tied correctly

This is another position that is best performed by someone who has no other responsibilities. It is not a position on which to scrimp, even if personnel are limited. The safety officer should have knowledge and experience in every aspect of the operation, including as a rigger, brakeman, and rope rescuer. They should be capable of spotting—and even predicting—potential errors and hazards almost as second nature.

Optional Additional Roles

Depending on the type and extent of the rope rescue operation, it may be advantageous to assign additional roles to help supplement the primary functions.

- Medical attendant(s): For a very ill, severely injured, or especially distressed subject, it may be wise to send a separate medical attendant to care for them so that the rope rescuer can focus on the technical aspects of the evacuation. The medical attendant is sometimes lowered on a separate rope system so that they need not become a member of the litter team and are not encumbered by the evacuation.
- Litter attendant(s): If a litter evacuation is necessary, litter attendants will be required to manage the litter. Depending on the slope angle, the subject's weight, and the system safety factor, the number of litter attendants may range from three to six. The rope rescuer can serve as one of these.
- Litter captain: During a litter evacuation, one member of the litter team should be designated as litter captain. This individual is responsible for directing the movement of the litter, communicating with the brakeman, and ensuring safety of the litter team.

- Radio attendant: Communications between the brakeman and the rope rescuer are of prime importance, but it is also important that the brakeman focus solely on the security and rate of speed of the rope running through the braking device. For this reason, having someone else—either a dedicated person, the site manager, or the rope handler—operate the radio is best for safety.
- Rope handler: A rope may be flaked, coiled, stacked, or fed out of a bag into the braking device. Without good rope management, kinks in the rope and tangles coming from the stack or bag could jam the brakes and slow the operation or bring it to a complete halt. The rope handler assists by feeding the brakeman the rope and removing kinks and tangles before they reach the brakes.
- Scouts: On the rare occasion when there are extra available rescuers, they can act as scouts to move ahead of the litter to clear a path or to warn the team of debris and other hazards. Scouts work a short distance ahead of the litter and maintain voice contact with the litter team to provide direction and information.
- Haul team: If a raising system will be required, one, two, or more rescuers will be required to pull on the haul system when it becomes necessary to raise the load. See Chapter 15, *Mechanical Advantage Systems* for more on raising systems.

Ambulatory Subjects

When there is no need to place the subject in a litter, and where the slope is not particularly steep, ropes may be used to help provide stability and security for a subject as they walk themselves to safety. At the most basic level, a **handline** can be set for this purpose. A handline typically consists of a secure anchor to which a fixed rope is attached for the subject (and perhaps rescuers) to grasp as they negotiate uneven terrain. Anchoring the handline at both ends, with a little tension between, offers the additional benefit of helping users to maintain a consistent path of travel.

Handlines are best used where there is minimal risk of injury should the user let go or fail to use it properly. If the subject is unable to walk facing downslope, the slope is likely too steep for a handline. Again, the angle at which this occurs will depend on several factors, including the experience, skill, and comfort level of the subject; presence of debris or loose footing along the path of travel; and environmental factors (visibility, moisture, etc.). Rescuers may or may not also need the handline, and in fact might even walk alongside as the subject grasps the line.

For steeper slopes, an ambulatory subject may be secured into a harness (manufactured or improvised) and connected to the end of a rope so that they can be lowered. If the subject is not wearing a seat harness, one can be improvised using an 18- to 22-foot (5.5- to 6.7-m) length of rope or webbing. There are many ways of tying an improvised full body harness; for the purposes of **SKILL DRILL 14-1**, this method incorporates both

SKILL DRILL 14-1
Tying an Improvised Full Body Harness

1 Choose an 18 to 22 foot (5.5 to 6.7 m) looped webbing sling. With the strands parallel, pass the sling around the subject's back with one bight protruding from beneath their arms on each side; hold these at chest level. Adjust the two strands of webbing behind the subject so that one of the strands lies high on the back, just beneath the armpits, while the other strand droops down behind the knees.

SKILL DRILL 14-1 (continued)
Tying an Improvised Full Body Harness

2 Reach between the subject's legs, grasp a bight of webbing, and pull it forward and up to meet the bights you are already holding at chest level.

3 Pass the two chest-bights beneath the lower bight so that they hold the lower bight up, just above waist level.

4 Stick the subject's arms through the resulting loops to create shoulder straps. Clip the shoulder straps together behind the neck with a carabiner, to keep them in place.

5 Attach a carabiner to the strand of rope at front waist and pull taut; the improvised harness will tighten throughout. Twist the carabiner several times to create a snug fit. When the harness is sufficiently tight, clip the carabiner back to the waist strand, close to the body.

leg loops and shoulder-straps, and that results in a high frontal attachment, thus offering optimum security. This type of improvised harness creates a connection point that is high enough of the subject's body to provide good security and helps prevent the webbing from slipping down over the hips when not loaded.

Ambulatory Subject Lower Method

The **ambulatory subject lower method** requires minimal person power to assist a subject down a slope while largely supported by rope. The accompanying rescuer helps to guide and direct the subject as the two of them, together, back slowly down the slope. This approach should be used with caution, and only with a subject who is relatively sure-footed. To perform this skill, follow the steps in **SKILL DRILL 14-2**:

1. Set a secure anchor and connect a braking device to the anchor.
2. Reeve a rope through the descending device and make a termination in the end of the rope.
3. Connect a harnessed rescuer to the terminated rope, using a carabiner to their ventral attachment.
4. Use a rope adjuster, such as a sliding hitch (Prusik) or rope grab to connect the subject to the rope above the rescuer.
5. As the rescuer and subject slowly back down the slope, release tension on the braking device to provide just enough friction to help control their descent.

Litters and Nonambulatory Subjects

Nonambulatory subjects typically require the use of a litter and a greater number of rescuers to carry it. Typically, the lower the slope angle, the more litter attendants are required; as the slope angle steepens, more of the load is taken on the rope and thus fewer litter attendants are needed. Although there may be as many as four or even six litter attendants, plus a subject in a litter, loading the business end of a low-angle evacuation system, the force applied to the system will rarely be anywhere near the combined weight of the load. Basic litter carries are covered in Chapter 13, *Patient Evacuation*.

Low-Angle Rope Rescue Systems

The foundation of any rope rescue system is the anchor system. Taking the time to carefully select the most appropriate anchorage at the outset of a roped evacuation will pay dividends in efficiency and security. The principles of anchoring for low-angle evacuation are essentially the same as for other types of rope rescue: the anchor systems must be of sufficient strength to withstand the highest possible force applied during loading, plus a safety factor. See Chapter 11, *Anchorages* for a detailed discussion on anchors.

Anchorages chosen for a low-angle evacuation should be positioned at the top of the intended evacuation slope, preferably back far enough from the edge where the slope begins to drop to permit sufficient working room for the rigging team. The anchor point should be in line with the anticipated evacuation route, and of sufficient strength to ensure an adequate safety factor. Anchor points are discussed in Chapter 11, *Anchorages*.

Low-Angle System Braking Devices

A braking device will be attached to the anchorage connector to serve as the primary means of friction for controlling the rate of speed at which the subject, litter, and/or rescuers are lowered down the slope. While nearly any braking device will suffice, rescuers should consider the different features and performance offered by different devices to determine which is most conducive to their intended operation. See Chapter 8, *Rescue Equipment* for more on braking devices.

Autolocking Devices

While autolocking devices have become the de-facto standard for many rescue organizations, some of these can be more than a little frustrating to operate in a low-angle slope evacuation where relatively low loads are transmitted to the brake. This is because most autolocking devices are designed to work within a specific range of force. With forces at the lower end of the spectrum, the device may not actuate properly and feeding rope through becomes quite difficult.

That said, for low-angle rescue systems where it is determined that a secondary rope system for belay is not required, the autolocking function of these devices does provide a measure of backup to the brakeman. Also, an advantage of bollard-type autolocking devices is that most are conducive to a quick and easy transition from lower to raise and back again, without the need to rig a completely separate system—again, a potential advantage where single-rope systems are concerned. Chapter 12, *Belay Operations* discusses belays in detail, while Chapter 15, *Mechanical Advantage Systems* discusses bollard-type autolocking devices in detail.

Non-Autolocking

Devices without a panic lock and without a hands-free lock can be advantageous in low-angle operations. The wide range of compatibility with different loads makes it easier for the brakeman to adapt to the constantly varying forces that are inevitable as the litter attendants jostle down the slope.

Adjustable Friction Devices

Particularly in the case of terrain that varies between steep and gentle, rescuers will find devices with easily adjustable friction (such as the brake rack) to be especially smooth. In many cases, there is simply just not enough force on the system to maintain a consistent speed. Most adjustable-friction devices are not auto-locking, so these should be used only by those who have been specifically trained in doing so.

Friction Wrap

One technique that rescuers can use in low-angle evacuation may find useful is the friction wrap (also called the tree wrap), in which turns of rope are taken around any substantial object (a post, stanchion, tree, etc.) to gain friction for controlling the load. This is particularly useful when there is a shortage of equipment for braking systems, or an abundance of substantial objects that are not easily damaged, and offers the added advantage of resolving both anchor and brake at once. This method is best used with bagged rope and should be practiced in training before being used in an actual rescue.

Rope

The rope runs through the braking device and then usually is attached to the head end of the litter for a low-angle slope evacuation. To maintain greater control over the operation, most rescuers prefer a rope with a minimum of stretch, such as a static kernmantle rope.

Static rope, classified as having less than 6 percent elongation at 10 percent of its minimum breaking strength, works best for low-angle evacuations. The reason for this is that stretchier alternatives, such as low-stretch rope, can introduce a lot of bounce as the load changes over the course of the lower due to different slope angles, inconsistent loading by litter team members, and the amount of rope length let out. A more static rope provides a firmer hold to facilitate the litter team's ability to maintain continuity in their loading, which in turn allows the brakeman to adapt accordingly. Because large impact loads are less likely during low-angle evacuations, the force-absorbing advantages that a low-stretch rope might otherwise offer over a static rope are not a factor.

The operations leader will determine whether or not a secondary, or backup, anchor or belay system is required, depending on risk (or the probability of failure plus the consequence of failure). If a secondary anchor is required, it should be further back from the edge than the main anchor, but as near as possible to being in line with the main anchor.

Litter

A rigid basket litter is typically preferred for low-angle rope rescue operations, both to protect the subject and to facilitate ease of handling. Using a rigid basket litter helps rescuers maintain an envelope of protection for the subject despite stresses and unequal loading and is strong enough to withstand the weight of the subject and rescuers while being supported by rope. Impact with obstacles is common in low-angle operations, and although these are rarely severe, a rigid litter does better to protect the subject from them.

Flexible litters may be used for low-angle evacuation, but require specialized rigging and handling by litter attendants, as they provide fewer options for holding, attaching, and carrying.

A subject who will undergo a low-angle evacuation should be carefully packaged into a litter in a manner that keeps them securely in place and prevents them from slipping down toward the foot end of the litter due to gravity. Simple straps over the subject will not do the trick. Some method of restraining the subject lengthwise is also necessary. Depending on injury, a subject may be able to help themselves by maintaining firm pressure against a foot loop that is secured to a litter rail. Other approaches might include a pelvic restraint or seat harness. Care should also be taken in packaging to ensure a patent (open) airway, including the ability to roll the litter onto its side if the subject begins to vomit. See Chapter 13, *Patient Evacuation* for a detailed discussion on packaging. A detailed discussion on managing litters in the low-angle environment is in the next section.

The litter is attached to the end of the rope, either directly or by means of a closedloop litter bridle.

Tying the Mainline Rope Directly to the Head of the Litter

If the distance of the low-angle evacuation is only one rope length and the rope does not have to be detached from the litter during the operation, the mainline rope may be tied directly onto the head of the litter. This is achieved by looping the rope several times

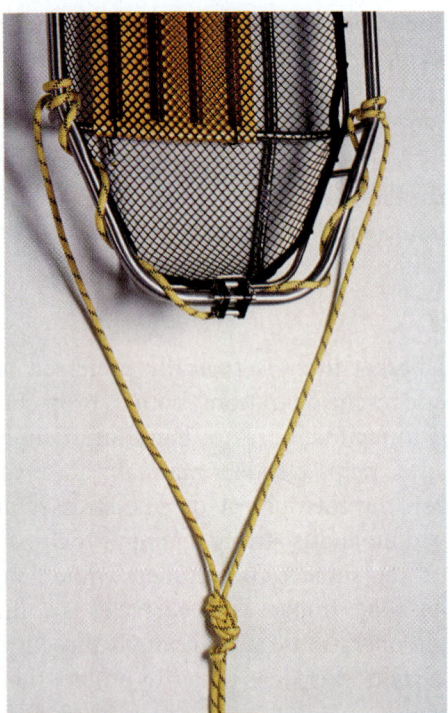

FIGURE 14-2 Direct mainline attachment to the litter.
Courtesy of Steve Hudson.

FIGURE 14-3 Tying the mainline to a litter that has plastic-covered head and end rails.
Courtesy of Steve Hudson.

around the head of the litter and then terminating with an end knot, such as the figure 8 follow-through knot (**FIGURE 14-2**). To tie the mainline rope directly to the head of the litter, follow the steps in **SKILL DRILL 14-3**:

1. At the end of the rope that is to be attached to the litter, measure off twice the distance between outspread arms (a total of approximately 10 feet [3 m]).
2. At this point in the rope, tie a simple figure 8 knot.
3. Run the rope around the head rail of the litter, using the upright structural bars on either side to help spread the forces evenly along the rail.
4. Bring the end of the rope back to the simple figure 8 knot and retrace it back through the simple 8, leaving sufficient tail for safety.
5. Center the knots so that both legs of the loops pull evenly on the litter rail.
6. Be sure to dress the figure 8 follow-through knot and pull it down tightly.

Note: This technique cannot be used on plastic litters on which the plastic material covers the rail at both ends of some models. **FIGURE 14-3** shows a technique for attaching a mainline to this type of litter. This technique uses clove hitches at each corner where the litter side rails meet the head end of the litter. Make sure to tie the clove hitches so that there is no slack between them in the part of the rope that is inside the litter.

> **TIP**
>
> Do not make the V of your tie-in too big, as this increases the hazard of it snagging on things and getting in the way during the evacuation. It is good practice to keep the angle at the figure 8 knot less than 120 degrees; although you would be unlikely to overload anything at this point, too great an angle can compromise the knot.

Tying a Closed Loop Bridle to the Head of the Litter

For low-angle evacuations involving transition from one lowering system to another, and when the litter must be freed quickly from the system at the completion of the evac, a litter bridle helps to facilitate quick removal and reattachment to the litter. In this case, the mainline rope is attached to the bridle simply by means of an end-knot or manufactured termination that is clipped with a locking carabiner (**FIGURE 14-4**).

- Monitor the bowline knot so that it does not "capsize" when being pulled over an obstruction such as a rock, tree, or building edge.

> **TIP**
>
> Always rig to the large, primary litter rail—usually, this is the top rail. Be sure to spread out the force evenly along the rail by wrapping the attachment around the rail several times. This also helps to stabilize the litter in use.

Low-Angle Litter Management

Litter management in low-angle roped evacuation systems is very different from management of a litter on level terrain. On level, unbroken terrain, the usual number of people carrying a litter is six (plus any additional personnel attending to the subject's medical needs). The full weight of the litter is on these individuals; they are completely supporting their own weight as they maneuver their way across the terrain.

Usually in low-angle slope evacuations, the rope will be tied to the head-end of the litter so that the subject is carried feet-first, with their head uphill. As noted earlier in this chapter, care must be taken to package the subject effectively so that they do not slip down in the litter. They should be secured firmly but comfortably into the litter. While the litter is still supported by the litter attendants, some additional lift can be achieved by their pulling it against the rope system.

Generally speaking, the steeper the slope, the fewer the number of litter attendants required. As the slope becomes less stable, or as it becomes less steep, more litter attendants will be required. The more litter attendants there are, however, the more they tend to get in each other's way. Too many litter attendants on a very steep slope can create undue stress on the system. There is a fine balance, then, between using enough litter attendants to do the job comfortably and safely, but not using too many.

As with most things in rescue, there is more than one appropriate way to lower a litter down a slope.

Low-Angle: No Tie-In

For a very mild slope where the rope is providing more peace of mind than actual security, the low-angle

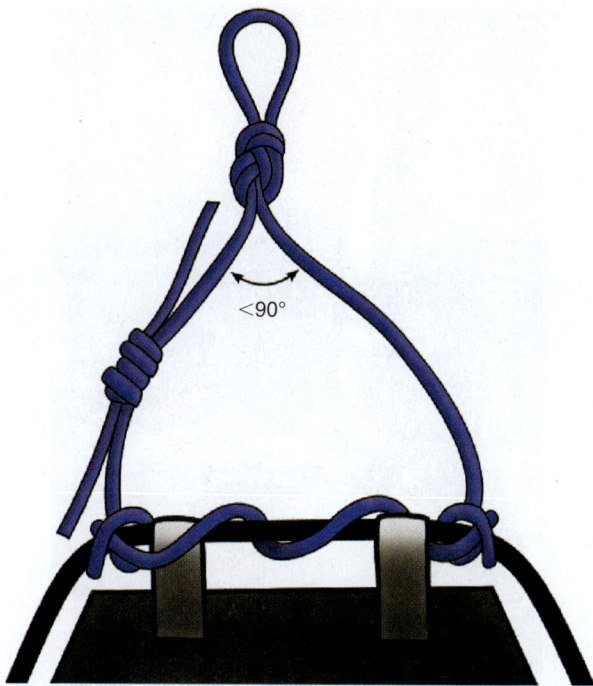

FIGURE 14-4 Tying a loop onto a litter.

To tie a closed loop bridle to the head of the litter, follow the steps in **SKILL DRILL 14-4**:

1. Take a length of rope or webbing about 6 feet (1.8 m) long. (Add additional rope if you will be tying a figure 8 on a bight in the loop as described in step 5, below.)
2. Run the rope or webbing around the head rail of the litter to spread the force around the rail.
3. Tie the two ends together with a grapevine (double fisherman's) knot, figure 8 bend knot, or some other type of joining known to have sufficient strength.
4. Adjust the grapevine knot so that it is off to the side and not in the center, where the mainline rope will attach.
5. Optionally, in the center of the loop farthest from the litter, tie a figure 8 on a bight, into which the main carabiner can be clipped.

For a more stable tie-in, start the wrap of the litter rail with a clove hitch and end it with a second clove hitch on the opposite side. This prevents the bridle from slipping around as the direction of the load shifts.

When the rope is tied directly to the head of the litter, generally a retrace-8 is the termination of choice. While a bowline knot may be used in place of the figure 8 follow-through, rescuers should take into account the following considerations:

- Make sure the bowline knot is tied correctly
- Back up the bowline knot with a safety knot, such as the double overhand backup (barrel) knot.

evacuation will largely represent a standard six-person carry with minimal variation other than the fact that there is a rope tied to the head of the litter. Performing a no-tie-in slope evacuation is quite taxing with fewer than six litter attendants.

Litter attendants align themselves two abreast, on opposite sides of the litter, facing downhill. In addition to lifting and carrying the litter, they will also need to exert downhill pressure to effectively pull against the braking system. The brakeman must be quite attentive to not over-brake the system, as this makes the job of litter attendants even more difficult. A nonautolocking braking device generally works best for this type of lower.

This method should be used with care, and only in situations where the litter attendants would be otherwise capable of effectively maintaining footing and carrying the litter even without the rope. In this case, a rope simply provides an added sense of security, somewhat akin to the handline described above for ambulatory subjects.

Low-Angle: Tie-In to Rope

Litter attendants can protect themselves from a fall by also clipping into the lowering system. One way of tying into the system for low-angle evacuation involves running pigtails down from the litter evacuation rope to each of the three or four litter attendants, who wear seat harnesses and clip into the ventral attachment. In this case, attendants face uphill (toward the brakeman) and clip into the pigtails with a carabiner attached to adjustable hitches or knots. In this case, they do not lean into the rope system, but they are secured to it in case they fall.

Using this method, the litter attendants grasp the litter rail with their hands and proceed as they would for a standard litter carry—with the exception that there are only four (or sometimes even three) of them, and they are walking backward. By pulling the litter downhill, against the rope system, they can leverage the rope system to achieve some assistance from the rope. These techniques may work well on smooth, gentle slopes that do not have broken terrain, but on steeper slopes with more rugged terrain, attendants could lose grasp of the litter rail, causing the team to drop the litter, along with the subject.

This method can be modified to incorporate three, four, or even five litter attendants. Those who use this system cite as a benefit reduced strain on the rope and rescue system. Alternative methods that involve litter attendants tying directly to the litter offer certain advantages in leverage, lift, and security.

FIGURE 14-5 Adjustable low-angle attendant tie-in.

Low-Angle: Tie-In to Litter

Another method of tying into the system for low-angle evacuation is for litter attendants to place a short lanyard between their seat harness connection and the litter rail. When litter attendants clip themselves directly into the litter rail, they can create additional leverage against the system by leaning back into the tie-ins so that their legs—rather than their arms—do most of the work. As litter attendants leverage against the rope system, the weight of the loaded litter is taken by the rope, braking device, and anchors. On a steep enough slope, and with experienced litter attendants, this essentially creates a hands-free lift.

Any number of methods may be used as **low-angle litter tie-ins**, including a pickoff strap, webbing loop, Prusik, or an **adjustable low-angle litter tie-in** (**FIGURE 14-5**). The tie-in can be attached to the litter top rail with a girth hitch, and then to the waist attachment point on the litter attendant's harness. Because each person's body proportions (e.g., trunk size and arm length) are different, a variable-length tie-in allows different rescuers to use the same tie-in, and (once attached) the litter attendant can quickly adapt to varying terrain and circumstances.

To construct the adjustable tie-in, follow the steps in **SKILL DRILL 14-5**:

1 To make a safety attachment, use an accessory cord 8 mm in diameter and about 6.5 feet (2 m)

long. Using a grapevine (double fisherman's) knot, tie the two ends together, forming a continuous loop.
2. Using a girth hitch, attach the end opposite the grapevine knot onto the litter rail.
3. Clip the end opposite the girth hitch to the litter attendant's seat harness.
4. Now, take a piece of accessory cord 7 mm in diameter and about 3 feet (0.9 m) long and tie it into a continuous loop using a grapevine (double fisherman's) knot.
5. Attach the loop over both strands of the safety attachment with a Prusik hitch.
6. Attach the other end of the adjustable loop to the litter attendant's harness using a separate locking carabiner.

The optimum length for a low-angle litter evacuation tie-in is one that lifts the litter off the ground as the litter attendant pulls back against the system with their legs, placing the litter at height where the litter attendant's hand can rest comfortably on the rail, arm bent. The trip downslope will be somewhat reminiscent of a team-rappel with the litter functioning somewhat akin to an oversized rigging plate as all litter attendants back down the slope while simultaneously pulling—hard—against the rope system via their litter connection.

Effective use of this method requires practice, teamwork, and cooperation. At the command of the litter captain, all litter attendants lean back into their tie-ins and allow the litter and rope system to take the weight. As with any rappel, stability can be improved by placing feet wide apart and as high as practicable on the slope in relation to the person's body.

A litter attendant who slips continues to hold onto the litter rail and pulls their body taut on the tie-in. The rope system and the other team members usually can keep the litter stable and prevent the attendant from falling. The litter team, in concert with the rope, acts as a sort of self-equalizing table and provides stable transportation for the subject.

If more than one litter attendant loses footing or the terrain becomes particularly treacherous, the team captain can call "Stop" to halt the descent temporarily. This gives the team the chance to regain its stability.

An attentive litter captain who gives firm, clear commands and monitors team members to optimize efficiency will help the operation go much more smoothly.

This method works best with four litter attendants, but can be modified to incorporate two, three, or even five litter attendants, depending on slope angle, weight of the subject, and other factors.

> **TIP**
> **Characteristics of Litter Management on Slopes**
> - The rope takes much of the litter's and the subject's weight.
> - Litter movement is controlled by the rope system.
> - Much of the attendants' weight is taken by the litter and the rope system.

Low-Angle Evacuation with a Litter Wheel

Any of the methods described previously—with or without tie-ins—may be adapted to incorporate the use of a litter wheel. This is particularly useful where personnel resources are limited and works best where terrain is relatively smooth. Large chunks of concrete, debris, or vegetation are not conducive to inclusion of a litter wheel.

When performing a low-angle evacuation with a litter wheel attached, the number of litter attendants required to lift the load is generally lower. However, the litter wheel changes the physics of the system, shifting the balance of the system and creating a destabilizing pivot point. The effects of any one litter attendant pulling too hard in one direction, or even falling, can be magnified by this pivot point.

As with any technical rescue system, it is imperative to practice methods for low-angle evacuation while using a litter wheel before putting this method to use in a real operation.

Practicing Low-Angle Litter Evacuation

Most rescue organizations will perform far more low-angle rescues over the course of a year than vertical evacuations, so teams should practice and become skilled in executing these. Systems for steep-angle rescues are loaded with the combined weight of several attendants plus the subject, but because the load is not hanging completely from the rope, the actual force on the system is often less than what people might assume. This is because the majority of the force is being transmitted into the ground through the feet of the litter attendants.

A foundational precept of performing rope rescue at the operations level of training and skill is that the rope system is intended to assist with the tasks of moving the load and negotiating obstacles through a steep slope environment, from one stable location to another.

The most advantageous means of providing security to operations-level rope rescuers and their subject(s) in this low-angle environment is arguably the low-angle tie-in to litter method. Therefore, any organization and/or individual functioning at the operations level should practice and become capable of executing such an operation.

Communications

Good communication and coordination, especially between the litter team and the brakeman, are essential to a successful rope rescue operation. Communication must be simple, clear, and to the point. Once the rope rescue operation commences, voice communications should be limited to a very few people—most notably, the litter captain and secondarily the brakeman. Any time a rescuer is on rope, that rescuer's communications take top priority.

If a radio is being used, the channel should be cleared and dedicated to the rope rescue operation until its completion. If radios are not available, a relay person may be stationed in a position where they can hear both sides of the system. Everyone else should keep quiet.

Another communication method is whistle blasts, which are often audible when no other form of communication is. SUDOT is a recognized whistle command system used in rope rescue (see ASTM Standard F1768, Standard Guide for Using Whistles During Rope Rescue Operations):

- **S**—Stop (one blast)
- **U**—Up (two blasts)
- **D**—Down (three blasts)
- **O**—Off rope (four blasts)
- **T**—Trouble! (continued long blast)

If verbal commands are intended to be used during a rope rescue system, the parties involved (particularly the brakeman and the litter captain) should agree to the sequence of commands that will be exercised. The person who is *on rope* (or, if a litter team, the litter captain) should be the one who provides the cues for moving, stopping, etc. By agreeing upon a pre-established set of words and their intended response, the brakeman and the rescuer can become more effective in the lowering operation.

Words selected for rope rescue operations commands should effectively communicate the appropriate information while altering cadence, intonation, and sound enough to prevent misunderstanding.

Ideally, communications should follow a statement-and-response pattern. One example of a good set of commands might be as follows:

Operations Leader: Roll call! (*general statement to entire site, and radio if applicable*)

Operations Leader: Safety ready?

Safety Officer: Safety ready!

Operations Leader: Brakeman ready?

Brakeman: Brakeman ready!

Operations Leader: Rescuer ready?

Rescuer: Rescuer ready!

Operations Leader: Litter operations commence when ready.

These commands may be used in any order, as dictated by need/situation. A similar series of commands may be used in raising operations. See Chapter 15, *Mechanical Advantage Systems* for more information.

The specific communications used matter much less than that the people using them on a given operation agreeing to them. Details may vary somewhat among rescue groups, but the basic premise remains: communications should be closed loop, and all those involved in the low-angle rescue agree on the particular sequence of commands and responses.

A Typical Low-Angle Lowering Operation

The following sequence outlines the basic elements of a low-angle lowering operation.

Preparation

Before the litter moves, all major elements of the low-angle evacuation system should be in place and prepared. The safety officer should verify the following:

- The subject has been medically assessed, treated, and packaged, and their condition is being monitored.
- All systems have been properly rigged.
- The rope has been properly attached to the litter.
- Secure anchors have been set and a braking device or descender attached to them.
- The rope is wrapped in the braking device or descender and locked off.
- The brakeman has the rope in hand and is ready to run the braking device.
- The rope handler is ready to feed the rope to the brakeman.
- The belay line is set in the belay device (if applicable).

- The belayer is ready to belay (if applicable).
- The litter attendants are attached to the litter with tie-ins and are ready to lift the litter.

When the operations leader senses that all systems are ready, they initiate the process as follows:

Operations Leader: Roll call! *Gets everyone's attention. Operations are about to begin. This is a sort of "quiet on scene."*

Operations Leader: Safety ready?

Safety Officer: Safety ready! (or, "Standby, X minutes!")

Operations Leader: Brakeman ready?

Brakeman: Brakeman ready! *The device is unlocked and in hand.* (Or, "Standby, X minutes!")

Operations Leader: Litter ready?

Litter Captain: Litter ready!

Operations Leader: Litter, on your command.

Litter Captain (directing the litter team to lift the litter): One, two, three, lift.

Litter Captain: Tension. *Litter team members remove slack from the rope by holding the litter downhill against the rope and leaning into their own tie-ins to apply tension to the system. The brakeman holds the braking system fixed. This pretests the system and ensures that everything is in order and that the litter team's tie-ins are correctly set.*

When the litter captain determines that the system is stable and the team is prepared, he commences lowering as follows:

Litter captain: Down slow. *The brakeman begins to let the rope through the braking system. The litter team leans into the system and moves downhill. (The litter captain may say, "Down faster" if they want to move downslope faster, or "Down Slow" if they want the litter to move more slowly.) If anything begins to go wrong (e.g., a kink slips past the rope handler and jams the brakes or a litter attendant begins to lose a boot), anyone can call, "Stop!"*

The rope handler is responsible for monitoring the amount of rope remaining to be fed into the braking device. At some point before lack of rope becomes critical, perhaps when around 20 feet (6.1 m) are remaining, he warns the brakeman that the rope is running out by calling or radioing, "Twenty feet of rope!"

The litter captain looks for a good place to set down the litter. When the litter has reached that place, the litter captain calls, "Stop!" The captain then directs the team to set the litter down. When the litter is secure and the team is in a stable position, the captain calls, "Off rope."

Evacuation of More Than One Rope Length

Sometimes, low-angle evacuations occur in areas where the litter must travel a distance down a slope further than the length of available rope. In this case, the operations leader has two choices: they can either extend the system by tying another rope onto the end of the first one, or they can establish a new anchor system downslope and move the system. In either case, close coordination is needed to ensure smooth and effective transitions.

Adding a Rope. Where a determination is made to simply tie another rope onto the end of the first one, it may be necessary to perform a knot pass. It may be possible to simply set the litter down and re-reeve the descender on the other side of the knot, but if not, a well-practiced team can perform a knot pass with minimal interruption to the operation by using a second anchor, a second brake, and a **transfer line** to temporarily control the load while the first device is detached and moved to the other side of the knot. A transfer line can be made using a 30–50 foot (9.1–15.2 m) piece of anchor rope, or simply by borrowing the unused and of the new (second) rope. The process for this is described in **SKILL DRILL 14-6**.

While this may sound complicated, with practice the transition can be very smooth and efficient, with minimal interruption to the lowering operation.

Transferring to a New Braking System. Another option is to establish a new anchor and braking system at a strategic location downslope and leapfrogging a new brakeman into place, ready to receive the load just before the first rope runs out. Ideally, this anchor and braking system would be set some 20 feet (6.1 m) before the original brake runs out of rope, but this is not always something that can be custom ordered.

Setting a new anchor and braking system can be resource intensive, consuming a significant amount of time, equipment, and personnel. It is best to avoid the operation having to come to a screeching halt to wait for rigging every time a switch is made. Two solutions to this problem are prerigging and leapfrogging.

If enough equipment and skilled personnel are available for anchor rigging, the anchor and brake systems can be rigged ahead of the litter team. This is called **prerigging**. Anchors should be set so that the litter just passes the new braking system by about 10 feet, where it can be set down while connection is made between the new braking system and the litter. To this end, a new rope can be used (if available) or the original rope

SKILL DRILL 14-6
Performing a Knot Pass

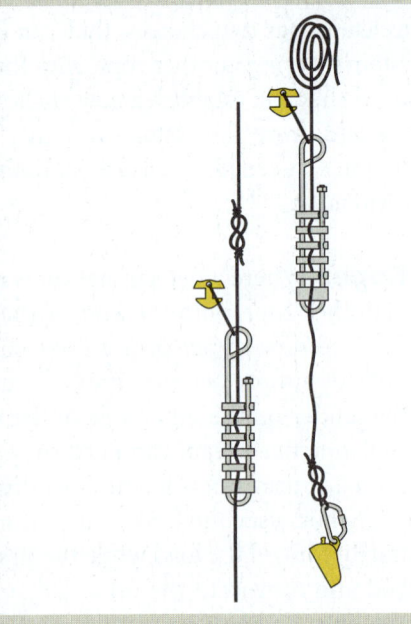

1 Tie the new, extension rope onto the tail end of the primary lowering rope, minimizing the size of the bend to reduce the chance that it will snag on something. Establish a second anchor system with an attached braking device slightly behind and in line with the first one.

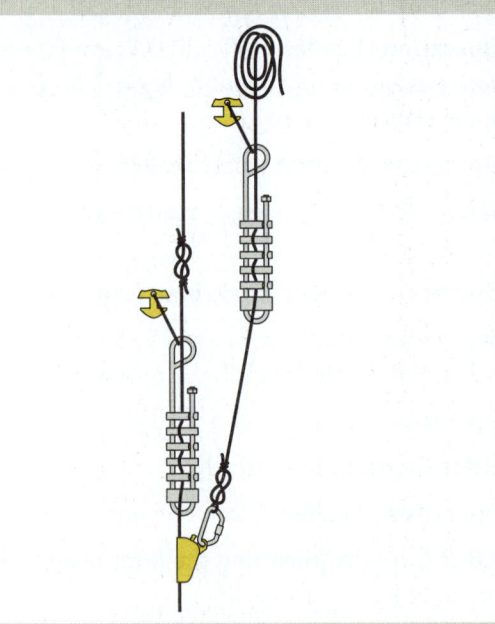

2 When the knot that joins the two ropes (primary and extension) is about 6 feet (1.8 m) from reaching the original braking device, attach the transfer line to the loaded line just below the braking device with a rope grab or sliding hitch system. Reeve the transfer line through the second braking device that is slightly behind and in line with the first one and lock it off.

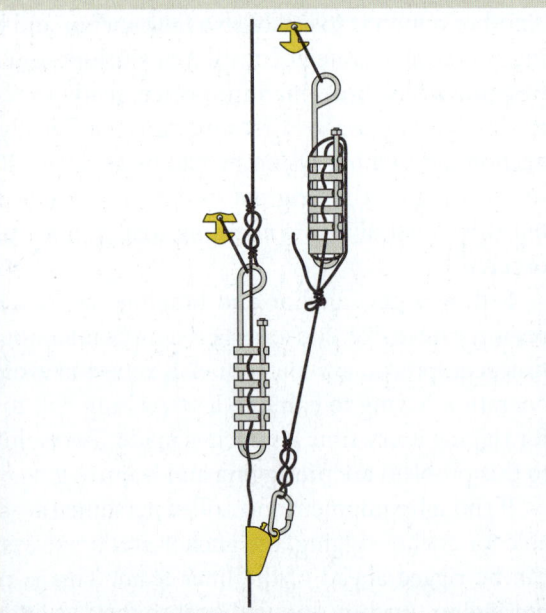

3 Continue to feed the original primary rope through the original braking device until all of the load is taken onto the transfer line with its locked off braking device.

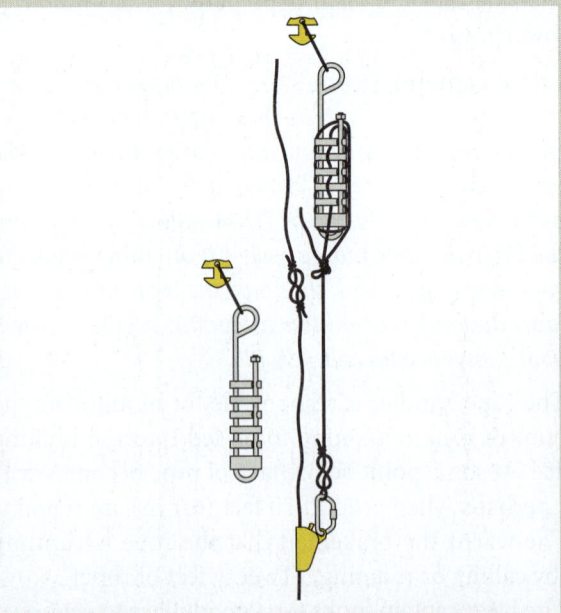

4 When there is sufficient slack in the system, detach the primary rope from the original braking device, then reattach it on the other side of the connecting knot and tie it off.

SKILL DRILL 14-6 (continued)
Performing a Knot Pass

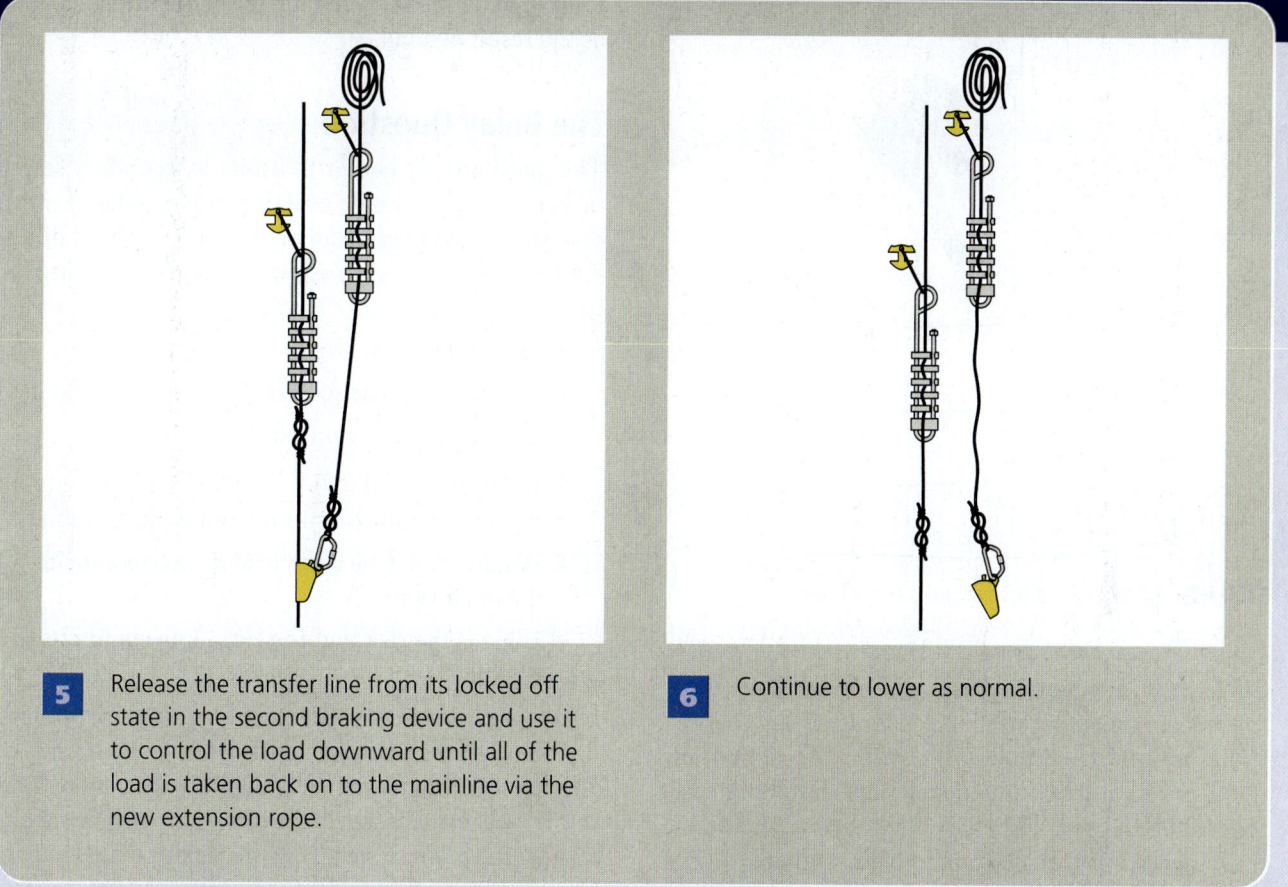

5 Release the transfer line from its locked off state in the second braking device and use it to control the load downward until all of the load is taken back on to the mainline via the new extension rope.

6 Continue to lower as normal.

that is already attached to the head of the litter can simply be released from the first braking system, pulled down the hill, and attached to the second one.

Another method, **leapfrogging** is the use of two sets of rigging teams and brake teams to alternate the rigging of anchors and the operation of the brake. While the first team operates the first set of brakes, the second team rigs the second set of anchors below the first team. After the litter rope has been detached from the first set of brakes, the first team derigs the first anchors and moves them down below the second team to the position for the third anchor and brake system.

Leapfrogging can be a very effective system and can speed the evacuation of a rescue subject. However, for it to be an efficient technique, the rigging teams must be skilled at anchoring, knowledgeable about the equipment, and self-reliant. If they do not have these qualities, the entire operation may be interrupted and the litter stopped as rescuers wait for the riggers to finish establishing the next brake and anchor system.

Safety Considerations in Low-Angle Rope Rescues

According to NFPA 1006 and 2500 (1670), operations-level rope rescuers are expected to steer clear of trying to perform rescue in an unstable environment—that is, an environment where they would normally not be able to travel without a rope. This means that the premise behind using a low-angle rope rescue system should be that it is the rescue itself that necessitates the rope, not just the slope angle or terrain.

Slope Safety Lines

That said, it is sometimes useful to incorporate a safety line on a slope to help rescuers more safely move up and down the slope for rigging, medical evaluation, litter rigging, and other tasks. The additional stresses associated with an ongoing rescue operation present challenges with route finding, footing, or dislodging

FIGURE 14-6 Low-angle slope safety lines.

debris from the slope, thus endangering themselves, the rescue subject, or other rescuers.

To prevent this problem, one of the first actions on the scene should be immediate establishment of a restricted area (sometimes called a hot zone) to reduce the danger to the rescuers and subject. Only personnel directly involved in the rescue should be allowed into this area. In particularly steep terrain, rescuers can establish personnel safety lines (**FIGURE 14-6**). These lines should be well anchored and rigged off to the side so that personnel can travel up and down them without endangering those below.

Once the safety lines have been established, they can be treated as handlines, or personnel can attach to them with a rope adjuster, such as an autolocking descender, rope grab, or sliding hitch, to provide security as they move up and down the slope. Note that handled ascenders made for climbing rope should not be used in situations where they might be impact loaded as the result of a fall.

The Belay Question

The question of whether to utilize a secondary safety, or belay, in a low-angle evacuation is a topic of much discussion and may be answered differently at different times depending on relevant factors, including the following:

- How steep is the slope?
- Is the footing particularly loose or treacherous?
- Is the slope icy or muddy?
- What would happen if the litter team fell?
- Are the mainline brake anchors questionable?
- Would a belay be beneficial or detrimental to the operation?
- What is the risk that the extra line could create a rockfall?

Belays are discussed in Chapter 12, *Belay Operations*.

The determination of need for belay can only be answered on the scene by trained, experienced personnel who will weigh the risk versus benefit of the second line. These decisions generally boil down to two questions:

1. What is the probability of failure?
2. What is the consequence of failure?

If the probability of failure is minimal, and if the consequence of failure is little more than a skinned knee or bruised ego, the potential benefit of the extra rigging may not be worth the increased demands and complexity on personnel, equipment, and time.

Summarizing Low-Angle Evacuations

Low-angle rescue operations are an extremely common form of rope rescue, and can be highly demanding in terms of both resources and need for close cooperation. Rescuers should become familiar with and practice sloped rope rescue techniques in the environment(s) in which they may be called upon to perform them. A well-tuned team that has low-angle rope rescue skills mastered will likely find graduation to vertical, or high-angle rope rescue, somewhat anticlimactic.

TIP

- Using good body position during low-angle evacuation will help reduce the workload and prevent injury of litter attendants.
- Litter attendants face uphill and lean back into the system.
- Attendants lean back against the tie-ins, letting their weight (and that of the litter) be taken by the rope system.
- Litter attendants' bodies are perpendicular to the slope.
- If the attendants are pulling hard enough with their legs, one hand on the litter rail should suffice.
- The team allows the rope system to determine the litter's rate of descent or ascent.

High-Angle Rope Rescue Overview

High-angle rope rescue, also called vertical rope rescue, refers to the controlled lowering of a rescue subject down a surface where rope is necessary for suspension. If the subject's injuries are severe enough, the lowering is done with the subject packaged in a litter. In some cases, when the subject is uninjured or only slightly injured, it may be possible to lower the individual without the litter, either alone or with a rescuer.

A misperception has developed among some practitioners that the NFPA dictates that operations-level rope rescuers are not permitted to perform high-angle rope rescue operations, and that these kinds of rescues must be deferred to their technician-level counterparts. This is not accurate. The differentiator between NFPA technician- and NFPA operations-level functions is not in the steepness of the incline nor the need for verticality, but simply in *how* the rescue is achieved.

According to NFPA 1006 and NFPA 2500 (1670), the operations-level rescuer is expected to be able to transfer a subject from one stable location to another, but is not expected to be capable of performing rescue skills in an unstable environment (for example, free hanging, with air under their feet). What this means is that the operations-level rescuer may be called upon to lower a subject from an elevated platform or structure through an unstable environment to ground level, but they are expected to use methods and techniques that do not require them to accompany the subject through said unstable environment.

Operations-level rescuers may also find themselves in need of being lowered (or raised) to reach a subject in a stable location. As long as the rescuer is not performing rescue skills or tending to the subject while they are in the unstable environment (e.g., hanging in free air or in a place where they would not be able to stand normally without assistance of a rope), this is still considered an operations-level skill.

The following sites are some of the environments in which a high-angle lowering may be used:

- Cliffs
- Buildings
- Industrial sites, either outside a structure or inside a vessel
- Construction sites, such as tower cranes
- Other structures, such as stacks, silos, wind turbines, or towers
- Cliff faces
- Vertical caves

Vertical lowering may take place on the outside of a structure, such as a building or stack, or on the inside, such as the interior of a silo or tank. As rescuers lower the load, it may run down against the side of a steep, vertical incline, such as the side of a building, structure, or cliff face, or it may be *free* (not touching the wall). This depends on the nature of environment where the rescue is taking place. If the top of the drop is overhung, the litter will naturally hang out away from the wall.

High-Angle Rope Rescue System

There are many correct ways to rig and operate a high-angle rope rescue system, but there are a few key points that these should always have in common:

- Primary anchor system
- Mainline (rope)
- Braking device
- Belay system
- Means of securing the subject to the system
- Optional tag-lines for the management of the load
- If the subject is not ambulatory, a litter and bridle may also be required (optional) (**FIGURE 14-7**)

At the operations level of response, the litter will typically be unaccompanied through the unstable portion of the lower, with load control provided by some

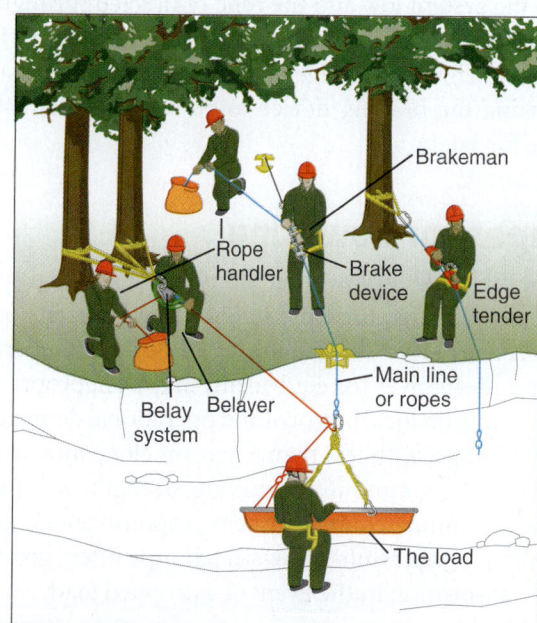

FIGURE 14-7 Overview of basic lowering system.

means such as tag-lines. The vertical system should be relatively simple and should achieve the stated goal of transferring the subject from one stable location to another.

Primary Anchor System

In selecting an anchor for a lowering system, consider the direction of travel, the mass of the load to be lowered, potential for belay, and likelihood of the system needing to be converted to raise. The anchor point for a lowering system should be built based on the following principles: secure, in line with the intended pull, and able to withstand the anticipated loads with a comfortable margin of safety. Note that the direction of pull may vary throughout the course of the rescue, especially if tag-lines will be used to help direct the movement of the load. See Chapter 11, *Anchorages* for more on anchoring systems.

The anchor should be set in line with and as close as reasonably comfortable to the subject, yet with enough working room available for the rescuer(s) at the top to work efficiently. A setback of at least 6 feet (1.8 m) from the edge to the braking device is a good goal, but of course the rescue site may not be conducive to such distance. In any case, there should be sufficient working space at the anchor to allow for the rescuer(s) to be comfortably positioned to package the subject, operate the braking device, and effect the rescue.

Keep the anchor relatively high in relation to the edge that the load will have to go over as it begins descent, but not so high as to make the brakeman's job harder. Rigging a slingshot-type system, with the braking system low and the rope redirected through a pulley block at a secure high point, can really help with loading. This makes edge negotiation easier while still allowing the braking device to be set back from the edge for safety.

Ropes for Lowering

Ropes selected for lowering should be of a design specifically manufactured and intended for life safety use. NFPA 2500 (1983) T (Technical) Rated or G (General) Rated alternatives (or equivalent) may be appropriate, depending on local protocol and operational demands. A static rope, with less than 6 percent elongation at 10 percent of its minimum breaking strength, will provide maximum control, system responsiveness, and minimal creep, while a low-stretch rope offers greater force absorption in the event of a dropped load.

Whichever ropes are used, they must be compatible in size and construction with other components of equipment presently on the market. In the past, all G-rated rope was at least of a ½ inch (12.5 mm) diameter, but with the recent proliferation of G-rated 7/16-inch (11-mm) rope, there is an increased hazard of having mismatched equipment in the field. Rescuers should be attentive to ensuring compatibility in every connection.

Braking Systems for High-Angle Lowering

The principles governing the use of a braking device for high-angle lowering are similar to those we have already discussed: the device imparts friction to the rope running through it. Some devices of these will be autolocking, some will have panic locks, some will offer adjustable friction, and some will be free-running. See Chapter 8, *Rescue Equipment* for further detail.

Any braking device used in lowering must provide an adequate margin of control over the load being lowered. This concept is not the same as *strength*. Strength refers to how strong something is, not how much friction something imparts. This question of friction is something that requires experience and discretion, as the same device will offer varying degrees of friction depending on several factors, including (but not limited to) the rope with which it is used, environmental conditions (wet/dry), wear and tear, amount of load applied, etc.

The braking device will be attached to the anchor, as for a low-angle operation, and with it a brakeman will control the rate at which the litter is lowered. Autolocking devices are a good choice for safety, particularly where the ability of the brakeman to hold the load might be in question.

Belays for High-Angle Lowering

Belays, discussed earlier as being an option for consideration in low-angle operations, are nearly always appropriate in high-angle rescue operations for professional rescuers. Keeping in mind the juxtaposition of probability and consequence, the goal is to ensure that the failure of any one point does not result in the complete failure of the rope rescue system. A belay is typically achieved by building redundant systems wherein if one system fails, another catches the load. This can mean a typical system involving a belay that is both apart from the main lowering system and on a separate anchor, or it can mean two separate lowering lines that provide back up to one another, a concept known as a dual tension rope system (DTRS).

If the main lowering line is run through a high directional anchor, rescuers must decide whether to also run the belay line through a high directional. In this case, it depends on how secure the high point is. Structural high points, such as an overhead beam or eye bolts, are often very secure and unlikely to be compromised. In such cases, where the high point directional anchor is unquestionably sound, matching the path of the mainline is advisable, as this reduces the amount of potential drop in the system in the unlikely event of a failure. However, if there is any question as to the integrity of the high point, such as it being part of an artificial high directional system where there is likelihood of failure, the belay line may be more secure if it is left at ground level. See Chapter 17, *Horizontal Systems* for a detailed discussion on high directional systems.

Rescuers who are also familiar with Occupational Safety and Health Administration (OSHA) mandates for fall protection in industrial and construction workplaces may wonder why this strong affinity for belay is not mirrored in requirements for coworker-assisted rescue. The reason is simple and practical.

Coworker-assisted rescue is performed in the moments following a fall incident, usually with prepackaged, prerigged, plug-and-play **rescue solutions** that incorporate high safety factors, autolocking connectors, and autolocking descenders, all of which dramatically reduce the *probability of failure* to the point where the risks of *probability of entanglement* and *difficulty in managing the system* far outweigh the potential risk of something breaking. Considering these concerns, along with the time and complexity that integration of a belay would introduce, implementation of a separate belay is often contraindicated in such situations.

Unlike the prepackaged systems often used in the workplace to assist a coworker, professional rope rescuers are more likely to be rigging piecemeal systems from individual components, customized to the various sites and situations to which they respond. The wide variety of incident responses they make, the propensity to mix and match equipment, allowance for nonautolocking connectors and equipment, the allowable variables in training and rigging, and the availability of more personnel to operate systems all swing the risk pendulum back over to the side of preferentially using a belay for professional rescue.

When a belay is used, it should be rigged so that it and the primary system do not interfere with one another or become entangled. However, they should be close enough together to prevent a dangerous pendulum should the main system fail and the belay be forced to catch the load. Potential drop distances should be managed to a minimum, but belayers must still remain aware of the potential impact, swing fall, and shock load.

For simplicity, this text prefers the use of a two-tensioned rescue system as the rescue system of choice. A two-tensioned belay is essentially nothing more than a mirrored replicate of the primary lowering system. As always, follow local protocols.

Two-Tensioned Rescue Systems

A two-tensioned rescue system (TTRS), also sometimes called a twin-tensioned rescue system, uses two anchored friction devices to control the two ropes at the same time, with both under load. A TTRS is a type of DTRS. If properly controlled, this can result in each rope supporting one-half of the load as it is being lowered. In reality, the two lines are rarely loaded equally, so the term *twin-tensioned* is a bit of a misnomer—thus the alternative term, two-tensioned rescue system, indicating that both lines are under load, albeit not necessarily at an equal level.

The fact that the ropes are not each carrying exactly 50 percent of the load is not especially important. What is important is that both ropes be loaded enough to mitigate excessive shock load to the other due to slack in the event that one line fails. In any dual tension system, neither rope is officially designated as belay, but if either line fails for any reason, the other will function as a belay. This differs from a traditional mainline system with an untensioned belay, where inherent system slack and rope stretch can result in higher impact forces in the event of a failure.

For maximum redundancy, the two rescue systems in a TTRS should be anchored separately and run by two separate belayers (**FIGURE 14-8**).

The downside of this approach as that the brakemen controlling the friction devices in a TTRS may not always operate the two systems at exactly the same speed, whether due to different experience levels,

FIGURE 14-8 A TTRS with maximum redundancy.

unequal edge friction between the two ropes, ropes with different elongation, or just the nature of friction in physics. In this case, the system that operates slower will end up carrying the entire load while the faster system will become slack.

One way to help reduce the operator-induced differences is for the same brakeman to operate both ropes. To do so, identical devices should be used for both systems, and they should be placed as close together as possible so that the brakeman can operate both handles (if so equipped) with one hand, and grasp the free ends of both ropes with the other hand.

Of course, the downside of this approach is that the shared anchorage eliminates redundancy in one of the highest risk areas of rigging (the anchor), and having only one brakeman likewise fails to protect against human error. This also protects only against operator-induced tension differences—it does not do anything for the friction-induced or rope-induced differences in tension between the two systems.

The only other option for balancing tension between the two systems is to slow the faster system until the slack in the slower system is caught up. This can become something of a seesaw exercise as the two systems strive toward equilibrium.

If two brakemen are used, it can be helpful to designate one of the two-tensioned systems as primary, or lead. The individual operating the primary system takes charge and sets the pace for the operation, while the operator of the second system focuses on following the lead so that the systems are balanced as equally as possible. This approach gives more structure to the operation and reduces confusion between the brakemen; it also helps eliminate confusion for rescuers who are more experienced with traditional mainline/belay systems and allows for more consistency in utilizing familiar terminology, as in the preceding table.

TIP

Some rescue teams use color coding to identify and express which rope serves what purpose in either mainline belay systems or two-tensioned systems. There are some pitfalls with using color codes to direct system movement commands, however. The first is that rescuers who are colorblind may not be able to tell the difference between the two rope colors. This can easily lead to confusion, which in turn can cause accidents or damage to the system. Another complication is that mutual aid rescue teams may not be familiar with your team's color-coding system, or with two-tensioned rescue systems. Training of all members of teams that may work together is the best way to resolve this, but that is not always possible.

Designating one of the two-tensioned ropes as the mainline and the other rope as safety is one way to differentiate and help reduce or eliminate confusion between rope systems.

Rigging a System for Lowering

Conceptually, the same kind of system can be used to lower a litter or a person. Although rescuers at the operations level are not expected to perform rescue skills in a free-hanging or other unstable environment, it is quite possible that an operations-level rescuer may need to be lowered (or raised) from one stable environment to another to reach a subject who is in need of care. This section will discuss systems and methods that may be used to lower a rescuer, a rescuer with a litter, or even to lower a litter or other load without a rescuer. These methods for lowering provide a foundational skill that will set the stage for lowering litters as well as raising loads. Working in suspension as a rescuer is discussed in Chapter 16, *Working in Suspension*.

To practice this skill, you will need the following:

- A lowering system
- A raising system
- A brakeman
- A secondary brakeman or belayer
- An operations leader
- A safety officer
- A rescuer

Choose a short vertical face (approximately 20 feet [6.1 m]) where the top breaks over gradually into a steep face. Establish two secure high-angle lowering system anchor points safely back from the edge previously. If possible, have the anchor points high off the ground to assist the rescuer in going over the edge. A high directional anchor between the main anchor and the load is a good solution for this. If there is any danger that the brakeman might be exposed to a potential fall, they should be secured to a safety system that is not part of the rescue system.

A dual-tension lowering system may be used to lower the load, offering all the same advantages of any other dual tension rescue system. In this case, for optimum redundancy two separate lines should be used, supported by two separate anchors, just as would be the case for a rescue system with primary and belay lines. If there is any risk that any of the persons operating the system might be exposed to a potential fall, they, too, should be tied into a safety line that is not part of the rescue system. Place edge protection to guard the ropes from abrasion if needed.

To practice a high-angle lowering of a rescuer using a DTRS, follow the steps in **SKILL DRILL 14-7**:

1. Both DTRS lines should be connected to the load. If the load is a rescuer, both lines may be attached to the same point on the rescuer's harness, or the main lowering line may be attached to the waist attachment while the belay line is attached to another full-strength attachment point (if available) such as the sternal attachment **FIGURE 14-9**.
2. If two brakemen are used, one brakeman is designated as "primary" and the other as "safety." Both brakemen hold their respective systems at a FULL STOP until otherwise instructed.
3. After confirming with the brakemen that the system is ready, pretension the system to make sure all parts are rigged correctly under load, that no lines will tangle, and that shock loading will not occur when the load goes over the edge. The brakemen hold tension on the brakes, while the rescuer(s) controlling the load pulls slack out of the system.
4. When all is ready the rescuer controlling the load gives the "Down slow" voice signal to the brakemen. Both brakemen slowly allow the rope to pass through their respective braking devices. If the load is a rescue, they lean back against the rope and move backward toward the drop. If the rope is not moving through either one or both of the brakes, the respective brakeman can help it along by reducing friction.
5. Practice the full stop, tie-off as follows:
   a. The practice rescuer calls, "Stop!"
   b. The brakemen stop the ropes from going through the brake.
   c. If the load is to be stopped for an extended period, the brakemen can lock off the devices in the manner described by the manufacturer's instructions.
6. Next, practice commencing descent:
   a. When ready, the rescuer controlling the load calls "Prepare to lower."
   b. While maintaining a firm grip on the brake side of the rope to keep the rope taut, the brakemen release the device from the locked off position; the primary brakeman announces, "Ready to lower on your command."
   c. When ready, the rescuer controlling the load calls "Down slow," and the process begins again.
7. If the load is a rescuer, they should practice getting off rope as follows:
   a. When the practice rescuer reaches the ground or other intended stable and secure position, they call, "Stop."
   b. The brakemen grip the rope firmly and actuate the brakes to prevent rope from passing through them.
   c. The brakemen maintain the load on belay until the practice rescuer relieves them by completing the belay cycle.
   d. If the rope is too taut for the practice rescuer to disconnect, they call, "Slack." The brakemen allow some rope through the brake devices to create slack in the mainline.
   e. If the practice rescuer has finished with the rope and is ready for it to be used for other purposes, they disconnect from it and signal to the brakemen, "Off rope."

Alternatively, a rescuer or other load may be lowered using a mainline with belay system. To execute a lower using a mainline/belay system, follow the steps in **SKILL DRILL 14-8**:

1. Rescuer positions themself near the top of the drop, as in Skill Drill 14-7, and connects both primary and belay lines to their harness.
2. Primary brake and belay brake operators hold their respective systems at a FULL STOP until otherwise instructed.
3. Rescuer calls "Brake, rescuer, on belay?" and primary brakeman confirms "On belay!" Belayer stays quiet but continues to hold tension.
4. Rescuer leans into the system to pretension and ensure that all parts are rigged correctly under load, that no lines will tangle, and that shock loading will not occur when the load goes over the edge.
5. When all is ready, the Rescuer gives the "Down slow" instruction to the brakeman. The primary

FIGURE 14-9 Practicing a high-angle lower of a rescuer using a DTRS.

brakeman slowly allows the rope to pass through the braking device while the belayer maintains a loose belay line through the belay device.

6. Rescuer leans back against the rope and moves backward toward the drop. If the rope is not moving, the top brakeman may need to help it along by reducing friction.

7. Next, the rescuer practices the full stop, tie-off:
   a. The rescuer calls, "Stop!"
   b. The top brakeman stops the ropes from going through the brake.
   c. If the rescuer is to be stopped for an extended period, both the brakeman and the belayer can lock off their respective devices in the manner described by the manufacturer's instructions.

8. Next, the rescuer practices commencing descent:
   a. When ready, the rescuer calls "Prepare to lower."
   b. While maintaining a firm grip on the brake side of the rope to keep the rope taut, the brakeman releases the device from the locked off position and announces, "Ready to lower on your command."
   c. When ready, the rescuer calls "Down slow," and the process begins again.

9. Next, the rescuer practices getting off rope:
   a. When the rescuer reaches the ground or other intended stable and secure position, they call, "Stop."
   b. The brakemen grip the rope firmly and actuate the brake to prevent rope from passing through.
   c. The brakemen maintain the load on belay until the rescuer relieves them by issuing the command "Off belay."
   d. If the rope is too taut for the practice rescuer to disconnect, they call, "Slack." The brakemen allow some rope through the brake devices to create slack in the mainline.
   e. If the rescuer has finished with the rope and is ready for it to be used for other purposes, they disconnect from it and signal to the brakemen, "Off rope."

Practicing a High-Angle Litter Lower

As previously mentioned, at the operations level, the high-angle litter lower is generally unattended. This means that some other means of controlling the load should be employed so that it does not careen off obstructions or spin out of control as it is lowered.

The operations leader should ensure prior to commencing the operation that both parts of the DTRS are rigged with a safety factor appropriate to the load that will be on the system.

The braking devices must be attached to the anchor systems such that the following are achieved:

- The brakemen are close enough to the edge to hear voice communication from the litter attendant, there is positive radio communication, or an edge attendant can relay communications between the brakemen and the litter attendant.
- There is enough room at the top between the brakes and the edge so that the litter can be rigged safely (and, if applicable, so that the litter attendant can also be tied in safely).

Litter Orientation

In some parts of the world, rope rescue operations are frequently performed with the litter in a vertical orientation, not unlike the system shown earlier in this chapter for a low-angle evacuation (**FIGURE 14-10**). For whatever reason, in the United States, vertical (or high-angle) rope rescue systems have evolved in such a way that the litter is typically oriented in a horizontal configuration unless there is a specific reason to do otherwise (**FIGURE 14-11**). A horizontal configuration, in which the subject lies as though on a cot, is usually a more comfortable and reassuring position for the subject and is also more conducive to medical care.

FIGURE 14-10 Subject packaged for vertical configuration.

CHAPTER 14 Lowering Systems

FIGURE 14-11 Subject packaged for horizontal configuration.
Courtesy of Skedco, Inc.

FIGURE 14-12 A litter spider.

A vertical litter orientation may be more prudent in a confined space rescue or when obstructions on the face require a small cross-section for the litter.

Securing the Subject in the Litter

During a high-angle rescue, the subject must be securely packaged into the litter, with appropriate consideration to injuries and medical care. If the subject is on a spine board, they must be securely packaged onto the board, which, as a unit with the individual, is connected to the safety lines. The subject should also be secured in the litter with lacing across the top to prevent shifting in the litter.

The subject should be wearing a harness, and a safety sling should run from the subject's harness to the master attachment point at the top of the litter spider. This safety sling is designed to catch the subject should they somehow become disconnected from the litter. Slack must always be left in the safety sling so that the subject is not pulled upward if the litter tilts.

Note that a subject packaged for rescue with the litter positioned in the vertical configuration will need some extra protection from the potential for slipping out the end of the litter. A seat harness, foot loops, or other methods for holding the body may be used, but consider carefully how these will be loaded when the litter is stood on end. It is natural for the body to want to slip down with gravity, so a subject who is packaged with the litter lying horizontally and then stood on end may shift significantly.

Attaching the Litter to the Main Rescue Line

The means by which a litter is attached to the rescue system will dictate its orientation during the evacuation, among other things. A different kind of litter bridle from that used for low-angle evacuation may be fashioned from various components to join the mainline lowering rope to the litter (**FIGURE 14-12**). For differentiation purposes, a litter bridle used for vertical evacuation is sometimes called a **litter spider**. A single-point litter spider for lowering a litter in a horizontal orientation should have a minimum of four legs, each supporting the litter at multiple points along the litter rail, with these collected back to a central point where it is attached to the mainline lowering rope. The connection where the litter harness and rope rescue systems come together is called the **master attachment point**.

Commercially made litter spiders that are adjustable for height and angle, and that incorporate tie-in systems for litter attendants, are available from a number of manufacturers. In use, the litter harness should be adjusted so that the subject rides in the litter slightly head up (unless medical reasons dictate head down). Riding head down adds to the subject's anxiety and disorientation.

Rigging Without a Litter Attendant

While operations-level rescuers should be capable of performing high-angle rescue operations, they should be capable of doing so without employing a litter attendant to ride the litter with the subject. In an operations high-angle rope rescuer scenario, the subject should be packaged into the litter at a stable location, attached to the rope rescue system, then lowered (or raised) from that stable location to another stable location where they can be removed safely from the system.

Of course, lowering a subject without an accompanying rescuer is fraught with challenges. Clearly this is not an option if the subject requires medical attention or management during the lower or raise. Beyond this, if the vertical segment is a wall, tower, cliff face, or other structure, preventing the litter from bumping and slamming into protrusions and obstructions along the way is a priority. If the vertical segment to be transited is free hanging, presumably the biggest problem would simply be limiting spin.

One way to help control a litter during an evacuation is by means of tag-lines. A **tag-line** is a non–load-bearing line that connects to the litter to help move it side to side to avoid obstacles, prevent swing-fall, or attain a more advantageous position. Tag-lines must be managed by tag-line attendants, who take rope in or let it out as the load is moved to manage it around obstacles or away from the structure, or to prevent spin. While tag-lines are simple to use, it is often difficult to control the load precisely.

Some rescuers erroneously consider tag-lines to be a technician-level rescue skill, perhaps because tag-lines are often associated with helicopter operations, and sometimes confused with methods used in more complex systems such as cross-hauls, guiding lines, and highlines. The primary difference between tag-lines and these other methods is that tag-lines should be managed by hand, with no auxiliary equipment to create mechanical advantage on the tag-line component. The reason tag-lines become a technician-level skill when haul systems are incorporated into them is because of the potential force amplification that can occur.

It is, however, good practice for operations-level rescuer to consider using directional pulleys, additional friction, or braking devices on tag-lines to help manage the load. In fact, the use of friction is often essential to help effectively control tag-lines. The caution here is to not use additional mechanical advantage that can exert undue forces on the system. Because tag-lines are intended, by design, to exert forces in opposition to one another, care must be taken to ensure that these forces are not so great that they compromise the safety of the system. This is the reason that the use of haul systems in tag-line operations is discouraged at this level of skill, but to instead rely simply on the manual hauling ability of one tag-line operator at a time.

Ropes used to construct tag-lines need not be G- or T-rated ropes, as long as they are not also intended to double as a belay or load-bearing line. In fact, use of a lighter weight line in the mm to 8-mm range may offer advantages in rope management and weight.

Tag-lines may be attached either to the bridle main attachment point, or to hard points on the litter. In the unlikely event that a tag-line will be expected to double as a haul or belay line, it is best attached to the rigging plate above the litter. Connection to the main attachment point rather than the litter itself offers an additional advantage in that it does not imbalance the litter when a tag-line is pulled. Tag-lines attached directly to the litter rail at the head-end or foot-end of the litter have more immediate, direct, and positive control—for example when negotiating edges; however, these may cause the basket to pull out of level, which can be disconcerting to the subject.

There may be one or more tag-lines, depending on the direction(s) in which the litter needs to be controlled. Most often, it is advisable to place multiple tag-lines in opposition to one another for optimum control. Tag-lines are typically controlled by personnel positioned beneath the load (outside the hazard zone), who pull on the lines manually to provide tension and deflect the litter's path. There is much to be said for the simplicity of nonanchored tag-lines, but if the consequence of a tag-line attendant accidentally letting go would be critical or catastrophic impact with something, additional friction (or an extra set of hands) should be considered.

Friction may be provided by means of an appropriate friction device, such as a braking device with an autolocking function. The device selected for controlling and securing tag-lines should be capable of quickly and seamlessly transitioning between applying and releasing tension so that there is no delay in response as needs change during the course of the lower. Use of nonautolocking devices is often appropriate here. Note that autolocking braking devices are typically rope-size dependent, so when using those be sure to select an appropriate device for the diameter of tag-line(s) that will be used in the system.

For optimum efficiency, use tag-lines attached to the ends of the litter, ensure good offset angle with the main system, and train regularly with them. To connect tag-lines to a litter, follow the steps in **SKILL DRILL 14-9**:

1. Create two litter spiders as described in Skill Drill 14-5; attach one to the head end of the litter and the other to the foot end.
2. Select two pieces of light rope 30 percent longer than the height of the intended lower to serve as your tag-lines.
3. Tie an end termination in each.
4. Use carabiners to connect one tag-line to each end of the litter.

Practicing High-Angle Litter Lower Techniques

Skills for lowering a litter should be practiced in whatever environment(s) where operations-level rescuers anticipate possibly needing to achieve a rescue. To rig the litter for a single point suspension high-angle lower, follow the steps in **SKILL DRILL 14-10**:

1. Rig an appropriate anchor at the top of a drop.
2. Stack the rope in a convenient, out-of-the way place, ready to feed into the braking device.
3. Attach a braking device to the anchor and insert the rope, leaving just enough room to reeve the rope into the device and attach it to the main attachment point on the litter spider. Be sure there is not enough slack in the line to allow the litter to break the plane of the edge.
4. If possible, run the mainline up to a change of direction pulley on a secure high anchorage.
5. Set the litter in a secure place.
6. Attach a litter spider to the litter, and connect it to the mainline.

To prepare a two-tensioned rescue system, follow the steps in **SKILL DRILL 14-11**:

1. Rig an appropriate anchor for the second system, aligned appropriately with the first lowering system. For clarity of communication, we will refer to the first system as primary, and to this second system as the safety system, even though the intent will be to maintain equivalent tension on both systems.
2. Stack a second rope for the safety system in a convenient place where it will not interfere with the main system, ready to feed into the safety brake device.
3. Attach a safety brake device to the safety system anchor and insert the rope, leaving just enough room to reeve the rope into the device and attach it to the litter, but not so much length as to allow the litter to break the plane of the edge. If a high directional anchor is used, and is unquestionably sound, run both lines along the same path through the high directional.
4. Attach the safety line to the main attachment point on the litter spider. Connect tag-lines to the master attachment point or to the ends of the litter.
5. Drop the tag-lines over the edge of the drop to tag-line attendants (NOTE: Each tag-line must have a person operating it, with clear title/designation to enable effective communications. If there is only one tag-line, the tag-line operator may simply be called Tag-line. If there is more than one, using terms like HEAD TAG, FOOT TAG, MAIN TAG, etc., or designating each line by rope color, will help facilitate coordinated movement.)
6. If appropriate, connect the tag-lines to anchored rope adjusters (such as a lightweight braking device with autolocking function) **FIGURE 14-13**.
7. Assign and create a safety system(s) for one or more edge attendants to help move the litter from its place of rest to a suspended state.

FIGURE 14-13 Connecting the tag-lines to the anchored rope adjusters in a DTRS.

To execute the lower with TTRS, follow the steps in **SKILL DRILL 14-12**:

1. Position edge attendant(s) to hold the litter as the subject is being packaged.
2. Package the subject into the litter that is properly secured by both main and safety lowering systems.
3. As there is no litter attendant, identify a litter manager who will direct the litter movement; be sure everyone knows who the litter manager will be. This must be someone who can see and/or who is in direct communications with the litter—either the main tag-line operator or the operations leader.
4. The operations leader initiates the roll call, as described earlier in this chapter.
5. The designated litter manager (main tag-line operator or operations leader) initiates the lowering sequence.
6. The primary and safety brakemen hold the system in place as the edge attendant(s) and tag-line operator(s) pretension the system.
7. A check is made by the safety officer to ensure that all parts of the system and all subsystems have been rigged correctly, no lines will tangle, and that slack is removed from the system.
8. Once readiness is confirmed, the designated litter manager calls for "Down slow."
9. The primary brakeman serves as lead brake, repeats "Down slow" to signify that he has heard, and signals the safety brakeman to begin movement.
10. The primary brakeman slowly lets rope through the primary braking device at the desired rate; meanwhile, the safety brakeman feeds rope to the secondary brake at a rate that strives to maintain equivalent tension with the mainline.
11. Edge attendants maneuver the litter over the edge while the tag-line operator(s) pull tension and the brakemen control the speed of the lower. NOTE: This part of the operation should move very slowly, at least until the litter reaches a suspended state. Attendant(s) and TAF-LINE operator(s) may need to pull hard on the litter while the brakemen reduce the friction to get the litter into suspension. However, once the litter goes over the edge, there will be greater weight on the system, and greater friction will be needed.

Control movement of the litter with the tag-lines, following these steps:

1. As the load is lowered, the tag-line attendants should practice pulling tension or giving slack through the tag-line braking device(s) to move the load away from obstacles or to direct its path.
2. The litter manager should practice calling for the litter to move in various directions by specifically directing the brakemen and tag-line operator(s) to stop, haul, and lower as appropriate.

Throughout the entire evolution, the litter manager should remain in command of the movement of the litter. Although anyone may call a stop at any time, it is the litter manager who should direct the rate and direction(s) in which the litter is lowered. Tag-line operations require clear, direct communication and close coordination.

As an alternative to the TTRS, the litter lower may be performed using a mainline with belay. To practice a litter lower with mainline and belay, follow the steps in **SKILL DRILL 14-13**:

1. Rig the litter for a single point suspension high-angle lower, following the steps in Skill Drill 14-10.
2. Rig an appropriate anchor for the belay system, aligned appropriately with the main lowering system.
3. Stack the belay rope in a convenient place where it will not interfere with the main system, ready to feed into the safety brake device.
4. Attach a belay device to the belay anchor and insert the rope, leaving just a little more slack than what is in the mainline.
5. If a high directional anchor is used, and is unquestionably sound, run the belay line along the same path as mainline from brake to litter, using a separate connection for the belay; if there is any question about the integrity of the high point, leave the belay line at ground level.
6. Attach the belay line to the main attachment point on the litter spider.
7. Connect tag-lines to the master attachment point or to the ends of the litter.
8. Drop the tag-lines over the edge of the drop to tag-line attendants (NOTE: Each tag-line must have a person operating it, with clear title/designation to enable effective communications. If there is only one tag-line, the tag-line operator may simply be called Tag-line. If there is more than one, using terms like HEAD TAG, FOOT TAG, MAIN TAG, etc. will help facilitate coordinated movement.)
9. Connect the tag-lines to anchored rope adjusters (such as a lightweight braking device with auto-locking function).

10 Assign an create a safety system(s) for one or more edge attendants to help move the litter from its place of rest to a suspended state.

To execute the lower with main and belay lines, follow the same steps as were used in Skill Drill 14-11, except that during the entire evolution, the belay operator should strive to maintain a slightly loose belay line rather than sharing half the tension with the primary system.

Aerial Ladder Slide Lowers

One alternative to tag-lines, particularly for urban structural rescue operations, is to employ an aerial ladder as a slide for the litter to follow. In effect, this method converts what would otherwise be a high-angle (or vertical) lower into a low-angle operation, with the ladder providing a track for the litter to slide down as the rate of lower is controlled by a rope system. This approach works best where an aerial ladder truck can be positioned adjacent to the structure with a safe working angle, and where adequate anchorages are available for the rope lowering system.

For this type of system, the subject should be packaged in a rigid basket litter that has a gap between the top rail and basket. You will need a properly constructed lowering system, similar to what might be used for a low-angle evacuation; an aerial ladder; and a pike pole. The pike pole will be used to hold the litter off the rungs of the ladder, allowing it to travel more smoothly down the ladder.

At least three rescuers are typically required to securely lift the subject into the litter, package them, and move the litter onto the ladder slide. One of these three will serve as a litter attendant. The subject should be secured into the litter using an adequate patient restraint.

Follow the steps in **SKILL DRILL 14-14** to perform an aerial ladder slide:

1 Position the ladder adjacent to the structure where the rescue will take place.

2 Establish a lowering system in a secure location on the structure.
Secure the lowering system to one end of the litter as you would for a low-angle evacuation (see Skill Drill 14-5).

3 Package the subject securely in a rigid basket litter.

4 Slide the pike pole between the top rail and the shell of the litter so that it lies across the short axis of the litter, at the end opposite to that which the rope system is connected.

5 The litter attendant should position themselves on the ladder, facing upward, a few steps down, yet within reach of the litter.

6 As the brakeman maintains adequate friction control on the rope, additional rescuers should lift and position the litter into place lengthwise on the ladder so that the end bearing the pike pole is nearest the litter attendant, and the end bearing the rope system is nearest the top.

7 At the command of the litter attendant, the brakeman slowly lowers the litter down the ladder; meanwhile, the litter attendant maintains grip on the pike pole and helps guide the litter down the ladder.

There will be some jostling to the subject, but the aerial ladder slide method is a convenient and low-stress method of simplifying a high-angle (vertical) operation. It is dependent on the ability to position the apparatus near the structure with an appropriate working angle, access to the subject, and availability of basic rope rescue equipment. The aerial operator must have adequate training in load capacities, tip angles, and device function to be competent and comfortable with the operation.

It is also possible to execute this technique with a ground ladder, but extra care must be taken to ensure security of the ladder before using these methods. Ladder methods are not covered deeply in this text because they are not explicitly called out by NFPA 1006 or 2500 (1670.) Specific instruction from a competent trainer should be obtained before using this approach.

After-Action REVIEW

IN SUMMARY

- A rescuer trained at the operations level is expected to be capable of moving a subject from one stable location to another.
- A stable environment is one in which a rescuer would be capable of maintaining their position and/or moving comfortably without a rope system but for the additional difficulties presented by the work they are doing.

- The operations-level rescuer is not expected to perform rescue functions in an unstable (vertical, free-hanging) environment.
- The difference between high-angle and low-angle rescues has less to do with a specific slope angle than with the type of system most appropriate for use given the circumstances at hand.
- The type of system will depend on multiple factors, including slope angle, foot friction, available resources, environment, and more.
- A low-angle evacuation may be appropriate for moving a subject down a slope where a simple litter carry without roped assistance would be hazardous.
- A low-angle evacuation can be resource intensive.
- Rope may be attached to the litter for a low-angle evacuation either by means of direct tie-in, or with a litter bridle.
- The core members of rope rescue operation are:
  - Someone to be in charge (operations leader)
  - Someone to assist the subject (rope rescuer)
  - Someone to rig the system (rigger)
  - Someone to control the brake (brakeman)
  - Someone to monitor safety (safety officer)
- Operations-level rescuers can perform unattended high-angle (or vertical) litter lowers.
- The litter may be positioned laterally or longitudinally to the slope for a vertical litter lower.
- Tag-lines may be used in a vertical litter lower to help control horizontal movement of the litter, and to keep it away from obstacles.
- An aerial ladder may be used to create a litter slide, effectively turning a vertical litter lower into a low-angle evacuation.

KEY TERMS

Adjustable low-angle litter tie-in A low-angle litter tie-in that can be adjusted.

Ambulatory subject lower method The process of helping an ambulatory subject help themselves through a roped evacuation.

Brakeman Rescuer responsible for operating a braking device in a rescue system.

Handline A fixed rope anchored at the top of a slope and grasped by a subject(s) and/or rescuer(s) for stability.

High-angle rope rescue Use of a rope to make a controlled lowering of a rescue subject down a slope that is steep enough that the load is dependent on the rope for support.

Leapfrogging The use of two sets of rigging teams and brake teams to alternate the rigging of anchors and the operation of the brake for an extended evacuation.

Litter bridle The mechanism by which a rope system is attached to a litter for the purpose of supporting and stabilizing the load.

Litter spider A type of litter bridle used for horizontal suspension of a litter for vertical evacuation; this type has many legs, which connect to different parts of a litter for the purpose of balancing it from several points simultaneously.

Low-angle evacuation A rescue over any inclined or rugged area where a litter must be carried and where it is difficult or dangerous to do so without the assistance of a rope.

Low-angle litter tie-ins Method used by rescuers to attach themselves to a low-angle litter lowering system.

Master attachment point The location at which the rope system is attached to the litter.

Operations leader The person in charge of the rescue operations.

Prerigging During a rescue operation that requires transfer of the load from one system to another, the act of setting up a subsequent system while waiting for the load to arrive.

Rescue solutions A complete, prepackaged system that can be used to execute a rescue.

Rope rescuer Rescuer primarily tasked with extrication and removal of the subject from hazard.

Rigger Rescuer responsible for building, analyzing, and adapting rescue systems in the field.

Scree evac A type of low-angle evacuation over a loose, rocky slope, typically in wilderness terrain.

Tag-line A non–load-bearing line that connects to the litter to help move it side to side to avoid obstacles, prevent swing-fall, or attain a more advantageous position.

Transfer line A system used to temporarily control a load during a knot pass.

Vertical evacuation Use of a rope to make a controlled lowering of a rescue subject down a slope that is steep enough that the load is dependent on the rope for support.

On Scene

1. You respond to reports of an injured subject on the track of a roller coaster. What factors will you use to determine whether to use low-angle methods to lower them down the track, or high-angle methods to lower them over the edge of the structure?

2. What is the best way to attach a rope to a litter for a low-angle evacuation, and why?

3. How many people would you need to staff a low-angle litter lower and what would their respective roles be?

4. How many people would you need to staff a high-angle litter lower at the operations level, and what would their respective roles be?

5. Talk through how to perform a system safety check for a high-angle litter lower at the operations level.

CHAPTER 15

Operations

Mechanical Advantage Systems

KNOWLEDGE OBJECTIVES

After studying this chapter, you should be able to:
- Summarize the principles of mechanical advantage. (**NFPA 1006: 5.2.15**, pp. 300–304)
- Calculate the total mechanical advantage of a system. (**NFPA 1006: 5.2.15, 5.2.16, 5.2.17, 5.2.18**, pp. 300–301)
- Describe the components of a simple mechanical advantage system. (**NFPA 1006: 5.2.15**, pp. 300–307)
- Describe the methods for rigging mechanical advantage systems. (**NFPA 1006: 5.2.15**, pp. 306–307)
- Describe how to operate a mechanical advantage system. (**NFPA 1006: 5.2.16**, pp. 310–311)
- Explain the methods of holding a load in place. (**NFPA 1006: 5.2.15, 5.2.16, 5.2.17, 5.2.18**, pp. 311–316)
- Describe the safety considerations when operating a hauling system. (**NFPA 1006: 5.2.15, 5.2.16, 5.2.17, 5.2.18**, pp. 316–317)

SKILL OBJECTIVES

After studying this chapter, you should be able to:
- Construct simple mechanical advantage systems. (**NFPA 1006: 5.2.15**, pp. 306–308)
- Rig a hauling system. (**NFPA 1006: 5.2.15, 5.2.17**, pp. 309–310)
- Direct the operation of a mechanical advantage system during raising and lowering of loads in a low-angle or high-angle environment. (**NFPA 1006: 5.2.16, 5.2.18**, pp. 310–312)
- Rig a compound and communicate effectively during a raising operation. (**NFPA 1006: 5.2.16, 5.2.18**, pp. 310–311)
- Construct a compound mechanical advantage system. (**NFPA 1006: 5.2.17**, pp. 313–316)
- Conduct a preoperational system safety check on a hauling system. (**NFPA 1006: 5.2.15, 5.2.16, 5.2.17, 5.2.18**, p. 316)

You Are the Rescuer

You are the leader of a rope rescue team that has just arrived at the scene of a construction accident. There is an injured worker at the bottom of a construction shaft. The worker does not have life-threatening injuries, but is unable to self-extricate and will need to be transported to a hospital. Your team is trained and experienced for the job, but you may need assistance from the workers on site.

1. How will you deploy your team in the most efficient manner and safely use the workers to assist?
2. How will you construct a haul system that is both efficient and safe?
3. How will you construct a belay (safety) system that will work well with the haul system?

Access Navigate for more practice activities.

Introduction

In this chapter, we will explore the concepts surrounding rescue by raising and mechanical advantage. With knowledge and practice, a rescuer can combine these techniques with those already learned to move a load in nearly any direction. These concepts may be employed on either low-angle or high-angle terrain, and in any environment—urban to wilderness. Chapter 14, *Lowering Systems* discusses rescue by lowering.

By the end of this chapter, rescuers will be able to choose equipment and rigging methods to raise a load using a little bit of pulling power, or a lot. They will be able to estimate the amount of pulling power exerted by a system, demonstrate the ability to construct and use a mechanical advantage system, and discuss safety considerations when employing such systems during rescue operations.

Principles of Mechanical Advantage

FIGURE 15-1 A direct pull raising system.

Not all litter evacuations have the luxury of being assisted by gravity; it is often necessary to fight gravity and raise the load to a higher elevation rather than lowering it. In fact, any time a rescue load is being moved in a direction with rope, it is wise to have a system in place for reversing the effect to move the load in the opposite direction, should this become necessary. This is for practical reasons, so that rescuers have complete control over the load in the event that something does not go as planned, as well as for safety. Without some means of opposing gravity, its effects could allow the load to be inadvertently lowered into an unplanned or less-than-ideal predicament.

Ideally, transitions from lower to raise—and vice-versa—should be made quickly and seamlessly. At the most basic level, raising a load to a higher elevation might be achieved simply by attaching a rope to a load and pulling on the uphill end of the rope to move the load. Obvious hazards notwithstanding, we will explore the mechanics of this type of raising system.

To effect a raise using a direct pull method, a rescuer would have to exert a force at least equal to that of the load (plus a little extra to overcome inertia and system friction) to raise it (**FIGURE 15-1**). Practically speaking, this means that it would take at least 100 pounds (45 kg) of force (plus a little extra to overcome inertia) to move a 100-pound (45-kg) load. In other words, the amount of force required to move the load represents a one-to-one (1:1) relationship with the load itself.

This simple example can be expressed as a ratio:

Load : Force required to move load

100 pounds : 100 pounds.

1 : 1

So the ratio is 1:1. Likewise, in this example, the rescuer would have to pull 10 feet (3 m) of rope in order to move the load 10 feet (3 m), again reflecting a 1:1 ratio.

Direction Change

While the math works, on a practical level it would be quite difficult for a rescuer, or even a group of rescuers, to move a 300-pound (136-kg) subject even a short distance using a direct pull. Sometimes, adding a **fixed pulley** at the anchor might permit the rescuer(s) to pull in a more convenient direction (**FIGURE 15-2**). When the pulley is anchored so that it remains stationary even as the rope moves, it does not add any mechanical advantage—quite the contrary, in fact. Some efficiency will actually be lost through the pulley's inherent inefficiency and friction in the system. Even so, by changing the direction of the rope pull, rescuers may find that the work becomes easier if it allows them to (for example) pull along the contour of a slope rather than straight up, or to work in a more open or convenient location. **Direction change** pulleys are also an important building block for haul systems that do offer mechanical advantage.

When using a pulley as a direction change, rescuers must rig carefully to avoid overstressing the system. Forces on the pulley and its anchor can be more than twice the weight of the load during a raise, particularly in the moment when a little extra force must be applied to overcome inertia. Edge friction can also increase the load on systems. The actual force on the system will depend on these and other factors, including the angles involved.

A fixed pulley added at the anchor to redirect the rope will not offer additional **mechanical advantage**, but it may make the direction of pull more convenient for the rescuer(s). The interior angle of the rope passing through a direction change should be kept as small as possible for optimum efficiency.

Counterbalance Haul System

When attempting to raise a litter up a steep slope, rescuers can optimize the efficiency of a 1:1 system with a change of direction by using the slope, and the effects of gravity, to augment their efforts. Known as a **counterbalance haul system**, this 1:1 raising system offsets the weight of the load as it moves up the slope with the increased pulling power of the **haul team** as they move down the slope (**FIGURE 15-3**). Care must be taken during use of such systems to avoid the risk of falls for the counterbalance rescuers, particularly in slope evacuations above a vertical drop-off.

In this system, the rope is attached to the head of the litter as for a steep slope lower, and is then passed through a direction-change pulley to make a 180-degree turn. As the haul team pulls the working end of the rope down the slope, the litter is pulled up the slope. A rope grab or Prusiks may be used at the anchor as a **progress-capture device (PCD)** to ensure

FIGURE 15-2 A direct pull with direction change.

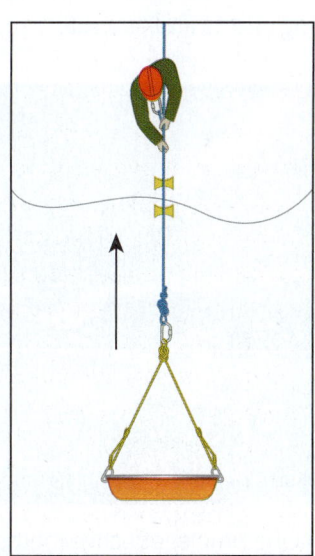

FIGURE 15-3 Elements of a counterbalance haul system.

that the load does not slip back down the slope after it us pulled up. Attaching haul team members to the haul rope, either with a rope grab or a knot, can help boost their pulling power, but can also increase potential risk of getting pulled off balance.

Strictly speaking, a counterbalance hauling system is also a 1:1 hauling system. Without considering loss caused by friction and other inefficiencies, the force used to haul is pretty much the same as the weight of the load being hauled. However, the key differentiator of a counterbalance haul system is that instead of the haul team going off to the side (as in a change of direction) or up the slope (as in a direct pull), the haul team takes advantage of gravity by going downhill. The load (i.e., the litter) still goes uphill.

Progress Capture

The direct-pull or counterbalance methods may be viable for moving light loads short distances, but because they rely heavily on brute strength of the rescuer(s), they can quickly become prohibitive for larger loads. A system that is totally reliant on the grip of rescuers' hands is at increased risk of dropping the load or letting it slip back down the face or incline. Incorporating a PCD into the system can offset this risk.

The term **progress capture** is used to describe the action of permitting the rope to move in an upward direction while preventing motion in the opposite direction. There are several ways to accomplish this end. One simple, effective method is to secure an anchored rope grab on the mainline, and to set or activate it whenever you want to hold tension (**FIGURE 15-4**).

Sometimes, the term **ratchet cam**, or simply cam, is used to refer to this device or function in a haul system. A PCD can also be used to give rescuers a rest between hauling efforts and as a safety redundancy.

Positioning the PCD

If possible, the PCD should be on a separate anchor from the hauling system. However, it should be close and parallel to the mainline rope. This helps prevent shock loading and reduces the slack that interferes with hauling system efficiency.

The PCD offers most protection and efficiency when it is positioned as far forward on the hauling system (toward the load) as possible. Because the PCD is anchored, it does not move as the load moves, but its location in a slack system will be different from its position when the system is under tension. Rig the PCD within easy reach of rescuers, close to the edge but not over it, and where it is not compromised by the edge even when under tension. Being the last hope to grab the mainline rope and prevent the system and the rescue load from falling should anything else in the hauling system fail (e.g., anchors, hauling team, haul cam, and pulleys), the PCD should be anchored securely.

When a PCD is rigged, it should be set so that, as much as possible, its anchor and anchor sling are in line with the rope the PCD is grabbing. If the PCD anchor is rigged too far off to the side of the rope it is protecting, the angle formed by the offset will allow too much slack into the system (**FIGURE 15-5**). This will result in a loss of efficiency during the operation, and also exposes the system to potentially dangerous shock loading in the event that it fails.

For systems that incorporate a direction-change pulley, the PCD mechanism can be placed behind the redirect, on the side furthest from the load. In this position, it experiences less force, but it does not offer protection against failure of the direction change, and it is a bit more difficult to manage.

FIGURE 15-4 One simple, effective method is to secure an anchored rope grab on the mainline, and to set or activate it whenever you want to hold tension.
© King Ropes Access/Shutterstock.

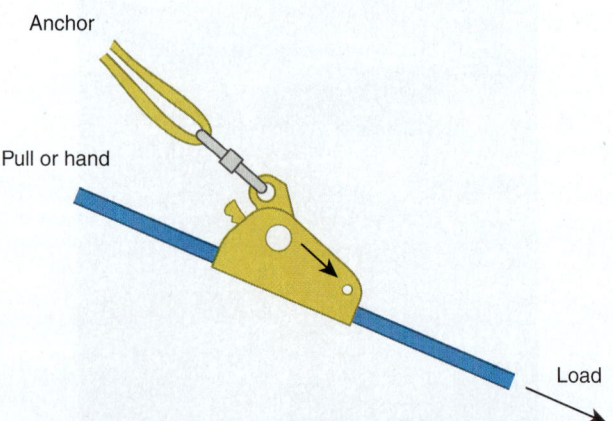

FIGURE 15-5 The PCD should be anchored in line with the primary system to avoid (1) dangerous shock loading of the PCD and its sling and anchor, and (2) optimize the amount of purchase gained by the hauling team when it sets the PCD to reposition for another bite.

A PCD must be managed during a raise to ensure that it stays in proper position, and that it activates to clamp the rope when needed. A rescuer may be assigned to serve as a PCD tender, with the sole responsibility of ensuring that the PCD grabs when it is supposed to, but does not grab when it is not supposed to. When a shell-type cam is used for progress capture, the cam tender should hold the shell from the back side, palm and fingers extended out of the way of the cam. In this position, fingers and gloves are at minimal risk of getting caught by the device, yet the PCD can be snapped into position quickly as needed. Use of a PCD tender can be effective, but requires extra personnel and introduces the potential for human error.

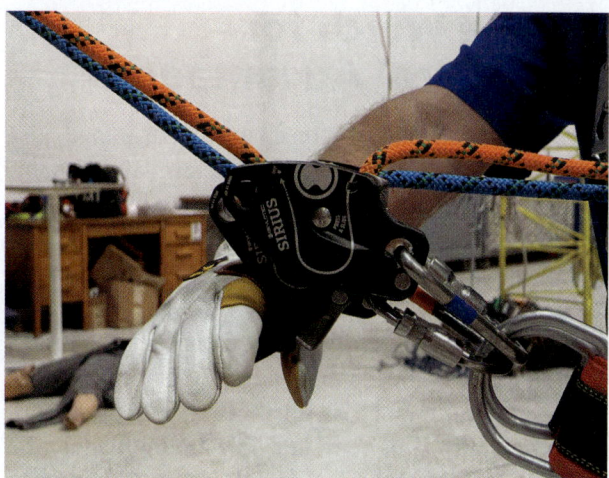

FIGURE 15-6 A bollard-style autolocking braking device.
© King Ropes Access/Shutterstock.

> **TIP**
>
> Two primary concerns must be addressed when rigging a PCD:
> 1. The PCD must clamp the rope when needed.
> 2. The PCD must not ride up the rope as the rope moves, because this could result in dangerous shock loading.

Some cam devices have a small hole at the load-end to facilitate an alternative, human-free method for managing a free-running cam. To use this method, a lightweight bungee (elastic) cord must be connected into the empty hole in the shell, and the other end of the bungee cord is anchored securely to a convenient spot forward of the anchor (toward the load). Minimal tension is required on the bungee cord to maintain position and prevent the ascender from riding up the rope and to keep the cam clamped on the rope. In a vertically suspended system, a small weight can be used in lieu of the forward anchor. The bungee cord is tied only to the shell and not to the cam itself. If the bungee is tied to the cam, the cam may not close when it should.

A rope grab is not the only device that can be used for progress capture. Another option is to use a bollard-style autolocking braking device in the place of the directional pulley, with the autolocking mechanism serving as the PCD. Bollard-style autolocking braking devices that have a fixed bollard are rather inefficient when used as pulleys; for larger loads, those with one-way rotating bollard-wheels offer greater efficiency (**FIGURE 15-6**).

Of course, greatest efficiency of mechanical advantage will be offered by a high-quality pulley with ball bearings. Organizations using this method often choose to rely on Prusiks for progress capture. Tandem

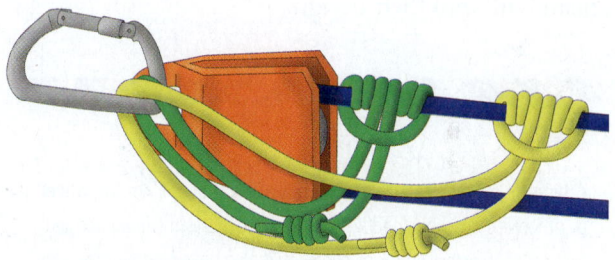

FIGURE 15-7 Tandem Prusiks against a Prusik Minding Pulley.

Prusiks, spaced about a hands-width apart, have been shown to offer more consistent and reliable performance than a single Prusik. Tandem Prusiks should be used with care, as compatibility between Prusiks and rope must be optimized for best performance. Some pulleys offer squared-off side plates designed to block a Prusik and hold it open when the pulley and Prusik are jammed together (**FIGURE 15-7**). This type of pulley is called a Prusik Minding Pulley (PMP).

Load Release

In keeping with the principle that any time a load is moved in one direction, rescuers should also be capable of moving it in the other direction, consideration must be given to how a PCD will be released if it catches inadvertently. The most obvious solution is to simply reactivate the raising method of choice to pull the rope in an upward direction to release the tension. In the event that this is not feasible (for example, with tandem Prusiks), having had the forethought to have incorporated a load-releasing (LR) hitch will be beneficial. Alternatively, a piggy-backed mechanical advantage system, which will be discussed later in this chapter, may be used to release the load. LR hitches are described in Chapter 12, *Belay Operations*.

Using a Winch for Mechanical Advantage

The easiest way to add mechanical advantage to a system is with a powered winch or similar mechanical device (**FIGURE 15-8**). Winches may be driven by human power or by an external power source. Any device used for raising human loads should be specifically approved by the manufacturer for that purpose. These should also be of a **fail-safe** design—that is, a design in which mechanical failure will not cause catastrophic failure (such as breaking the rope or dropping the load.) This may be achieved by means of a kill switch that trips if the winch is overloaded, or a force at which rope will slip on the drum. The fail-safe should not permit sudden release of the load. Winches may, but do not always, have a progress capture mechanism built into their design.

> **TIP**
> A fail-safe mechanical winch design is a one that will not freewheel or otherwise drop the load if the raising mechanism should accidentally fail. A winch used for rescue should also have reverse capability.

As with any rescue system, loads should be able to move in either direction. Powered devices can be difficult to control, and it is all too easy to apply too much power at the wrong time. If a litter is jammed or a hand is caught between a hard surface and a litter, the powered winch may not slow down until after damage has been done. Especially when raising, the amount of force used for the raise should not be so great as to cause damage to equipment or injury to persons if the load became jammed on something, and there should be some means by which rescuers are warned of the jam. A properly rigged system would then be activated to lower the subject slightly so that they may be dislodged from the jam before continuing to raise. A winch used for rescue should not only have a fail-safe design, but also means to quickly switch from raising to lowering.

When rescuers practice using winches and understand their potential and limitations, powered winches can help perform high- and low-angle rescue raising operations safely and efficiently.

> **TIP**
> - Never use a powered winch for hauling a human load unless it has been specifically designed for that purpose.
> - Always follow the manufacturer's instructions when using a rescue winch.
> - Practice with rescue winches in realistic conditions before using them in a real rescue.

Mechanical Advantage Systems

Where a 1:1 system does not provide enough pull-power, and where powered winches are inadvisable or unavailable, mechanical advantage can be created using pulleys and other equipment to reduce the amount of force that a rescuer must exert in order to move a load. Commonly referred to as a **haul system**, mechanical advantage can be applied by rescuers using some basic physics principles to move a relatively heavy load with relatively little force. Depending upon the types and configuration of pulleys and rope, varying amounts of mechanical advantage can be produced.

The amount of mechanical advantage produced by a given system is calculated by formulas and is expressed as a ratio. A ratio is a mathematical expression that is used to make comparisons between two values. In the case of mechanical advantage, the ratio compares the amount of pulling power to the weight of the load. The second number in a mechanical advantage ratio is always a 1, and represents the weight of the load. The first number in the ratio represents the amount of pulling power the haul system is capable of applying as compared with the weight of the load.

In our earlier discussion, we pointed out that when a rescuer must apply force equal to that of the load to

FIGURE 15-8 A powered winch.

move it, the ratio is 1:1. In this case, there is no advantage. However, if a rescuer is able to create a mechanical advantage system wherein the amount of pulling power is twice that of the load, we would call the system a two-to-one system:

$$2 \times \text{Pulling Power} : 1 \times \text{Load Weight}.$$

In other words, a 200-pound (91-kg) load could be moved with a calculated 100 pounds (45 kg) of force. This would be called a 2:1 (two-to-one) mechanical advantage. This calculated mechanical advantage is referred to as theoretical mechanical advantage (TMA). It is referred to as *theoretical* because the actual amount of pulling power we actually get from the system will be less than the calculated amount because some of the energy is lost through pulley friction, rope rubbing or dragging, rope stretch, and other inefficiencies. The resulting mechanical advantage, after friction and other losses, is known as the actual mechanical advantage (AMA). This text will refer to the TMA in hauling systems, but rescuers must keep in mind that AMA will always be a bit less.

FIGURE 15-9 A simple in-line haul system.

TIP

The actual amount of mechanical advantage offered by a hauling system will depend on efficiency in rigging and components. Direction of pull, quality of components, and limiting friction in the system will all help improve AMA.

Simple Mechanical Advantage Systems

A simple mechanical advantage system is one that uses just one line to reeve through all of the pulleys used, with all of the moving pulleys traveling in the same direction as the load. When the mechanical advantage system is rigged directly on the mainline that is already supporting the load, it is called an **in-line haul system** (**FIGURE 15-9**).

An infinite number of pulleys may be used in a simple mechanical advantage system, but there are good reasons to use as few as possible. First, as the number of pulleys in a system increases, the amount of rope that will need to be pulled through the system also increases. Second, each of the moving pulleys in a system reduces the speed at which the load travels, as compared with the rate at which rescuers are pulling. Third, if rescuers have too much pulling power, they might not notice if the load were to get hung up on something. There are other reasons as well, including the amount of equipment required, time to set up, number of resets, increased friction, and other factors. Therefore, rescuers should consider using only as much mechanical advantage as absolutely necessary to move a load.

2:1 Simple Systems and Beyond

It is possible to create mechanical advantage to reduce the amount of force that the rescuer must exert in order to move the load with the introduction of just one pulley, placed at the load itself. A pulley affixed to the load is known as a **traveling pulley**, because it travels with the load. Unlike the fixed pulley described earlier on direction change, a traveling pulley *does* add mechanical advantage. With one end of the rope anchored, then passed through the pulley at the load, the rescuer(s) pulling upward on the end of the rope benefit from a 2:1 TMA.

A disadvantage of this setup is that the rescuer(s) must now pull 20 feet (6.1 m) of rope to move the load 10 feet (3 m). Another disadvantage is that the direction of pull (i.e., upward) may not be the most ergonomically efficient approach. In this case, changing the direction of pull by adding a fixed pulley at the anchor, as described in the earlier section on direction change, may make the rescuers' job easier (**FIGURE 15-10**). The pulley at the anchor is stationary, merely offering a change of direction and no additional mechanical advantage. When adding a pulley, it is always best to anchor it independently from any other pulleys or connectors, even if it is attached to the same anchorage. This helps to prevent binding or jamming.

A quick and easy way to determine the amount of mechanical advantage in a simple mechanical advantage system is to count the number of lines actually supporting the load. If four lines now support the load, it is a 4:1 mechanical advantage. To summarize,

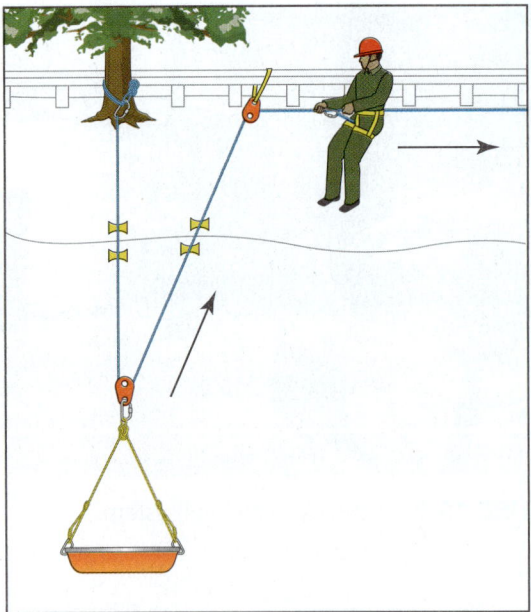

FIGURE 15-10 2:1 with direction change at the anchor; lines supporting the load are numbered.

travelling pulleys supporting the load provide additional mechanical advantage, while fixed pulleys added to the anchor do not.

Rigging a 2:1 Mechanical Advantage with Direction Change

To rig a 2:1 mechanical advantage with direction change, follow the steps in **SKILL DRILL 15-1**.

TIP

Whenever the first pulley in a simple system is placed at the load, the mechanical advantage system will always be an even numbered system (i.e., a multiple of two). In a simple mechanical advantage system, when the line is first attached to the load (without a pulley), the mechanical advantage will be an odd number.

SKILL DRILL 15-1
Rigging a 2:1 Mechanical Advantage with Direction Change

1. At the top of a slope, set an appropriate anchor. Attach an adequate life safety rope to the anchor.

2. Connect a litter bridle to the head of a litter, as for a lower, but with a pulley attached.

3. Maintaining tension on the rope attached to the anchor, pass it through the pulley at the head of the litter and pull it back up slope toward the anchor.

4. Create a second anchor near and in line with the first, and attach a pulley for direction change and a progress capture device. Note: If the original anchorage is of sufficient strength and performance, it may be used again for the additional pulley(s). Using a separate connection to the anchorage will help keep rigging clean and prevent jamming.

5. Place a PCD on the rope that travels upward toward the fixed pulley when hauling, to prevent that rope from traveling downward when released.

6. Pass the rope through the directional pulley and PCD.

In theory, with a 2:1 mechanical advantage hauling system, the force needed by the haul team to move the litter system is only half the force needed in the 1:1 mechanical advantage hauling system previously examined. However, with the 2:1 mechanical advantage hauling system, the haul team must travel twice as far. A direct pull down slope will be more efficient than a pull off to one side, but terrain features may or may not permit this. Care must be taken with rigging to ensure that the counterbalance team is not placed at risk of being pulled over the edge or at risk of a serious fall.

Converting a 2:1 Mechanical Advantage to 4:1

To convert a 2:1 mechanical advantage to 4:1, follow the steps in **SKILL DRILL 15-2**.

With a 4:1 mechanical advantage, moving a 200-pound (91-kg) load would require only 200 / 4, or 50 pounds (23 kg), of effort (system inefficiencies notwithstanding). Additional mechanical advantage may be achieved by replicating this pattern of alternating direction change and mechanical advantage pulleys ad infinitum, but this quickly becomes terribly inefficient and rope/equipment intensive.

3:1 Simple In-Line Systems and Beyond

All of the mechanical advantage systems discussed up to this point have been even-numbered ratios (2:1, 4:1, 6:1). Odd-numbered mechanical advantages are also possible, but to create these the rope must be tied off at load, rather than anchored at the top, with the first pulley in the system placed at the anchor.

As you will recall from the previous discussion on direction change, one fixed pulley at the anchor does not result in any mechanical advantage, but only adds a change of direction to the 1:1 system. Moving the load will require an input of force equal to at least 100 percent of its weight.

However, by attaching a traveling pulley to the load we do realize a mechanical advantage. Remember,

SKILL DRILL 15-2
Converting a 2:1 Mechanical Advantage to 4:1

1 Connect a second pulley to the litter bridle. Grasp the free end of the rope as it exits the directional pulley at the anchor. Then, draw it down toward the litter and pass it through the second pulley at the head of the litter.

2 Run the rope back up slope toward the anchor and through another directional pulley. Move the PCD to the strand of rope that enters the final directional pulley.

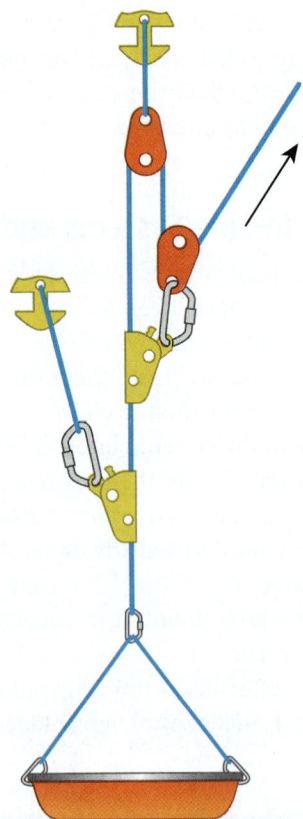

FIGURE 15-11 3:1 mechanical advantage.

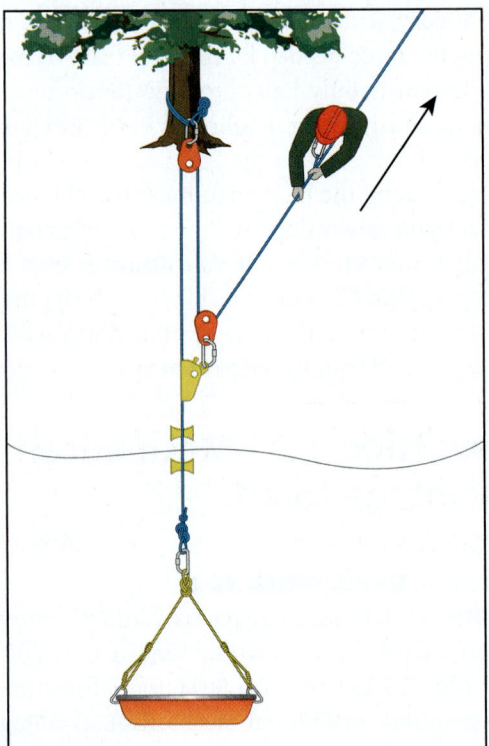

FIGURE 15-12 3:1 hauling system.

traveling pulleys in a simple mechanical advantage system always add mechanical advantage (**FIGURE 15-11**).

By counting the lines that directly support the load, we find that this is a 3:1 mechanical advantage system. In other words, for a given amount of force exerted on the system, three times that force will be applied at the load. Moving a 300 pound (136 kg) load would require 100 pounds (45 kg) of force. Rescuer(s) will pull 3 feet (0.9 m) of rope to move the load 1 foot (30.5 cm).

Load : Force required to move load

300 pounds : 100 pounds

3 : 1

Amount of Rope Pulled : 3 × Distance Load Will Travel

TIP

In a simple haul system, whenever the first pulley is placed at the anchor, the result is an odd-numbered mechanical advantage. The value of the mechanical advantage is equal to the number of pulleys behind the final traveling pulley, plus one.

Again, with odd-numbered systems it is possible to continue adding an infinite number of fixed pulleys and traveling pulleys in series to create direction change and mechanical advantage respectively, but the efficiencies of doing so are quickly exhausted. The 3:1 mechanical advantage system, also sometimes called a Z-Rig because the configuration of the system causes the rope to resemble a sideways Z shape, is perhaps the most common raising system for rescuers to build because it provides a good balance between mechanical advantage, ease of construction, rope length, and internal system friction (**FIGURE 15-12**).

To rig an in-line 3:1 (Z-rig), follow the steps in **SKILL DRILL 15-3**. This skill drill requires the following equipment:

- One rescue rope to serve as mainline and haul system
- Three locking carabiners
- Two pulleys
- One rope grab for PCD
- One rope grab for haul cam
- A separate belay system appropriate for the load to be hauled

In this configuration, the direction of pull must be back toward the anchor and parallel to the other

SKILL DRILL 15-3
Rigging a 3:1 (Z-Rig) Hauling System

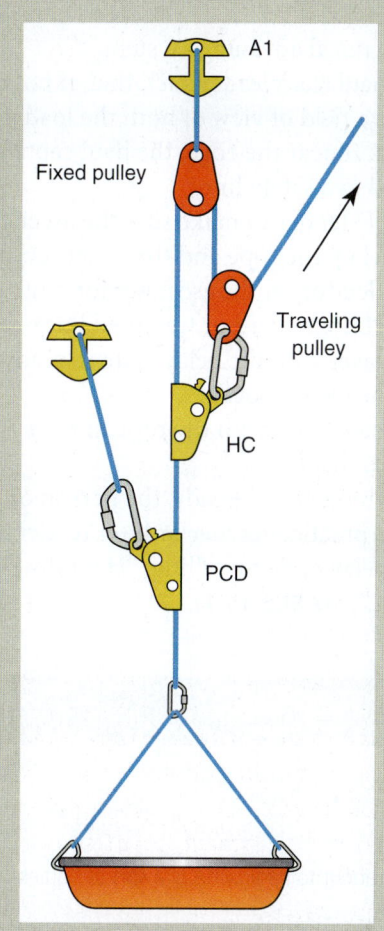

1. Set an anchor point above and in line with the load. Attach an anchor sling with a locking carabiner and pulley attached.

2. Thread the mainline rope onto the pulley and lower an end of the rope to the load.

3. Set a separate anchor point and establish a PCD (with LR hitch, if applicable). This anchor should be close to where the main rope will run but slightly off to the side so that the PCD and its sling do not tangle with the rope.

4. Thread the PCD onto the rope. After the rope has been loaded, make sure the PCD does not create much of a bend in the mainline rope, which could mean a much higher load for the haul team.

5. At a point just above the PCD, place the haul cam. (If a general-use ascender is used for this purpose, the arrow should point toward the load.) Clip a locking carabiner into the haul cam and attach a pulley.

6. Bring the loose end of the main rope back in the direction of the load and run it through the pulley on the haul cam.

7. Reverse direction again with the upper end of the rope, bringing it back toward the first anchor and parallel to the other two strands. This is the hauling end of the rope. Remove all rope slack from the hauling system.

strands of the rope. If it is off to the side, some efficiency will be lost. If, because of limited space or some other reason, the haul team must go off in another direction, set a directional deviation, using a fixed pulley. In this way, the only advantage lost is the small amount accounted for by the friction of the directional pulley. With the PCD proximal to the load as described in this method, there is less opportunity for slack to build in the mainline during a haul, thereby adding security by reducing potential drop distance on the mainline in the event of a problem with the haul system. However, it is also acceptable to place the PCD near the anchor and the haul cam closer to the edge. This latter method may result in fewer resets. A competent rescuer on site will determine which approach best fits the circumstance.

In-Line Haul System with Braking Device PCD

Some bollard-type braking devices may be conveniently integrated into an in-line haul system by using them in place of the anchor pulley. Simply extend the loose part of the rope toward the load, add a pulley at the load, and pull the loose end back toward the anchor—thereby creating a (admittedly lower-efficiency) in-line 3:1 haul system. Bollard-type braking systems are discussed in Chapter 8, *Rescue Equipment*.

With the rope bent around the bollard in the braking device, it effectively takes the place of that pulley, and the inherent one-way braking action serves to capture progress without the need for extra equipment. Some braking devices actually have a moving bollard for improved efficiency in this configuration. To rig an in-line 3:1 with an integrated brake, follow the steps in **SKILL DRILL 15-4**:

1. Set an anchor point above and in line with the load, with a bollard-type autolocking braking device attached. Connect the end of the rope to the load, pull it taut, and thread it into the braking device. Rig a pulley onto the load (or onto the rope just above the load).
2. Pull the loose end of the rope back toward the load and run it through the pulley. Reverse direction again with the upper end of the rope, bringing it back toward the anchor and parallel to the other strand. This is the hauling end of the rope.
3. Remove all rope slack from the hauling system and pull to raise the load; check to be sure that the autolocking braking device captures progress properly.

An in-line 3:1 built in this manner is particularly conducive to switching easily between raising and lowering the load.

Operating a Mechanical Advantage System

To operate any type of mechanical advantage system, follow the steps in **SKILL DRILL 15-5**:

1. Rig any mechanical advantage system.
2. Position the haul team leader such that, if possible, they have a field of view of both the load and the haul team. If near the edge, the haul captain should be tied to a safety line.
3. Position a PCD tender to make sure the ascender does not travel up the rope and that it sets on the rope as intended (or set the ascender for "automatic" with a bungee cord or weighted line).
4. Position a belayer with the belay system, ready to establish the load on belay.
5. Position the haul team with the rope in their hands, ready to haul.
   When all the rescuers are ready, the person on the load (e.g., practice rescuer or litter tender) initiates the belay cycle by calling, "Haul slow" (or "Haul fast.") (**TABLE 15-1**).

TABLE 15-1 Communications for Hauling

Stage of Operation	Command	Response
Roll call	"Roll Call!"	(Gets everyone's attention. Operations are about to begin. This is a sort of "quiet on scene.")
	"On Belay?"	"Belay Is On!" (System is functional and operating) or "Standby!" (Give indication of time needed.)
	"Mainline Ready?"	"Ready on Mainline!" (Haul team is in position.) or "Standby!" (Give indication of time needed.)
	"Litter Captain Ready?"	"Ready on Litter!"
	"Safety Ready?"	"Ready on Safety!"
Tensioning the system and movement	"Load the System."	"System Loading" (Action: Mainline is initially tensioned to feel resistance.)
	"Haul or Raise on Mainline!" (slowly)	Haul—set—slack (to be determined and used by haul team. DO NOT REDO ROLL CALL with each reset of mechanical advantage. Operation carries on.)

*Rescuer needs: Commands should be specific to the needs of the tender, such as speed, belay, and indication of progress (e.g., "Haul on Mainline Slow" or "Raise on Mainline Fast").

Stage of Operation	Command	Response
Termination	"Load at Top." (or at changeover point)	(Action: Stop—set—reset for final haul, as necessary.)
	"Load on the Ground!"	(No reply needed, but all rescuers remain at devices until told off main and off belay.)
	"Off Mainline!"	(Load is no longer on mainline and can be disconnected.)
	"Off Belay!"	(Load is free and clear of drop zone and can be disconnected from the belay.)
Additional commands	"Stop!"	(Action: Can be given by anyone; it means "freeze.")
	"Stop, Stop, Why Stop?"	(A question asked to the group when an unexpected long pause has occurred.)
	"Tension"	(Action: Too much slack exists; take up excess rope.)
	"Slack"	(Action: too much tension; need more rope. Give amount when appropriate, e.g., "Slack belay 2 feet.")
	"Prepare to Change Over—Rig for Lower!"	(Action: Initiate change from haul to lower.)
	"Reverse Haul!"	(Action: Used to initiate the reversal of the mechanical advantage system to lower the load)
	"Rock!"	(Safety issue: Given when any object is falling or dislodged)

6. The haul captain relays the signal to the haul team: "Haul slow" (or "Haul fast").
7. Moving in unison, members of the haul team should grip the rope firmly and pull. Greater pulling force can be generated by leaning into the rope and using leg power, rather than relying on back and arms (**FIGURE 15-13**). A Prusik or rope grab may be applied to the rope to improve grip as long as it is easily removed/adjusted as needed.
8. Before the moving pulley contacts the fixed pulley, the haul captain calls, "Set," and the haul team immediately stops hauling and slowly eases back so that the PCD catches. The PCD tender makes sure the PCD has set.
9. The haul team leader calls, "Slack." The haul team instantly reverses direction and moves back toward the load with the rope still in their hands. As they do so, the PCD tender resets the system by grasping the carabiner attached to the haul cam and pulling it back down on the rope as far as practicable.
10. When the haul team reaches its original position, it is ready to begin another cycle.

FIGURE 15-13 Using leg power can assist in hauling operations.

Piggyback Rope Systems

In rescue terms, a **piggyback rope system** is any rope system that is used to exert force on another. In the preceding examples of haul systems, we have discussed in-line haul systems that are created by

FIGURE 15-14 3:1 Haul system piggybacked onto a mainline lowering system.

FIGURE 15-15 1:1 hauling system with progress capture.

simply converting the mainline to a haul system by reeving it through a series of pulleys. However, it is also possible—in fact, is even more common among some rescue organizations—to maintain the mainline as a separate system from the haul system by using a separate rope to create the haul system and using a rope grab, known here by its function as a **haul cam**, to attach the haul system to the mainline (**FIGURE 15-14**).

Piggybacking the haul system onto the mainline offers a high level of versatility, and allows rescuers to apply virtually any kind of a haul system to a rescue scenario, even after the load is out of reach. It is not unusual for a rope rescue operation to involve 100 feet (30 m) or more of rope, but rarely will rescuers have this amount of space to effect a haul. More often, the working area at the top of a rescue site is no more than 100 to 500 square feet (9.3 to 46.5 m^2), and the distance a load can be raised during any one evolution will be limited by the available working space. In such situations, a haul cam is placed on the mainline as proximal to the load (or edge) as space permits, and thenceforth this cam is treated as "the load."

During a raising operation, the haul cam must be managed by a **PCD tender** who has a good field of view and can move the cam forward on the line as opportunity permits. If there is a shortage of personnel for this operation, the haul team leader may

be stationed at the haul cam to manipulate it, but it is better if one person can be assigned specifically to the job of PCD tender. In operation, the haul team will raise the load as much as possible; as much tension as possible is taken up in the mainline and progress captured; and then the PCD tender slides the haul cam down the rope, toward the load, as far as possible to "reset" the system in readiness for another haul (**FIGURE 15-15**).

TIP

Alternative Approach: Placing the PCD to the Rear in a Z-Rig

In certain cases, it may be impossible to place the PCD forward of a Z-rig hauling system, such as when no anchors are available for attaching the PCD. When a Z-rig (3:1 mechanical advantage) is used, the PCD can be placed in back of the hauling system. This can be done with a rope grab, or if a Prusik minding pulley is used for the rear pulley, Prusiks can be added to make a PCD out of the rear pulley.

Compound Systems

Simple mechanical advantage is all that is required for the vast majority of rescue work. However, in some cases, particularly where working space, resources, or personnel are limited, there may be some benefit to creating a **compound mechanical advantage system** in which one haul system is devised to piggyback onto another haul system, thereby multiplying force as one pulls on the other. This is also sometimes referred to as a **stacked mechanical advantage** system.

To calculate the TMA of a compound system, simply multiply the first digit of the mechanical advantage ratio of the first system by the first digit of the mechanical advantage ratio of the second system. The product of the two systems is the TMA of the compound system. By pulling on a 2:1 mechanical advantage system with another 2:1 mechanical advantage system (2 × 2 = 4), you can achieve a 4:1 compound mechanical advantage. Pulling on a 3:1 with a 2:1 (3 × 2 = 6), we get a 6:1; pulling on a 4:1 with a 2:1 (4 × 2 = 8) gives an 8:1, and so on. To rig a 4:1 system, follow the steps in **SKILL DRILL 15-6**. The minimum equipment requirements for a 4:1 (piggyback) hauling system are as follows:

- One mainline rope
- One hauling rope (50 to 100 feet [15.2 to 30.5m], depending on the space available for the haul)
- Three locking carabiners
- Two pulleys
- Two rope grabs (one to act as a PCD and one for hauling)
- Separate belay systems appropriate for the load to be hauled

SKILL DRILL 15-6
Rigging the 4:1 System

1. Attach the mainline rope to the load. Secure the upper end so that it does not accidentally slip over the edge.

2. Establish a strong anchor point for the PCD, close to where the main rope will run but slightly off to the side so that the PCD and its anchor sling do not tangle with the rope. Attach an LR hitch to the anchor point. Into the end of the LR hitch, attach a locking carabiner and clip a general-use ascender to it (or tie a Prusik to the rope and clip the end of the sling into the carabiner).

3. Establish an anchor point for the hauling system. This anchor point should be well back from the edge and in line with the load to allow room for the haul system. Attach a sling to the anchor point and clip a locking carabiner to it.

4. Find the center of the haul rope (the shorter line). At the center, tie a figure 8 on a bight knot and clip it into the locking carabiner on the anchor sling.

5. After the figure 8 on a bight knot has been tied, the hauling rope will have two strands. Bring one leg of the haul rope back in the direction of the load. About two-thirds of the way down this leg of the haul rope, create a bight and clip a pulley into it. Into the pulley, clip a locking carabiner and attach a rope grab to it; the rope grab must be attached on the mainline rope above the PCD (between the PCD and the top end of the mainline rope). If the rope grab is a general-use ascender, the arrow on the shell should point toward the load. This is the point where the hauling system piggybacks onto the mainline rope.

6. On the same leg of the haul rope to which the rope grab has been attached, find the free end of the rope. In this end, tie a figure 8 on a bight knot. Clip a locking carabiner into it and attach a pulley to it.

7. Take the second leg of the haul rope and bring it down to the pulley just clipped into the first leg of the haul rope. Thread the second strand through the pulley.

(continues)

SKILL DRILL 15-6 (continued)
Rigging the 4:1 System

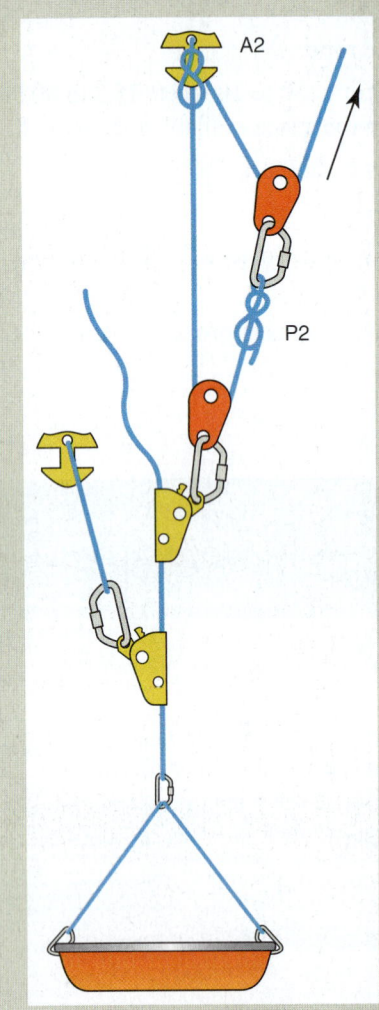

8. Pull the end of the second strand back toward the anchor for the hauling system. This is the end the haul team pulls.

9. Commence hauling. Before the pulleys at P2 and A2 come together, the haul captain calls, "Set." The haul team immediately stops hauling and slowly eases back so the PCD catches. The PCD tender makes sure the rope grab sets.

10. The haul team leader calls, "Slack." The haul team instantly reverses direction and moves back toward the load with the rope still in their hands. As they do this, the person attending the haul cam resets the system. When the haul team reaches its original position, it is ready to begin another cycle.

Keep in mind that the more pulleys there are in a system, the greater the frictional energy loss, so the "more is better" philosophy quickly erodes in this scenario. In fact, sometimes compound systems can be used to optimize the AMA that can be achieved with a limited number of pulleys. For example, the same two pulleys used to build a 3:1 system could be used to construct a compound mechanical advantage system of 4:1 (**FIGURE 15-16**).

Compound systems are easiest to construct if separate ropes are used for each set of mechanical advantage. This can quickly eat up a lot of rope, and in fact the total amount of rope that must be pulled through the system is quickly multiplied. The potential advantages offered by having greater pulling power and fewer pulleys to induce frictional loss must be considered against the potential disadvantages of having to pull more rope through the system. To rig a compound 4:1 piggybacked to a mainline system, follow the steps in **SKILL DRILL 15-7**.

Stacking systems in a compound haul system requires numerous anchors, multiple ropes, and the ability to envision the outcome. It can quickly become a tangled mess. When two systems of unequal mechanical advantage are combined to create a compound system, riggers are faced with the additional challenge of having to accurately assess and plan travel distances, as the two systems will move at different rates from one another.

CHAPTER 15 Mechanical Advantage Systems

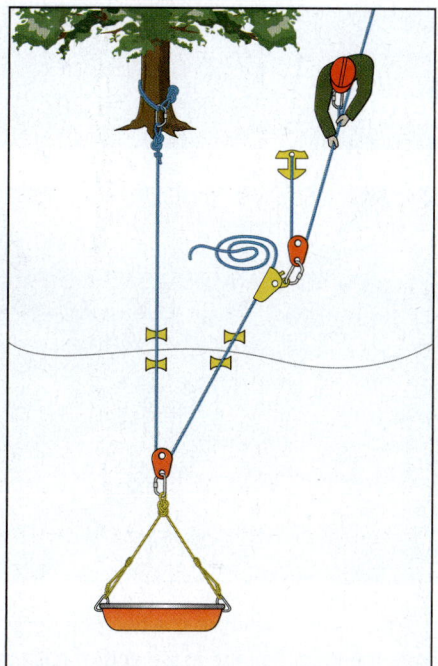

FIGURE 15-16 A 4:1 compound mechanical advantage system uses the same number of pulleys as a 3:1 simple mechanical advantage system.

It is possible to build even more **complex mechanical advantage systems** in which traveling pulleys move in the opposite direction from the load or even from one another, but these are generally more useful for training than practical application. The difficulties associated with estimating pulley travel distances and speeds are not conducive to smooth and efficient rescue operations.

Two-Tensioned Rescue Systems

Two-tensioned rescue systems (TTRS) may warrant some additional considerations when raising a rescue load. When two autolocking bollard-type brake devices are used for the TTRS, the easiest solution by far is to utilize an in-line 3:1 mechanical advantage with integrated braking device PCD, as described earlier in this chapter, on each of the two lines. In this case, each of the two lines may be readily converted to a raising system simply by reeving the free end of the rope through pulleys at or near the load, and pulling. **SKILL DRILL 15-8** discusses rigging a TTRS inline 3:1 raising system. TTRS are discussed in Chapter 14, *Lowering Systems*.

SKILL DRILL 15-7
Rigging a Compound 4:1 Piggybacked to a Mainline

1 Set an anchor with an autolocking braking device at the top of a slope. Attach a mainline to the autolocking braking device. Connect the end of this rope to a litter with a bridle, as for a steep lower.

2 Set a rope grab onto the mainline below the braking device, to act as a haul cam, and connect a pulley.

3 Anchor a haul line at the top of the slope, near the braking device anchor. Pass the haul line through the pulley at the haul cam.

4 Attach a haul cam with a pulley to the slack end of the first haul line.

5 Anchor a second haul line at the top of the slope, alongside the first. Pass the haul line through the pulley at the second haul cam. The haul team can now move the load with a pull.

SKILL DRILL 15-8
Rigging a Two-Tensioned Inline 3:1 Raising System

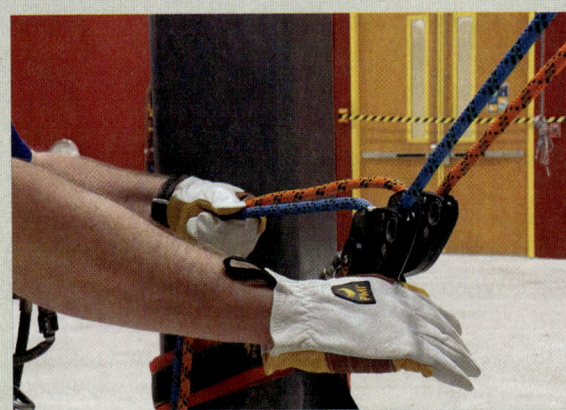

1 Build a TTRS using two bollard-type autolocking braking devices (one on each line). Connect the ends of both ropes to the load, pull taut, and thread them into the braking devices. See Chapter 14, *Lowering Systems* for more on building a TTRS.

2 Rig a pulley onto each of the two lines at or just above the load. Pull the loose end of each rope toward the load and pass them through the pulley on their respective lines. Reverse direction again with the upper ends of the ropes, pulling the free ends back toward the anchor and parallel to the load line. These are the hauling ends of the ropes. Remove all rope slack and pull both systems simultaneously to raise the load; check to be sure that the autolocking braking devices capture progress properly. To convert the system back to lower, simply remove the pulley and associated equipment at the load-end and commence lowering.

Safety Check

Before operating any haul system, a safety check should be performed by a competent rescuer. The safety check should be performed by someone other than the person who rigged it, even if the person who rigged it is an expert; having multiple eyes looking at the system and multiple brains thinking through it improves the odds that errors will be caught and corrected!

The best way to perform a safety check is to start at one end of the system and work toward the other, observing and evaluating all aspects of the system along the way. The person performing the safety check should not re-rig the system or make any changes; if errors are found they should stop, inform the safety officer, and not proceed until the errors are corrected. When restarting, the safety check should be started over completely from the beginning, not just picked up from the point where an error was found or correction made.

The following things should be considered as part of the safety check:

1. Appropriate anchorage(s) that is correctly rigged This specifically requires a visual and tactile check to ensure that all carabiners are correctly locked.
2. Proper use of equipment, in accordance with manufacturer instructions
3. Rigging appropriate to anticipated direction of pull (preload to verify)
4. All knots properly tied
5. Edges protected
6. System sufficient to perform the desired function
7. Proper safety factor
8. Operational commands understood among participants
9. Appropriate redundancy (failure of any one point will not result in failure of the entire system)

10. All loose items at the top secured to prevent them from being kicked or dropped over the edge
11. All personnel wearing proper PPE
12. Everyone at the edge protected by a belay or fall arrest system

The individual performing safety checks should be skilled and knowledgeable in construction and operation of haul systems, and should be capable of analyzing performance of a system to determine its adequacy. They should have full authority from the authority having jurisdiction to stop operations until concerns have been alleviated.

> **TIP**
> - Never substitute voice haul commands unless there is an overriding reason and all team members were informed before the change.
> - Use only crisp, distinct terms that have no chance of being mistaken for other words. For example, do not replace "Stop" with "Whoa," which easily can be mistaken for "Slow" or, even worse, "Go."
> - All commands should be given over the radio or loudly enough so that everyone—litter team, belayer, and rope handlers—can hear them.

Operating a Haul System

Haul team members should not be selected on the basis of brute force ability, but instead according to the following criteria:

- Ability to follow commands
- Ability to react quickly
- Sensitivity to the feel of the haul rope
- Teamwork ability

Raising systems can create extraordinary forces on the rope rescue system. Unnoticed problems quickly can result in catastrophic system failure. Personnel constantly must monitor for potential problems, including, but not limited to, the following:

- Knots on moving rope that jam in cracks
- Broken gear that causes system failure
- Systems reaching their limit
- Pinned arms and legs
- Loads getting caught beneath overhangs or during edge transitions

Good communication must be established between team captains and the haul team. Also, the haul team must keep in mind that what might seem a reasonable speed for them might feel very fast to those being hauled (i.e., the rescue subject and the litter tenders). Therefore, the haul team must pull slowly unless told to do otherwise, and must be prepared to stop at a moment's notice.

Communications in Raising Operations

Commands and voice communications for moving a litter, whether up or down, should always be initiated by the litter captain. Everyone else remains quiet. If verbal commands are intended to be used during a rope rescue system, the parties involved (particularly the brakeman and the litter captain) should agree to the sequence of commands that will be exercised. The person who is "on rope" (or, if a litter team, the litter captain) should be the one who provides the cues for moving, stopping, etc. By agreeing upon a pre-established set of words and their intended response, the brakeman and the rescuer can become more effective in the lowering operation.

Example:

Rescuer: Belay ready?

Brakeman: Ready!

Once the litter captain has called for a raise, the haul team leader should take over the responsibility of initiating actions until such time as the litter captain calls again for a different action. During hauling evolutions, the haul team leader may use a concise set of terms like HAUL—SET—SLACK to efficiently communicate actions to the haul team. In this case, with haul team leader and haul team in close proximity to one another, the haul team need not repeat the commands; their execution of those should suffice to show they have understood.

- Haul: Haul team pulls on the haul system.
- Set: Haul team stops pulling and holds the line taut; meanwhile, the ratchet cam attendant ensures that the ratchet is set.
- Slack: Haul team fully releases all tension on the haul line; haul cam attendant rapidly resets haul cam.

With a firm understanding of roles and responsibilities, and a grasp on communications, a team should practice haul system skills until they are fully adept.

After-Action REVIEW

IN SUMMARY

- There are two general approaches to hauling systems: mechanical power and human power. Due to the shear force required to raise a load, utilizing mechanical advantage to reduce the amount of human power required is essential.
- One of the simplest hauling systems used in low-angle evacuation is the 1:1 mechanical advantage hauling system.
- With a 2:1 mechanical advantage hauling system, the force needed by the haul team to move the litter system is only half the force needed in the 1:1 mechanical advantage hauling system, but the haul team must travel twice as far.
- Other types of force-multiplying hauling systems include 3:1 and 4:1.
- A quick and easy way to determine the amount of mechanical advantage in a simple mechanical advantage system is to count the number of lines actually supporting the load.
- Although force-multiplying hauling systems can make it easier for haul teams to move a litter, all such systems have some drawbacks.
- Communications methods and language should be agreed in advance.
- Before operating any haul system, a safety check should be performed by a competent rescuer. The safety check should be performed by someone other than the person who rigged it.
- Haul team members should not be selected on the basis of brute force ability, but instead according to the following criteria:
  - Ability to follow commands
  - Ability to react quickly
  - Sensitivity to the feel of the haul rope
  - Teamwork ability
- Communications methods and language should be agreed in advance.

KEY TERMS

Complex mechanical advantage systems Mechanical advantage systems in which traveling pulleys move in the opposite direction from the load or even from one another.

Compound mechanical advantage system A system in which one haul system is devised to piggyback onto another haul system, thereby multiplying force as one pulls on the other; see also *stacked mechanical advantage*.

Counterbalance haul system A procedure for hauling that uses a 1:1 ratio mechanical advantage, employed by placing a pulley at an anchor a haul team at one end of the rope that moves in a direction opposite to the load using the haul team's own weight to offset the load.

Direction change A method of deviating the rope path by creating an angle, usually via a pulley placed into the system.

Fail-safe The failure of a device or system will not endanger life or property.

Fixed pulley A pulley that is rigged into a system in a position where it will remain stationary at a specific location.

Haul cam A camming device used to grip a rope for the purpose of hauling.

Haul system An arrangement of ropes and pulleys and connectors that is arranged in a manner to provide mechanical advantage for the purpose of raising a load under control.

Haul team The group of individuals who provide the power to raise the load.

In-line haul system A manner of arranging ropes, pulleys, and connectors directly on the primary support line so that mechanical advantage is provided.

Mechanical advantage A force created through mechanical means including, but not limited to, a system of levers, gearing, or ropes and pulleys usually creating an output force greater than the input force and expressed in terms of a ratio of output force to input force.

PCD tender The person assigned to tend the haul cam, ensuring that it sets when appropriate and that it does not set when it should not.

Piggyback rope system Any rope system that is used to exert force on another.

Progress capture The act of preventing a rope from traveling in a direction so that a load is maintained in a static location.

Progress-capture device (PCD) A rope-grab device, general-use ascender, or hitch placed on the rope in a hauling system to prevent the rope (and the load) from slipping back down as the haul system is reset. Also commonly referred to as the ratchet.

Ratchet cam A device that permits rope to travel through in one direction but not the other, commonly used as a progress capture device.

Stacked mechanical advantage A system in which one haul system is devised to piggyback onto another haul system, thereby multiplying force as one pulls on the other; see also *compound mechanical advantage system*

Traveling pulley A pulley that travels with the load.

On Scene

1. For a given rescue operation, how will you determine how much pulling power you might need in a raising system?

2. List the items of equipment you would need to create a 3:1 (Z-Rig) using the fewest components possible.

3. Given sufficient anchor points, two ropes, two single-sheave pulleys, and carabiners, what kind(s) of mechanical advantage systems might you be able to create without resorting to using carabiners in place of pulleys?

4. What is the difference between theoretical mechanical advantage and actual mechanical advantage? How much does it really matter?

5. What are some advantages and disadvantages of prerigged mechanical advantage systems?

CHAPTER 16

Operations

Working in Suspension

KNOWLEDGE OBJECTIVES
- Recognize the risks and signs and symptoms of suspension-induced injury. (pp. 322–323)
- Describe the role of the third man in rope rescue. (pp. 323–326)
- Describe the methods a rescuer on a rope can use to negotiate an edge. (**NFPA 1006: 5.2.19**, pp. 326–327)
- Describe rigging a litter for a high angle lower or raise. (**NFPA 1006: 5.2.12, 5.2.13**, pp. 327–330)
- Explain how to tend a litter moving over an edge. (**NFPA 1006: 5.2.21, 5.2.22, 5.2.23**, pp. 330–335)
- Describe the methods of attaching a litter to a lowering system. (**NFPA 1006: 5.2.21, 5.2.22, 5.2.23**, pp. 327–330)
- Explain the duty of an edge attendant. (**NFPA 1006: 5.2.14, 5.2.19**, pp. 324, 331, 334, 335)
- Describe rigging to a horizontal high line. (pp. 335–336)

SKILL OBJECTIVES
- Navigate an edge while suspended. (**NFPA 1006: 5.2.19**, pp. 326–327)
- Configure a litter attachment for a horizonal lower. (**NFPA 1006: 5.2.12, 5.2.13**, pp. 328–330)
- Configure a litter attachment for vertical lower or raise. (**NFPA 1006: 5.2.12, 5.2.13**, pp. 332–334)
- Manage a litter while suspended. (**NFPA 1006: 5.2.21, 5.2.22, 5.2.23**, pp. 331–332)
- Direct a litter raising and lowering operation in a high-angle environment. (**NFPA 1006: 5.2.23**, pp. 332, 335)

You Are the Rescuer

Your organization responds to a local manufacturing plant, where a worker has fallen to a lower level. The subject is motionless and unresponsive. The operations leader decides that the best course of action will be to quickly lower a rescuer to the subject from an upper level. The rescuer will check the condition of the subject, package them into a litter, and connect them to a rope rescue system. The litter will be lowered to ground level without an attendant.

1. What skills will the rescuer who is lowered initially need to have?
2. What communications methods will the rescuers use?
3. What challenges will the rescuer face as they negotiate the edge?

 Access Navigate for more practice activities.

Introduction

As a basic function, being lowered and raised by rope is an operations-level skill. The foundational skills of managing yourself while suspended in the vertical are prerequisite to the next step: that of a suspended rescuer making their own progress up and down rope, as well as managing a suspended subject. Managing a subject while suspended is considered to be a technician-level skill. An operations-level rescuer is expected to be capable of moving a subject from one stable location to another, but is not expected to accompany a subject on rope, nor to be able to ascend and descend on rope themselves. Another way to state this is to say that an operations-level rescuer may be lowered or raised on rope from one stable location to another, but an operations-level rescuer does not manage their own progress (ascent/descent) on rope and does not manage a subject while on rope.

Developing a knack for functioning effectively while being lowered takes time and practice—especially when it comes to working in a free-hanging condition, with no surface on which to exert friction to help control body position. Whereas ascending and descending require strength and stamina, suspended work requires coordinated communication and the ability to anticipate needs. There is something disconcerting about not having direct control of your own movement.

Suspension-Induced Injury

Concerns over the hazards and potential for suspension induced injury have risen to prominence in recent years. Various terms are used to describe the phenomenon, including suspension injury, harness trauma, suspension trauma, suspension intolerance, and others. In this text, we will use the term **suspension-induced injury**.

Various instances of suspension-induced injury have been noted over the years, but it was not until the 1980s that the concept gained traction, largely because of research performed by the Federation Francaise de Speleologie in an effort to determine clues as to the cause of death in apparently healthy, otherwise uninjured, suspended cavers who had expired while awaiting rescue. Before the project was halted due to perceived risk to test subjects, researchers did confirm a clear correlation between motionless suspension and severe physical response. Some years later, the National Institute for Occupational Safety and Health (NIOSH) in the United States performed tests related to harness suspension, confirming the findings of previous studies and emphasizing that keeping the legs elevated and moving is a key to survival.

While anyone suspended in a harness is potentially at risk of experiencing the effects of suspension-induced injury, hanging upright and motionless from an attachment point, especially the dorsal attachment point, appears to be a key factor in the slowing of circulation in the legs, leading to shock (**FIGURE 16-1**). Other factors, such as weight, overall physical fitness, harness fit, fall distance, and other medical history, may affect a person's risk level, but do not seem to have a direct impact as to whether a person experiences the signs and symptoms of suspension-induced injury.

A suspended person who is conscious and who can move their hands and feet has a distinct advantage over an unconscious subject. Rescuers working while suspended in a harness should stay active while on the rope, flexing their feet and legs periodically either against a vertical surface or in a footloop

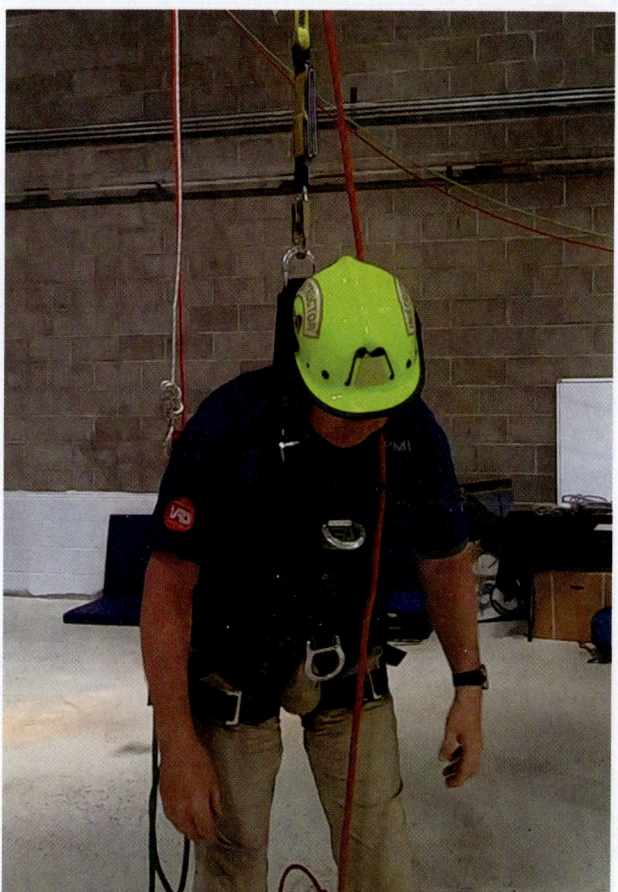

FIGURE 16-1 Motionless suspension can be hazardous.

(or etrier), and even keeping their legs elevated, if possible. Indications of increased difficulty in breathing while suspended, a growing discomfort in extremities, or tingling or numbness in the legs should be taken as an early warning sign to get off the rope and stretch a bit to avoid injury.

A person who is suspended motionless, particularly after a significant fall and/or in a compromised environment, is in imminent danger, and this represents a true emergency. Personal escape becomes increasingly difficult as the condition progresses, and the subject may eventually lose consciousness.

Prompt rescue should be provided to anyone who is suspended in a harness, particularly if they are unconscious or unable to move. Rescuers responding to a person who is experiencing suspension-induced injury may observe symptoms including swelling of the face and hands, puffiness around the lips and eyes, shallow breathing, and extreme lividity (purple skin). On the other hand, symptoms may not be seen. An absence of symptoms does not necessarily mean an absence of the condition, and subjects should be closely observed during and after an incident.

Treatment for the condition was once believed to be to keep the subject upright, in a seated position, even after they are on the ground. However, more current protocols suggest that treating for shock, including providing high-flow oxygen and fluid replacement, is the preferred approach. Follow the local protocols as crafted by a medical director and the level of medical training of the rescuer.

Third Man Operations

A suspended rescuer role in which support functions are performed without direct responsibility for the subject is sometimes called the **third man**. This term, derived from a phenomenon in which an unseen being is sensed as intervening in a critical moment of distress, can be traced to the poem *The Waste Land* by T. S. Eliot.

In rope rescue vernacular, the term *third man* is used to describe a rescuer whose role it is to access the subject quickly, assess the situation, help with packaging and preparation, and generally anticipate and assist with the many tasks required. They are typically not the primary rescuer, nor the rescuer who takes direct contact and control with the subject.

The third man is, paradoxically, usually the first person over the edge in a rope rescue operation. Equipped with basic technical and medical gear to stabilize and assess the subject, they set the stage for the rest of the operation with the resources and evacuation recommendations they request. The third man need not have extensive medical skills but should be capable of providing basic life support assistance to the subject. More important, the third man should have a strong understanding of rope rescue systems and be capable of securing the subject to prevent further harm (**FIGURE 16-2**).

When accessing the subject from above, the third man must first prioritize their own safety. Although there may be a sense of urgency to reach the subject, taking the time to evaluate site hazards, don proper personal protective equipment (PPE), select appropriate equipment to carry, establish appropriate position relative to the impending operation, and set up an appropriate lowering/raising system will set the stage for a successful rescue.

When the third man reaches the subject, their initial role is to assess the subject's physical predicament and medical condition. They should provide relevant information back to site management and the primary rescuer to assist with ongoing evacuation and rescue planning. While the third man may initiate urgent medical interventions, the most urgent of interventions involves readying the subject for extrication and transport.

FIGURE 16-2 The third man should have a strong understanding of rope rescue systems and be capable of securing the subject to prevent further harm.

FIGURE 16-3 Radio chest harness.
Courtesy of Steve Hudson.

To the best of their ability, the third man should prepare the subject for the next phases of the rescue. In addition to immediate lifesaving medical measures, this includes having them ready for arrival of the primary rescuer and, if applicable, a basket litter for transport. This may include providing physical protection or shielding if required, determining where the litter should be positioned, and having the subject completely ready for loading into the litter by the time the primary rescuer and litter arrive. If the subject is suspended on rope or unable to move themselves, the third man may need to create a pickoff system (described later in this chapter) to help extricate and move them into the litter.

When the primary rescuer arrives, the third man's role becomes that of an assistant. Their role then is to support the efforts of the primary rescuer in any way needed. If a litter is being used, the third man may even continue to be lowered and raised alongside the litter so that they may continue their support role. Particularly in the case of a lower with obstructions, it can be very helpful to have an extra set of hands to help maneuver the patient.

In some cases, it may be prudent to run an adjustable tether between the third man and the primary rescue system or litter, to help keep them close. If such a contraption is used, it should be set in such a manner that it can be quickly detached as needed.

Communications

With system operation at the discretion of a brakeman or haul team, good communication between the rescuer who is on rope and the person controlling their progress is key. If possible, direct voice communication between these two individuals is preferable.

When conditions such as distance, line of sight, or ambient noise preclude direct communication, a relay person may be employed. Where available, an edge attendant can serve double duty with this assignment. Over longer distances, radios may be beneficial.

Many rope rescue professionals choose to use a radio secured to their body by way of a radio chest harness to facilitate more efficient radio communications. A radio chest harness may integrate with the full-body rescue harness, or it may be a completely separate configuration of straps. In either case, it should be secured in a location where the rescuer can easily hear the speaker and key the mic, as needed (**FIGURE 16-3**).

Some rescuers use voice-activated radio systems for the hands-free convenience they offer, but these have some clear disadvantages, including a tendency to pick up and transmit unwanted sounds, including ambient noise, rescuer grunts and mutterings, and private conversation.

Regardless of the transmission source and method, voice communications between rescuer and brakeman should be limited to a few simple, predetermined commands. If there are to be multiple rope rescue operations occurring simultaneously—such as a third man system and a litter system—be sure to assign each system an identity, and to preface each radio communication with that identity. For example, the third man system may use language such as:

Third man brake, third man rescuer, on belay?

While the litter system communications may progress as follows:

Litter brake, Litter, on belay?

Using different radio channels for the different systems is also considered good practice, but once a rescue is underway, operators should not need to change channels to communicate. Setting your radio at a primary communications channel while continuing to monitor other relevant channels is one way to help ensure adequate communications. When a rescue operation is in a critical phase, all other radio traffic should be held. See Chapter 12, *Belay Operations*; Chapter 14, *Lowering Systems*; and Chapter 15, *Mechanical Advantage Systems* for further discussion about on-scene communications.

> **TIP**
>
> When more than one rope is used for lowering, such as in a situation with a belay or third man, line management efficiency can be improved by using ropes of different colors. This helps rescuers to communicate more succinctly and clearly about needs. For example, a rescuer might simply call, "Red rope down 1 foot." Check before commencing the operation to ensure that everyone has the same understanding of which ropes are which color.

During the operation, no conversation or communications should be taking place except between the suspended rescuer and the brakeman or haul team captain. Commands should be crisp, effective, and to the point. Communications should be discussed and agreed in advance. What words or terms will be used for which actions?

The specific verbiage used may vary depending on local protocol or preference. Of most importance here is that the brakeman and the suspended rescuer have the opportunity to identify and agree in advance to the sequence of communications and the words that will be used. In any case, the suspended rescuer should be in primary control, and all communications should be repeated by the hearer (i.e., "closed loop communications") to ensure precision and accuracy.

Connecting to the Rope Rescue System

Typically, the third man (or other suspended rescuer) makes connection to the lowering rope by clipping into their harness waist attachment with a locking carabiner. Attaching to the waist attachment is generally more effective than clipping into the sternal connection on the harness, as the latter tends to place the center of gravity below the connecting point, thereby reducing mobility. Likewise, the use of a locking carabiner for the connection (rather than a direct tie-in) also improves mobility, allowing for quick attachment to or detachment from the rope. If a secondary system is used for belay or backup, it may well be attached to the sternal, dorsal, or even the same waist attachment point. A belay line is generally kept slack during operation, and should not interfere with mobility (**FIGURE 16-4**).

As with other systems, the choice of whether or not to belay a suspended rescuer is a matter of probability versus consequence of failure. With human error being the most common failure mode in rescue operations, use of an autolocking descender or progress capture system can mitigate this hazard, making a secondary belay system arguably less imperative.

FIGURE 16-4 Rescuer tie-in.

At the same time, risks associated with the complexity of employing a secondary system should be considered—especially taken in context with the incident at hand, where the scene may be cluttered by other lines such as a litter lowering/raising system, edge lines, and even the subject's rope(s).

When possible, a safety officer should check the connection of the rescuer prior to their being suspended on rope, as well as the rigging and operation of the raising or lowering system to be used. Although any rescuer should take first responsibility for their own safety during an operation, assigning a safety officer who can check the rigging before the system is loaded, and who can monitor the system as it is loaded, is very helpful. In small team operations, it may be necessary for the safety officer to perform some other function as well, such as belay or communications.

Negotiating Edges

Lowering Over an Edge

A rescuer whose progress is being controlled by another faces the challenge of needing to direct their own progress, while at the same time being able to react and respond appropriately to sudden changes in rope control. With communications agreed in advance, the next step is simply to execute those communications smoothly and accurately to avoid misunderstanding.

As previously noted, the person being lowered should be the initiator of any actions or movement of the system. There are exceptions, of course—for example, if something just does not feel right to the brakeman or belayer, or if anyone on site observes or anticipates a hazardous situation brewing. In these cases, the word "STOP" should be announced loudly and clearly, so that the brakeman, haul team, and belayer (if applicable) bring the system to a complete halt until the situation is examined and corrected.

It is generally best to avoid descending (or ascending) straight in line with a subject, to avoid dropping debris, objects, or yourself onto them. When initiating the descent, the suspended rescuer should rig into their system well away from the edge. When everyone is ready, the rescuer can start the process by verifying that the belay is on.

With the brakeman and belayer in place, the safety officer should analyze the system and declare it ready for use. As always, the brakeman should respond to or even repeat commands as they are heard to ensure harmony. The rescuer commences the process and remains in command throughout. When the rescuer is ready, they should pull backward on the system, hard, toward the edge, and call for the system to begin moving. When the rescuer reaches the edge, they may call for a slowdown or even a stop so that they can carefully negotiate the edge.

Getting over the edge is the most difficult part of any suspended rope operation. A sharp edge is more difficult to get over smoothly than a sloping one. In fact, if a sloping edge is gentle enough, a slowdown may not even be required. However, for a sharp or undercut edge, the rescuer may wish to call for a full stop just as their heels overlap the edge. The most graceful way to get over an edge is to use a high point anchor or multi-pod to keep the horizontal path of the rope above the height at which it connects to the rescuer, but with practice these skills can be mastered even without a high point.

Most rescuers will choose to attach at the waist, which allows good balance when using the legs to pull backward. With the rope taut, and the edge behind, keep your feet shoulder width apart as you allow the harness attachment point to apply as much force as possible in the direction of pull. Leaning into the system, call for the brakeman to lower slowly.

As the rope begins to move, the rescuer keeps their feet still and bends at the waist over the edge. When the rescuer is in a seated position, with feet at about waist level, the rescuer can begin taking small steps backward. The rescuer should keep their body in a seated position, feet in front, throughout the entire descent. When the rescuer is ready, they can call for a faster descent to the brakeman.

A good brakeman will be responsive to commands, and will maintain a consistent speed unless requested otherwise. Likewise, a skilled rescuer will perform smoothly and avoid any uncontrolled or loading or swing.

If getting over the edge as described previously is too difficult for a rescuer, they can instead downclimb from a seated position. To negotiate an edge in this manner, initiate the operation as described previously, but when the rescuer reaches the edge, call for a full stop and then sit down with their legs dangling over. Then, the rescuer will roll over from the seated position onto the belly, and brace against the edge with their knees. Alternatively, an etrier can be hung over the edge and used as a ladder to downclimb.

Once over the edge and in a stable position with knees against the vertical surface, the rescuer calls to be lowered down very slowly. As the rope begins to move, the rescuer extends their legs to press against the vertical surface with their feet instead of their knees. Once on their feet, the rescuer keeps their feet at or above waist level to maintain maximum stability, and proceeds with the lower.

Raising Over an Edge

Raising over an edge is similar to lowering over an edge, but requires practice to execute smoothly. Slowing down before getting to the edge will help to prevent dragging the rescuer nose first when they reach the point at which the rope travels in a horizontal position.

As soon as the rescuer can see up and over the edge, there is a natural tendency for them to want to shift into a more vertical stance and "climb" over the edge. Resisting this urge and maintaining a seated position, with legs horizontal, for as long as possible will pay dividends, making it much easier to get up and over.

Landing

As the suspended rescuer reaches their landing zone, they should call for the brakeman (whether raising or lowering) to move more slowly. If rescuer and brakeman are out of sight of one another, calling distances helps. For example, the rescuer may call out, "10 feet from landing."

As they approach the final landing zone, the rescuer should survey the situation to ensure safety and to more specifically plan their approach. It may even be necessary to call for a full and complete stop while still some distance away, and take time to observe. If the subject is ambulatory, it may be necessary to speak to them to calm and direct them. In any case it is best to intentionally stay out of their reach so that they do not lunge toward the rescuer.

The rescuer should call out progress as they near the end of the lower. Communicating progress as the rescuer reaches their destination can be instrumental in success of the entire operation:

RESCUER: Five Feet to Landing; Four Feet; Three Feet; Two Feet; One Foot; Stop.

Even though the rescuer may have called that they have reached the landing area, the brakeman should not go "off belay" until specifically directed to do so. This helps to ensure the safety of the rescuer in the event that they need to anchor themselves at the location for protection, or if the subject makes some sudden motion that could create a hazard.

Upon reaching the landing area, the rescuer should determine whether to stay attached to the line on which they were lowered, secure it for later use, or release it.

Moving a Litter While Suspended

The operations-level rescuer may be called upon to transport an empty litter from one stable location to another by rope, for example to reposition it nearer the subject in preparation for a rescue lower. In this case, the rescuer will attach to the litter in much the same way as they might if tending a loaded litter, but without the added degree of difficulty created by a heavy litter or by the need for patient care. Managing a litter down a vertical surface is arguably easier if rigged in a horizontal configuration (**FIGURE 16-5**).

It is also possible to lower a vertically oriented litter, which may be necessary in a confined space or where obstructions require a narrow profile. Moving a litter in this configuration is a little less graceful an endeavor, but quite feasible (**FIGURE 16-6**).

FIGURE 16-5 Litter in a horizontal position.

FIGURE 16-6 Litter in a vertical position.

Horizontal Litter, Single Attachment Point

The simplest method of configuring a litter for vertical lower is to attach a four-legged litter bridle, sometimes called a **litter spider**, to the top rail of the litter at four equilateral points to create a stable platform of the litter.

With the litter spider, each of the legs meets at a main attachment point (MAP) a couple of feet above the litter. Rescuers may choose to use a manufactured litter spider for this purpose, or may devise one out of webbing or rope. Although a quick and simple litter spider can be created from a bowline on a coil with four loops, with each of the loops serving as a litter spider leg, a much more adaptable and versatile version can be created relatively simply. Features such as adjustable legs, tie-in points for litter attendant, and subject safety tether are nice to have.

Most rescuers who construct their own litter spider make them from rope because it is easier to handle and allows for a greater variety of knots.

To utilize an adjustable rope spider, follow the steps in **SKILL DRILL 16-1**. The following materials will be needed:

- One rigging ring or rigging plate (as the main attachment point)
- Two lengths of static kernmantle rope, each 12 feet (3.7 m) long and at least ⅜ inch (9.5 mm) in diameter.
- Four lengths of Prusik material, each 4 feet (1.2 m) long. The Prusik cord must be of appropriate diameter to grip the static kernmantle rope; for example: 7-mm (0.28-in.) Prusik cord on ⁷⁄₁₆-in. (11.1-mm) rope, or 8-mm (0.31-in.) Prusik cord on ½-in. (12.7-mm) rope.
- Four locking carabiners; note: if the litter to be used does not have adequate rigging point, carabiner gate openings should be large enough to fit over the litter rail.

Using the Litter Spider

Regardless which type of litter spider is used, the rescue system rope can be attached to the MAP either

SKILL DRILL 16-1
Constructing the Litter Spider

1. At the midpoint of each leg of static kernmantle rope, tie it to the rigging ring using a retrace figure 8. When held at the ring, the four strands of rope should be of approximately equal length.

2. Hitch one length of Prusik material to each of the legs, ensuring that the remaining strands of Prusik material are the same length.

3. Tie the ends of each Prusik cord to its host spider leg with a grapevine knot, so that it creates a loop that may be adjusted by sliding the Prusik hitch up or down the host rope.

4. Clip a large locking carabiner to each of the four loops. NOTE: While a locking carabiner can be used at the MAP, a ring or plate is more secure and avoids the risks associated with three-way loading of carabiners.

SAFETY TIP

Some litters have connecting points on the rails; attachment to other litters is made by clipping directly over the large rail. Not all large locking carabiners fit easily over litter rails. Before purchasing carabiners for this use, measure the diameter of the rail and check the specifications of the carabiner's manufacturer or distributor.

the litter, using a second MAP, or rigging the belay to the spider legs with a carabiner. Some organizations choose to connect both the primary and secondary lines to the main attachment point about 6 feet (1.8 m) from the end, and then tie the remaining tail directly to the head of the litter. As with any belay, the authority having jurisdiction (AHJ) should make this determination based on the risks and the consequences if the risks impact the rope rescue system.

with a direct loop or with a carabiner. The advantage of tying directly into the rigging ring is that this reduces complexity and opportunity for failure. However, a carabiner allows for faster connection/disconnection, if this is a priority. To attach the mainline to the litter, follow the steps in **SKILL DRILL 16-2**.

The spider legs should not be tied directly into the litter rail because rope or webbing may abrade through when rubbed over the face of the vertical surface. Also, carabiners give greater flexibility for attaching or detaching from the litter rail. This may become important during subject loading.

If a belay line is to be used, it may be tied or connected directly to the main attachment point in the same manner as the primary rescue system. Some rescuers may be reluctant to connect the belay to the main attachment point, as this then creates a potential single point for failure. If, upon consideration, probability warrants use of other alternatives these may include attaching the belay directly to a point on

TIP

The belay line is never tied directly onto the subject in the litter. If the belay line caught the subject directly, it would pull on the person's harness while the subject supported the remainder of the load (litter and tender[s]).

Carabiner gates should be set inward toward the center of the litter. This helps prevent the lock nut on the locking carabiner from being rubbed open or damaged on the face of the cliff or wall. The gates also should be oriented so that the locking nuts close with gravity and cannot vibrate to an unlocked position.

The rescuer may also be attached to the MAP with an attendant rig, also sometimes called a pigtail. While a fixed lanyard will work for this, an adjustable system is preferred. One way to make a versatile attendant rig out of rope is described in **SKILL DRILL 16-3**.

SKILL DRILL 16-2
Attach a Rescue Line to a Litter

1. Tie the rescue line to the main attachment point with a retrace-8. NOTE: Alternatively, if a carabiner is to be used form an end knot in the rescue line, connect a carabiner, and clip it to the main attachment point of the spider. Keeping the litter legs organized and free of twists, attach the carabiner at the end of each leg to a rigging point on the litter, in accordance with litter manufacturer instructions.

SKILL DRILL 16-3
Crafting the Attendant Rig

1. Use a retrace-8 to tie the static kernmantle rope to the main attachment point. Use a barrel knot to attach one carabiner to the other end of the static kernmantle rope, and clip it to the attendant harness.

2. Attach the short webbing loop and the footloop to the ascender with a carabiner.

3. Cam the ascender onto the rope.

4. Attach the other end of the short webbing loop to the attendant's harness.

The materials needed include following:
- Static kernmantle rope at least 10 mm (0.4 inch) in diameter and 6 feet (1.8 m) in length
- Ascender or Prusik
- Short webbing loop
- Footloop or etrier
- Three carabiners

The attendant rig should be of sufficient length to permit freedom of movement, allowing the attendant to move all the way around the litter as needed. In this role, the rescuer is called a litter attendant, even if the litter they may be attending does not have a subject.

Rescuers should always consider their own safety as being of primary importance. Before committing themselves to the rope rescue system, the litter attendant should make a final check of all rigging—even if there is a safety officer already assigned to do the same.

Commencing a Lower

Litter movement should always be at the discretion of and on the command of the litter attendant. If more than one litter attendant is utilized, one should be deemed litter captain and take command.

After confirming with the brakeman that the system is "On Belay," the litter attendant will call for "Tension," which simply means that the brakeman should hold the system stationary. This affords the litter attendant a moment to lean into the system to remove any slack, ensuring that the direction of pull is as anticipated, no lines are tangled, and to make any last-minute adjustments to body position.

With practice, the litter attendant will be able to accurately predict the best length for their attendant rig so that they are most effective after getting over the edge. When suspended vertically, the attendant should hang so that their legs extend just below the litter so that they can push against the vertical surface on the other side of the litter. This means that the top litter rail should be somewhere between belly button and chest height, depending on the rescuer. The attendant can make a good estimation while still on the horizontal surface by lifting the litter by the closest rail and tilting it so that both the attendant rig and the spider legs are taut.

No further changes should be made to the rigging or to anything in the rope rescue system after it has been inspected by the safety officer or while the system is on belay. If there becomes a need to make such an adjustment, the rope rescue system should go off belay until all is satisfactorily ready. Before going back on belay, the system should be re-evaluated by the safety officer. When the litter attendant is ready, litter movement will commence on their command.

Sometimes it can be a bit of a challenge for the litter attendant to move the unloaded litter horizontally across a surface to reach the edge of the planned vertical drop. The brakeman may need to feed rope through the brake device as the litter attendant leans back hard, pulling the litter. Once the litter and attendant are over the edge, there will be greater weight on the system and the brakeman will need to apply more friction.

As previously noted, getting over the edge is usually the most difficult part of any lowering or raising operation. This is true with or without a litter. Close communication and teamwork between the litter attendant and brakeman are crucial to a successful outcome. The litter attendant should feel free to call for a "Stop" at any time, and the brakeman should be alert and responsive.

The process is quite similar to that of lowering a lone rescuer, except in this case the rescuer happens to be toting a 20–50 pound (9.1–22.7 kg) litter and gear. It is important to have all of the slack out of the rope system before the litter attendant tries to get over the edge. Otherwise, when gravity takes hold, the load will drop, possibly shock loading the system and equipment, not to mention giving a fright to rescuers.

Holding the litter by its nearest top rail so that the litter harness legs are taut, the litter attendant leans back hard into their spider connection to maintain constant tension on the system. The brakeman lowers very slowly as the litter attendant backs toward the edge. When the attendant's heels are just over the edge, they will lean back into the system, knees bent and feet widely spaced. As the line is lowered, the litter attendant slowly transitions to a seated position, just below the litter.

If personnel are in ready supply, positioning one or two edge attendants at the transition can be very helpful. Edge attendants should be on safety lines that permit them to just reach the edge, but not to fall over it. The edge attendant(s) can help the litter attendant to carry the litter across the horizontal surface, maintain adequate tension on the system, and provide balance as the litter goes over the edge.

Undercut edges and 90-degree transitions can be among the most difficult to get over. If the standard approach proves too challenging, it may be possible to set the litter on the edge of the drop, maintaining as much tension as possible on the rope system, while the litter attendant maneuvers into place around the head or foot of the litter.

Raising a litter over an edge is also quite similar to the methods used for a raise of a lone rescuer—again, the difference being the 20–50 pound (9.1–22.7 kg) appendage. It is especially important to approach the raise over an edge slowly, and with caution. Likewise, the litter attendant must be especially diligent to continue pulling back and out over the edge as

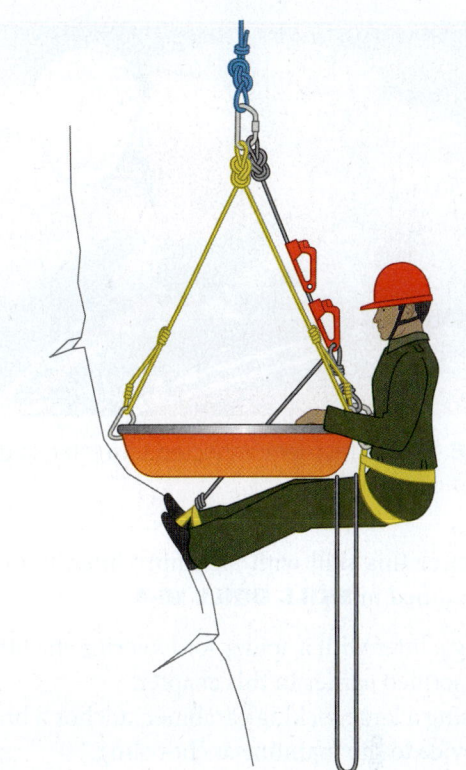

FIGURE 16-7 The proper litter attendant position.

much as possible for as long as possible, to avoid a nose-first drag.

Body Management with Litter Movement

The litter attendant is most functional if they are able to assume a natural, seated position, legs extended, with the litter just a few inches above their lap. The litter should not be resting on the attendant's legs, as this could create a hazard (**FIGURE 16-7**).

The attendant grasps the litter with both hands, either at the nearest top rail or at the closest rail underneath, leveraging with their feet to pull the litter away from the vertical surface. Taking small, backward steps down the vertical surface, the attendant continues to prevent the litter from bumping or snagging against it as they move.

If, at any point, the litter attendant wishes to stop—or, when they have reached their desired location—they simply request that the brakeman "Stop!" If the stop will be for an extended period of time, the brakeman can tie off the braking device. When the litter attendant is ready to move again, they commence lower once again using the same commands as before.

Practicing the skill of managing a litter while suspended should first be performed with an empty litter, before attempting to manage a loaded litter.

FIGURE 16-8 A rescuer connected to a litter spider using an attendant rig.

To practice this skill with an empty litter, follow the steps outlined in **SKILL DRILL 16-4**:

1. Rig a litter with a spider for lowering the litter, as described earlier in this chapter.
2. Using a large locking carabiner, anchor a braking device to the mainline anchor sling.
3. Connect the end of the mainline to the litter spider main attachment point as described earlier in this chapter.
4. Attach a belay system to the main attachment point in accordance with AHJ requirements.
5. Connect a rescuer to the spider using an attendant rig (**FIGURE 16-8**).

Personnel requirements and positions should be similar to those used for basic lowering systems, except that there is a rescuer attached to the litter. Execute the lower as follows:

Operations Leader: Roll call! (*general statement to entire site, and radio if applicable*)

Operations Leader: Safety ready?

Safety Officer: Safety ready!

Operations Leader: Brakeman ready?

Brakeman: Brakeman ready!

Operations Leader: Rescuer ready?

Rescuer: Rescuer ready!

Operations Leader: Litter operations commence when ready.

From this point, communications will be primarily between the rescuer and the brakeman. Any team member can call, "Stop."

Vertical Litter Configuration

A litter can also be rigged to hang in a vertical orientation, either to avoid obstacles, to get through a confined space, or for whatever reason. High directionals make the vertically rigged litter much easier to manage, but they are by no means mandatory.

When using a vertically oriented litter, it is particularly important to pay close attention to securing the patient in such a way that they will not slide down toward the foot end when the litter goes vertical. This may be accomplished with foot loops, a seat harness, or other means.

The litter bridle for a vertical rig needs only to have two points, one on each side of the head of the litter. Bridle systems used for low-angle litter rigging often serve well for vertically rigged high-angle rope rescue systems. Keeping the attachments wide, on either side of the litter, will help improve stability. However, this presents a new challenge in that when the main attachment point of the litter is at the edge transition, the weight of the remains well below the edge. Therefore, it becomes particularly challenging to get the litter over the edge when lowering without shock loading the system, and it becomes nearly impossible to get the litter over the edge at all during a raise. (See Chapter 14, *Lowering Systems* for more on low-angle litter rigging.)

If a high directional is not going to be used, rigging the system in advance with a pike and pivot in mind will make all the difference.

Pike and Pivot Litter Bridle

The **pike and pivot bridle** is used for the vertically oriented litter only during edge transitions. During the remainder of the rope lowering and raising operation, a standard two-point bridle (or direct tie-in) to the head of the litter will suffice. The pike and pivot litter bridle should be rigged in advance, and stored on the litter for when it is needed. To create a pike and pivot bridle, rescuers will need a 25-foot (7.6-m) piece of 8-mm or 9-mm (0.31- or 0.35-in.) cord, and a carabiner. Tie a knot (butterfly or figure 8 on a bight) into the middle of the cord so that it creates a V shape, and clip into it with a locking carabiner.

To store the bridle for later use, attach it loosely to the main attachment point of the litter bridle, but be sure to clip it behind the mainline so that when it is employed it will be nearer the ground than the primary rope rescue system and belay. Clip this in such a way that it can be disconnected while the primary system is under load. Run the two loose ends of the bridle cord behind the litter, and tie them to the litter rail, one on each side, at a point that will be at about waist level of the subject when packaged into the litter. Select a tie-in point between two uprights so that these cannot slide up or down. This is your pike and pivot bridle. It should be just long enough to rest loosely when not in use, but not long enough to drag or snag on obstructions. The equipment needed to rig a pike and pivot litter bridle include the following:

- One 25-foot (7.6-m) piece of 8-mm or 9-mm (0.31- or 0.35-in.) cord
- Locking D carabiner

When it comes time to employ the system, rescuers will also need a short, 50-foot (15.2-m) rescue line of a chosen diameter (11 or 12.5 mm [0.43 or 0.5 inch]) and a separate device for lowering/raising of this part of the system. Tie a figure 8 knot into the end of the rope, and rig the other end into the anchored raising/lowering system. To rig a pike and pivot litter bridle, see **SKILL DRILL 16-5**:

1. Tie a midline loop knot (butterfly or figure 8 on a bight) into the middle of the cord.
2. Clip into the loop created by the midline knot with a locking carabiner.
3. Clip the locking carabiner loosely to the main attachment point of the litter bridle, behind the mainline.
4. Run the two loose ends of the bridle cord behind the litter, and use an end knot to tie them to the litter rail, one on each side, between uprights at a point that will be at about waist level of the subject when packaged into the litter.

Pike and Pivot Lower

To lower the vertically oriented litter using the pike and pivot system, unclip the pike and pivot bridle from the MAP and connect that carabiner to the knotted end of the short lowering system. There is no need to tension this part of the system until the litter and attendant reach the edge of the drop.

As the foot of the litter nears the drop, the rescuer begins to take tension on the pike and pivot system. The rescuer slides the foot of the litter out over the edge so that it protrudes into the air. While maintaining tension on the pike and pivot bridle with it right at the edge of the drop, the rescuer extends the primary system at the main attachment point so that the vertically oriented litter begins to tip foot down over the edge. The litter attendant and edge attendants (if present) can help maneuver the litter into this configuration. When the litter is vertically oriented, on the command of the litter attendant the entire litter can be lowered from this pike and pivot bridle until the head of the litter reaches the edge.

When the head of the litter is at the edge, take the load back onto the primary system and loosen the pike and pivot system. When the load is fully on the primary system, the rescuer calls for an all-stop so that the pike and pivot lowering/raising system can be released and disconnected. The rescuer stores the V-point of the pike and pivot bridle at the main attachment point, as before, in case it is needed later. Follow the steps in **SKILL DRILL 16-6** to negotiate the edge using the pike and pivot during lowering.

SKILL DRILL 16-6
Negotiating the Edge Using a Pike and Pivot

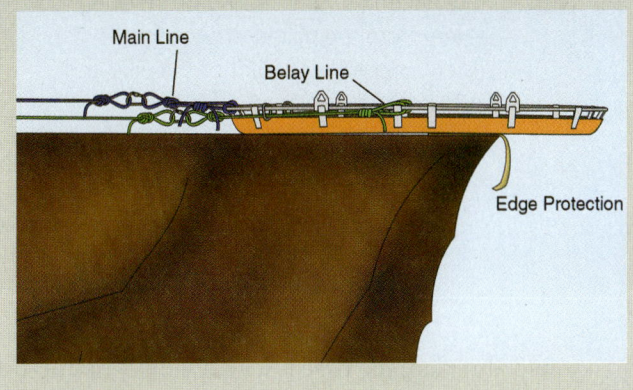

1. Rig a litter for lower in a vertical orientation with a main lowering system attached to the bridle.

2. Prepare a secondary lowering system for use with the pike and pivot.

3. Attach a pike and pivot bridle as described in Skill Drill 16-5.

4. Begin lowering operation using the main lowering system.

5. As the litter nears the edge, connect the pike and pivot bridle connection to a secondary lowering system.

(continues)

SKILL DRILL 16-6 (continued)
Negotiating the Edge Using a Pike and Pivot

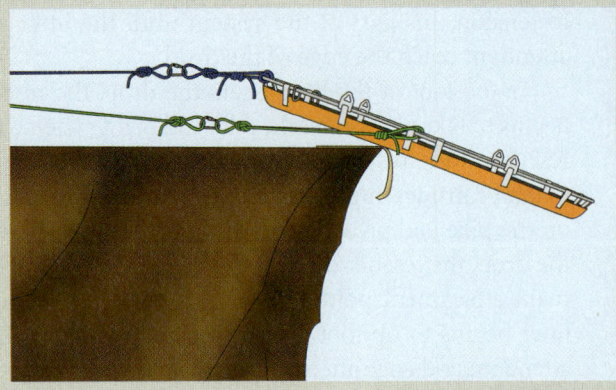

6. As the foot of the litter nears the drop, stop the secondary lowering system with the pike and pivot bridle connection point just at the edge of the drop, while continuing to lower on the main lowering system.

7. Slide the foot of the litter out over the edge so that it protrudes into the air.

8. While maintaining tension on the pike and pivot lowering system, extend the primary lowering system at the main attachment point so that the vertically oriented litter begins to tip foot down over the edge.

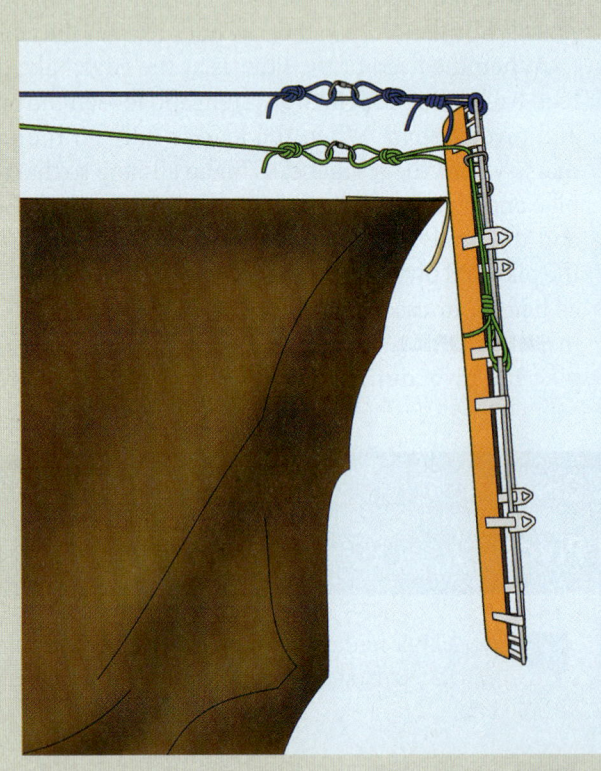

9. When the litter is vertically oriented, litter attendant calls for the entire litter to continue being lowered by the secondary system (while keeping the main system loose) until the head of the litter reaches the edge.

10. When the head of the litter is at the edge, litter attendant calls for the primary system to stop while the secondary system continues lowering so that the load is taken back onto the primary system.

11. When the pike and pivot system is loose and the load fully on the primary system, litter attendant calls for an all-stop.

12. An edge attendant or litter attendant can now disconnect the secondary lowering system from the pike and pivot lowering system. Store the V-point of the pike and pivot bridle at the main attachment point in case it is needed later.

Pike and Pivot Raise

During a raise with the litter in a vertical orientation, the pike and pivot can be used to help get the litter up and over the edge. Without it, the litter would stall just below the edge, with no way to get it the rest of the way other than to lift it manually—a task that would be difficult at best, and perhaps even impossible. Such an approach would also place the rope rescue system at risk of shock load with the litter in its most precarious position.

This maneuver works best when the pike and pivot bridle is pre-attached to the litter. While it is possible to configure the bridle as an afterthought, the necessity of rigging behind/below the mainline and belay make this especially difficult.

With the pike and pivot bridle pre-attached and loosely at the ready at the MAP, simply raise the litter until it is just a couple of feet below the edge transition. Either the litter attendant or an edge attendant needs to be able to reach the MAP, but be sure to stop before the MAP is jammed up against the transitional point of the edge. Lower the short raise/lower system to the litter. With the load still taken by the primary raise/lower system, the assigned person should disconnect the pike and pivot carabiner from the MAP and connect it to the short raise/lower system. Upon command of the litter attendant, the top will begin raising again, but this time with the short system. Raise as far as possible with this system, until the attachment points at the sides of the litter are as high as possible at the edge.

It may be necessary to temporarily loosen the primary raising system (the one attached at the head of the litter) to prevent contradictory forces. Before the primary system is slacked, the litter attendant should carefully climb up and over the edge so that their weight is off the system. They should, however, remain at the edge to assist in positioning the litter and should remain tied-in for safety.

When the pike and pivot bridle is as high as it will go on the edge, hold it steady and once again begin to tension the primary system so that the head end of the litter begins to tilt horizontally toward the ground. Here the litter attendant and/or edge personnel can assist in pivoting the litter over the edge as appropriate. When the heaviest part of the litter is over the edge and the litter is near horizontal, properly secured rescuers can manhandle it the rest of the way.

The pike and pivot should be practiced first with an empty litter, as this technique becomes increasingly difficult with additional weight in the litter. To perform this skill, follow the steps in **SKILL DRILL 16-7**:

1. Rig a litter for raise in a vertical orientation with a main raising system attached to the bridle.
2. Prepare a secondary raising system for use with the pike and pivot system.
3. Attach a pike and pivot bridle to the litter as described in Skill Drill 16-5.
4. Begin the raising operation using the main raising system
5. When the top of the litter is two feet below the edge transition, just within reach of either the litter attendant or an edge attendant, connect the pike and pivot bridle to the secondary raising system.
6. Litter attendant calls for the secondary raising system to begin raising, so that the litter is lifted by the pike and pivot bridle. Do not raise the primary raising system at this time; in fact, if it comes under tension, it may be necessary to loosen it a bit.
7. When the pike and pivot bridle attachment points at the sides of the litter are as high as possible at the edge, the litter attendant should carefully climb up and over the edge so that their weight is off the system.
8. While holding tension on the secondary raising system, begin to haul on the primary raising system. The head end of the litter will tilt horizontally toward the ground.
9. Edge attendant(s) and litter attendant assist in pivoting the edge until the heaviest part of the litter is over the edge and the litter is near horizontal, then carry the litter away from the edge to a secure location.

Suspended Work on A Highline

Working in suspension along a horizontal system differs from working in a vertical system in many ways, not the least of which is the complexity involved in the system. Horizontal systems are typically built and operated by technician-level personnel, although operations-level personnel will likely help build the system and may well be moved across it to reach a rescue site or subject. A rescuer at the operations level is not expected to move themselves along the system, nor to manage a patient during movement, but they may find themselves in a position to be moved from one stable location to another by riding the line of a horizontal system.

Rigging to a Horizontal System

Rigging to a horizontal system is really as simple as clipping a pulley onto the track line. Typically a Kootenay, or Knot Passing Pulley, is used for the track line. The rescuer is connected to the pulley with a sling that runs between their seat harness and the pulley. The sling is usually just long enough to preclude any temptation toward grabbing the track line (in which case fingers could be pinched and injured in the pulley) with locking carabiners at both the pulley end and at the harness end. Still, the rescuer riding the highline should wear gloves to help prevent rope- or pulley-induced injuries. Often, a slightly longer sling will be clipped as a "backup" from the rider directly to the track lines so that it may trail the pulley.

Highline pulleys generally have multiple tie-in points in addition to the central notch where the rider will be clipped in. Tag-lines are generally connected, one fore and the other aft, to help provide control and a means of hauling and lowering the load. An operations-level rescuer will generally not be managing a rescue subject while riding a highline, but at times multiple rescuers may be pulled across a line simultaneously. In this case, the other person may be attached directly to the same pulley with a separate sling.

Riding with a Litter

At times, an operations-level rescuer might be asked to transport an empty litter across a highline (**FIGURE 16-9**). Simply put, when attaching a litter to a highline system, it should first be rigged with a litter spider, just as for a vertical lower. The spider is connected to a highline pulley at the MAP, just as an individual might be. Methods of attaching the litter to the track line are discussed in Chapter 17, *Horizontal Systems*.

As with a litter spider for vertical lower, the attendant rig (also called a pigtail) consists of a short (6-foot [1.8-m]) length of 9- or 10-mm (0.35- or 0.4-in.) rope connected from the MAP to the waist harness attachment point of the attendant. A short sling on an ascender, along with an etrier, complete the setup and add versatility.

When riding with a litter across a highline, the position of the litter attendant is quite similar to that used for vertical lower: litter at about chest height, legs beneath the litter.

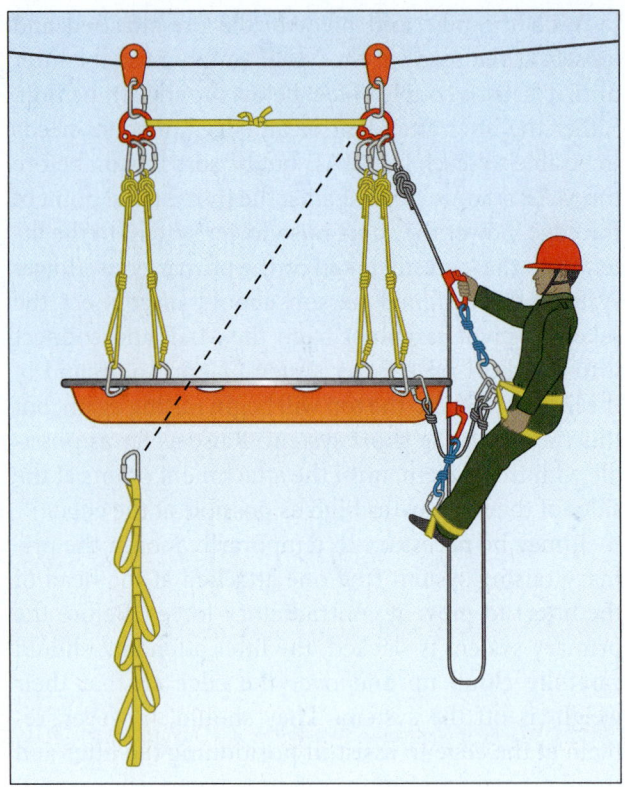

FIGURE 16-9 Litter and attendant attachment for highline.

After-Action REVIEW

IN SUMMARY

- Rescuers at the operations level may work in suspension as they are lowered or raised from one stable location to another.
- Operations-level rescuers are not expected to manage a subject while suspended, nor to manage their own ascent/descent on rope.
- In order to prevent suspension-induced injury, rescuers should move their legs and stretch regularly and unresponsive subjects should be rescued as promptly as possible.
- In rope rescue vernacular, the term *third man* is used to describe a rescuer whose role it is to access the subject quickly, assess the situation, help with packaging and preparation, and generally anticipate and assist with the many tasks required. They are typically not the primary rescuer, nor the rescuer who takes direct contact and control with the subject.
- All commands should be discussed and agreed in advance before rope operations begin.
- Getting over the edge is the most difficult part of any suspended rope operation. A sharp edge is more difficult to get over smoothly than a sloping one.
- If a sloping edge is gentle enough, a slowdown may not even be required.
- For a sharp or undercut edge, the rescuer may wish to call for a full stop just as their heels overlap the edge.
- The most graceful way to get over an edge is to use a high point anchor or multi-pod to keep the horizontal path of the rope above the height at which it connects to the rescuer.
- Raising over an edge is similar to lowering over an edge, but requires practice to execute smoothly. Slowing down before getting to the edge will help to prevent dragging the rescuer nose first when they reach the point at which the rope travels in a horizontal position.

- As soon as the rescuer can see up and over the edge, there is a natural tendency to want to shift into a more vertical stance and "climb" over the edge. Resisting this urge and maintaining a seated position, with legs horizontal, for as long as possible will make it much easier to get up and over.
- As the suspended rescuer reaches their landing zone, they should call for the brakeman (whether raising or lowering) to move more slowly. If rescuer and brakeman are out of sight of one another, calling distances helps.
- As they approach the final landing zone, the rescuer should survey the situation to ensure safety and to more specifically plan their approach. It may even be necessary to call for a full and complete stop while still some distance away, and take time to observe the situation.
- Upon reaching the landing area, the rescuer should determine whether to stay attached to the line on which they were lowered, secure it for later use, or release it.
- The operations-level rescuer may be called upon to transport an empty litter from one stable location to another by rope.
- Managing a litter down a vertical surface is arguably easier if rigged in a horizontal configuration
- Horizontal systems are typically built and operated by technician-level personnel, although operations-level personnel will likely help build the system and may well be moved across it to reach a rescue site or subject.
- At times, an operations-level rescuer might be asked to transport an empty litter across a highline.

KEY TERMS

litter spider An arrangement of straps and connectors used to connect to a litter so that it is appropriately balanced and stabilized and to provide a means by which a litter can be connected to a rescue rope; also called a litter bridle.

pike and pivot bridle A method of rigging into a litter using straps and connectors to facilitate the pike and pivot technique.

suspension-induced injury Physiologic shock response resulting from the effects of hanging motionless in a harness for a period of time.

third man A rescuer who is assigned to access the subject quickly to provide assessment and assistance with patient care and rescue tasks.

On Scene

1. Under what circumstances might an operations-level rescuer need to be suspended and moved on a roped system?

2. How much of a risk to rescuers is suspension-induced injury, and how can it be avoided?

3. Under what circumstances might a vertically oriented litter be preferred for rescue operations?

4. How might techniques for negotiating an edge while being lowered differ from negotiating the same edge while being raised?

Chapter Opener: © Jones & Bartlett Learning. Courtesy of Loui McCurley; On Scene siren: © Bildgigant/Shutterstock.

SECTION 3

Technician

CHAPTER **17** **Horizontal Systems**

CHAPTER **18** **Personal Vertical Skills**

CHAPTER **19** **Pickoff and Litter Management**

CHAPTER **20** **Special Rescue Disciplines for Additional Training**

CHAPTER 17

Technician

Horizontal Systems

KNOWLEDGE OBJECTIVES

After studying this chapter, you should be able to:

- Differentiate fixed line systems from dynamic directional systems. (pp. 342–343)
- Describe the use of a track-line or highline system. (**NFPA 1006: 5.3.5, 5.3.6**, pp. 343–344)
- Describe the components of a track-line system. (**NFPA 1006: 5.3.5, 5.3.6**, pp. 343–344)
- Describe the purpose of a tag-line. (**NFPA 1006: 5.3.5, 5.3.6**, p. 345)
- Describe anchorage and anchor system considerations with horizontal systems. (**NFPA 1006: 5.3.5, 5.3.6**, pp. 345–350)
- Describe the purpose and components of carriage systems. (**NFPA 1006: 5.3.5, 5.3.6**, pp. 348–349)
- Describe the purpose and components of high directionals. (**NFPA 1006: 5.3.5, 5.3.6**, pp. 350–351)
- Describe the forces in a track-line or highline system. (**NFPA 1006: 5.3.5, 5.3.6**, pp. 354–357)
- Identify safety considerations for a horizontal rope rescue systems. (**NFPA 1006: 5.3.5, 5.3.6**, pp. 352, 354–357)
- Describe the procedure for rigging a horizontal or sloped rope rescue system. (**NFPA 1006: 5.3.5**, pp. 358–360)
- Describe methods of attaching a litter to a horizontal system. (**NFPA 1006: 5.3.5**, p. 357)
- Explain use of a guiding-line system. (**NFPA 1006: 5.3.5, 5.3.6**, pp. 362–364)

SKILL OBJECTIVES

After studying this chapter, you should be able to:

- Calculate necessary tension on a line. (**NFPA 1006: 5.3.5, 5.3.6**, p. 355)
- Construct a horizontal rope rescue system or highline. (**NFPA 1006: 5.3.5**, pp. 358–360)
- Conduct a preoperational system safety check on a horizontal rope rescue system. (**NFPA 1006: 5.3.5, 5.3.6**, p. 352, 360)
- Attach a litter to a horizontal rope rescue system or highline. (**NFPA 1006: 5.3.5**, pp. 357–358)
- Move a load across a horizontal rope system. (**NFPA 1006: 5.3.6**, pp. 360–361)
- Communicate effectively during a highline operation. (**NFPA 1006: 5.3.6**, p. 361)

You Are the Rescuer

Your organization has been called to a local amusement park where a tracked ride has come off its tracks. There are 24 people stranded in the ride 120 feet (36.6 m) off the ground. Rescuers can climb the tracks to reach the stranded subjects but getting them to ground is another story. They cannot simply be lowered straight to ground due to a water feature below. Dry ground is 50 feet (15.2 m) horizontally away from the vertical drop.

1. What system will you use to transport the subjects to safety?
2. Which equipment will you need for the system you will build?
3. Why did you choose this system over the other alternatives available to you?

 *Access Navigate for more practice activities.*

Introduction

Where a chasm exists and a bridge is not in place, rope may be used to move a load horizontally between two points. In fact, NFPA 1006, *Standard on Technical Rescue,* and NFPA 2500 (1670), *Standard for Operations and Training for Technical Search and Rescue Incidents and Life Safety Rope and Equipment for Emergency Services,* require that technician-level rope rescuers are capable of moving a suspended rescue load along a horizontal path to avoid an obstacle using a suitable anchorage and appropriate system within its limitations. In keeping with the performance-based requirements approach, NFPA does not, however, specify which particular systems should be used, nor exactly how this is to be achieved.

There are two commonly used concepts for achieving horizontal movement in rescue. One is to tension a rope horizontally between the two points to create a **track-line** system for rescuers to move equipment and personnel. Such systems are commonly known as a **highline**, tyrolean, telpher, or guiding-line. The other commonly used concept is that of a **dynamic directional system**, which involves two or more systems applying force in opposition to one another. Some of the more commonly used dynamic systems include reeves, cross hauls, and skate blocks.

A good understanding of vectors is essential to safe rigging of systems where forces may be applied from multiple directions. Many technical rescue teams regularly practice setting up and using systems for multi-directional movement of a load, not because these are the most commonly used systems for rope rescue, but because rigging these complex systems is a good test of many different rope rescue skills. See Chapter 10, *Principles of Rigging* for more on vectors.

Track-line systems and dynamic directional systems may be used independently, or even combined

TIP

Prerequisites for Performing the Skills in This Chapter

Before attempting the activities described in this chapter, you must have demonstrated competency in the following skills:

1. Use and care for rope.
2. Use and care for other equipment needed in the high-angle environment.
3. Tie correctly and without hesitation the knots necessary for safe, effective work in the vertical environment.
4. Apply the principles of anchoring and rig a safe and secure anchor point.
5. Accurately assess potential forces in a belay situation and correctly rig and operate an adequate belay system.
6. Apply the principles of belaying by using a Münter hitch or a personal belay device to belay another person.
7. Demonstrate how to release a jammed system under load, utilizing either a load-release hitch, haul system, or alternative means other than the primary lowering/belay device.
8. Apply the principles of high-angle lowering systems: correctly rig the elements of a high-angle lowering system and be capable of safely assuming the role of litter attendant, brakeman, belayer, rope handler, and edge attendant.
9. Apply the principles of hauling systems: determine the mechanical advantage (MA) required for a hauling system; correctly rig any of the elements of a hauling system; rig for a 1:1 MA hauling system, a 2:1 MA hauling system, a 3:1 MA hauling system (Z-rig), a 4:1 MA hauling system (piggyback rig), and compound versions of MA systems; and safely and confidently assume the role of PCD attendant, haul captain, haul team member, and belayer.
10. Accurately assess system strengths and loads.
11. Demonstrated experience over time as an operations-level rescuer is a key part of becoming a competent technician-level rope rescuer.

for maximum versatility. Rescue systems with horizontal capabilities may be used for the following:

- Bypass an obstacle such as a swiftly flowing river
- Span a hazard such as machinery or equipment
- Avoid a difficult area, such as collapsed building rubble

Horizontal rescue systems pose a few unique hazards due to the fact that forces are acting in several directions at once. Perhaps more than any other rope rescue system, highlines have the potential to overstress rope, equipment, and anchor points, which can result in failure of the system. The rigging and use of highlines require a thorough knowledge of the potential forces involved.

In addition to the forces involved, there is also a time factor. Horizontal systems require a great deal of teamwork and communications to set up. When they first attempt the task, teams sometimes find rigging horizontal systems to be very time-consuming. When time is crucial, even experienced rescuers often find other techniques preferable.

Track-Line Systems

Track-line systems may be built in a horizontal or sloping configuration. A **horizontal highline** is one that is suspended between two points that are nearly on the same level. A **guiding-line** is essentially just a sloping highline, in which one of the two points is much higher than the other (**FIGURE 17-1**). Control of a load when using a track-line system is typically achieved with mechanical advantage systems for raising and lowering systems. Special rigging, such as a reeve (discussed later in this chapter), can be added to a track-line system to allow loads to be raised or lowered from the span.

The basic elements of a highline system that may be used in a rescue are shown in **FIGURE 17-2**. In this operation, a rescue team is moving a litter with a rescue

FIGURE 17-1 Guiding-line.

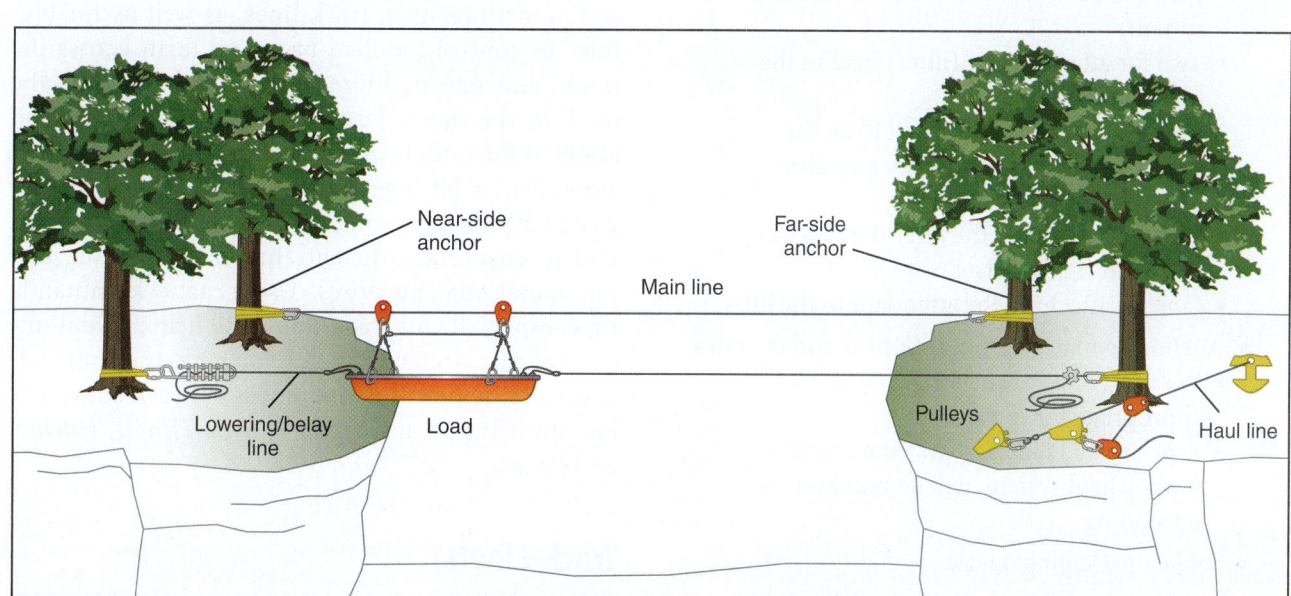

FIGURE 17-2 Elements of a highline.

subject from left to right on a horizontal highline. In many cases, it would be desirable to include a litter attendant. However, for simplicity, this is not shown in this illustration. (Litter attendants for highlines are discussed later in the chapter.)

Rigging a Track-Line System

The equipment needed to rig a track-line system will include at least the following:

- One track-line rope. The total length will include:
  - The length of the gap to be bridged, plus
  - The length from the edge to the anchor point on each side, plus
  - The length needed to tie onto anchorages on each side
- One lowering/belay rope (or control line). The total length will include:
  - The distance from the lowering device on the near side to the point on the far side where the load will be derigged, plus
  - The amount of rope needed to tie to the load, plus
  - 20 feet (6.1 m) to spare
- One tag-line/haul rope (or control line). The total length will include:
  - The distance from the point where the load is rigged on the near side to the belay device on the far side, plus
  - The length needed to attach to the load, plus
  - 20 feet (6.1 m) to spare
- 17 locking carabiners (minimum):
  - Four (at least) for anchoring (single-line system)
  - Two for attaching the (litter) load to the highline
  - Two for attaching the pigtail from the lowering/belay and tag-lines to pulley carabiners
  - Eight for attaching the litter harness to the litter and rigging plate
  - One for attaching the attendant to the litter
- Anchoring materials for both near and far sides
  - Webbing or rope anchorage connectors appropriate for the site
  - A pre-rigged haul system, or materials to make a haul system, will be required for tensioning
- Subject packaging device
  - A litter, harness, or some other means of connecting the subject to the system

Spanning the Gap

Getting one end of the track-line to the far side of the chasm to be spanned can be one of the most difficult parts of setting up a horizontal or a sloping highline. Simply throwing a line is generally not feasible. Instead, a lightweight leader line is launched across the span using a throwing device or line gun. Once the leader line is across, the ropes to be used as track-lines can be attached to the leader line and pulled across by rescue crew members who have made their way to that side. Some organizations have begun to utilize drones for transporting the leader line across a span, but this method should be used only when adequately trained personnel and sufficient equipment are available.

In rare cases, when no alternative exists, rescuers may choose to enlist the help of capable bystanders, or even the subject, to receive the leader line, pull the track-line over the span, and even establish a tensionless anchor or other simple termination.

If a line-throwing device is not available, depending on what type of gap is being spanned, it may be possible to lower one end of the track-line rope to ground teams on the near side so that they can transport the end of the rope to the base of the structure on the other side. From here, far-side rescuers can lower a line to which the ground team can secure the end of the track-line rope so that it may be pulled up to the far side anchor point.

Ropes for Highlines

Highlines consume a great deal of rope, and other equipment as well. Rope is needed for at least one, and sometimes two, track-lines, as well as for tag-lines to control the load back and forth across the track, and one or more haul systems. Ideally, the track-line(s) should be a single length of rope with no knots in the middle of the span. Choosing ropes that are at least a bit longer than the anticipated need is wise, both to accommodate for miscalculations and also to ensure a sufficient amount of working line for operational purposes. Using static kernmantle rope, especially for track-lines, will help prevent uncontrollable stretch from developing in the system. If necessary, review the differences between static and low stretch rope in Chapter 9, *Ropes, Knots, Hitches, and Bends*.

Track-Line(s)

The track-line is the line that supports the major portion of the weight of the load in the highline.

In many cases the track-line can be a single line. However, under circumstances of high loading, it may be preferable to use two ropes in parallel to create a double line system. The track-line(s) needs to be longer than the span itself, because the anchor points are typically set back some distance from the edges of the span. In addition, there must be sufficient length to allow for an appropriate amount of sag (also known as catenary) in the track-line rope to prevent overstressing of the rope, other equipment, and anchor systems.

Tag-Line(s)

A tag-line is a non–load-bearing line used to connect the load with either side of the highline for control purposes. A tag-line is typically attached either directly to the load or to the track-line pulley to which the load is attached. A tag-line may be used simply for the purpose of maintaining contact with the load, or it may double as a lowering line, a haul line, and/or belay line. When used in this manner, it is often referred to as a control line. It is usually best to rig systems so that the tag-lines on both sides can be used for lowering and raising, as needed.

Lowering Line. When used as a lowering line, a tag-line is run through a braking device that is attached to an anchorage near the terminated end of the highline, runs from the higher of the two end anchor points, and is connected to the load. It is used primarily to control the speed of the load as it runs down the slope of the track-line, typically by running the lowering line through a braking device (**FIGURE 17-3A**). In a horizontal highline, this lowering effect usually is necessary only until the load reaches the center of the track-line (**FIGURE 17-3B**). At that point, because of sag and stretch in the rope, the load starts "uphill" toward the far side (**FIGURE 17-3C**). After this point, the lowering/belay line functions more as a belay, if at all.

In a steep-angle highline, if the load is traveling from the upper point to the lower one, the lowering/belay line acts primarily as a lowering line until the load is quite near, or perhaps even reaches, the bottom. The amount of friction required for control will vary. In a steep-angle situation, the lowering line requires a braking device with a great deal of friction and control, but in a less steep system less friction is required. In fact, given the wide range of friction needed for tag-line control, autolocking devices may be particularly difficult to use for this purpose.

Haul Line. The haul line is a tag-line that is also used to haul the load. Beyond the center point of a horizontal highline, the load must be pulled upward to overcome the sag in the system. Depending on the situation, this can be quite strenuous and may require the use of a mechanical advantage system. In a steep-angle highline, a haul line is needed to bring the load from the low anchor point side to the high anchor point side.

Belay Line. A belay line is any line that protects against potential failure. Belays are discussed in detail in Chapter 12, *Belay Operations*.

Sometimes rescuers rely on a second track-line for belay; other times, tag-lines (i.e., lowering and raising lines) are used to achieve belay. When relying on tag-lines for belay, it is extremely difficult to achieve a level of protection that would not impart at least some damage to the load in the event of track-line failure. However, given a low probability of failure, it may suffice to prevent total failure by rigging the lowering and raising systems in such a way that they would protect the load from dropping to the ground if the track-line might fail. To qualify as a belay, a line must incorporate a means of securely grabbing the rope and transmitting the force to an adequate anchorage.

Where a significant distance is to be spanned (more than 100 feet [30.5 m] or so), tag-line hangers, also called festoons, are a useful tool for supporting the tag-lines on either side of the load, just under the track-line. Avoiding slack in the tag-lines can make them easier to manage and help prevent them getting caught up in obstacles or drooping in the water. Festoons can be made by girth hitching a short loop of accessory cord to the tag-line and securing it at regular intervals to the track-line with a lightweight, nonlocking carabiner.

Anchorages

Anchorages chosen for a highline will depend at least in part on environment, as this will dictate what anchorages are available. Rescuers are unlikely to find sturdy trees on the tops of buildings, or a structural truss in the wilderness, so rescuers' ability to analyze anchorages in the environment where they are likely to operate is essential.

Ideally, track-lines should be anchored separately from tag-lines, as this provides optimum security. Where a structural anchorage of unquestionable reliability exists, it is certainly reasonable to anchor multiple lines to the same edifice. However, rescuers will find that anchoring tag-lines behind—yet in line with—track-lines is most advantageous, as this maximizes working space.

346 Rope Rescue: Principles and Practice

FIGURE 17-3 A. Controlling the speed of the load by **A.** Lower. **B.** Load at center. **C.** Raising the load.

The tensioning system for a highline should also be positioned distal to, but in line with, highline anchor points. Again, maximizing the working area in which rescuers will operate helps to accommodate rigging as well as ensure rescuer safety.

Far-Side Track-Line Anchorage

The far-side anchor system is the point to which the track-line rope is attached at the side of the operation opposite where the system is being controlled. This is usually where the subject is being transported *from*. This may be on the opposite side of a river or chasm, or in the hot zone of a rescue. When possible, a team of rescuers should gain access to the far-side anchor system location prior to setting up the system. In rare cases, it may be necessary to direct a subject or bystanders to assist in establishing an initial far-side anchorage using equipment and instructions provided by rescuers. The far-side anchor system must be at least as strong as the near-side anchor system.

In a steep-angle highline, the lower anchor point receives less stress while the load remains near the top. However, as the load approaches the bottom, the far-side anchor point is subjected to greater stress (**FIGURE 17-4**).

Any anchoring method may be used for the far side, as long as it provides sufficient security and adequate safety factor for the operation (**FIGURE 17-5**). If an appropriate anchorage is available, tensionless anchors are a good choice for far-side anchor systems because they are easy to rig, quick to establish, and strong. At least three wraps of rope should be used for a far-side tensionless anchor.

Near-Side Track-Line Anchor/Tensioning System

The near-side anchor system is usually nearest arriving rescue personnel. It ordinarily is the point from which operation of the highline is initiated and controlled, and is the side *to* which the subject is being transported. If a highline system is to be tensioned (it usually is), termination is typically to a tensioning system on the near side, where there is greater accessibility to equipment and personnel resources.

For a single track-line system, an inline haul system with progress capture may be used, keeping in mind that if the progress capture is left in place during operation of the highline, the integrity of the connection between the progress capture device and rope will most likely become the weakest link upon which the safety factor should be calculated (**FIGURE 17-6**). An advantage of this approach is that it permits tension to be easily adjusted once the load is mid-span. If maximum

FIGURE 17-4 Forces on a steep-angle highline.

strength is required for the highline, a piggyback haul system may be used to pull tension on the rope, while the rope is anchored using some other means (tensionless hitch, webbing anchorage connector, etc.).

Highlines may be rigged with two track-lines to provide a greater margin of safety, particularly for heavy loads or taut highlines. Double track-lines may be rigged as two parallel highlines rigged beside each other (including sag and tension). By using the same bombproof anchorages on either end, track-lines can be kept very close together. Assuming the ropes are made with the same elongation characteristics, the tension will be nearly the same in each line as they elongate together under load. All else being equal, for any given load, adding "mass" in the track-line, whether by using a thicker rope, or by using multiple ropes, will typically result in less sag. Parallel track-lines may be tensioned using side-by-side single track-line tensioning systems (**FIGURE 17-7**).

For more equilateral track-line tension, a **Flying W tensioning system** may be used to pre-tension both track-lines as one (**FIGURE 17-8**).

In this method, each line is anchored independently (adjacent to one another) at the far side, while at the near side the "W" shape formed by the haul rope creates a balanced 2:1 mechanical advantage on each of the track-lines with a single tensioning system. Once the track-lines are tensioned, they are tied off directly to anchor points. Thus rigged, the lines are independent of one another, each providing backup to the other. The haul system can be easily reattached to the track-lines as needed to adjust tension once the highline is loaded.

Carriage System

Pulleys are used in various places throughout a highline. In addition to being used to create a tensioning system for the track-line(s), a pulley may be used to attach the load to the track-line rope, acting as a traveling pulley to carry the load across the track-line. A traveling pulley is used to create a **carriage system**

FIGURE 17-5 Any anchoring method with sufficient safety factor may be used.

FIGURE 17-6 Single track-line tensioning system with integrated progress capture.

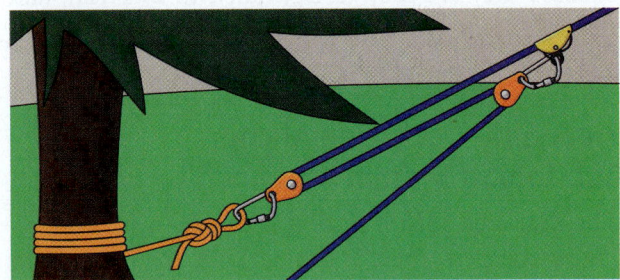

FIGURE 17-7 Parallel track-lines may be tensioned using side-by-side single track-line tensioning systems.

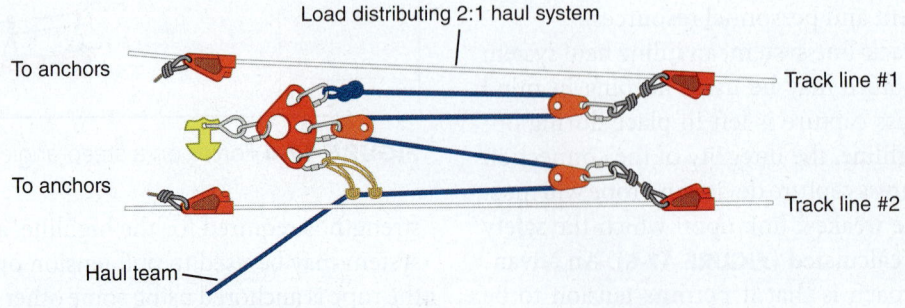

FIGURE 17-8 Flying W tensioning system.

by which the load travels along the track-line with minimal friction.

Especially with higher loads, tandem pulleys offer advantages over single pulleys in that they create less of a bend at the tangent point in the rope. By spreading the force a further distance along the rope, point stress is reduced and hauling is made easier. The *tandem* can either be two single pulleys rigged in line or a specialized tandem pulley manufactured with two sheaves in line. When two pulleys are used, it is important that pulleys be set so that they travel in a straight line along the rope. Otherwise, they may torque and create drag.

A special large pulley, sometimes called a **highline pulley**, Kootenay carriage, or knot-passing pulley, can be used for the traveling pulley on the track-line (**FIGURE 17-9**). The sheave of a highline pulley is larger in every dimension. Although it is not a tandem pulley, the larger sheave helps spread the load further along the track-line, and also accommodates double track-lines when needed. A highline pulley also typically has additional attachment points for tag-lines.

Tag-Line Anchor Systems

Tag-lines, whether or not they are used for lowering, raising, or belay, should be anchored independently from the track-line(s), with sufficient strength to perform adequately with an appropriate margin of safety. Especially if they are expected to function as a belay, these anchorages should be of sufficient strength to withstand a failure of the track-line system. Anchoring tag-lines to their own anchorages beyond—but in line with—track-line anchor points yields best results.

> **TIP**
>
> **Avoid Overstressing the System**
> A highline system must never be stretched very tight and then loaded. This could result in overstressing and failure of the rope, other equipment, or anchorages.

High Directionals

Anchor systems rigged low are typically more secure than anchor systems rigged high. At the same time, in the case of highlines, the track-lines should be well above and clear of the edge when not loaded—otherwise, when loaded, the subject(s) or rescuer(s) may not clear the edge. The system can be anchored low and suspended high by using a high directional near the edge, either by rigging pulleys into available high rigging points at the rescue site, or by rigging an artificial high directional (AHD) near the edge. This is an optional feature, not absolutely required for all highlines, but very useful. In fact, high directionals have become popular in all types of rescue rigging, including vertical evacuation, confined space, and more. They can be particularly useful for highline applications.

Site-built systems can be rigged by constructing an overhead anchor point near the edge, for example from a higher work level or in a nearby tree, from which to suspend one pulley for each track-line (**FIGURE 17-10**).

Alternatively, where there is a scarcity of available overhead anchorages, an artificial high directional (monopod, bi-pod, tripod, quad-pod) may be carried into the site and rigged to provide a high point from which to suspend the pulley(s).

High directionals should be considered a convenience rather than a necessity, and rescuers should be

FIGURE 17-9 A highline pulley.

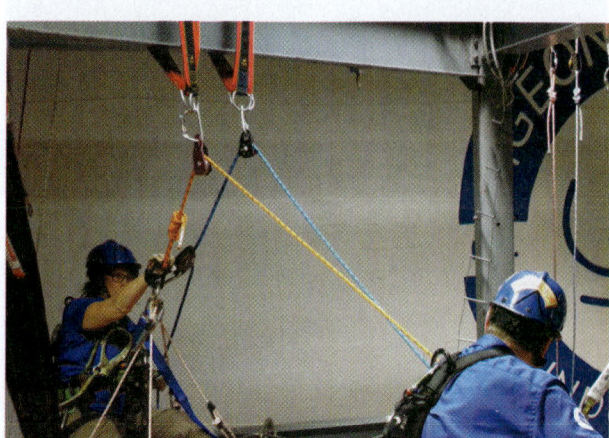

FIGURE 17-10 Site-built high directional.

capable of rigging a highline with or without the aid of such devices. Use of a high directional introduces a whole new range of parameters, angles, and forces, which are a specialty unto themselves.

Parts of a High Directional

One piece of equipment that is often used to create an artificial high directional are multi-pods. Multi-pods are a type of portable high anchorage, and depending on the number of legs may be configured as a monopod, bi-pod, tripod, or quad-pod, and in some cases may be used in a horizontal configuration across a span (**FIGURE 17-11**). Some devices are adaptable between two or more of these configurations, while other devices are preset as one. While not specifically called out in NFPA 1006, nor in the training requirements of NFPA 2500 (1670), rigging portable high anchorages is generally considered to be an advanced skill. As such, industry best practice is to defer rigging of such anchorages to experienced rope technicians. When used to create a high directional, the portable high anchorage is often referred to as a portable high directional (PHD). The most common configuration of PHA used as a PHD is a tripod. There being various brands and models of such devices, the specific features and components of these types of devices vary somewhat, but there are a few key parts that most portable high anchorages share.

Head

The part of the portable high anchorage where primary connection occurs is called the head (**FIGURE 17-12**). Generally the high point of the device, the head will contain certain connection points that are approved for rigging with rope, either directly or with a pulley, carabiner, or other component.

With certain portable high anchorages, the angle of the head may be adjusted to accommodate different terrain or directions of pull; keeping the head level even when the terrain is not helps to keep rigging clean and direct the load more precisely along the supporting legs, thereby reducing the likelihood that the device will tip dangerously. The height of the head is typically determined by legs.

Legs

A multi-pod may have one leg (monopod/gin pole), two (bi-pod), three (tripod), or even four (quad-pod) legs. These provide greatest strength when in compression, but are less secure when pulled sideways. Generally speaking, more legs will provide greater strength, but the wider the legs are spread and the higher they are extended, the weaker the portable anchorage will be.

Usually of a tubular design (either square or round), the legs of most portable high anchorages may be dismantled so that they may be transported in two, three, or more sections. They are often adjustable in height, allowing rescuers to customize the portable high anchorage to fit the ground, head angle, and situation. Always follow manufacturer's instructions for rigging the components together, as some configurations will be stronger than others. Similarly, the maximum height specified by the manufacturer should not be exceeded. When increasing the height of a tripod or similar device, extending the lower leg section first, so that the majority of tube overlap is nearest the head, typically results in greatest strength. The head is attached to the uppermost part of the legs by some means of clamp.

FIGURE 17-11 Portable high anchorage.

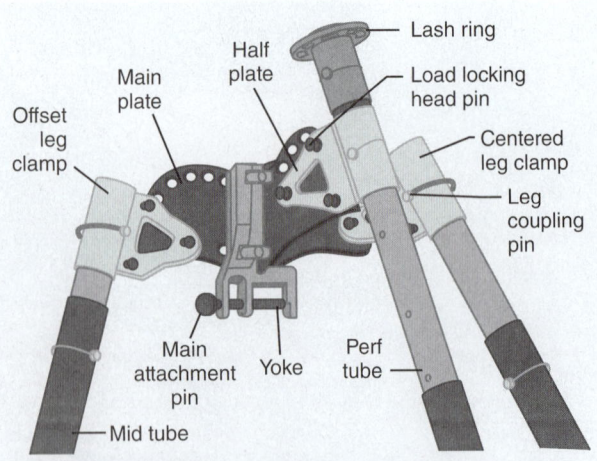

FIGURE 17-12 The head of the portable high anchorage.

Feet

At the bottom end of the legs, a foot of some sort provides direct interface with the ground. Feet may be flat, curved, round, pointed, or almost any shape. Sharply angled feet with pointed ends can be used to dig into rugged surfaces; round-shaped or curved feet offer great versatility, but limit direct contact with the surface; flat feet work well on flat roofs, and may even be screwed or bolted into place; and some feet are conducive to lashing around rails or other surfaces.

Hobbles

The integrity of the legs is foundational to the overall strength of the portable high anchorage. When two or more legs are used in a multi-pod, a downward force exerted on the head between them will want to spread the legs further apart. As the angle between legs widens, the strength of the device is reduced. While the interface between the legs and head is generally the primary means of maintaining leg position, inclusion of even the smallest amount of support at the widest end of the legs helps to reduce movement and preserve the integrity of their position. This can be achieved with short lengths of cord, cable, strap, or chain between them (**FIGURE 17-13**).

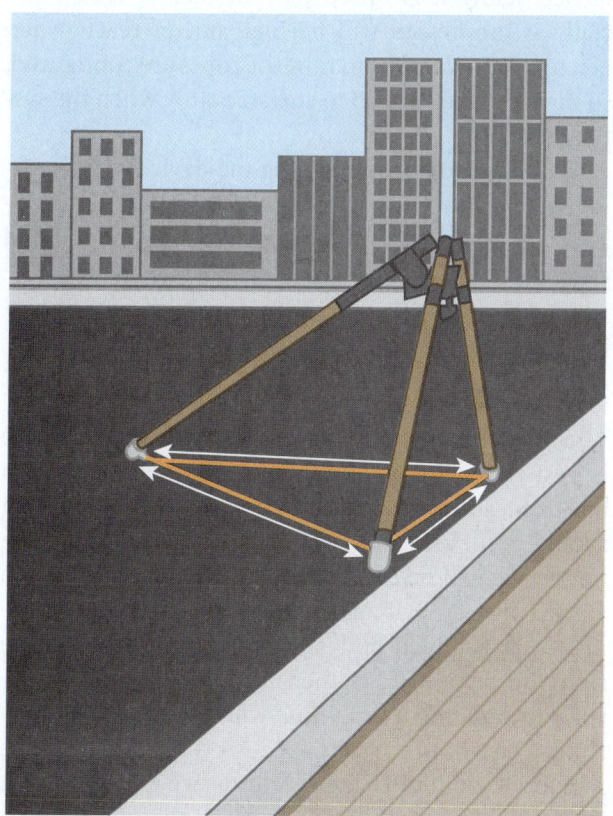

FIGURE 17-13 Hobbles.

Hobbles are exposed to relatively moderate forces during normal use, so super–high-tenacity material is not essential. More important is that the material is easy to work with, and offers easy connection to the legs or feet. If hobbles of fiber material are used, take care to prevent damage from moving rope during use.

Hobbles should be rigged and tensioned only after positioning and adjusting all of the other elements of the multi-pod. They should be snugged just enough to prevent movement but not made taut.

Additional Rigging Points

Most multi-pods are fitted with lash rings, winch brackets, and other accessories that can be positioned at various points along the legs to provide additional rigging points. These are used for various purposes, including safety lines and back-ties. Because these are removable and adaptable in position, users must take special care to ensure that the direction of pull and forces exerted are appropriately offset and balanced so that their rigging does not exert a twisting action on the device.

Configurations

Tripod

Perhaps the most common configuration of a portable high anchorage is the symmetric tripod. Such devices are widely available as one-piece, preconfigured assemblies, and may also be created from modular systems. They are frequently used to provide a high point anchorage directly above a hole or hatch. A symmetric tripod is set so that all three legs are of equal height and equidistant from the hole that it is providing access to. This helps to ensure that all three legs are equally loaded in their strongest configuration.

When used to provide a high point over an edge, a tripod may be rigged with two legs parallel to the edge with the third leg behind to provide stability and greater opportunity to secure the device into place (**FIGURE 17-14**). This configuration is sometimes known as an edge A, and may even be used without the third leg. Having a third leg that can be extended so that the system leans either over, or away from, the edge, helps to better align the system for the desired direction of pull. This is called an easel leg. Some systems allow this third leg to float freely, a configuration known as a lazy leg. While the lazy leg allows greater variability in position, being able to lock this third leg into place makes the tripod much easier to manage during setup and raising and also offers more stability when back-tying and securing the device into place.

FIGURE 17-14 A symmetric tripod.

Quad-Pod

Quad-pods are not as widely utilized as tripods, probably because they are not as widely available. These are prized for the greater strength and stability that comes with the fourth leg, but perhaps more importantly they can be set with a narrower profile and greater height (as well as greater strength) than a standard tripod.

Bi-Pod

As the name suggests, a bi-pod has two legs for support. Sometimes referred to as an **A-frame**, these may be used with the two legs perpendicular to the working area, or with the frame in profile to the working area (also called a **sideways A-frame**. These are most useful in narrow areas such as catwalks or rocky ledges.

Monopod

Also referred to as a **gin pole**, the monopod is the lightest high anchor system configuration and, perhaps surprisingly, quite strong. These do require back-ties for stability but are highly adaptable to tight spaces and difficult locations, and when properly rigged can support significant amounts of weight.

Horizontal Span

Some monopod devices can be configured to rest on existing supports to span a horizontal chasm such as trenches, catwalks, and pits. This use should be attempted only where specifically permitted by the manufacturer, as shear strength of tubes is generally much lower than compressive strength. Rigging a portable high directional for a horizontal span is quick, and it is particularly advantageous in industrial settings where edges around the area to be accessed are either nonexistent or are not conducive to the point loading induced by legs/feet.

Using Multi-Pods

The easiest way to erect a multi-pod in almost any configuration is to first assemble and adjust all of the components with the device lying flat on the ground; then, raise it into position and make necessary adjustments; next, secure the device into its desired location with back-ties; and finally, make final adjustments to the system.

Assembly

Assembly begins with the legs, either one, two, three, or four. These should be assembled with appropriate feet and legs set to the desired height. If the head is removable, it can be attached next. Some systems have an adjustable head. While the angle can be estimated and preset with the device lying on the ground, adjustment may be necessary once it is raised into position.

At this point, the rescue system may be rigged into the multi-pod as well. This requires some vision and finesse. Being able to envision the finished product before commencing rigging is essential to avoiding creating a tangled mess. Rigging the system into the device before it is raised pays dividends later, especially if the device will be high out of reach when erected. Pay attention to how the ropes are configured, so that they are aligned to run smoothly when the system is raised.

Before raising and standing the device up, do a final system safety check to ensure that all of the components and connections are properly configured and set.

Raising and Securing the Portable High Anchorage

The fully constructed portable high anchorage device is best raised to an upright position by lifting from the front and tilting towards the rear leg(s). Raising should occur as near as possible to the location where the device will be used, yet while taking great care to ensure the safety of all rescuers. For safety, the device may be assembled and pre-rigged at a safe distance from the exposed edge, then carried to its point of use prior to standing it up. When working near an unprotected edge, both the portable high directional and the rescuers should be belayed or otherwise secured with a safety system to prevent falling—or, for that matter, any unwanted dynamic event. This may require securing at the head as well as the foot ends of the device. If applicable, back-ties may double as safety lines.

A tripod-type system placed over a manhole and loaded in a straight-down direction will be fully supported by the legs of the device, with little chance of tipping. Imagine drawing lines between the legs of the tripod to form a triangle; the area inside this triangle is known as the footprint. As long as the load and the rescue system remain inside the footprint, the tripod will remain stable. Any application of force from another direction, however, can destabilize it and cause it to tip. To prevent tipping, a portable high directional can be secured into place using back-ties. Back-ties as used relative to anchorages and anchor systems are discussed in Chapter 11, *Anchorages*.

The primary purpose of a back-tie is to provide security and stability to an anchor system by reinforcing one with another. When applied to a portable high directional, multiple back-ties are needed to exert oppositional pressure against one another so that the portable high anchorage itself does not shift. Back-ties may be attached to the head of the portable high anchorage, or to legs, as long as they are rigged in such a manner that they provide tension in equal and opposite directions.

Back-ties are usually created using rope or cordage, along with some sort of progress capture device (PCD). Tension may be applied manually, or using pulleys for mechanical advantage. Many rescuers find small, pre-rigged haul systems to be advantageous for this purpose. Tensioning back-ties used in opposition to one another can be tricky, in that a load must be established in one direction so that tension can be applied equally in the opposite direction. This can be achieved by multiple team members pulling simultaneously, or by one rescuer who simply works their way around the device making small adjustments until all is sufficiently tensioned. PCDs are discussed in Chapter 15, *Mechanical Advantage Systems*.

As the portable high anchor system is used, rescuers may find that things need readjusting as the system settles in. In fact, most portable high anchorage devices will have a bit of play in them to accommodate this settling in, as it allows the device to find balance as it conforms to the terrain and rigging. Some rigging, including back-tie tension, may need to be adjusted during use of the tripod to accommodate settling in, rope stretch, and small variations in loading.

Managing Forces

As the device is raised into position, consider the outcome: specifically, be intentional about where and how resultant forces will be applied once the system is loaded. Resultants are the sum of all of the vectors acting on a given object. Like all vectors, a resultant will have both magnitude and direction. In this case, the vector with which rescuers should be most concerned is the force that will be imparted on the high directional when it is loaded. See Chapter 10, *Principles of Rigging* for more on resultants.

While portable high anchorages are occasionally used with a load suspended directly beneath the center of the head, when used as a portable high directional, they are usually rigged with a pulley hanging from the center of the head so that rope can be effectively lifted and redirected off the edge.

In this configuration especially, rescuers must ensure that the resultant lies within the footprint of the device. The size and shape of the PHD footprint will be influenced by the number of legs, how widely they are spaced, and the use of back-ties (**FIGURE 17-15**).

The resultant in a PHD can be visualized by determining the magnitude and force in which the pulley is drawn, relative to the footprint. The pulley is typically pulled in a downward direction by the rescue load at one end, and the controlling system (raising, lowering) at the other. The resultant, then, is the bisection of this angle. Some rescuers attach a small laser pointer to the pulley in this position to help visualize the resultant. Note that both the rescue load and the control system can be set up and used outside the footprint as long as the resultant force remains inside.

If the resultant is outside the footprint, the PHD will be at risk of tipping. This can be mitigated in one

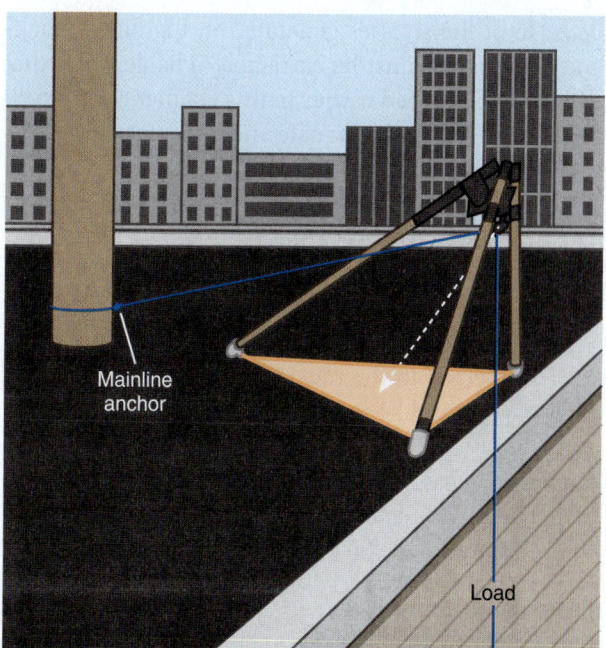

FIGURE 17-15 PHD footprint/resultant.

of two ways: either by adjusting the resultant or adjusting the footprint. The quickest and often simplest approach is to adjust the resultant by managing the direction from which the load is controlled, or that from which it is applied. This can often be achieved by simply making a small adjustment to the raising/lowering system location, or by applying a change of direction near ground level either within or very close to the footprint.

If changing the resultant is not readily feasible, changing the footprint is the next best course of action. Adding a pair of back-ties (especially toward the front of the device) is a quick way to widen or shift the footprint to resist tipping.

When erecting and using the PHD, visualize the lines as though they were loaded before you load the system—and then when you load, do so with caution—revisiting and confirming any assumptions that may have been made while setting up. Likewise, care must be taken when transitioning from raise to lower, or from lower to raise, to ensure that this simple adjustment has not caused the resultant to shift outside the footprint.

Forces in a Highline System

Highline forces can be predicted based on the weight of the load, the geometry of the triangle formed by the lines connecting track-line anchor systems, and the load, given the sag in the system.

The Anchorages

Because of the stresses generated in highlines, track-line anchorages must be extremely reliable. Note that a PHD, as described earlier in this chapter, is not itself considered to be an adequate anchor point for a horizontal system. The PHD simply serves as a change of direction to raise and redirect track lines to facilitate its use.

In a horizontal highline, the two track-line anchorages opposite one another are subjected to similar stresses. In a steep-angle highline, the upper anchorage is subjected to the most stress, much as an upper anchorage would be in a lowering system.

Webbing or rope anchorage connectors (or both) should be used as needed and appropriate for the site. Highlines exert great force on the system. High-strength tie-offs should be used with multipoint anchor systems or other methods to build bombproof anchor systems. Trees or vertical columns used as anchorages should be back-tied and pretensioned as needed to provide additional strength.

The Load

The load is the total mass, including personnel and equipment, that will be riding on the highline at any given time. For example, this may be a litter with the rescue subject and possibly a litter attendant. In other cases, the load may be one person, a rescue subject attached to a rescuer, or equipment.

If the load is to be just one person, typically connection between the load and the track-line is made with a vertical sling between that person's harness and the pulley on the line. For multiple people riding the line together, such as a rescuer with a subject, the second person should also have a direct connection to the track-line, rather than just to the first person, yet with some means to ensure that the rescuer does not get too far from the subject. The two different people may be connected via separate track-line pulleys, but being both connected to the same track-line pulley can be easier to manage. A highline pulley or a rigging plate can help to separate the lines.

Litter loads typically require some form of litter bridle to balance and ride smoothly along the track-line. A litter bridle may incorporate either one or two pulleys to ride along the track-line. Tag-lines should be attached to the pulley or the rigging plate for optimum control. Attaching and moving suspended loads is discussed further in Chapter 16, *Working in Suspension*.

The Sag

Sag is the term used to the curved shape the unloaded line takes when suspended between two points. The more accurate engineering term for this is catenary. Too much sag, and the line will droop down into the very obstacle rescuers are trying to avoid. By tensioning the rope, the line becomes straighter, making it easier for rescuers to pull the load across.

But a tighter line also puts more force on the opposing anchorages, and adding a load to a pretensioned line increases the force even more. And so the question becomes, "How much tension is too much?"

Analyzing the System for Force

There are mathematical formulas that can be used to solve for force in systems where there are the same number of equations as variables, but neither rope materials nor highline systems are constant, and most rescuers are not mathematicians. Even with a solid understanding of the force multipliers working on a rescue system, the forces in a track-line are difficult to predict because of the nonlinear and changing geometry of the track-line. Angles and resulting forces will change significantly as the load travels across the span, making accurate real-time calculation unrealistic. For this reason, it is especially important for rescuers to be capable of analyzing the effect of sag and estimating forces with relative accuracy. Force multipliers are discussed in Chapter 10, *Principles of Rigging*.

Pre-Tension Guidelines

When building a system, rescuers must decide how much tension to put into the line before it is loaded. **FIGURE 17-16** shows the forces that can be present in a highline system when the track-line has been stretched tight, with no visible sag. As shown in Figure 17-16B, a 200-pound (90.7-kg) load at the middle of a 100-foot (30.5-m) tight span can create tremendous stress. Conversely, Figure 17-16C shows how forces can be reduced simply by introducing more sag into the track-line.

Loads on a highline are directly proportional to the geometry of the angles formed by the lines connecting the anchor systems and load. The shallower the system, the higher the load. Keeping this in mind, there are two commonly accepted rules of thumb that can be used as a practical guide for pretensioning horizontal highline systems: the 10 percent sag rule and 10 percent loaded sag rule.

10 Percent Sag Rule (Pre-Loading)

One conservative method of tensioning a highline, the 10 percent rule, suggests that the center of the *unloaded highline* should sag a vertical distance equal to about 10 percent of the span for every 200 pounds (90.7 kg) of expected load and every 100 feet (30.5 m) of span in the rope. This sag should be calculated based on the total weight of the load, and should take into consideration the total length of the span between the two supports, including the anchor systems—not just the width of the gap to be bridged. The rule refers to the amount of sag in the system *before* the load is applied.

For example, the formula for determining sag on a 100-foot (30.5-m) span with a 200-pound (90.7-kg) load (1L) would be calculated as follows:

$$1L \times 100 \text{ feet } [30.5 \text{ m}] \times 0.1 = 10 \text{ feet } [3 \text{ m}]$$

Therefore, a 10-foot (3-m) rope sag would be required. The two variables in the 10 percent rule are (1) the weight of the load and (2) the length of the rope span. If either one of these changes, the amount of sag also must change. As an example, sag for moving a 200-pound (90.7-kg) load (1L) across a 200-foot (61-m) span would be calculated as:

$$1L \times 200 \text{ feet } [61 \text{ m}] \times 0.1 = 20 \text{ feet } [6.1 \text{ m}]$$

A 20-foot (6.1-m) rope sag is required for this load. Similarly, the sag required for a 200-foot (61-m) span expected to transport a 400-pound (181.4-kg) load (2L) would be:

$$2L \times 200 \text{ feet } [61 \text{ m}] \times 0.1 = 40 \text{ feet } [12.2 \text{ m}]$$

In this case, a 40-foot (12.2-m) rope sag would be required for the 400-pound (181.4-kg) load. The 10 percent rule is a particularly conservative approach to determining the amount of pre-tension for a highline. According to a research study by Stephen Attaway and colleagues, the 10 percent sag rule consistently resulting in a static system safety factor (SSSF) well in excess of 10:1. A horizontal line represents the worst case scenario for highline forces; when determining sag for a sloped line, the 10 percent rule would be especially conservative, but also very difficult to apply, as the amount of sag would be tricky to visualize.

10 Percent Loaded Line Sag Rule

A relatively simple-to-use formula for estimating the force in a highline when the load is center-span is:

$$\text{Tension} = \frac{\text{Load} \times \text{Span}}{4 \times \text{Sag}}$$

With the load at center-span, the hypotenuse (long sloping span) of each of the two the triangles formed on either side of the load is assumed to be approximately half the distance of the span, with rope tension applied proportionally.

TIP

The 2:1 pre-tension rule is another method for estimating appropriate pre-tension in a horizontal highline. This one suggests that pulling about twice the force that one person can pull horizontally will result in sufficient tension to offer a 10:1 SSSF for a 2-kN (450-lb) load.

This rule is challenged by two variables in particular: the load is not always 2 kN, and elongation varies with different rope materials and diameters. In short, the 2:1 rule is not a very conservative approach, and research has shown that it does not consistently yield satisfactory results. For this reason, in this text we will not recommend this as a viable method.

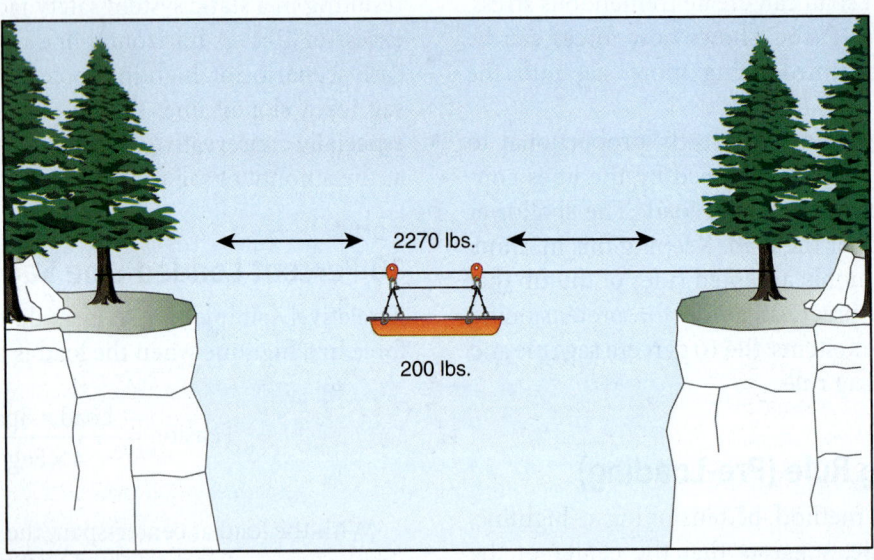

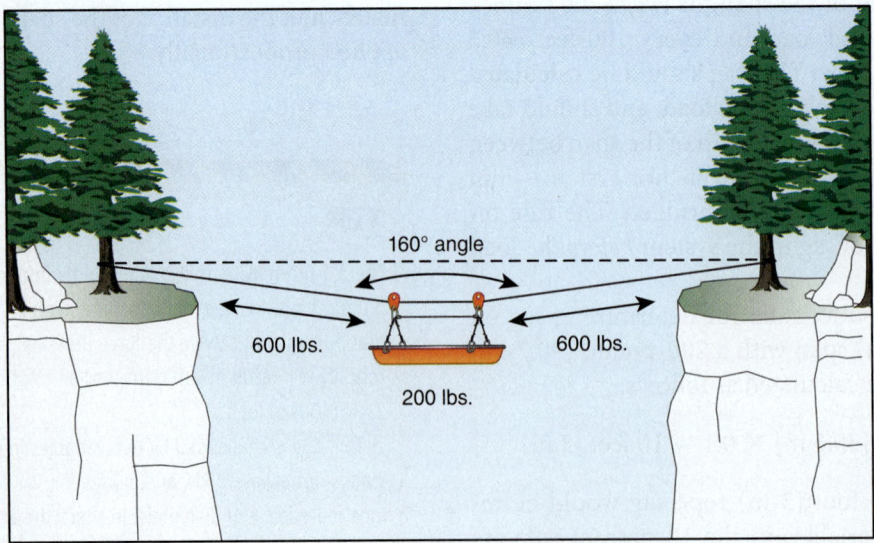

FIGURE 17-16 Line tension and mass of the load determine forces in a highline.

Based on this method, Richard Delaney, in his text, *Physics for Roping Technicians,* suggests that one approach to field calculation is to use as a foundational assumption the fact that a highline with a sag of 10 percent has a load magnification factor of 2.55. In other words, a 100-kg (220-lb) mass applied to the center of a highline with a sag of 10 percent applies a force equivalent to a static load of 255 kg (562 lb) to each anchor system.

Loaded Line Guidelines

Pre-tensioning aside, what really matters most is the actual amount of force on the highline with a load on the line. It bears repeating that for span lines with equal height anchor systems, the worst case for span line tension occurs when the load is at mid-span. Rescuers should bear this in mind, and rig so that the maximum potential force when the load is mid-span yields a safety factor acceptable under the parameters of the authority having jurisdiction (AHJ).

> **TIP**
>
> The 12:1 loaded line rule of thumb assumes a significant amount of sag at the outset of an operation, and provides guidance for rescuers to pull additional tension into the track-line after the load is at the center of the span. In this case, the tension in the system is controlled by limiting the amount of force used to haul on the system to the equivalent of 12 rescuers. Stated differently, for 7/16-in (11.1-mm) rope, the product of the number of people pulling and the mechanical advantage of any tensioning system used should not exceed 12. This might be achieved with 12 people using a direct pull method, 4 people pulling on a 3:1 (4 × 3 = 12), 6 people pulling on a 2:1 (6 × 2 = 12), etc. For the purposes of this rule, *people pulling* is defined as the per-person hand gripping strength on the rope providing the input power to the mechanical advantage system, not their full weight or with the haul team tied into the haul system using their weight and legs to pull.
>
> Attaway et al.'s research study confirmed the veracity of this method (haul friction notwithstanding) so long as the load is in the center of the span at the time of tensioning, and the haul line is co-located with the highline anchorage. The study also noted that for a rope with 9,000-lbf (40-kN) strength, the rule may be adapted to an 18:1 loaded line rule—meaning that the product of the number of people pulling and the MA of any tensioning system used could be as much as 18.
>
> It is important to note that teams using the 12:1 loaded line rule for sloping highlines may be mistakenly motivated to apply the 12:1 haul while the load is still quite near the anchor system; this does not produce the desired results, because the tension on the track-line will increase as the load travels further from the anchor system.

While this text discusses highlines in context of a 10:1 SSSF, this should be no means be assumed to infer that this is the *correct* safety factor for highlines. This value is used for discussion purposes only. Determination of an appropriate safety factor is at the discretion of the AHJ.

Attaching a Load to a Highline

It is not uncommon for a single rescuer to cross a highline prior to sending the rescue load/subject across. Putting a rescuer on the line first helps to achieve initial tension, and may be essential to transferring a tag-line and/or equipment across the span. **FIGURE 17-17** shows a highline load consisting of only one person.

The essential elements of securing this load include the following:

- An appropriate carriage system, in this case a Kootenay HX Pulley.
- A support sling hanging from the carriage system.
- A rescuer wearing an adequate life safety harness, to which the support sling is attached.
- Tag-lines attached to either side of the carriage system.

Some organizations choose to add a second support sling for redundancy, loosely attached to the track-line with a carabiner, adjacent to the carriage system. This protects against the unlikely failure of the carriage system.

The person riding the highline should wear gloves. This is to prevent rope burns should they need to grab a rope while moving, to enable them to grasp the rope and pull hand-over-hand if necessary, and to prevent

FIGURE 17-17 One-person load on a highline.

injuries to fingers in the moving pulley. If the person transported on the highline is a rescue subject, it may be wise to make the support sling long enough so that the subject cannot reach the highline and injure their hands. To rig a highline, see **SKILL DRILL 17-1**.

Rigging a Litter to a Highline

Attaching a litter to a highline requires the use of a litter spider to suspend the litter below the carriage on the track-line. Depending on the type of spider used, either one or two carriage systems may be required.

A rigging plate can be useful here to help keep lines separated, especially if a litter attendant will accompany the litter across the track-line. The rigging plate provides a central tie-in point that can effectively be pulled from either direction without twisting the system. The rigging plate can also help facilitate a reeve, if one of these is used. To rig a litter to a highline, follow the steps in **SKILL DRILL 17-2**.

TIP
Dual track-lines often are a good idea, but involve more anchorages and good rope management skills. Two parallel 7/16-in. (11.1-mm) ropes offer more strength (and usually less elongation under a given load) than one 1/2-in. (12.7-mm) rope. When utilizing this approach, special wide-sheave pulleys are recommended, as they are easier to manage than double-sheave pulleys. Dual ropes can flip over each other, and the large-sheave pulley can continue on much more easily with less chance of jamming.

Regardless of whether one track-line or two are used, as the length of the highline increases, the tag-line tends to droop well below the track-line and can become a line management problem. Short tag-line hangers, or festoons, can be made from cordage about 1 foot (0.3 m) long.

As a point of caution, rescuers using a highline system should maintain constant awareness of weather conditions. High winds, for example, can entangle lines or add weight to the systems.

SKILL DRILL 17-1
Rigging a Highline

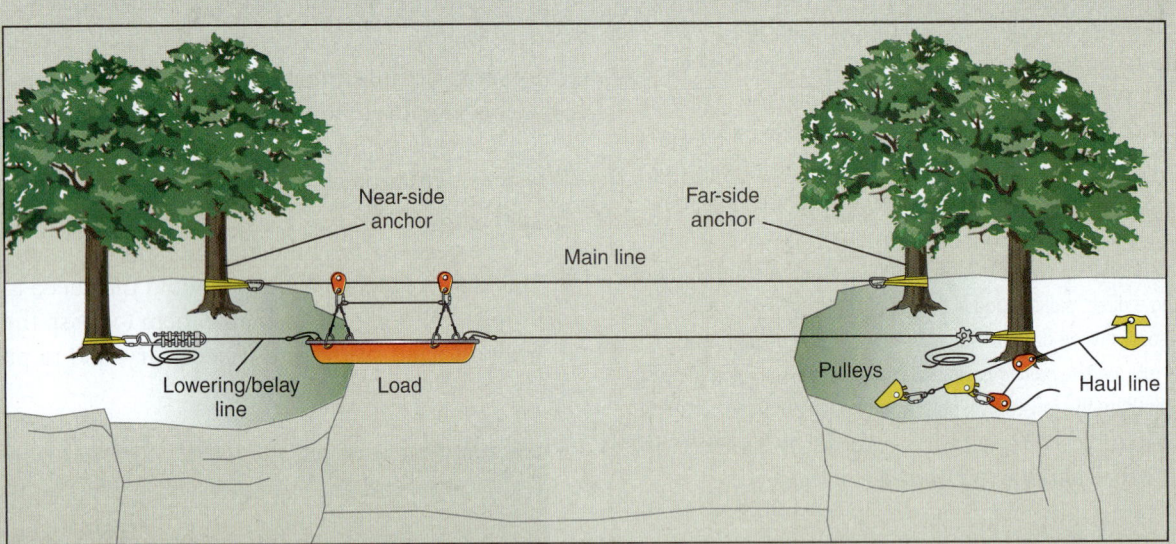

1 All personnel must be thoroughly briefed beforehand on the steps to be followed and on communications. Radios are very helpful. Select an appropriate site. It should have:

- As narrow a span as possible
- Room on both sides to rig and de-rig the load and for personnel to get on and off
- Track-line anchor systems available that are very strong and secure and high enough so that the load can get over the edge without dragging
- A tensioning system anchor system distal to, but in line with, the near side track-line anchor system
- Sufficient tag-line anchor systems on both sides

Insert a team of rescuers to the far side of the site. They should carry sufficient equipment to establish an adequate far-side track-line anchorage, and to create a tag-line system capable of both raising and lowering.

SKILL DRILL 17-1 (continued)
Rigging a Highline

2 Transport the far-side end of the track-line rope over the span, as described earlier in this chapter—usually with some sort of leader-line. Depending on the circumstances and why the chasm must be spanned in the first place, this may be one of the more difficult parts of the operation. The presence of hazardous environments, physical barriers, great distances, or extraordinary heights or depths can all make this act more difficult. Anchor the track-line(s) at the far side. A tensionless anchorage connector works well here, but any adequate anchoring method may be used.

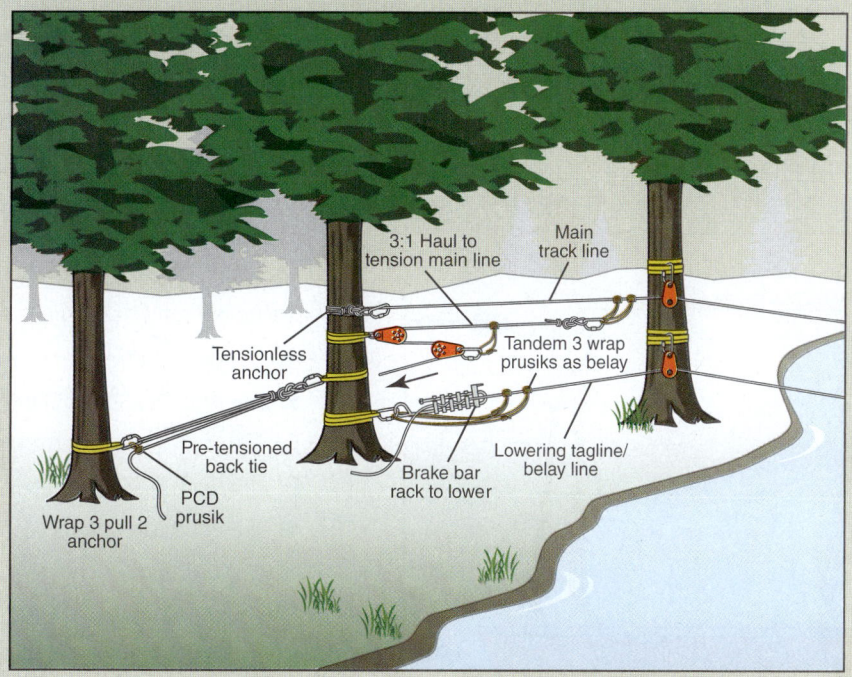

3 Establish an anchor system for the track-line(s) at the near side. The anchor system should be far enough back from the edge to permit sufficient working room between it and the edge.

(continues)

SKILL DRILL 17-1 (continued)
Rigging a Highline

4 Tension the track-line(s) by pulling from the near side to take slack out of the track-line rope until it is stretched horizontally across the span. Use an appropriate pre-tension method in accordance with AHJ guidelines.

5 Establish tag-lines at both near and far side. Both ends should have capability to both brake and raise. Have the safety officer(s) check the system at both ends; what they will check depends on the specific system being set up.

6 Attach a rescuer to the line with a pulley and lanyard. This rescuer should trail a tag-line from the near side. This tag-line may be used to help control the speed of the rescuer as they cross the downhill portion of the span.

7 When the rescuer is mid-span, refine the tension in the track-lines using your method of choice. From this point, the rescuer will need to pull themselves, hand-over-hand, up the uphill side of the system until they reach the far side. Note: While the rescuer is at mid-span, with the track-lines at their steepest angle, it may be feasible for far-side rescuers to attach a tag-line to the track-line(s) with a carabiner, and lower it down the sloped part of the highline toward the rescuer. Once the rescuer and the tag-line meet, the rescuer can attach the tag-line to their track-line pulley so that far-side rescuers can pull them the rest of the way up the slope using an adequate haul system.

8 Having transferred the carriage system to the far side, both tag-lines can now be attached to the carriage system so that control may be achieved from either side.

SKILL DRILL 17-2
Moving a Load Across a Highline

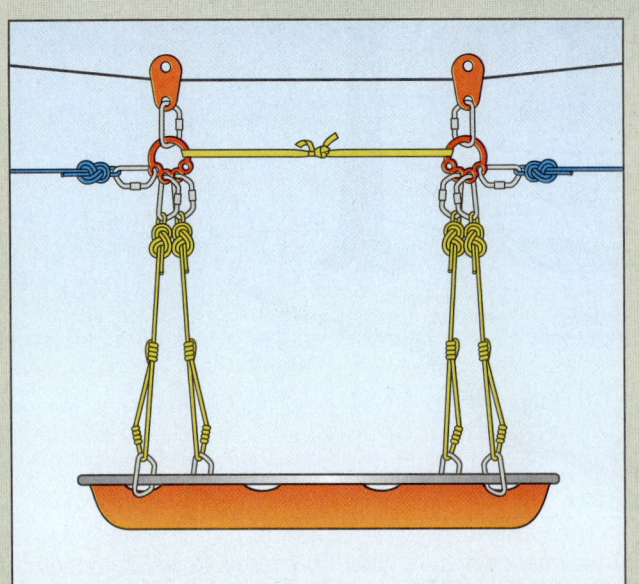

1 Attach the litter to the carriage. Before beginning movement, recheck the rigging and anchor systems for track-line and tag-lines. Make sure all those involved in the rescue are ready, including:

- The person attached to the load
- The brakeman on the lowering/belay line
- The belayer on the far-side tag-line
- The edge attendants

SKILL DRILL 17-2 (continued)
Moving a Load Across a Highline

2 When all personnel are ready, the litter attendant or person attached to the load begins the voice signals; they call, "On belay."

Near-Side Tag-line/Brakeman: "Near-side belay on."
Far-Side Tag-line: "Far-side belay on."
Litter attendant: "Down slow."

The near-side brakeman begins allowing rope through the brakes on the near side, while the belayer on the far-side tag-line begins taking in slack. It may be prudent to use a line-holding device (e.g., rope grab, Prusiks, or other belay device) at one or both ends. This decision should be based on the potential and consequences of the rope reversing direction and traveling quickly out of control.

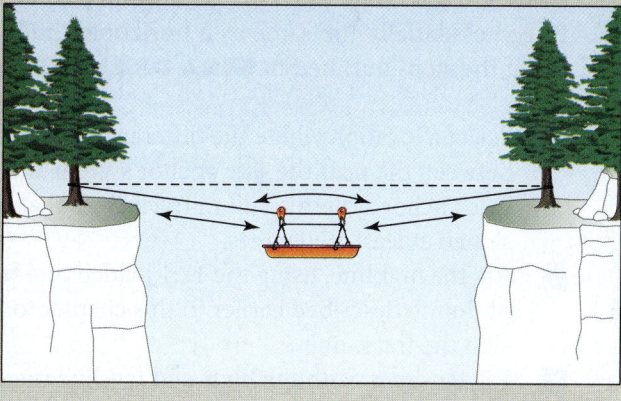

3 As the load nears the center of the trackline, tension should be monitored to ensure that it is within parameters established by the AHJ. Adjust as needed.

4 Once the load has reached the far side and is in a secure position, the litter attendant or person attached to the load calls, "Stop." The lowering/belay line brakeman stops feeding rope.

When the load is in a secure position on the far side, the rescuer managing the load calls, "Off belay."
Near-Side Tag-line/Brakeman: "Near-side off belay."
Far-Side Tag-line: "Far-side off belay."

> **TIP**
>
> The designations *near side* and *far side* are established by rescue leaders at the outset of the operation and remain such for the duration. If inexperienced rescuers mistakenly begin to refer to near side or far side as relates to themselves, or even to the subject, confusion will ensue. Once an end is designated near side and the other end designated far side, those designations should remain constant until that operation is complete.

Guiding-Lines

In some cases, a rescue site requires a subject to be moved both horizontally and vertically to clear obstacles or reach safety (**FIGURE 17-18**). Where there is a clear path from the high point where the subject is located to a lower point where safety awaits, both functions can be performed simultaneously simply by rigging the highline in a sloping configuration. Such systems, called guiding-lines or **sloping highlines**, may be used to clear obstacles or simply to avoid a ground carry.

While these terms are used synonymously in this text because they are so substantially similar in design and construction, a key differentiator between a sloping highline and a guiding-line system is that the sloping highline generally incorporates both a lowering system and a tag-line at the bottom end, while a guiding-line system eliminates this need.

A sloping highline or guiding-line system is highly dependent on the simultaneous operation of a lowering system to maintain integrity. Given the presence of a full-strength lowering or raising system operating concurrently with the sloping highline, rescuers may be lulled into a false sense of security thinking that the strength or integrity of the guiding-line may be reduced. Nothing could be further from the truth. Any time a load is pulled out of the fall line, there is swing fall potential and perhaps even a chance to drop the load if the guiding-line is compromised. In fact, a guiding-line should really be considered as two full-strength systems—a lowering system and a sloping highline—operated at the same time, in harmony.

Equipment needed to rig a guiding-line system includes all the same equipment required for a horizontal highline, plus all the equipment needed to rig a lowering/raising system A guiding-line system is rigged essentially the same as a horizontal highline, using the steps outlined in **SKILL DRILL 17-3**:

1. Select a location where the difference in height between the near the side anchor system and far side anchor system result in track-lines that are sloped at least 15 degrees.
2. Rig the highline, using the 12:1 loaded line rule of thumb described earlier in this chapter to tension the track-lines.
3. Rig tag-lines, with the high-end tag-line rigged as you would for a typical lowering system, including capability for both lower and raise.

Track-line tension may be fixed or variable in a guiding-line system. With a variable tension system, the track-line is tightened just enough to clear the obstacle and then relaxed when the obstacle is cleared. Alternatively, the guiding-line system may be tensioned at the outset to transfer a rescuer or load from one end to a midline point above the subject. By relaxing the guiding-line while the load is part way across the span, the rescuer can be lowered to the subject's location so that they may be attached to the system,

A

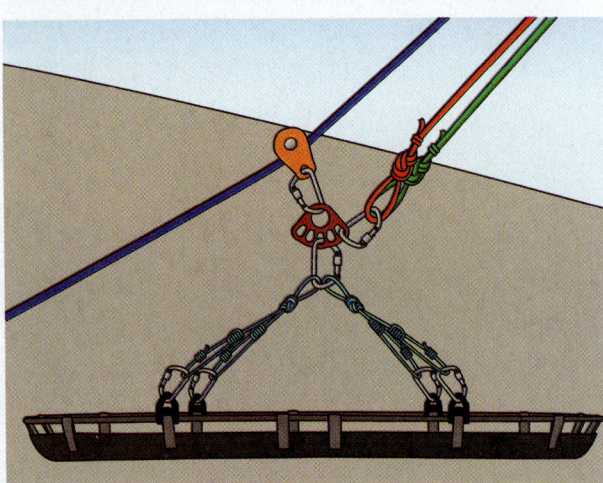

B

FIGURE 17-18 A. Guiding-lines may be used to clear obstacles such as trees, rocks, or debris. **B.** Attach control lines directly to the main attachment point, where the litter spider meets the guiding-line.

FIGURE 17-19 Relaxing a guiding-line to pick off a subject.

then the system is once again tensioned to lift the load and transfer it into place (**FIGURE 17-19**).

Guiding-lines are a viable, useful method for deflecting a load away from obstacles during a raising or lowering operation.

Dynamic Systems

With a solid understanding of simple horizontal highline rigging, more involved and creative systems for achieving horizontal movement can be considered. As we will see later in this chapter, forces in a simple horizontal system are typically greater than in some of the more creative systems, but the number of moving parts and variables in complex horizontal systems make these worthy of a little extra practice.

Care should be taken to practice these methods under qualified instruction, and to ensure that appropriate load ratios and safety factors are applied. Rescuers should always be aware of the potential for unexpected dynamic loading events, and should be prepared to respond accordingly.

English Reeve

Reeve is a broad term used to describe the act of threading rope through pulleys to create a haul system. Used in context with highlines, the term *reeve* generally refers to a specific configuration for lowering and raising the load from the carriage system as it travels across the track-lines.

The **English reeve** and the **Norwegian reeve** are both particular types of arrangement involving ropes and pulleys rigged to facilitate vertical movement of the load at virtually any point across the span of a highline. They may be used in conjunction with either a horizontal or sloping highline system to lower and raise the load (**FIGURE 17-20**).

Both the English reeve and the Norwegian reeve involve a 6:1 compound MA. The key difference between the two is that the line for the English reeve extends all the way across the highline system, just like the track-line, while the line for the Norwegian reeve terminates on one side at the carriage.

Reeve lines are in addition to, not instead of, tag-lines. In this case, tag-lines are still used to move the line horizontally across the span, while the reeve is used solely for the purpose of raising and lowering the load—for example, to lower a rescuer to snatch a stranded subject mid-span. When a load is to be moved horizontally, the reeve line(s) should be secured to prevent them from moving. Likewise, when a load is to be moved vertically, the best control is gained with the tag-lines tied off.

Reeves can be difficult and time consuming to rig and operate, and should be used only after extensive practice.

Skate Block

The **skate block** system provides an interesting alternative to the guiding-line. In this dynamic system, one continuous rope is used to form both the track-line and the lowering system. These systems are particularly useful where rescuers are positioned at ground level, with the subject(s) to be rescued being on a high point, within view of and with a clear path between rescuers and subject. The majority of the work of the operation takes place at ground level, with just one or two rescuers required to ascend to the height of the subject(s) to secure a high anchorage and connect the subject.

A single-line skate block system can be built with a minimal amount of equipment, which at a minimum must include the following:

1. A rope whose length is at least three times the height that the subject is on the structure
2. Anchorage connectors and materials for rigging above the subject

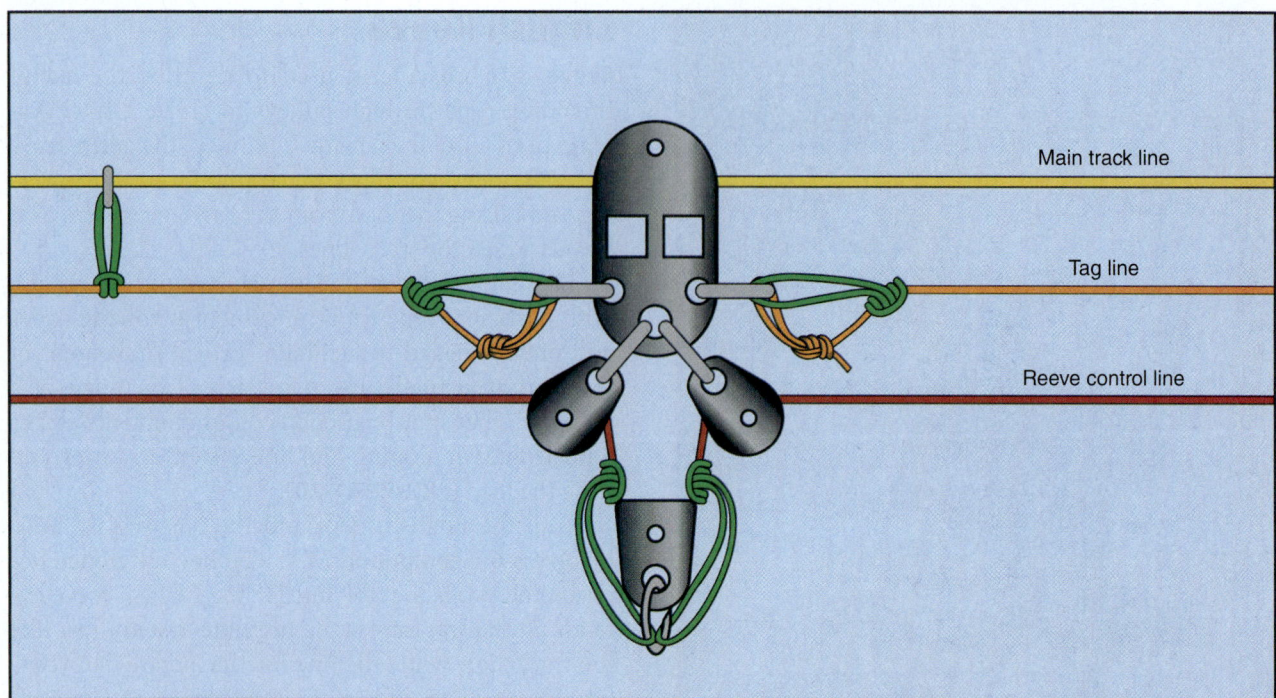

FIGURE 17-20 English reeve.

3. One pulley for rigging at the high anchorage above the subject
4. A ground anchorage (and materials for rigging) that is about as far away from the structure as the subject is high on the structure
5. An autolocking braking device that will function in both a lowering configuration and as a progress capture for an inline haul system
6. A pulley and rope grab to make an inline Z-rig
7. A pickoff strap and track-line pulley for connecting the subject to the track-line

To create a skate block system, one rescuer proceeds to the location of the subject, trailing one end of the rescue rope that will become the mainline (**FIGURE 17-21**). After establishing a high anchor point and pulley at least 6 feet (1.8 m) above the subject, the mainline is passed through the high pulley, and the terminated end of the mainline is attached to the subject. If the subject is wearing a full-body harness, attaching this line to their waist attachment is preferred.

Meanwhile, the ground team establishes an anchorage at a distance about as far away from the subject as the subject is high (a truck is often a good choice of anchorage here) and attaches an autolocking braking device, through which the free end of the mainline is reeved.

At the command of the rescuer on the structure, all available slack is removed from the mainline and the

FIGURE 17-21 The top rescuer runs the mainline through a high anchor point with pulley and connects the terminated end to the subject.

subject is raised slightly, either by means of an inline haul system or by vectoring the mainline.

The rescuer on the structure attaches a pulley with pickoff strap (i.e., skate block) onto the angled track-line just below the anchor point. When the subject has been raised high enough to remove them from their perch and to within reach of the track-line, the rescuer connects the pickoff strap to the subject's harness (sternal connection, if available.) On command, the brakeman at the ground anchor point slowly feeds rope through the braking device. As the load descends, their connection to the angled track-line simultaneously moves them away from the structure.

Forces in this method are self-limiting, with tension in the line corresponding directly to the applied load. Regardless of the weight of the load, length of the lower, or angle of skate, the subject will typically reach ground level at a point about one-third of the way between the ground system and the structure.

Opposing Systems

One other way of achieving horizontal movement with a rescue load is to create multiple haul systems that act in opposition to one another. In the high-angle safety and rescue world, this method is sometimes called a cross-haul.

The matter of balance is always at play when using opposing systems, and ensuring that operation of the systems is well-coordinated to achieve the desired result is key. All systems used should be capable of both raising and lowering, and should be easily transitioned between those two actions. Each should also incorporate backup/belay as appropriate to protect against failure in the anticipated direction(s) of pull.

Moving a load with opposing systems requires good cooperation and communication among team members. With practice, a team should be able to utilize two or more systems in opposition to one another to effectively raise a load from one location, move it to another location (perhaps even through obstacles) and then set it back down at a predetermined location.

After-Action REVIEW

IN SUMMARY

- Horizontal systems may be used to transport rescuers, rescue subjects, and equipment across an area that presents a barrier to customary rescue operations.
- Fixed directional systems, also called highlines, are constructed of a rope line suspended between two points to create a track-line, across which a load may be transferred.
- Dynamic directional systems are those in which moving systems are used to move a load in the horizontal plane.
- Drawbacks on highlines include stress on the equipment, lengthy setup, and difficulty getting initial personnel and rope across.
- Dynamic directional systems can be complex to rig and operate, and can impart challenging directional forces on a system.
- In a fixed directional system, the track-line supports the majority of the weight of the load.
- The near-side anchor system in a directional system is ordinarily the point from which operation of the highline is initiated and controlled.
- The far-side anchor system is the point to which the track-line rope is attached opposite the near-side. This anchor system must be at least as strong as the near-side anchor system.
- The load is the mass, including personnel and equipment, that will be riding on the highline at any given time.
- With higher loads, tandem pulleys are preferable to single pulleys because tandem pulleys create less of a bend in the rope, and they spread the load along the rope.
- A tag-line may be used simply for the purpose of maintaining contact with the load, or it may double as a lowering line, haul line, or belay.
- Reeves, cross-hauls, and skate blocks are all different types of dynamic directional systems.
- Essential elements of a one-person load are a support sling that runs up the rigging plate under the pulley and a rigging plate clipped into the pulley with a large locking carabiner.
- A critical factor in the rigging of a rope for a highline is determining the proper amount of sag in the track-line rope.
- According to the 10 percent rule, the center of the unloaded highline should sag a vertical distance of about 10 percent of the span for every 200 pounds (90.7 kg) of expected load and every 100 feet (30.5 m) of span in the rope.

- Anchorages are the foundation of any horizontal system.
- If the highline cannot be rigged well above and clear of the edge using existing natural anchorages for high directional pulleys, an artificial high directional may be considered.
- Guiding-lines can be used to clear obstacles such as trees, rocks, or debris.
- A reeve system is a particular type of arrangement involving ropes and pulleys rigged so as to facilitate movement of the load.
- Highlines may be rigged with two track-lines to help provide a greater margin of safety for heavy loads or taut highlines.

KEY TERMS

A-frame Another name for a bi-pod (also called a sideways A-frame).

Carriage system A traveling pulley used on a track-line to enable a load to travel with minimal friction.

Cross haul Multiple haul systems that act in opposition to one another to achieve horizontal movement.

Dynamic directional system A type of track-line that involves two or more systems applying force in opposition to one another.

English reeve One of several rigging systems that can be added to a highline to control the load from either side.

Footprint The area between the imaginary lines between the legs of a tripod that form a triangle.

Festoons Short cordage hangars used along a highline to suspend and control the droop of control lines beneath the track-line.

Flying W tensioning system Device used to pre-tension two track-lines as one.

Gin pole A monopod.

Guiding-line A low- to medium-tensioned line used to clear obstacles such as trees, rocks, or debris. Often has variable tension; the tension is tightened just enough to clear the obstacle and then released when the obstacle is cleared. Also known as a sloping highlines.

Haul line A type of tag-line that the haul team uses for pulling on a load; with highlines, it is used to transport the load across the highline.

Highline A system of using a rope suspended between two points to move people or equipment over an area that is a barrier to a rescue operation. Also known as a *telpher* or *tyrolean*.

Highline pulley A special large pulley used for the travelling pullet on a track-line, also called *Kootenay carriage* or *knot-passing pulley*.

Horizontal highline A highline in which the two suspension points are nearly on the same level.

Kootenay carriage A modified knot-passing pulley that is used on highlines as the track-line pulley for single and double track-line systems or as a high-strength tie-off.

Lowering/belay line The line attached to a load on a highline that is used to control the load from the near-side point.

Norwegian reeve One of several rigging systems that can be added to a highline to control the load; in this case, control occurs from just one side of the highline, allowing rescuers to raise or lower a load from the carrier at any position along the full length of the highline.

Reeve A broad term used to describe the act of threading rope through pulleys to create a haul system.

Skate block A rescue system in which one continuous rope is used to form both the track-line and the lowering system for a load.

Static system safety factor (SSSF) The ratio of the overall rope system strength divided by the anticipated load when it is not moving. It is based on the weakest link on the system, not the strongest.

Steep-angle highline A highline in which one of the suspension points is considerably higher than the other.

Tag-line A line attached to the load on a highline that is used to maintain contact or control of the load from either side.

Tandem pulleys Two in-line pulleys on the same rope. Used in highlines to stabilize a load and to distribute its weight along the rope.

Track-line The line (or lines) that provide the support of the load

REFERENCES

Attaway, Stephen, Beverly, J. Marc, Mills, Mike, Weber, Erin, Belknap, Evan, and Attaway, Nancy. "Track-lines and Guiding-lines: Forces Based on Point Loads in a Catenary." Presentation at the International Technical Rescue Symposium, Albuquerque, NM, November 8-10, 2013.

Delaney, Richard. *Physics for Roping Technicians.* 2018. RopeLab. Blue Mountain, Australia. https://rigginglabacademy.com/wp-content/uploads/2018/07/Physics_For_Roping_Technicians.pdf. Accessed November 3, 2021.

On Scene

1. When might you choose to use a guiding-line instead of a highline?

2. Using the 10 percent loaded line sag rule, how much force would be applied to a highline anchor point if a 125-pound (56.7-kg) mass is applied to the center of a highline with a 100 foot (30 m) span and 12.5 ft (3.8 kg) of sag?

3. When using a 3:1 mechanical advantage system to tension a highline, what is the maximum number of rescuers that can haul on the system and still stay within the 12:1 loaded line rule?

4. How would you go about keeping the resultant inside the footprint of an AHD used at the edge of a drop?

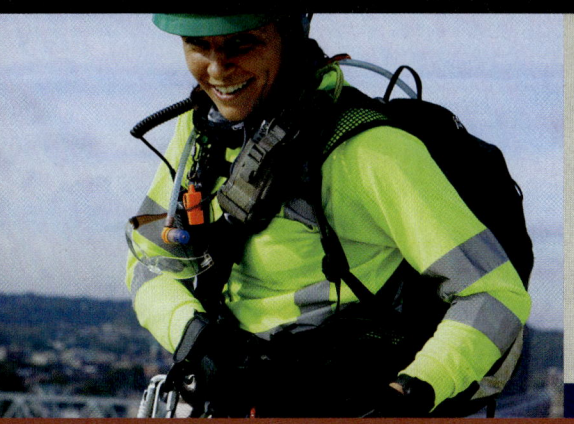

CHAPTER 18

Technician

Personal Vertical Skills

KNOWLEDGE OBJECTIVES

After studying this chapter, you should be able to:

- Describe the mechanics of rappelling to descend a rope. (**NFPA 1006: 5.3.10**, pp. 372–373)
- Describe the techniques of descending a rope. (**NFPA 1006: 5.3.10**, pp. 373–384)
- Describe how to set up a descending rope system. (**NFPA 1006: 5.3.10**, pp. 373–374)
- Describe how to initiate descent. (**NFPA 1006: 5.3.10**, pp. 375–376)
- Explain the methods of descending over edges. (**NFPA 1006: 5.3.10**, p. 376)
- Describe the process of setting up and using a brake bar rack. (**NFPA 1006: 5.3.9, 5.3.10**, pp. 378–379)
- Identify the methods of locking off on a rope. (**NFPA 1006: 5.3.9, 5.3.10**, pp. 382–384)
- Describe the methods of ascending a rope. (**NFPA 1006: 5.3.9**, pp. 384–385)
- Explain the considerations for selecting an ascender. (**NFPA 1006: 5.3.9**, pp. 385–388)
- Describe how to create a Prusik ascending system. (**NFPA 1006: 5.3.9**, pp. 385–388)
- Explain the use of mechanical ascenders. (**NFPA 1006: 5.3.9**, p. 388)
- Describe how to construct an ascending system. (**NFPA 1006: 5.3.9**, pp. 389–393)
- Describe the mechanics of ascending a rope. (**NFPA 1006: 5.3.7, 5.3.9**, pp. 391–406)
- Describe the safety considerations for ascenders. (**NFPA 1006: 5.3.9**, pp. 384–389)
- Explain the techniques for working while suspended on a rope. (**NFPA 1006: 5.3.9, 5.3.10**, pp. 391–406)
- Describe the methods of maneuvering between ropes. (**NFPA 1006: 5.3.9, 5.3.10, 5.3.11**, pp. 386–388)
- Describe the methods of maneuvering around obstacles on a rope. (**NFPA 1006: 5.3.9, 5.3.10, 5.3.11**, pp. 406–407)
- Explain the methods of self-belay. (**NFPA 1006: 5.3.9, 5.3.10, 5.3.11**, pp. 406–407)
- Describe the steps that can be taken to self-rescue when stuck on rope. (**NFPA 1006: 5.3.11**, p. 407)

SKILL OBJECTIVES

After studying this chapter, you should be able to:

- Throw a rope. (p. 374)
- Practice rapelling. (**NFPA 1006: 5.3.10**, p. 377)
- Thread a brake bar rack and rappel. (**NFPA 1006: 5.3.10**, pp. 378–381)
- Utilize a lock off during a descent and lock off a brake bar rack. (**NFPA 1006: 5.3.10**, pp. 382–384)

- Release a locked-off descender and release a brake bar rack during descent. (**NFPA 1006: 5.3.10**, pp. 383–384)
- Attach a Prusik to a host rope and utilize Prusiks as ascenders. (**NFPA 1006: 5.3.9**, pp. 386–388)
- Use a mechanical ascender. (**NFPA 1006: 5.3.9**, p. 414)
- Rig an ascending rope rescue system. (**NFPA 1006: 5.3.9**, pp. 391–392)
- Perform edge transitions, including the kneeling transition method. (**NFPA 1006: 5.3.9, 5.3.10**, pp. 395–396)
- Perform a changeover. (**NFPA 1006: 5.3.9, 5.3.10**, pp. 398–404)
- Negotiate around obstacles while on a rope. (**NFPA 1006: 5.3.7, 5.3.9, 5.3.10**, pp. 404–406)
- Use self-rescue techniques to address jammed or otherwise disabled equipment. (**NFPA 1006: 5.3.11**, pp. 406–412)

You Are the Rescuer

You are at the scene of a high-angle rope rescue, at the bottom of a cliff where three individuals riding a large all-terrain vehicle (ATV) went over the edge in the dark. The ATV fell some 50 feet (15.2 m) down. One victim is deceased, one has significant torso injuries, and the third has a simple leg fracture. The two injured patients have been hauled to the top and transported, while the body of the deceased subject has been hauled to the top. Now, you and the other rescuers must get to the top. Since it is a sheer cliff and the only walk-up access is some 1.75 miles (2.8 km) away, your only option is to ascend on the fixed ropes that are still hanging. Since you were on the initial rigging team at the top, you know that all the hanging ropes were well anchored and padded at the edge.

1. What equipment will you need?
2. How will you attach your ascending system to your body?
3. How will you get over the edge to the top of the cliff?

Access Navigate for more practice activities.

Introduction

In accordance with the National Fire Protection Association (NFPA) Professional Qualifications for technical rope rescue, a rescuer at the technician level should be capable of performing a range of vertical skills in the high-angle environment to help enable them to reach a subject and/or perform rescue skills while suspended. At this level, a rescuer may be required to move a subject from an unstable (mid-air, etc.) environment to a stable one. While some of this information may be useful to the operations-level rescuer who is being lowered or raised from one stable environment to another to gain access to a subject, it is generally not advised for rescuers to control their own ascent or descent, or to serve as a litter attendant in the vertical environment, until they are well-versed and skilled at the technician level. Personal vertical skills require practice and can take years to master.

According to NFPA 1006, *Standard for Professional Technical Rescuer Qualifications*, a technician-level rescuer must be able to descend a fixed rope while attached in a manner that will not allow them to fall and be able to stop at any point on the fixed rope system to work or to rest. Likewise, they must be able to ascend a fixed rope system with at least two points of contact, possess the ability to stop to work or rest on rope, and be able to convert from ascent to descent and back again (**FIGURE 18-1**). Any rescuer who intends to control their own direction and speed while ascending or descending rope, or who will serve as a litter attendant for a vertical evacuation, should be capable of maneuvering comfortably in virtually any direction while on rope to optimize safety and efficiency, and for self-rescue. This may include ascending, descending, passing obstacles, or even transitioning from one rope to another.

The question of whether to require a complete secondary system for belay, or backup, for personal vertical systems (i.e., ascending, descending) is answered at the discretion of the authority having jurisdiction (AHJ). This determination should be made

FIGURE 18-1 A technician-level rescuer must be able to ascend a fixed rope system with at least two points of contact, to stop to work or rest on rope, and to convert from ascent to descent and back again.

after weighing the probability and consequence of failure. The higher the risk of failure, either of equipment or rescuer, and/or the higher the consequence in the event of such failure, the more prudent it will be to incorporate a secondary system for backup protection. That said, secondary systems also necessarily involve more equipment and complexity, which in themselves can add further risk, so these additional resources should be justified. It is not uncommon for skilled rescuers to safely and appropriately use single rope technique (SRT) in a rescue environment. Belay methods are discussed in detail in Chapter 12, *Belay Operations*.

To allow for variations in how any given AHJ approaches the question of belay, the techniques described in this chapter do not include specific instructions for backup, or belay. If a secondary system is preferred, whether a second line with manual belay or a rope access backup, refer to appropriate instructional materials for use of those systems and equipment.

Single Rope Technique (SRT)

The term **single rope technique (SRT)** describes an approach to vertical ropework in which one rope is used for ascending and descending, without a second rope for backup safety or belay. This is not to be confused with the term *stationary rope technique*, which itself is a term coined by arborists (workers who care for trees) as an alternative to the term fixed rope technique. A fixed rope (or stationary rope depending on your terminology preferences) may be used for either SRT or in a dual rope system. The reference to fixed or stationary simply means that the rope is anchored at one end rather than dynamically operated, as would be the case with a lowering or raising system.

NFPA 1006 requires that rope rescuers maintain two points of contact with their system when ascending, and that they be connected in a manner that prevents accidental disconnection when descending. This is not to be confused with the idea that there must be two separate ropes, a concept that is not mandated by NFPA. The standard merely says that the system must be set up to prevent disengagement, and that two points of contact must be maintained when ascending. As we will learn later in this chapter, when ascending there are, by design, usually already two points of contact on the rope. In the case of descending, simply maintaining a rope grab or Prusik on the rope either above or below the braking device can provide additional security when using SRT. Of course, especially when only a single rope system is used, extraordinary care must be taken to protect the rope from potential damage and to nullify the probability of failure.

Descending

Descending, which is simply the act of using friction to control the speed at which rope rescuers slide down a rope, is a necessary skill for operating safely in the vertical environment. This is the first step toward developing vertical competency.

Known colloquially as rappelling, this may appear to the inexperienced to be a spectacular act. However, the ability to rappel should not be taken as indication that a person is a skilled and knowledgeable technician-level rescuer. This elemental personal skill is merely one aspect of rope work and must be used in combination with other skills to be useful in technical-level rescue operations. In fact, even the term rappelling is most commonly used to refer to recreational activity, while the term descending is the preferred term in professional application.

All things being equal, and if sufficiently trained personnel are available, being lowered is often a better choice than descending because it allows the rescuer to maintain hands-free ability to work with the subject while all of the lifting and lowering is done by other team members. See Chapter 8, *Rescue Equipment* for a detailed discussion on rope systems equipment, including harnesses, connectors, and different types of rappelling devices.

> **TIP**
>
> Before attempting the activities described in this chapter, rescuers must first demonstrate the ability to properly:
> 1. Select an appropriate life safety rope and related equipment for a defined purpose (Chapter 9, *Ropes, Knots, Bends, and Hitches*)
> 2. Use and care for rope and other high-angle equipment (Chapter 8, *Rescue Equipment*)
> 3. Tie correctly and without hesitation the knots necessary for safe, effective work in the vertical environment (Chapter 9, *Ropes, Knots, Bends, and Hitches*)
> 4. Apply the principles of anchoring and rig a secure anchor system (Chapter 11, *Anchorages*)
> 5. Belay a load (Chapter 12, *Fall Protection and Belay Operations*)

Importance of Control

When it does become necessary to use descending methods for access into or egress out of a rescue site, control is paramount. Often, the very idea of rappelling conjures up visions of outdoor enthusiasts rapidly bouncing down rock faces, swinging wildly in the air and yelling. This is nothing like what the professional rescuer will do. An important sign of a person's competence in descending is control. Evidence of control includes the ability to control the descent with minimal physical effort, without bouncing or shock loading the system, and to manage speed effectively to reduce heat buildup between the rope and descender.

The rescuer should be able to stop at any point on the rope with complete confidence and control and tie off or otherwise secure the braking device so that they can use their hands for managing a subject or for other work. Rescuers should also be capable of controlling their descent and continuing to function effectively even if their body should become oriented in an unusual way, even to the point of inversion, as a result of terrain or operational conditions.

How Descending Works

The basic idea behind any descending method involves the use of friction against a host rope to slow the rate of descent of an attached mass. At its crudest level, the body itself can be used as a friction device on the rope, as with the **body rappel (Dulfersitz rappel)** and the **arm rappel (guide's rappel)**, discussed later in this chapter. Especially in very steep conditions, these techniques can be unpleasant to use due to extreme compression as well as frictional heat buildup during descent.

Most modern techniques involve the use of braking devices, also called **descenders**, which are attached to the rappeller by means of a carabiner attached to their harness. The rope runs through the device to create friction so that the heat and pressure are directed into the device and not to the rappeller's body. A well-designed rappel device also offers more control of the descent than does a body rappel.

With most rappel devices, the rate of descent is controlled by pulling tension on the part of the rope below the device. This increases friction by increasing rope tension and pressure on metal parts of the device. The rescuer who is descending usually exerts this controlling action with the dominant hand (the right hand for a right-handed person). This hand is known as the **brake hand** (**FIGURE 18-2**).

The rescuer's other hand cradles the rope above the device or (with some descenders) the device itself. This hand does not support body weight by grasping the rope. With most descenders, this hand is known as the **guide hand**, which helps balance the person on rope. The guide hand can also be used to help the person maneuver around irregular and hazardous areas on a structure or rock face. In some rappel devices, it may also help control the descender.

FIGURE 18-2 Using the dominant hand as a brake.

> **TIP**
>
> Anyone learning to descend must maintain absolute control of the descent and must ensure that provision is made to protect against a catastrophic fall. Among the ways of ensuring this control are the following:
>
> - Avoid rapid, bouncing descents that can lead to loss of control, damaged rope, overloaded anchorages, and injuries.
> - Use a separate belay, as determined by the authority having jurisdiction (AHJ).
> - Learn under the guidance of a qualified, experienced instructor.
>
> When descending, make sure your rope is long enough for you to reach a safe location at the end of the rope.

Descending Techniques

With few exceptions, recreational rappelling methods are generally not suited for professional rescue. Equipment and techniques used for rescue operations should be purpose-designed, and a competent person should ensure that components are compatible, systems are rigged appropriately, and personnel are adequately trained.

At a minimum, a rescuer will require a rope, anchorage, harness, descender, and connectors for descending (**FIGURE 18-3**). A pair of sturdy rope gloves and a helmet are also advisable. Regardless of which descender is used, methods for descending are quite consistent. Attaching the descender to the waist attachment of the harness will help the rescuer to maintain an ergonomic stance. Attaching the descender to the sternal attachment causes the body to want to hang like a sack of potatoes, with the center of gravity too far below the attachment to the rope to facilitate control; this can also result in the descender extending out of reach. A dorsal or rear attachment is also not recommended as this places the device out of visual range and away from safe operation. Keeping the device within reach is essential for safety.

Setting Up the System

Any adequate life safety rope may be used for descending. A static rope will offer greater stability and less creep, especially for heavy or varying loads, and while a dynamic rope offers more force absorption and forgiveness in the event of a fall, it also results in a bouncier ride. Especially when using rope with a lot of stretch, be sure to protect sharp edges and keep an eye on contact points with the structure, as repeated rubbing can abrade or damage the rope.

The anchor system should be set far enough back from the edge to permit adequate working space. While anchoring low can attain more strength (especially from natural anchorages), an anchor point that is at least waist high offers greater stability for the person on rope as they transition over the edge.

Another alternative is to anchor low, and then place the rope through a high directional near the edge. If a separate belay or backup line is to be used, it should be anchored separately from the mainline for optimum redundancy. The anchor point should be set in line with the desired fall line, or intended rappel path: near the subject, but not so close that the rope might strike the subject, nor where they can reach out and grab the rope as it hangs.

Either an autolocking or a free-running manual descender may be used for descending. Regardless of which descender is used, it should be used fully in accordance with manufacturer's instructions. It is generally best for the rescuer to place the device on the rope while facing the anchor point, so that their back is to the drop they will descend. This helps to ensure that the device is oriented correctly on the rope. The rope rescuer should familiarize themselves with the device first by reading product instructions, then by practicing putting the device on and off the rope, locking it off, and releasing it on a flat or low-angle surface before using it for steep or vertical rappel (**FIGURE 18-4**). Types of descenders are discussed in Chapters 7, *Hazard Specific Personal Protective Equipment* and 8, *Rescue Equipment*.

Connectors used in the anchor system and at the rappel device should be unquestionably secure, with an adequate factor of safety and of appropriate size and proportion for the intended load and direction of pull. Size and proportions should complement the

FIGURE 18-3 At a minimum, a rescuer will require a rope, anchorage point, harness, descender, and connectors for descending.

FIGURE 18-4 Practice with equipment on level ground before moving to a vertical environment.
Courtesy of Steve Hudson.

task to which it is assigned. Bigger is not always better, and aluminum may be a better choice than steel in many cases; matching the connector material with the material to which it is attached is considered good practice. Care should be taken in rigging to prevent cross-loading or three-way loading, as these can result in damage or failure at loads below the rated strength. Use of autolocking connectors at critical points, such as at the anchor point and at the harness, can also help prevent failure.

The rescuer must first decide whether to drop the free end of the rope down the rappel route prior to descent, or to carry it with them while descending. There are advantages and disadvantages to each approach. If the rope will be thrown down the slope before descent, first anchor the top end of the rope to an appropriate anchor point then stack the excess rope on the ground next to the edge with the end of the rope to be thrown coming off the top of the stack. A stopper knot should be tied into the furthest part of the free end of the rope to help prevent the rescuer from rappelling off the end of the rope in the event that it is shorter than the drop. A barrel knot, in-line figure 8, or figure 8 on a bight works well for this purpose.

Throwing the Rope

Rope should be removed from its coil or bag prior to throwing and flaked or stacked in a location that will be conducive to paying it out over the edge. Removing the rope from its bag or coil before throwing affords the rescuer an opportunity to quickly inspect the rope to ensure that it is in ready condition and that it has not been damaged during storage (for example, by chemicals or rodents.) Removing the rope from the bag also facilitates a cleaner throw.

Throwing a bagged rope increases the probability that it will become tangled, jammed, or caught on something on the way down—thereby creating a potential hazard.

The person throwing the rope should stand beside the stacked rope, near enough for a clean throw but far enough to ensure that they will not be caught by any of the loops; they should be secured to a nearby anchorage for safety. With six to eight loops of rope coiled loosely in each hand, yell "Rope" loudly. Pause briefly (to give people who are in earshot time to get clear), holler, "Rope," again, and throw—first the loops from the throwing hand, and then tossing the other loops after, and finally letting any remaining rope feed from the stack on the ground. To throw a rope, follow the steps in **SKILL DRILL 18-1**:

1. Anchor the rope to a secure anchor point in line with the intended fall line.
2. Stack the rope near the edge, with the end to be thrown coming off the top of the slack. Stand beside the stacked rope, being sure to not get tangled in it.
3. Clip a lanyard from your harness to a secure anchorage.
   Pick up the end of the rope and tie a stopper knot.
4. Starting at this end, lightly coil six to eight loops of rope into your throwing hand.
5. Take six to eight additional loops of rope in your other hand.
6. Yell "Rope!" and count to five. Yell "Rope!" again, and throw the loops from your primary throwing hand, tossing the loops from your other hand immediately thereafter.

TIP

Under some circumstances, rescuers may want to consider NOT throwing a rope down before them, such as the following:

- In tactical operations in which a hostile person below the rappeller could grab the rope (and thereby control the rappeller)
- When a very frightened and unpredictable person is below the rappeller
- When unstable rocks could be knocked loose by the rope and fall on people below
- When the rappeller wants to prevent the rope below them from being damaged or cut by falling debris and rock
- When the rope could become tangled or jammed, and the rappeller could not retrieve it

While dropping the free end of the rope down over the edge before descending may at first seem the easiest approach, there are potential hazards associated with this method. For starters, even when best practices are employed, there could be a risk that the rope will become snagged or tangled on itself or in a protrusion on the way down. Clearing entanglements can be challenging and risky for the rope rescuer and can slow the operation down. Also, especially in the case of a long drop, the weight of a significant amount of rope hanging below can create excessive tension on the braking side of the descender, making descent difficult or even impossible. In such cases, it is possible to descend only by lifting the rope so that it does not impart so great an effect on the descender.

Descending with a Bagged Rope

An alternative solution to throwing the rope is to carry the free end of the rope with you in a bag, feeding it out as you descend. Throwing a bagged rope (while still in the bag) is not recommended. If the rope does not feed cleanly from the bag as it falls, it could jam, become tangled, and even impart an undesirable shock load to the anchor system when the mass stops suddenly.

While the bagged rope may be heavy, the bag can be connected to a utility ring on your harness or even to the same connector as your descender and/or seat board. This method helps to avoid entanglement, prevents the subject from grabbing the rope, and makes it easier for the rappeller to change direction of travel during descent (**FIGURE 18-5**).

Most any descender will work with either method, but keep in mind the effect that the method selected for rope management will have on friction. If descending with a bagged rope, controlling descent may be more difficult due to the increased load (weight of rescuer plus weight of the bagged rope). If the rope has been thrown down first, the rope weight will instead be hanging beneath the descender, imparting potentially significant friction on the device, even before employing a brake hand. This will result in a slower and more difficult descent near the top of the drop, increasing in speed as the rescuer nears the bottom. Either of these situations can create a hazard if it catches the rescuer unaware.

> **TIP**
>
> 1. It is difficult to estimate the length of a bagged rope accurately. When rappelling from a bagged rope, therefore, the danger exists that the rescuer could rappel off the end of the rope. Always either tie a stopper knot in the bottom end of a bagged rope or tie the rope end to the bag to avoid rappelling off the end of the line.
> 2. Rappelling with a bag attached to the body requires special care to prevent rope tangles that could jam in the descender. The rescuer also must be very careful that the bag does not jam in the descender. These are particular problems when rappelling from helicopters, a situation in which a jammed rappel device could leave a rescuer stranded on the rope hanging from the helicopter, possibly leading to severe injury or death.

Initiating Descent

In practice, a safety attendant or second rescuer should usually be stationed at the top of any slope or structure where someone is descending. While a rescuer is always primarily responsible for their own safety, the role of the safety attendant is to double check all aspects of the system and to monitor the system as it is loaded to help ensure safety. The safety attendant may also serve as a belayer if this is the method being used by the rescuer for backup. If an active belay is selected, communication between the safety attendant and/or belayer should be crisp, effective, and to the point. If voice communications are used, radios may be necessary to compensate for long distances or noisy environments.

When a rescuer is completely connected to their system, the rescuer should allow the safety attendant to double check their connections, and the system as a whole. If the rescuer is preparing to be belayed,

FIGURE 18-5 Descending with a bagged rope.

the following series of commands can be used to communicate:

Rescuer: On belay?

Belayer: Belay on.

Rescuer: Descending.

Belayer: OK.

Rescuer (*upon reaching the bottom and clearing the system*): Off belay.

Belayer: Belay off.

Getting over the edge is the most difficult part of any descent. In general, a sloping edge will be easier to negotiate than a sharp one, and an undercut edge will be especially difficult to get over. This is because the feet are integral to keeping the body away from the surface the rescuer is descending. With an undercut edge, there is no place for the rescuer to put their feet. To help facilitate a smooth edge negotiation, the rescuer should position the anchor point as high as possible in relationship to the edge (at least waist height, and preferably higher) and ensure that it is back away from the edge.

To descend over a 90-degree edge, the rescuer stands facing the anchor point with the feet shoulder width apart just at the edge of the drop. The rescuer tensions the descender and leans gently back into the system. Keeping the feet still, the rescuer feeds rope slowly through the descender so that they feel cradled by the seat portion of their harness. Keeping their feet still, the rescuer continues feeding rope through the descender as while bending at the waist, allowing the glutes to precede the feet until they are in a seated position, with their feet at about waist level. This is roughly the position that the rescuer will maintain throughout descent.

When transitioning over a very difficult edge, such as an undercut one, it may be necessary to descend until the rescuer's feet are above their harness attachment point, perhaps even as high as their head, before moving their feet. As a general rule, keeping the stance wide and legs high offers the greatest stability. Rescuers should not jump, bounce, bound, or use other uncontrolled methods.

The most important part of negotiating an edge, and of descending in general, is to avoid uncontrolled or shock loading of the system. For particularly difficult edges, or if the rescuer is not comfortable with a standing method, an edge may be negotiated by tensioning the descender, then sitting with the legs hanging over the precipice and climbing down until the rope is fully loaded (**FIGURE 18-6**). An etrier or footloop may be connected at the edge and used to down-climb until the system is tensioned. Of course, this is not the most graceful method, but it does help

FIGURE 18-6 Negotiating an edge from a seated position.

prevent falling and impact-loading the rope rescue systems. All of these methods take practice and skill. See Chapter 16, *Working in Suspension* for more on alternate methods of negotiating an edge.

Practicing Descending

Practice descending (rappelling) using the device preferred by your AHJ. First, practice the steps on a horizontal or low-angle surface, then try it on a vertical surface. Follow the steps in **SKILL DRILL 18-2**.

Rappelling with a Brake Bar Rack

Rappelling with a **brake bar rack** is somewhat different from rappelling with an autolocking device. While these should be used only by trained and experienced technician-level rescuers, there is nothing like a rack for heavy, variable loads or for long drops. If there is any possibility that you might be called upon to rappel with a brake bar rack, be sure to practice these skills first under the supervision of a competent trainer. Although autolocking devices are more in vogue these days, brake bar racks are covered here simply because they are still used by experienced technician-level rescuers. To thread the brake bar rack, see **SKILL DRILL 18-3**.

To rappel with the brake bar rack, follow the steps in **SKILL DRILL 18-4**.

Locking Off the Midline

When a rescuer descends, it is often for purposes other than just reaching Point B from Point A. Usually there is a task to achieve along the way, whether it is caring for a subject, setting rigging, or other tasks. In such cases, the rescuer will want to secure themselves to the mainline so that they can work hands-free, and not have to maintain a brake hand on the braking device. This is called **locking off**.

SKILL DRILL 18-2
Practice Rappelling

1. Don an appropriate harness with a waist or sternal attachment.

2. Connect your descender to the waist or sternal attachment with an appropriate connector.

3. Connect the standing end of the rope to an appropriate anchor point.

4. Thread the life safety rope through your descending device (see your specific device instructions for further information).

5. Verify that the descender is properly connected to your harness or evacuation seat using a locking carabiner.

6. Attach a belay, vertical lifeline, or other secondary backup system to your harness, in accordance with AHJ requirements. *NOTE: It is acceptable to use the same harness attachment as for your primary system.*

7. Position yourself at least 2 feet (61 cm) from the edge over which you intend to descend.

8. Facing the anchor point, pull the rope backward through your descender to remove any slack, creating tension between your harness and the anchor system.

9. If you intend to carry the rope with you as you descend, while still facing the anchor point attach the bag to your system as appropriate; if you intend to drop the free end of the rope down ahead of you, do so.

10. Check that the area of descent is clear (and, if applicable, that the entire length of rope is free from obstruction, tangling or hazards). Ask the safety officer or belayer to double check your system.

11. Once the safety officer or belayer has confirmed readiness of the system, if applicable ask, "On belay?" to your belayer, or announce to the safety officer, "Ready to descend."

12. On the affirmative reply from your belayer or safety officer ("Belay on," or "Descend" respectively) gently lean into the system so that your weight is supported by the rope, connected to the anchor system. At this point you should be leaning into the system but still on a safe surface, not yet over the edge.

13. Hold tension on the tail of the rope with your brake hand; if using an autolocking descender, activate the operating mechanism.

14. Relax tension on your brake hand until you are able to allow rope to travel through the descender.

15. Walk backwards toward, and then over, the edge, adjusting your hand tension to maintain full control of your rate of descent at all times. NEVER LET GO OF THE ROPE WITH YOUR BRAKE HAND!

16. Negotiating the edge is the most difficult part of descending. See "Edge Transitions" to review. As you negotiate the edge, take care to control the path of your rope as close to the fall line as possible. Lay the rope where you want it as you go and pad it as necessary. Once the loaded rope is pressed against the surface, it is very difficult to move and is subject to abrasion or damage.

(continues)

SKILL DRILL 18-2 (continued)
Practice Rappelling

17 Once over the edge, *sit* in your harness, legs extended forward near waist level. Keep your legs straight, but without locking your knees. A stance with feet shoulder width apart will offer good control.

18 Descend steadily by walking backward down the slope or structure. Maintain balance and control, taking care to avoid abrasion, sharp edges, falls, bouncing, or other hazards.

19 Once you have reached the bottom, unclip your harness connection from the descender, notify your belayer/safety officer that you are clear ("Off belay" or "Off descent," respectively) and move to a safe location.

SKILL DRILL 18-3
Threading a Brake Bar Rack

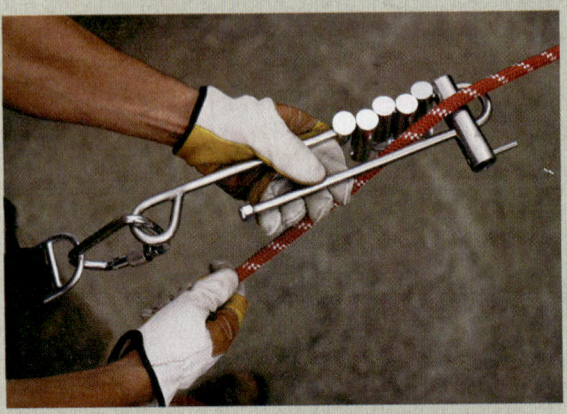

1 Attach a rappel rope securely to the anchor point. Clip the rack into your seat harness carabiner and lock the carabiner (with the gate toward your body).

NOTE: If your seat harness carabiner is in a horizontal plane, attach the rack to it with the short leg of the rack down. If you have a seat harness carabiner in a vertical position, attach the rack with the short leg toward the brake hand (e.g., the right hand on a right-handed person).

2 Stand facing the anchor point with the rappel rope on your brake hand side. Hold the rack out in front of you in your guide hand.

SKILL DRILL 18-3 (continued)
Threading a Brake Bar Rack

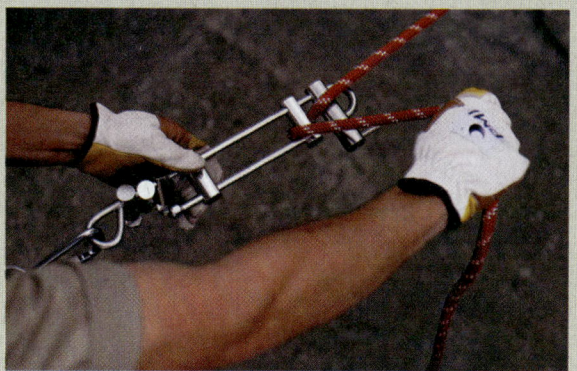

3 Disengage all bars except the top one on the rack.

NOTE: Do this by sliding them one at a time toward the bottom of the rack (toward the eye). Squeeze the two legs of the rack together with one hand and, with the other hand, flip back each bar.

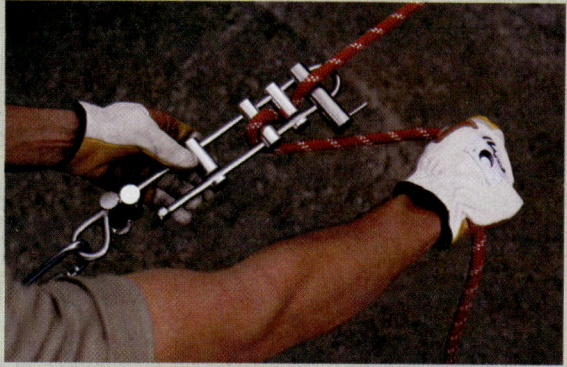

4 Pick up the rope with your brake hand. Guide the rope over the top bar, then down between the two legs of the rack, pulling out any slack.

NOTE: Do not pass the rope between the top bar and the bend of the rack. This results in pinching of the rope, making the descent harder to control and causing excessive wear on the rack.

5 With the other hand, clip the notch of the second bar to across the span of the rack near the bottom and slide it up to trap the rope between it and the top bar.

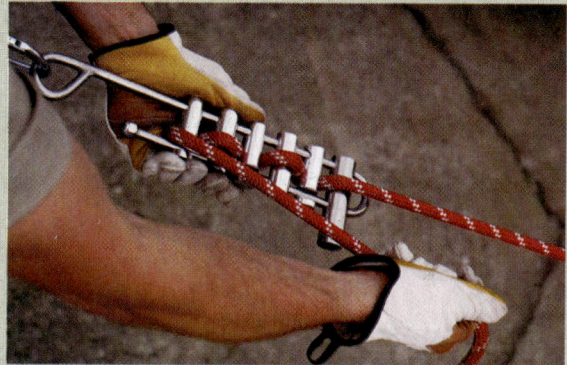

6 Pass the free end of the rope back between the legs of the rack and across the second bar and snug it into place by pulling the rope end upward, toward the anchor point.

NOTE: The rope must be on the side of the bar opposite the notch to hold the bar in place on the rack frame.

7 Repeat the process with the remainder of the bars until all six have been clipped in.

SKILL DRILL 18-4
Descending with a Brake Bar Rack

1 Attach the brake bar rack to a rope as described previously, with all six bars engaged, and ensure that it is properly clipped to your harness. Grasp the rope with your brake hand below the rack and off to the side. Rest your guide hand on the bars of the rack, holding the bar ends between the thumb and fingertips. Lean back into the system.

2 With your brake hand, pull the rope away from you (toward the anchor point), jamming the bars together at the top end of the rack. This is known as the quick-stop position. Bring the rope back to the normal rappel position. With your guide hand, grasp the bottom bar on either side of the rack and push it (along with the other bars) toward the top of the rack. This is the stop position for the guide hand. By jamming the bars together in this manner, toward the top of the rack, you increase the friction on the rope and add another element of control.

SKILL DRILL 18-4 (continued)
Descending with a Brake Bar Rack

3 Now use the guide hand to pull the bars one by one back toward you, spreading them out along the length of the rack. Meanwhile, ease your grip on the rope with the brake hand. As friction between the bars and the rope is reduced and the load on the device is increased, you may feel the rope begin to move through the rack. Use your brake hand to help control the movement.

4 If you have not moved, disengage the bottom bar; this is accomplished by lifting the tail end of the rope off the last bar, sliding that bar down toward the bottom end of the rack, and squeezing the legs of the rack together so that the bar may be unclipped. Let the bar slide down the rack toward the eye and out of the way.

5 Now spread the remaining bars apart along the length of the rack; this reduces the friction even more.

6 If you still have not moved, remove the fifth bar (which is now on the bottom) in the same way that you disengaged the sixth bar. Spread the remaining bars along the length of the rack.

7 To add friction, reverse the process described previously by clipping the bars back in to gain friction. Use your guide hand to squeeze the legs of the rack together and clip the bars in, one at a time, at the bottom, lacing the rope back between them. This is the same process used when you initially laced up the rack on the rope.

With autolocking descenders, stopping and locking off descent is relatively easy. When the activation handle of the device is released, it automatically jams itself on the rope. Most devices require some additional action for security; check the manufacturer's product instructions for specifics about each device.

For nonautolocking devices such as a brake rack, jamming the rope into a fail-safe position and then preventing it from coming loose will do the trick. Locking off midline with a brake bar rack is more challenging than with an autolocking device, so if this is a device you use, practice and become adept in all aspects of its use before employing it on an actual rescue. To lock off the descent, see **SKILL DRILL 18-5**.

When unlocking the descender, always keep a firm grip on the braking part of the rope, with no slack in

SKILL DRILL 18-5
Lock Off the Descent

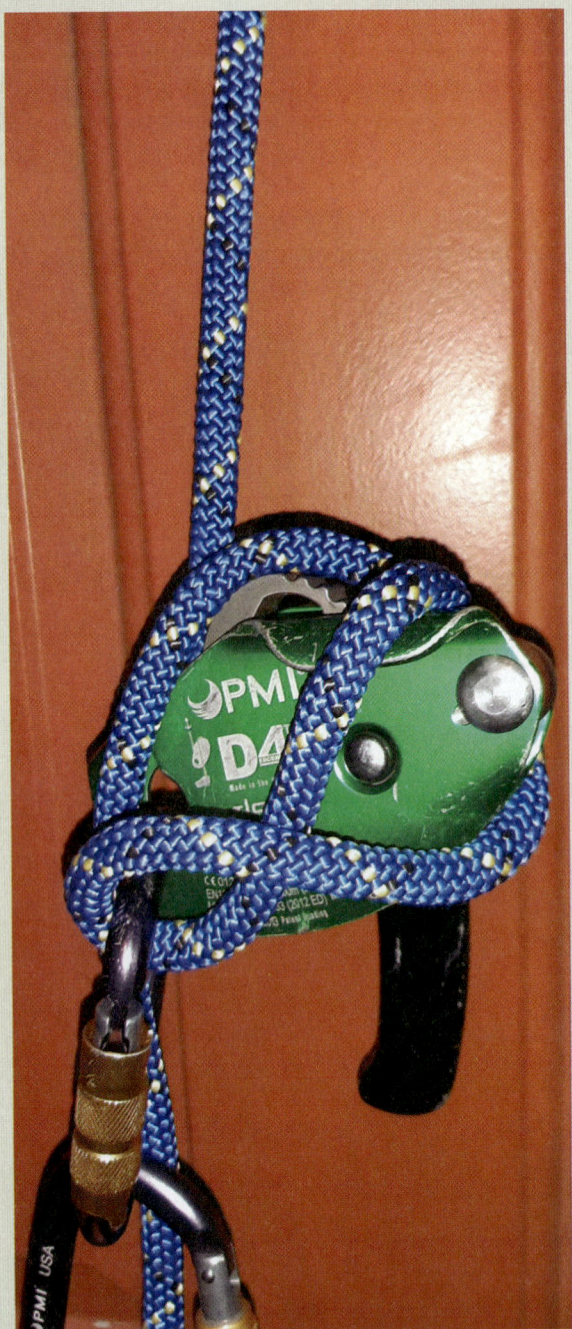

1 Begin descent using your autolocking descender. Stop your descent, activating the automatic lock-off for that device. For added security, use your brake hand to pass the trailing part of the rope between the device and the taut rope so that it is pinched in place there.

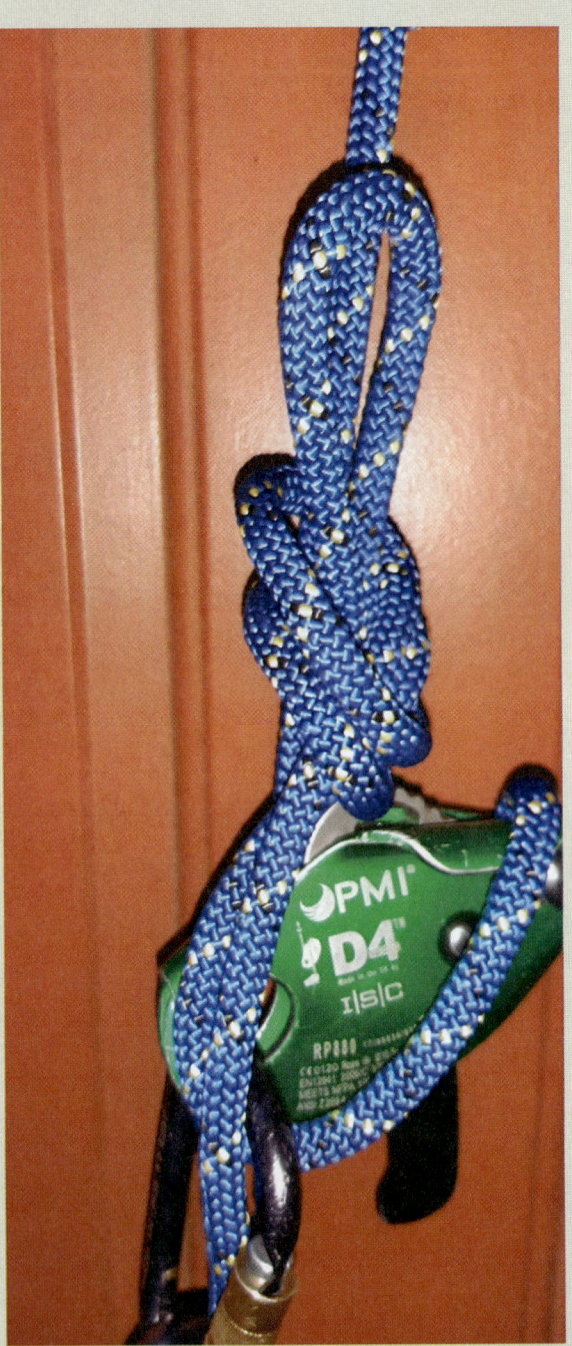

2 For even greater security, pass a bight of the trailing rope through the carabiner that connects the descender to your harness, and use the bight to tie an overhand knot around the tensioned host rope where it exits the top of the descender. Cinch the overhand knot firmly against the top of the device.

NOTE: There must be no slack in the tie-off, or else slippage could occur.

the rope between your hand and the device. To release a locked-off descender, follow the steps in **SKILL DRILL 18-6**:

1. To unlock, reverse the locking process. Untie the overhand knot, while maintaining tension on the rope with your brake hand.
2. Maintaining a firm grip on the braking part of the rope, pull the bight of rope back through your carabiner, releasing it so that you are holding just one strand.
3. Keeping tension with your brake hand, unravel the rope from between the tensioned line and the top of the device, returning to the normal rappel position. Resume normal descent.

To lock off a brake bar rack, see **SKILL DRILL 18-7**.

SKILL DRILL 18-7
Lock Off a Brake Bar Rack

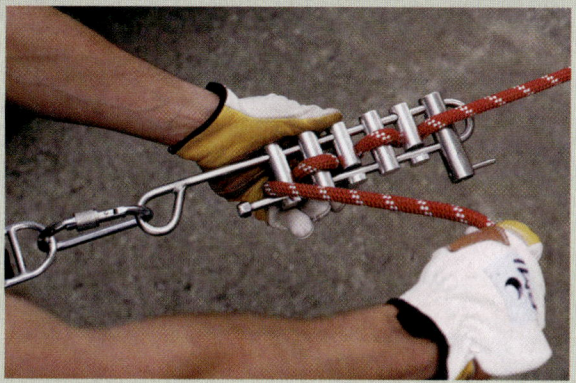

1. Begin descent using a brake bar rack with all six bars engaged. With the rope between rack and anchor system taut, pull the rope away from you toward the top of the rack, to engage the quick stop position.

2. With your brake hand, pull the rope over to the side of the rack and across the top hyper bar (if so equipped) between the rack frame and the pin at the end of the hyper bar so that the rope runs across the top bar. If your rack does not have a hyper bar, simply run the rope between the top of the rack and the tensioned line.

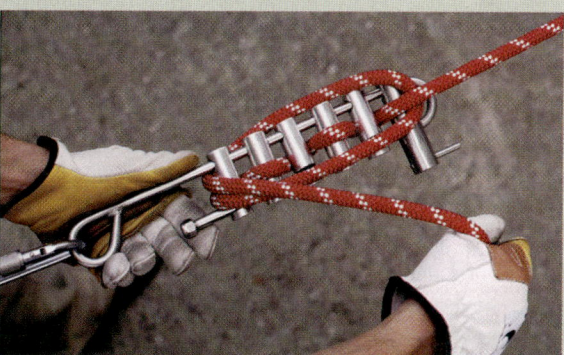

3. Pull the rope firmly in a downward direction, toward the eye, so that all the bars are jammed together.

4. Pass the trailing part of the rope through the open leg of the frame so that it runs between the two legs of the rack; pull it back in the direction of the anchor point so that it compresses against the bottom bar to help keep all the bars jammed together.

(continues)

SKILL DRILL 18-7 (continued)
Lock Off a Brake Bar Rack

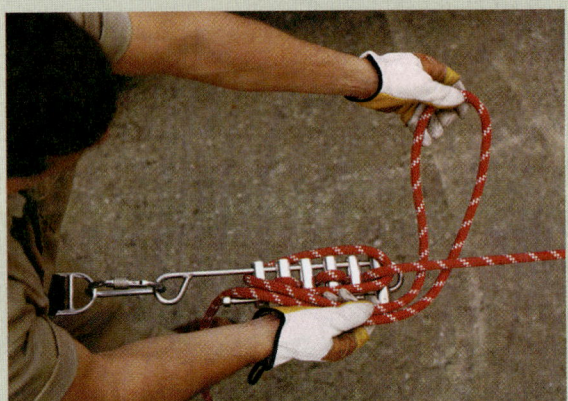

5 The rack should now be firmly in a *stop* position. To *lock* it into this mode, form a large bight in the trailing rope; use your guide hand to make the bight, with your brake hand at the apex, so that you can maintain a firm grip with your brake hand.

6 Treating this bight as one rope, use it to tie an overhand knot around the tensioned host rope where it exits the top of the rack. Cinch the overhand knot firmly against the top of the rack.

NOTE: There must be no slack in the rope running over the bar or any space between the bars.

To release a locked-off rack, follow the steps in **SKILL DRILL 18-8**:

1. To unlock, reverse the locking process. Untie the overhand knot, while maintaining tension on the rope with your brake hand.
2. Maintaining a firm grip on the braking part of the rope, release the bight so that you are holding just one strand.
3. Pass the trailing part of the rope back through the open leg of the rack so that it no longer runs between the legs.
4. Keeping tension with your brake hand, unravel the rope from the hyper bar (or from between the rope and the top of the rack and back to the normal rappel position.
5. Resume your guide hand's normal position of cradling the bars. If necessary, pull the bars apart with your guide hand until the decrease in friction allows you to rappel again.

When unlocking the rack, always keep a firm grip on the braking part of the rope, with no slack in the rope between your hand and the device.

Whatever method is used, the goal is to prevent any potential for descent to occur.

> **TIP**
>
> **Prerequisites to Performing the Skills in this Section**
> Before attempting the activities described in this section, you must have demonstrated that you can properly do the following:
> - Select, use, and care for rope
> - Select, use, and care for other equipment needed in the high-angle environment
> - Tie correctly, and without hesitation, the required knots described in Chapter 9, *Ropes, Knots, Hitches, and Bends*
> - Apply the principles of anchoring and rig a safe, secure anchor system

Ascending Basics

The opposite of descending is **ascending**. A person who descends a rope must also be capable of ascending that same rope, if for no other reason than to self-extricate in the event of an emergency. More than that, though, ascending affords the technician-level rescuer greater freedom to travel up and down rope effectively and efficiently to help achieve their mission. For maximum operational agility, a rescuer must be able to change over from descent to ascent, and back again, while on rope. This procedure, known as a **changeover**, requires thorough knowledge of necessary equipment and an innate use of skills that comes only through practice of the required techniques.

Rope rescuers ascend using friction hitches or mechanical devices called **ascenders**. When properly secured, these grip the rope firmly and can be made to slide up the rope but hold fast when pulled in a downward direction.

Choosing an Ascender

The most basic type of ascender is the friction hitch. There are several kinds of friction hitches, but the most commonly used type is the **Prusik hitch**. Every rescuer should be familiar with and capable of using a Prusik hitch, but these are not the most efficient choice for ascending—especially long distances.

Mechanical ascenders, which work by an offset camming action that presses the cam against the rope, are a more effective tool for ascending. The number and type of ascenders required will depend on the type of ascending system being utilized. In most cases, ascenders are attached to the climber's body by slings, which are connectors made of webbing or rope. The slings may be connected by various means to the seat harness, the chest harness, and the climber's feet.

The actual ascending process involves alternating actions on the part of the climber. First, the climber rests their weight on the first ascender as it grips the rope, keeping the body weight off the second ascender; the climber then moves the second ascender up the rope. Body weight is shifted to the second ascender and taken off the first ascender, which is then moved up the rope. The climber ascends the rope by repeating this cycle.

For most ascending activities, at least two ascenders are required. The use of three ascenders increases the margin of safety.

Ascending with Prusiks

Learning to ascend with Prusiks provides a good foundation for rescuers. Knowledge of this skill allows a rescuer to improvise an ascending system with available rope or cord, even when mechanical ascenders are not available. Improvisation of a friction hitch can be a lifesaver in an emergency situation.

FIGURE 18-7 Prusik hitch.
Courtesy of Steve Hudson.

A Prusik hitch works more efficiently when constructed of a rope or cord that is smaller in diameter than the host-rope to which it is attached (**FIGURE 18-7**). Choosing the optimum diameter differential will depend on the needs of the AHJ, specifically the overall strength required, the amount of slip preferred, and the desired ease of releasing. Fine tuning can be achieved by varying the number of wraps around the host rope.

Choosing a Cord for Prusiks

As a general rule for life safety rope diameters, the diameter of the Prusik cord should be at least 2.5 mm (0.1 in.) smaller than the diameter of the mainline rope, with optimum being around 4 mm (0.16 in.). The greater the range between Prusik and host rope, the more readily the Prusik will grab. If the two are very similar in diameter, the Prusik will slip rather than grip. If the range between diameters is very great, the Prusik will grip tightly without slipping, and will be more difficult to release.

With most life safety ropes being 10 mm (0.4 in.), 11 mm (0.43 in.), or 12.5 mm (0.5 in.), this usually works out to a 6- to 7-mm (0.24- to 0.28-in.) diameter Prusik for 10-mm (0.4-in.) rope, a 7- to 8-mm (0.28- to 0.31-in.) Prusik for an 11-mm (0.43-in.) rope, and an 8-mm (0.31-in.) Prusik for a 12.5-mm (0.5-in.) rope. Avoid extremes. The Prusik cord must be strong enough to support the intended load with a proper safety margin, but it should not be so large that it is hard to get the hitch to set.

There is more to cordage than just diameter. Material and construction lend to performance characteristics such as flexibility and elongation, which in turn affect the performance of the hitch. Age and wear are also factors. Performance of a Prusik is highly dependent on the relationship between itself and the host rope, so a given Prusik will not have the same performance when paired with a different rope.

Creating a Prusik Ascending System

At least two **Prusik loops** are required for ascending. One will be attached to the rescuer's harness waist attachment, the other will form a footloop to use as a step. Each can be made by forming a continuous loop by tying together the two ends of an approximately 6-foot (1.8-m) length of Prusik cord, using a grapevine (double fisherman's knot). The resulting loop will be about 2.75–3 feet (0.8–0.9 m) in length, depending on how much cord is used to form the knot.

TIP

Construction of Prusik cord is a compromise between two features:
1. A softer braid cord grips better but is more difficult to loosen and move up the rope; also, it wears out faster.
2. Harder braid cords are easier to loosen and move up the rope but do not set as well.

Establishing a Prusik ascending system begins with anchoring a rope at the top of the vertical or steep slope to be climbed. This forms the host rope. Attach both Prusik loops to the host rope, one above the other. To attach a Prusik to a host rope, follow the steps in **SKILL DRILL 18-9**.

The Prusik hitch may be used for a variety of purposes. For example, it can be used to tie onto a rope to hold rope pads in place. You should practice tying the

SKILL DRILL 18-9
Attach a Prusik Loop to a Host Rope

1 Hold the loop between your two hands, with the grapevine knot in your dominant hand. Hold the loop at about eye level and pass the grapevine knot behind and around the host rope, pulling it through the opening of the loop held in your non-dominant hand.

2 Repeat this action two more times, with each successive wrap lying inside the previous one, so that three wraps are formed around the host rope.

SKILL DRILL 18-9 (continued)
Attach a Prusik Loop to a Host Rope

3 Pull the hitch taut by pulling on the cord on the lower side of the grapevine knot; this will ensure that the grapevine knot is not at the apex of the bight.

4 Inspect the hitch to ensure that the coils of the Prusik hitch are even and parallel around the host rope, and when the loop is pulled downward, the grapevine knot is on one side or the other of the curve in the bight.

5 Test the hitch to ensure that it holds by grasping the end of the loop and pulling it downward; it should not move. Test the hitch to ensure that it can be slid upward on the rope. Grasp the coiled portion that is wrapped around the rope and use your thumb to loosen the cross-piece. Then, with the same hand, slide the hitch upward.

Courtesy of Steve Hudson.

Prusik under a variety of conditions, and you should be able to tie it with one hand. For ascending, the top Prusik should be connected to the waist attachment of the climber's seat harness using a carabiner. When the climber sits, the Prusik will hold firmly. When the climber releases tension by standing, the Prusik can be slid upward on the host rope. The lower Prusik will be used as a footloop.

> **TIP**
> Never press down on the top of a Prusik or other friction hitch unless you want it to slide down. In a life-safety situation, downward pressure could release the hitch, causing a fall, which could result in severe injury or death.

If the Prusik grips too aggressively on the rope, try taking out one of the wraps. A two-wrap Prusik hitch is adequate for most personal-use vertical applications, although the holding power of a three-wrap Prusik works well on rope that is new or slippery from mud or ice, or when greater weight is involved.

Ascending with a Prusik

Attach the upper Prusik to the climber's seat harness. The lower Prusik will be used as a footloop. If the Prusik is not long enough to suit the climber comfortably, a short loop of cord or webbing can be added with a girth hitch or carabiner, to extend its length.

This forms the most basic, foundational ascending system using Prusiks. Where a second rope for backup safety is not utilized, it may be prudent to

TIP

The dimensions of the Prusiks described in the previously described system must be fine-tuned to the preferences and ergonomics of the rescuer. One approach is to use a trial-and-error method of carefully tying, retying, and adjusting the Prusik lengths until they are just right. The other option to consider (especially for the footloop) is to tie a very short Prusik to grip the rope, and then add a webbing or cordage sling to achieve the desired length.

It is a good idea for rescuers to always carry a couple of Prusik loops that can be used quickly in unexpected situations such as to ascend short distances, as a self-belay at the edge of a drop, or for other unforeseen needs.

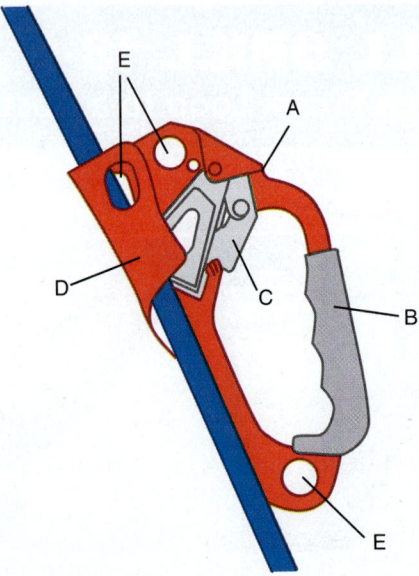

FIGURE 18-8 Typical technical-use handled ascender **A.** Frame. **B.** Handle. **C.** Safety lever. **D.** Nose. **E.** Tie-in points.

add a carabiner and short sling between the footloop and seat harness for extra security. To ascend using a Prusik ascending system, follow the steps in **SKILL DRILL 18-10**:

1. Attach the upper Prusik to your seat harness and sit down.
2. Raise the lower Prusik as high as practical so that you can step into its loop.
3. Stand up. Slide the upper Prusik as high as possible on the host rope. Sit down. Repeat these steps.

Ascending with Mechanical Ascenders

Mechanical ascenders are preferred by most technician-level rescuers as they are easier, more efficient, and more convenient than hitches. Some mechanical ascenders have handles, others do not. Although the parts of some models may not look exactly like the ones shown here, these devices all work in essentially the same way (**FIGURE 18-8**).

Parts of a Mechanical Ascender

The frame is the structure to which the parts are attached; it is the part of the ascender that mostly determines the device's strength. The frame may be made of extruded, stamped, or plate aluminum, or, in some cases, from cast aluminum. If the ascender has a handle, it may be an integral part of the frame, or it may be attached to the frame with rivets or bolts. In some designs, the handle is molded to fit the contour of the hand and can be comfortably used with gloves or mittens.

Most mechanical ascenders have a safety lever to help prevent the ascender from coming off the rope accidentally. The lever prevents full downward movement of the device's cam when it is in the locked position. The nose forms the inside channel into which the cam pushes the rope so that the ascender stays on the rope. Most ascenders have tie-in points at the bottom so that the climber can attach a sling to suspend themselves below the device. Some ascenders have additional tie-in points at the top. These are used in certain ascending systems in which the ascender is pulled along as the climber advances up the rope.

Handled ascenders may be of a right-handed or left-handed configuration. To determine whether an ascender is right-handed or left-handed, turn the ascender so that the opening for the rope between the nose and the cam is facing you. On a left-handed model, the handle will be to your left; on a right-handed model, the handle will be to your right.

Ascender Slings

To be used safely and efficiently, ascenders must be attached to the climber's body with **ascender slings**. These slings may be made of either webbing or rope. Many prefer rope for the following reasons:

- An appropriately designed rope may abrade less easily than webbing.
- Rope works better than webbing if the sling must go through a roller device (these are used in certain ascending systems).

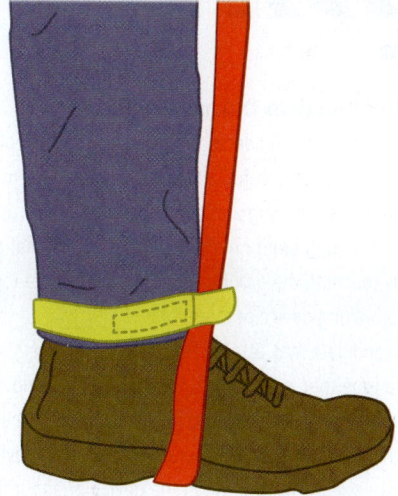

FIGURE 18-9 Chicken loop.

If you decide to construct your ascender slings from rope, you may be able to use line of a smaller diameter than that normally used for the mainline rope. A sling made of 7- to 9-mm (0.28-in. to 0.35-in.) diameter rope may be appropriate as long as it has an adequate safety factor. Most of those experienced in the use of ascenders prefer slings constructed of static rope. Static rope tends to stretch less than dynamic rope and therefore transfers the energy involved in ascending directly to the ascenders rather than absorbing it. The length of the slings depends on factors such as your body proportions and the type of ascending system you use.

Because the designs of ascender tie-in points differ, the methods of connecting the slings to the ascender likewise vary depending on the brand and design used. Consult the manufacturer's instructions for specific information.

When connecting the sling to an ascender, remember that a sharp bend in a rope diminishes the rope's strength. If the ascender's tie-in point is wide enough, the sling can be tied directly into it using a figure 8 follow-through knot. If the tie-in point is narrow, such that it would create too sharp a bend in the sling, the sling should be clipped into a carabiner or screw link, which then is clipped directly into the carabiner tie-in point.

Footloops

When climbing with ascenders, it is necessary to have some means by which to step up each time the ascender is moved up. Any loop of webbing or cordage may be used for this purpose, although there are commercially made steps called footloops and etriers available as well. A footloop is just that—a single loop for the foot—while an etrier has multiple steps and is able to function as something of a soft ladder as needed.

With all of the moving around that comes with ascending, it can sometimes be difficult to keep the foot in a footloop. A **chicken loop** can be used to secure the footloop around the ankles to help prevent the feet from slipping out of the stirrups (**FIGURE 18-9**). These are typically constructed of webbing that is stitched or securely tied into a loop that is larger than the ankle but smaller than the boot. By passing the footloop beneath the chicken loop before stepping into it, the footloop is better held in place.

Creating an Ascending System

Two or more ascenders are required create an **ascending system** to move up the rope effectively. There are numerous variations of acceptable

---

TIP

Personal ascenders can and do fail when misused. The following conditions are the most common means by which these ascenders fail.

Frame Breakage
Some very old ascender frames were constructed of cast aluminum; these can crack or break when subjected to the high stress of being dropped. Cast aluminum frames can crack under their paint, leaving no outward sign of damage. Although most of these are now out of circulation, these may still be found among rope technicians who pride themselves in very old gear.

Rope Damage
Rope damage can occur when the teeth of the cam tear the rope sheath. This typically occurs under very high loads, although sheath tearing has been known to occur with as little force as 800 pounds (362.9 kg). Ascenders with sharp-toothed cams should never be used in situations involving more than one person's body weight, or where a significant fall or high-impact force may occur.

Rope Slipping Out of the Ascender
The nose of the frame is designed to keep the rope in a secure channel, but when the device is pulled away from the rope or torqued sideways, rope can slip out, particularly if the nose is bent. This is most likely when ascenders are used on an angled rope, such as when ascending a sloping highline or traversing a rope along a ledge. Clipping a safety carabiner across the ascender and the rope will not prevent the ascender from slipping from the rope, but at least connection to the rope will be maintained in the event of a failure.

Previously owned ascenders of any type may have been subjected to stresses that could result in failure. For this reason, do not use a previously owned ascender unless you know its complete history.

ascending systems, each made unique by the number and kinds of ascenders used, the combination of methods by which they are attached to the climber's body, and the parts of the body to which they are attached. Each ascending system has its advantages with regard to safety, ease of use, and speed and ease of movement. However, no one system offers all these advantages, and each system has at least one drawback.

Especially at the outset of your training in ascending, you should use ascending systems that combine safety and ease of use. Some hallmarks of this type of ascending system include the following:

- The system should require more use of your legs and feet and less use of your arms and hands. Your legs are stronger and have greater stamina than your arms.
- The system should hold you upright on the rope with your body weight over your legs. This requires less arm strength, encourages use of the legs, and contributes to safety.
- The system should be able to support you in a sitting position while you are on the rope. Ascending is a tiring activity that often requires short periods of rest, possibly in a sitting position.
- The system should have attachments to the seat harness (and in some cases to a chest harness). Some systems have attachments only to the feet and depend on arm and body strength to hold the climber upright. These are risky systems, sometimes called "death rigs."

Whatever system you choose, it is important that it be fine-tuned to your body height and build. With a fine-tuned system, rope ascending is no more work than climbing a ladder. Conversely, a system that is poorly fitted and out of adjustment quickly exhausts even the most physically fit individual.

Customizing the Slings

You should tailor the slings in the ascending rope rescue system for your own use, using the following guidelines:

- Your primary ascender should be attached to your harness with a locking carabiner. It should ride on a short sling, very close to your body so that when it is attached to the main rope it is just long enough to allow working space, usually allowing the ascender to go no higher than shoulder level. Alternatively, if you plan to do a lot of ascending, using a chest ascender in this position makes a lot of sense. A chest ascender is typically hard rigged directly to your harness in a sternal position.
- Your second ascender will become a step ascender. This should be the highest ascender on the host rope and will have both a footloop and a safety lanyard attached.
  - The footloop should be tuned for your dominant foot (e.g., right foot for right-handed people). The foot loop should be long enough to allow the ascender to be attached just above your seat attachment when it is at its highest.
  - A safety lanyard should also be attached to the step ascender and connected to a secure

> **TIP**
>
> **Safety Practices During Training**
>
> To help prevent candidates from getting too high off the deck when learning to ascend, a floating practice system can be created by running a very long static rope through a direction-change pulley at a high point on a cliff or structure so that one end is just barely touching the ground. The other end of the rope is rigged into a braking device and tied off.
>
> A candidate may attach their ascending system to the part of the rope that is just barely touching the ground and begin to ascend. When they are just a few feet off the ground, a technician-level rescuer operates the braking device, permitting rope to travel through the device at about the same rate of speed as the person on rope is ascending. In this way, the candidate can be kept just high enough off the deck to use the ascending system but low enough that a fall would not cause injury. This system is sometimes referred to as a dreadmill.
>
> When the candidate has completed the ascending cycle, the trainer operating the braking device lowers the candidate back to the deck so that they can get off the rope. The candidate always must stop ascending while there is still enough rope to lower them to the ground. If there is any chance of injury by falling during this practice, add a belay rope.
>
> The operator of the braking device must be thoroughly experienced in its use as a lowering device. The floor under the practice rope should be covered with a mat. A basic two-ascender system can be created using two ascenders, two lanyards, a footloop, and four carabiners (along with a properly anchored host rope.) Because the proportions of individual bodies differ, no specific dimensions for the sling attachments can be shown.

attachment on your harness. This will serve as a backup in case the primary attachment comes off the rope. The safety lanyard must be long enough to allow the ascender to be attached above your primary ascender when it is at its highest, but not so long that it the ascender might extend out of your reach while you are sitting in your harness (**FIGURE 18-10**).

- All connections and knots must be contoured well, dressed, and pulled down tightly. *Note: A second footloop ascender with an attached harness safety lanyard is a very handy tool for the technician-level rescuer to carry on their harness. This setup, known as a* **QAS (Quick Attachment Safety)**, *provides smooth transitions for changeovers and edge negotiation. Alternatively, the QAS may be left on the rope at all times, as a third ascender. This is particularly good practice for technician-level rescuers using single rope techniques.*

FIGURE 18-10 Basic two-ascender system.

Basic Ascending System

To rig a basic ascending system, follow the steps on **SKILL DRILL 18-11**. Rig a practice system rope using an appropriate life safety rope and anchorage(s). If using a dreadmill, make sure that someone is attending the lowering device and that initially it is locked off tight.

Other Ascending Systems

The two-ascender system is just one example of an ascending system. There are dozens of different ascending systems from which one can be chosen, based on specific factors:

- The distance that must be traveled up the rope
- The climber's physical strength
- The stamina of the user
- Differences in physique (e.g., a body that is apple or pear-shaped, thin or fat)
- The speed required for the climb

With proper research and experimentation, each person can find the type of ascending system that is just right for them.

Ascending over an Edge

Usually, the most difficult maneuver in ascending is going over the edge at the top of a cliff or building. As a general rule, and just as in rappelling, the higher up the rope is anchored, the easier it is to go over an edge. Negotiating edges is covered in detail in Chapter 16, *Working in Suspension*.

Edge with a Gradual Rollover

The specific technique for getting over an edge depends on the particular nature of the edge. Getting over an edge with a gradual rollover can be easier than getting over a sharp or undercut edge if you follow the steps in order:

1. Ascend until the topmost ascender is about to make contact with the terrain or structure you are climbing.
2. Push yourself away from the wall using one hand and your feet.
3. Raise the topmost ascender above the contact point.
4. Take care not to get your fingers caught between the hardware or rope and the wall.
5. As you move upward and your weight is on the ascender above the contact point, the other ascender should follow more easily.

SKILL DRILL 18-11
Rigging a Basic Ascending System

1. With one hand, reach up and pull the mainline rope taut. Attach the primary ascender for the seat harness as high up as you can get it (still within reach) on the tensioned rope. Adjust the lanyard so that it is well within reach, no higher than shoulder level.

2. Attach the step ascender with attached footloop to the rope; it should rest just above the seat harness ascender when your leg is straight. Adjust the length of the footloop so that the ascender is still just within reach when your knee is bent, as for a step.

3. Attach a safety lanyard between the step ascender and your harness.

4. Pre-test the system by alternately loading the primary ascender and the step ascender to make sure they hold and that they are positioned properly. If possible, have another person help you get started by holding down the ascending rope close to the ground. This makes getting started easier because there may not be enough rope weight initially to allow the ascenders to pass smoothly over the rope; otherwise, you may have to hold the rope down yourself. Once you have gotten far enough off the ground, the rope weight will allow the rope to slide through the ascenders more easily as they are pushed up.

5. Standing on the ground, push your primary ascender up as high as possible and then sit down in your harness so that you are supported by the ascender.

6. With one foot in the footloop, bend your knee (to relieve pressure against the footloop) and push the step ascender up the rope as far as possible.

7. Stand up in the footloop, raising your seat primary ascender as high as possible on the host rope, and then sit down in your harness so that you are supported by the primary ascender.

8. Continue the cycle by repeating steps 6 and 7 until you have reached the desired height or run out of energy. If using a dreadmill, the brake operator can simply lower you to ground when the training evolution is finished.

Undercut Edge

Getting over a sharp or undercut edge can be more challenging, but with a bit of practice this, too, can be done gracefully. A second footloop and ascender (QAS) is quite useful here.

1. Ascend until the step ascender (highest ascender on host rope) is about to make contact with the wall.
2. Without jamming the two ascenders together, bring the primary ascender up as far as possible, so that one hand can reach the host rope above the edge.
3. Try to work the step ascender over the edge by pushing away from the wall with one hand and with your feet or knees.
4. If this is impossible and you have a QAS, place the QAS above the edge and place your weight on the QAS while moving your main system over. If you do not have a QAS, tie off short into the host rope while moving your system up over the edge.
5. Verify that the original ascending system is properly reattached and begin ascending again. Until you are in a safe, secure position and in no danger of falling, never have fewer than two ascenders securing you to the rope.

Tying Off Short

Tying off short is a safety procedure that involves tying directly into the mainline rope midline to achieve an additional attachment. This procedure is used anytime the climber wants a secure attachment to the rope, particularly when moving past a knot or transitioning over a difficult edge with only two ascenders.

Developing Vertical Skills

Descending, ascending, and negotiating edges represent just the tip of the iceberg of information and skills that a technician-level rescuer must master before working on rope. Becoming adept in ascending and descending requires practice. The more skilled and comfortable you become in managing your ascent and descent, the safer and more efficient you will be in performing rescue skills while working on rope. Technician-level rescuers should be able to apply these principles to ascend and descend safely, confidently, and under control at all times. They must be able to pass knots and other obstacles, as well as negotiate rolling edges, undercut edges, overhangs, parapet walls, and even open structures. They should be able to tie off whatever system they are using, operate hands free of the rope, and return to a safe, controlled rappel. They should be able to perform all of these skills safely and comfortably in whatever light (or dark) conditions, weather environments, and/or terrain in which they might be expected to work. With a solid understanding of the basics, work on these finer points of vertical skills to become more graceful and proficient on rope.

Effect of the Rope Angle

A factor that significantly affects the degree of difficulty involved in rappelling over an edge is the angle the rope forms from the rappeller to the anchor point. The effect ranges from greatest difficulty with a horizontal angle (the anchor point is level with or lower than the rappeller) to least difficulty with a vertical angle (the anchorage is above the rappeller). A vertical angle is rare. Most of the time, a compromise must be found: getting the rope angle as high as possible while maintaining a safe, secure anchor point (**FIGURE 18-11**).

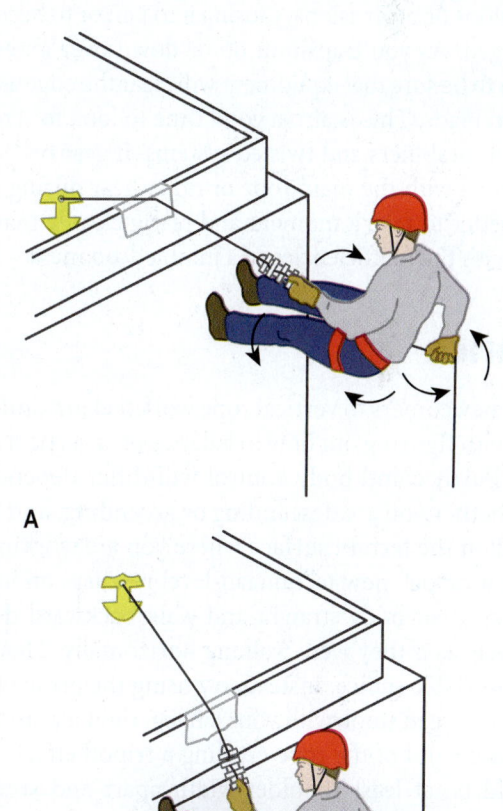

FIGURE 18-11 Effect of rope angle. **A.** Low anchor point. **B.** High anchor point.

Adding a directional deviation at a high point will help facilitate edge transitions, but adds complexity and introduces new questions, such as "does the belay line go high, with the mainline, or low, on the ground?" This, and more, is discussed in further detail in Chapter 17, *Horizontal Systems*.

Equipment Management

Whether ascending or descending, a technician-level rescuer who effectively manages their gear will generally be safer and more efficient in their work. Always placing your spare gear in approximately the same location on your harness is a good habit that will help you to function smoothly. As you rig, perform changeovers, and manage the patient, take care to ensure that your systems are neat and orderly. Avoid unnecessarily crossed lines, entanglement of ropes and lanyards, and crossed systems.

An edge transition, whether on ascent or descent, is a prime opportunity for equipment to get caught on obstructions or protrusions, causing it to jam or to become damaged. As you transition up or down over an edge, watch to be sure that equipment will clear the edge as the system loads. This is also a good time to look for cross-loaded carabiners and twisted systems. If gear becomes entwined with the main rope or other gear during use, it is better to take a moment and re-rig cleanly than to deal with the inefficiencies of a jumbled-up mess.

Stance

Most newcomers to vertical rope work feel uncomfortable with figuring out how to balance on a vertical surface. Balance and body control will differ depending on whether you are descending or ascending, and also based on the terrain surface where you are working.

In a rappel, new technician-level rescuers often try to keep their body straight and walk backward down the face, as if they were walking horizontally. This is a very unstable stance. Instead, try using the principle of a three-legged stool, with your left leg, right leg, and the anchored end of the rope creating a tripod effect. Feet should be at least shoulder-width apart and straight out in front of you, almost as though you are sitting on the floor. With your center of gravity low, the rope can do its work of supporting the vast majority of the load (your body) while you maintain stability with your legs.

For ascending, the optimum body position is quite different from that while rappelling. An ascending system works best when you are relying totally and completely on the rope system; in fact, a free-hanging rope is much easier to ascend than one against a vertical surface. The more you are able to center your body around the rope, the more stable you will be, and the easier will be your ascent.

When ascending, maintain as upright a position as possible on the rope, feet as close to directly beneath you as possible. Although your feet may stick out in front of you when you are sitting in your harness, when it comes time to take a step, tuck your feet under your butt and stand straight up. You will know you are at optimal efficiency if your rope is *quiet*, not swinging wildly to and fro, and if you feel balanced. Even if you are using only one footloop, placing the free foot on top of the foot in the etrier can help this positioning.

Edge Transitions

As with any journey, the first step in rappelling is arguably the most difficult. Technician-level rescuers should become very comfortable with edge transitions both on descent and ascent, learning to not only minimize fall potential, but also to perform these maneuvers gracefully. If high directionals are frequently used in the response area, practicing edge transitions with these in place is of course an obvious requirement. However, technicians should also practice edge transitions without high directional help, as this will only serve to improve their skills. To practice a standing transition, follow the steps in **SKILL DRILL 18-12**.

Good body position begins at the edge, just at the outset of rappel. Practice this method of attaining good body position on rappel.

If you just cannot quite get the hang of the standing transition, try the kneeling transition method. To practice the kneeling transition method, follow the steps in **SKILL DRILL 18-13**.

Undercut Edges

An undercut edge is one that is overhung so far back that your legs cannot reach the wall as you start your rappel over the edge, or one that does not permit you to brace with your legs as you make ascent over the edge. An undercut edge presents the technician-level rescuer with a special problem. It requires an advanced technique, which you should attempt only after you have developed full confidence and skill in controlling both your ascent and your descent systems.

Rappelling from an undercut edge involves keeping your feet on the edge while lowering the rest of your body until your head is well below your feet and the edge of the overhang. If your head does not clear the overhang, you could be in for a potentially painful injury to the face. Only after you are certain that the rappel device and your torso are far enough below the edge to clear it

SKILL DRILL 18-12
Performing a Standing Edge Negotiation

1 Connect to your descending system of choice, and an appropriate belay if called for.

2 Face the anchor point. Back up, slowly letting slack through your descender until you are standing with the balls of your feet on the edge, feet spread at least shoulder width apart, leaning backward into the rope.

3 Imagine that something is pushing you at your waist, so that your glutes are slowly being thrust back out over the drop and opposite the anchor point. Without moving your feet, lower your glutes until your feet are nearly at waist level.

4 Shuffle your feet in small steps backward, down the wall, maintaining a stable width.

5 Slowly feed rope through your descender to maintain the tripod effect, glutes low, feet wide, and rope supporting the majority of your weight. Continue to the bottom of the descent.

SKILL DRILL 18-13
Performing a Kneeling Transition Method

1 Connect to your descending system of choice, and an appropriate belay, as determined by the AHJ. Face the anchor point. Back up, slowly letting slack through your descender until you are as near to the edge as possible, rope taut.

2 Still facing the anchor point, get down on your knees at the edge of the drop with your toes hanging over it.

3 Lean back as far as possible, and push with your legs as you slowly feed rope through your descender, getting your glutes back away from the edge.

4 Slide over the edge on your knees. Your toes will hit the wall, and you will stabilize as the rope comes down on the edge.

5 Continue to rappel backward, with your feet shoulder width apart against the wall and your torso parallel to the face.

---

do you step off the edge. This often results in a forward pendulum. Your feet must absorb the shock of the forward motion against the vertical face (if there is one).

On ascent, the concern is that the final step up will inevitably put a bit of extra slack in the system, thereby introducing the potential for shock loading an ascender in the event of a fall. To prevent this, it is advisable to change over to a locked-off descender when just beneath the lip, as a descender can typically withstand a higher shock load force than an ascender. Hanging an anchored, separate rope loop, sling, or daisy chain over the edge makes a useful step. Before starting over the edge, always remove as much rope slack as possible to help reduce potential for shock loading of the system.

Gaining Extra Friction When on Rappel

A rappel should be done only when the rappeller has friction, and therefore control, to spare. At times, a rescuer might find themselves with less friction control than they might have expected due to variables such as carrying a heavy load of equipment, loading your system with the weight of another person, rappelling on wet rope, or even rappelling on a different diameter or material of rope from what you are accustomed to. In such situations, there are a few techniques that may be used to create additional control.

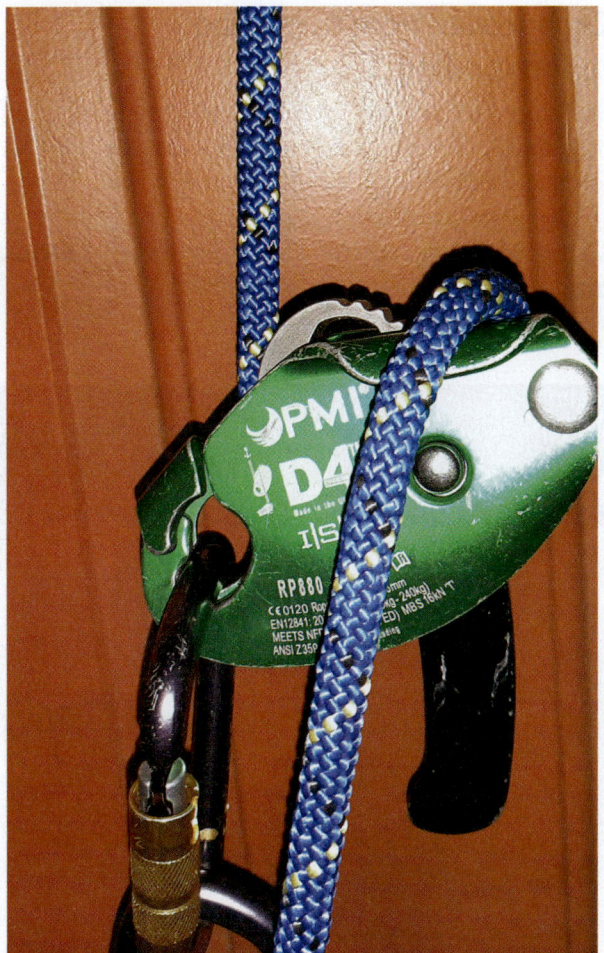

FIGURE 18-12 The first course of action if you feel a lack of control is to create as much bend as possible as the rope enters the device.

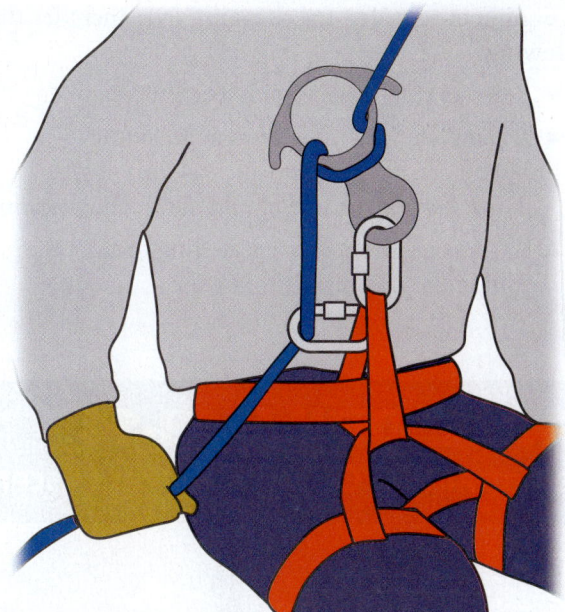

FIGURE 18-13 Gaining extra control with a spare carabiner.

Gaining Extra Control with Rope Angle

The angle at which the rope enters the device you are using will have some influence on the amount of control generated. If the rope enters the device in a straight line, there will be less friction than if the rope bends as it enters the device. The first course of action if you feel a lack of control is to create as much bend as possible as the rope enters the device (**FIGURE 18-12**).

Gaining Extra Control with a Spare Carabiner

To use a spare carabiner to gain extra friction, choose a substantial point on your harness to clip an extra carabiner. Side positioning rings, if you have them, are a good choice, but gear loops are typically not strong enough for this function. In a pinch, with no other options available, the friction carabiner can be added to the harness tie-in point at the waist, but this is not optimal. Place the friction carabiner on your brake hand side (**FIGURE 18-13**).

Clip the rope into the carabiner between your brake hand and the device, and immediately redirect the rope with your brake hand to create as sharp a bend as possible over the friction carabiner. The simple act of adding a friction point and bending the rope sharply over it will add a modicum of control to your descent. Note that some autolocking descenders, such as the Petzl ID, are required to have this extra carabiner in place whenever a heavy load is being lowered.

Gaining Extra Control with the Body

One method that may be implemented quickly to help prevent an out-of-control situation is to extend your brake hand on the rope far enough to bring the rope sharply against the thigh, and even part way around it. However, this is not a recommended technique for extended use, for two reasons:

1. The running rope may damage the material it runs over, whether seat harness, clothing, or body part.
2. This maneuver leaves no extra friction to spare in case it is needed.

Changeovers

A changeover occurs whenever a technician-level rescuer switches from an ascending mode to a descending mode or from a descending mode to an ascending mode while still on the rope. It is a skill that adds to vertical competency and is particularly useful in rescue operations where ropes are required to reach the subject. Changeovers are also important in emergencies, such as when you rappel to the end of a rope and find that it does not reach the bottom.

Equipment needed for changing over includes the following:

- A descending device, with its own carabiner
- A complete ascending system, including carabiners. *NOTE: Carabiners should not be shared between ascending and descending systems.*
- Harness gear loops or a gear sling to store the equipment not currently in use

Changeover from Ascending to Descending

For this exercise, use the floating practice system (dreadmill) described earlier in this chapter. This affords trainers the opportunity to safely lower the person on rope to ground should they have difficulty. To perform this exercise, follow the steps in **SKILL DRILL 18-14**.

SKILL DRILL 18-14
Performing a Changing Over (Ascending to Descending)

1 Before you start, make sure a competent person is attending the lowering device, which initially must be locked down tight. Begin ascending the rope using your ascending system of choice. When you reach the point where you want to change over, stop your ascent, placing your full weight on your primary (sternal) ascender.

2 Remove your rappel device from your equipment sling and attach it properly to your seat harness tie-in point with a locking carabiner. Attach the rappel device to the slack rope below your primary ascender. Remove all slack in the mainline rope between the descender and the primary ascender and securely lock off the descender.

SKILL DRILL 18-14 (continued)
Performing a Changing Over (Ascending to Descending)

3 Position your dominant leg ascender at about head-level (or just far enough so that when you put your weight on it, it removes the weight from the seat harness (top) ascender.) Shift your weight to the foot ascender and remove the weight from the seat harness ascender. Remove the seat harness ascender from the rope.

4 Sit down so that the rappel device takes your weight. Remove weight from your foot ascender by lifting your foot.

(continues)

SKILL DRILL 18-14 (continued)
Performing a Changing Over (Ascending to Descending)

5 Remove all remaining ascenders from the rope.

6 One by one, remove the ascender slings from your body. To secure them, wrap them around the ascender to which they are attached and clip them into your equipment sling. Unlock the descender and proceed with descent.

Courtesy of Steve Hudson.

Changeover from Descending to Ascending

To changeover from descending to ascending, follow the steps in **SKILL DRILL 18-15**.

Passing Obstacles

The ability to move proficiently between ropes, from point to point on a rope, or past obstacles is vital to personal safety during rescue operations. Regardless of environment, technician-level rescuers typically need to move around to accomplish their task, and seldom will circumstances accommodate a straight-line operation. At times, maneuvering into place may be as simple as briefly opposing the fall line with leg pressure or a short pendulum. At times, however, the rescuer may encounter various obstacles along the rope path, ranging from midline knots to direction change or re-anchored ropes. Negotiating such

SKILL DRILL 18-15
Performing a Change Over (Descending to Ascending)

1 Set up your ascending equipment before you leave the ground. Be sure you have at least one ascender with a footloop and safety lanyard to use as your step ascender, and one ascender with a short lanyard (or a pre-rigged chest ascender) to use as your primary ascender. Begin descending a rope using your descending system of choice. Stop the descent at the point where you want to begin the changeover. Lock off the descender securely.

2 Attach your step ascender to the host rope at about shoulder level. Connect its safety lanyard to your harness attachment point and move it upward on the rope so that the footloop will permit you to step up 10 to 12 in. (25.4 to 30.5 cm) when ready.

(continues)

SKILL DRILL 18-15 (continued)
Performing a Change Over (Descending to Ascending)

3. Attach your primary ascender to your harness and to the host rope between your locked-off descender and your step ascender. If you are using a hard-rigged chest ascender as your primary, you will need to step up into your footloop to accomplish this.

4. If you plan to use a third ascender (QAS) on the rope as part of your ascending system, attach it now. Step up in the footloop of your step ascender and push your primary ascender up the rope far enough so that when you sit back down your descender is no longer under tension.

SKILL DRILL 18-15 (continued)
Performing a Change Over (Descending to Ascending)

Courtesy of Steve Hudson.

5 When all the ascenders are securely attached, unlock the rappel device and remove it from the rope, then remove it from your harness attachment point and place it on your gear loop.

obstacles really involves nothing more than performing a series of changeovers, as described in the previous section.

Rope-to-Rope Transfers

When it is necessary to transfer to one set of ropes to another, ascending and then changing over to the decent system may be the least strenuous option.

To changeover from ascent, follow the steps in **SKILL DRILL 18-16**:

1 Secure the second set of ropes to your harness with a connector and start to ascend. Once you reach your goal height, changeover to descent. After locking the descender, attach your ascending system to the second line. Do not pull away from the fall line.

2 Climb down the first line until you feel the ascending system become tight on the second line. You may ascend the second line and climb down on the first line until you are suspended vertically on the second line. NOTE: Carefully control your horizontal movement so that you avoid entering swing fall potential or introducing excessive angles between the two lines.

3. As soon as you are suspended from the ascending system on the second line, you may remove the descending system from the first line and maneuver on the second line.

Passing a Knot

As a first step in passing obstacles, a technician-level rescuer should learn to pass a knot in the descent line. Passing a knot is a very real potential, as this may become necessary on long drops where multiple ropes are tied together, or in situations where a knot has been used to isolate a section of damaged rope. Once mastered, this process will transfer conceptually to passing other obstacles, such as directional deviations and mid-point anchorages. To pass a knot, follow the steps in **SKILL DRILL 18-17**:

1. Descend until the descender is just a few inches above the knot. Perform a changeover into your ascending system, continuing to stay above the knot (and your descender), but do not remove the descender from your harness attachment point, as it will be used again shortly.
2. Remove descender from the rope, then reattach it to the rope just below the knot. Use your ascenders to down climb until your lowest ascender is just slightly above the knot (within a couple inches).
Perform another changeover, back onto the descender, snugging the descender up as close as possible to the bottom of the knot.
3. When positioning the step ascender for the final transfer back into descent mode, be sure to keep the device as low as possible, so as to avoid it being out of reach when you sit down into your descender.
4. Remove ascenders and proceed with descent.

The biggest difference between a knot pass and a standard changeover is the height differential between the system above the knot as compared with the system below the knot. With practice, the technician-level rescuer will learn to keep equipment a concise distance apart, without impacting the knot. To pass a knot in an access line while ascending using an ascending system, follow the steps in **SKILL DRILL 18-18**:

1. Ascend up to the knot using your preferred system. Perform a changeover from ascending system to descending system, as previously described, but do not remove the ascending components from the harness attachment points, as they will be used again. Remove the step ascender from below the knot and replace it on the host rope above the knot.
2. Step up into the step ascender, pulling out as much slack as possible to snug it up closely to the knot. Lock off the descender.
3. Use the step ascender to step up high enough to connect your primary ascender to the working line at least a couple inches above the knot. Sit back into your harness, with your weight supported by the primary ascender.
4. Remove the descender from the access line. Continue ascending on ascent system to desired height.

Passing a Deviation

Deviations are points at which a rope is redirected to a new fall line, typically simply by passing the line through a carabiner or connector that is attached to a secure anchor point. There is typically no knot at a deviation; instead, the rope runs freely through the connector device. This method is typically used only when the new fall line is relatively near the original one.

The hardest part of passing a deviation on descent is reaching the deviation point itself. A body descending will naturally follow the inherent fall line. Because, by definition, a deviation creates a new fall line, the rescuer must have some means of pulling themselves across to the deviation anchorage. If the deviation has been rigged with this in mind, a large, bulky knot will have been tied in the ends of the ropes. In this case, all the rescue technician has to do is descend until about eye level with the deviation, then pull the ends of the rope up through the connector until the bulky knot jams in place. Then, using the bulky knot as a *stopper* against the connector, the technician-level rescuer pulls themselves over to the deviation anchorage and clips directly into it with a safety lanyard.

Once connected to the deviation anchor point, the technician-level rescuer is no longer at risk of an unsafe pendulum. Now, they may simply remove the anchorage connector from the rope below their descender and replace it back onto the rope above their descender. To pass a deviation on descent, follow the steps in **SKILL DRILL 18-19**.

SKILL DRILL 18-19
Passing a Deviation on Descent

1. Descend until the deviation anchor point is at about eye level. If a knot has been tied in the ends of the rope, pull it until it jams in the anchorage connector.

2. Use the rope to pull yourself to the deviation anchor point. Attach a safety lanyard from your harness directly to the anchor point.

3. Unclip the anchorage connector from the working line below your descender. Reattach the anchorage connector back to the working line, but now above your descender.

4. Unclip your direct connection from the deviation anchor point. Continue descent.

Passing a Deviation on Ascent

When passing a deviation from below, it is prudent to carry an extra anchorage connector for optimum security. Follow the steps in **SKILL DRILL 18-20**:

1. Ascend up to the deviation anchor point. Connect harness waist attachment point directly into the deviation anchor point with a safety/positioning lanyard to maintain position.
2. Attach a new anchorage connector (preferably a sling, slightly shorter than the original one) into the deviation anchorage.
3. Attach the new sling to the ascending line with a carabiner, just below the rescuer.
4. Unclip the safety/positioning lanyard from the deviation anchorage. While maintaining a firm grip on the rope below the deviation, remove the ascending line from the original anchorage connector sling.
5. Slowly feed out the tail of the rope to move horizontally into the new fall line.

Passing an Intermediate Anchorage Point

At times, an intermediate anchorage point is established into a drop to create permanent sag between the length of rope above and the length of rope below that anchorage point. This differs from the deviation in that the rope is firmly attached to the intermediate anchorage point with a knot, whereas in a deviation, the rope simply runs freely through a directional connector.

Intermediate anchor systems (also sometimes called *rebelay* from caving vernacular) can have anchorages that are very far apart, both vertically and horizontally. Intermediate anchor points with short distances between them are often used to permit multiple rescuers to be on different lengths of the same rope at the same time, to distribute the weight of the rope, to reduce the bounce effect on long drops, and to generally reduce the inconveniences associated with very long ropes. These can be negotiated in much the same way as a knot pass or rope-to-rope transfer. Intermediate anchor points with greater distances between them can be a bit more challenging but are especially useful for creating a path for horizontal movement across a span.

Still, passing a long rebelay involves nothing more than using a series of changeovers to progress through three different sets of ropes. Keeping both your descending system and your ascending system attached to your harness at all times will help facilitate this. To pass an intermediate anchor point, follow the steps in **SKILL DRILL 18-21**:

1. Using a properly rigged system, ascend up the first set of ropes to the main anchor point. Perform a rope-to-rope transfer from the vertical access line to the near side of the loop between the main anchor system and the intermediate anchor system. This should leave you suspended in your descent system from the near side of the rebelay loop.
2. Collect the far side of the rebelay loop and perform another a rope-to-rope transfer, attaching the ascending system to it. Ensure that it facilitates travel in an upward direction, toward the intermediate anchor point.
3. Using the descending system on one set of lines and the ascending system on the other set, maintain consistent tension between the two to avoid swing fall while descending/ascending across the span.
4. Continue this process until suspended from the far side of the rebelay loop in your ascending system directly below the intermediate anchor point.
5. Perform another rope-to-rope transfer, this one from the far side of the rebelay loop to the final set of vertical lines. Ascend or descend this set of ropes, as applicable to circumstance.

Of course, depending on circumstances, the technician may not need to move all the way through an entire system as described. If using just a portion of a system as described, the sequence may be truncated and modified as needed. The key takeaway is to ensure that both sides of the interior loop are used for the transition. For training purposes, if a candidate is capable of moving through all four legs as described, it is quite likely that they will be capable of any variation on the system.

Getting Off the Rope

As you approach the ground, watch for obstacles and hazards. Pay attention to footing, the presence of other rescuers and bystanders, and what the conditions are at the location. Because most ropes stretch at least a bit, completing a rappel involves a bit of finesse. With any rappel device, it is easier to get off the rope if you have a small amount of slack to work with. A quick trick for gaining this slack is to keep your weight on the rope and rappel device even as your feet touch the bottom; rather than standing up straight away, squat down before stopping the rappel. Then, when you stand upright, you will have some slack in the rappel rope to help you in unlacing the rappel device.

Troubleshooting

Self-Belay Techniques

When it is not possible or practical for a rappeller to be belayed by another person, self-belay techniques may be an option, especially when a manually operated, nonautolocking descender is used with single rope technique. Arguably, autolocking braking devices need no secondary means of safety.

Self-belay techniques in SRT systems are typically based on the use of some type of rope-grab device on the main rappel line. Usually, the rappeller is attached to the device by means of a short sling attached to their harness.

Using a Prusik Safety

A traditional means of rappelling self-belay has been to use a Prusik safety, which uses a Prusik hitch on the rope above the braking device, with the other end of the Prusik loop connected to the rappeller's harness. The theory is that should the rappeller's descent run out of control, the Prusik knot can be tightened on the main rope to stop the fall.

A variation on this system, known as an autoblock, involves placing a very short Prusik on the line below the descender, where the rappeller's hand is in contact with it during descent. Either of these methods may be considered appropriate by the AHJ, depending on just what they are trying to protect against. Some precautions can help the rappeller use the Prusik safety as intended:

- Practice using the Prusik safety at a safe height so that your emergency actions become automatic.
- Make sure the Prusik is tight enough and properly dressed so that it will catch when needed.

Secondary Backup Systems

In some cases, the AHJ may determine that it is most appropriate to rig a separate vertical lifeline for belay. In this case, rescuers may use rope grabs and lanyards designed for fall arrest as protection against a fall in the event of mainline failure. The advantages and disadvantages of a separate vertical lifeline must be weighed carefully. Although the phrase *safety backup* sounds inviting, the complexity and extra challenges involved with operating a secondary system may not always be justified. See Chapter 12, *Fall Protection and Belay Operations* for more on belays.

Self-Rescue

Occasionally, a technician-level rescuer will get stuck on rope. This can happen if a body part, an item of clothing, or other debris gets trapped in the descender or ascender they are using. Nearly any device, whether autolocking or manual, is susceptible to such an occurrence. Rescuers who may still be using a **figure 8 descender** also can become stranded when the host rope becomes girth hitched around the device. Girth hitching of the figure 8 immediately stops the rappel and prevents further descent.

A rescuer with good upper-body strength may be able to extricate themselves forcefully by lifting their body weight from the stuck condition. However, it is easier and usually safer to set an ascender or Prusik on the rope above the stuck device and step into an attached sling to release the weight. The best solution to this problem, of course, is prevention:

- Tuck in shirttails and other loose clothing.
- Keep hair trimmed or tied back and tucked into the helmet.
- Keep flabby body sections (e.g., underarms, stomachs) away from rappel devices.
- Prevent debris from entering the rope channel

In the unfortunate event that jamming does occur, having practiced extrication techniques beforehand will reap dividends. Avoid using a knife to extricate yourself from a jammed device, as it is difficult to avoid damaging or cutting the rope completely, or doing other irreparable damage, while using a knife under duress. For example, an attempt to cut hair out of a device could end up inflicting significant wound to the attached scalp. If you are trying to cut jammed material from a descender, you probably are under pressure, physically unbalanced, and, possibly, also in pain. Therefore, it is extremely difficult for you to cut away the offending material without also touching the rope with the knife.

Usually, the best means of escaping such a situation is to take your weight off the offending device. Having learned to perform changeovers from descent to ascent, and back again, is the foundational skill applicable here. At the very least, a technician-level rescuer should be prepared to attach an ascender or Prusik hitch above the jammed device, with a footloop to step up into. If extrication will require much energy, a second ascender or hitch can be added below the footloop ascender, to be used as a waist attachment. Note, this is easier said than done when also dealing with an entrapped appendage or other disruption, so practice one-handed and under pressure. To extricate an obstruction from a jammed rappel device, follow the steps in **SKILL DRILL 18-22**.

Emergency Descent Systems

Emergency situations may arise in which a rappel is necessary, but the person has no descender. There are possible solutions to this problem, some involving equipment and others involving only body friction. Because of the physical discomfort and the potential dangers associated with body rappels, these techniques are rarely considered suitable for vertical operations. That said, they are good skills to know for personal safety in low-angle environments.

Arm Rappel

The arm rappel is sometimes used for short distances on low-angle slopes (**FIGURE 18-14**). The user prepares for the arm rappel with their body perpendicular to the slope, with the anchored rope behind them. With both arms extended, the rappeler wraps the rope around each arm a couple of times, gripping the rope with both hands. As the rappeler slowly moves sideways down the slope, they adapt their hand grip to control friction and rate of descent.

FIGURE 18-14 Arm rappel.

SKILL DRILL 18-22
Extricating an Obstruction from a Jammed Rappel Device

1 Begin a rappel. Assume that the rappel device is jammed. Lock off the rappel device securely.

2 Place an ascender on the host rope above the rappel device; this will become your primary ascender.

SKILL DRILL 18-22 (continued)
Extricating an Obstruction from a Jammed Rappel Device

3 Use a short lanyard to attach the ascender to the front attachment point of your seat harness.

4 Place another ascender on the host rope, above the first ascender. This will become your step ascender. Attach an etrier or footloop to this ascender and push it up the rope. Use the step ascender footloop to step up 8–12 in. (20.3–30.5 cm), simultaneously moving your step ascender up high enough to take the tension off your descender. Settle your weight on the ascenders, removing the weight from the rappel device.

(continues)

SKILL DRILL 18-22 (continued)
Extricating an Obstruction from a Jammed Rappel Device

5. Remove the obstruction from the descender. Replace the descender on the rope and lock it off so that there is no slack in the mainline rope. Step up into the step ascender, removing your weight from the primary ascender. Remove the primary ascender from the rope. Sit back down into your harness, shifting your weight back onto your descender. Remove the foot ascender from the rope. Unlock the rappel device and continue the rappel.

TIP

Use the arm rappel only on short, low-angle slopes; it does not produce enough friction to control full body weight adequately in a completely vertical situation. Because of the potential for rope abrasion injuries on the arms and hands, use this technique only when you are wearing long-sleeved shirts and gloves.

Body Rappel (Dulfersitz)

Although the Dulfersitz is reported to have been used for vertical rappels, the practicality of this technique is questionable at best, and potentially disastrous. A significant amount of friction is created when the rope is wrapped around the body, often too much for a low angle rappel. At the other end of the spectrum, on a steep or vertical rappel, this technique can cause considerable pain to the hand and body, possibly severe enough to prompt the rappeller to let go of the rope to relieve the pain. To perform a body rappel, follow the steps in **SKILL DRILL 18-23**.

The body rappel technique presents at least two grave hazards:

1. The rope could come unwrapped from the rappeller's leg. Consequently, the rappeller would lose friction with the rope and may fall to the ground. This is a particular danger on a vertical face.
2. Rope abrasion and pressure can cause bodily injury, particularly to the crotch and shoulder. Thick padding in these areas is recommended.

TIP

Due to the higher probability of failure when performing a body rappel, it is prudent to use a separate belay when practicing these techniques! One danger of the body rappel is that on high-angle rappels, the rope can become unwrapped from the leg. To prevent this, in practice the rappeler must keep the wrapped leg kept lower than the unwrapped leg and the upper body.

Always practice the body rappel first on a low-angle slope, gradually working toward steeper slopes to determine your limitations. It should not be used as a general practice, but rather only in case of emergency, when no other alternatives exist.

SKILL DRILL 18-23
Wrapping the Rope for a Body Rappel

1 Straddle an anchored rope, facing the anchor point. Grasp the slack part of the rope from behind with your dominant hand and wrap it from back toward the front around the hip on the same side as the rappeller's dominant hand.

2 Extend the rope across the front of the chest and then to the back over the opposite shoulder.

3 Pass the rope down across the back to the braking (dominant) hand; the free or braking end of the rope should end up on the same side where the rope passes over the hip.

Carabiner Wrap

For vertical situations, it is always better to use methods that involve equipment, even if that equipment is a makeshift harness and nothing more than a carabiner. The carabiner wrap, which involves wrapping a seat harness carabiner with several turns of the rappel rope to create friction, is a method that was popular many years ago. However, the carabiner wrap rappel is no longer considered a satisfactory and safe technique for rappelling, for the following reasons:

- If the rope wraps are not put onto the carabiner correctly, they can spiral out of the carabiner gate, resulting in a free fall.
- The rope wraps can bear on the carabiner gate and break it.

Münter Hitch Rappel

An alternative to the carabiner wrap may be the Münter hitch rappel (**FIGURE 18-15**). As with any rappel, this technique must first be practiced on level ground, then on a moderate slope with a rappel, and finally on a short drop with a rappel, before moving on to real scenarios. Because the Münter hitch rappel twists the rope mercilessly, it is a limited-use technique best kept in reserve for emergency use only.

FIGURE 18-15 Münter hitch rappel.
Courtesy of Steve Hudson.

After-Action REVIEW

IN SUMMARY

- Single rope technique (SRT) describes an approach to vertical ropework in which one rope is used for ascending and descending, without a second rope for backup safety or belay.
- Descending is the act of using friction to control the speed at which rope rescuers slide down a rope, is a necessary skill for operating safely in the vertical environment. This is the first step toward developing vertical competency
- An important sign of a person's competence in descending is control.
- All descending methods involve the use of friction against a host rope to slow the rate of descent of an attached mass. Either the body or a braking device (descender) may be used.
- At a minimum, a rescuer will require a rope, anchorage, harness, descender, and connectors for descending. A pair of sturdy rope gloves and a helmet are also advisable.
- Getting over the edge is the most difficult part of any descent.
- The brake bar rack descender offers greater friction, and therefore greater control, than most descenders. It also allows the friction level to be changed after the rappel has begun.
- Ascending a rope is the opposite of descending. It involves the use of mechanical devices or friction hitches to climb a fixed rope safely and efficiently.

- Ascending is accomplished through the use of rope grab devices, called ascenders.
- Types of ascenders include friction hitches and mechanical hitches.
- Friction hitches were the first type of rope grab devices used in ascending. For the most part, they have been replaced in ascending by mechanical devices.
- When selecting rope for a Prusik hitch, consider the rope's diameter, stretch, and construction.
- Tying off short is a safety procedure that involves tying directly into the mainline rope to ensure an additional attachment. It is used during ascending when the danger exists that a climber may lose attachment to the rope.
- Whether ascending or descending, a technician-level rescuer who effectively manages their gear will generally be safer and more efficient in their work. Always placing your spare gear in approximately the same location on your harness is a good habit that will help you to function smoothly.
- Usually, the most difficult maneuver in ascending is going over the edge at the top of a cliff or building. As a general rule, the higher up the rope is anchored, the easier it is to go over an edge.
- Changing over means switching from an ascending mode to a rappelling mode or from a rappelling mode to an ascending mode while still on the rope. It is a skill that adds to vertical competency and is particularly useful in emergencies.
- The ability to extricate hair or clothing from a jammed rappel device without using a knife is an essential skill.

KEY TERMS

Arm rappel (guide's rappel) A type of rappel in which the rope wraps around both outstretched arms and across the person's back. The technique is sometimes used in sloping terrain. It does not give enough control for vertical situations.

Ascenders Rope grab devices used to ascend a fixed rope or, with specific types of ascenders, to create a hauling system. Ascenders are also known as *ascending devices*.

Ascender slings Attachments, usually webbing or rope, that connect a climber to the ascenders.

Ascending A means of traveling up a fixed line using either mechanical devices or friction hitches attached with slings to the climber's body.

Ascending system An arrangement of two or more ascenders used for traveling up a rope.

Body rappel (Dulfersitz rappel) A type of rappel that uses the body as friction by running the rope through the legs, across one hip, over the opposite shoulder, and to a braking hand. Because of the discomfort involved and the potential injury to body parts, this rappel has largely been supplanted by other techniques.

Brake bar rack A rappel device that consists of a series of short metal bars fixed to and sliding along a U-shaped metal rack with an eye at one end for attachment.

Brake hand The hand that grasps the rope to help control the speed of descent during a rappel. The dominant hand (e.g., the right hand in a right-handed individual) usually is the brake hand.

Carabiner wrap A rappel technique that uses several rope wraps around a seat harness carabiner to create friction and control the descent. It generally is not considered a safe or secure technique for rappelling.

Changeover To transfer from an ascending mode to a rappelling mode or from a rappelling mode to an ascending mode while on rope.

Chicken loop A safety loop that fits around the ankle to secure the ascender sling and to prevent the foot from slipping out of the sling should an upper connection fail and the climber fall backward.

Descender A rappel device that creates friction by means of a rope running through it; it is attached to a rappeller to control descent on a rope. Most descenders can also be used as a fixed brake lowering device. Also called a descent control device.

Descending The act of using friction to control the speed at which rope rescuers slide down a rope.

Figure 8 descender A commonly used descender made roughly in the shape of a figure 8.

Guide hand The hand that cradles the rope to help balance the rescuer. The nondominant hand (e.g., the right hand in a left-handed individual) is usually the guide hand.

Locking off The technique of jamming a rope into a descender or tying off securely so that the rescuer can stop the descent and operate hands free of the rope.

Prusik hitch A type of friction hitch used in ascending and belaying. The term Prusik also is used by some individuals as a verb, meaning to ascend, even when mechanical devices are used (e.g., to Prusik a slope).

Prusik loops Continuous loops of rope in which a Prusik hitch is tied.

QAS (Quick Attachment Safety) A short sling attached on one end to the climber's seat harness and on the other end to an ascender, which can be easily attached with one hand to a secure point. Also known as a *quick attachment point*.

Single rope technique (SRT) An approach to vertical ropework in which one rope is used for ascending and descending, without a second rope for backup safety or belay.

Tying off short A safety technique that creates an extra point of attachment during ascending by tying the person directly into the mainline rope.

On Scene

1. You need to access a vehicle about 50 feet (15.2 m) down the slope at a car-over-the-edge scene, but you have only a rope and no other gear. How do you do it?

2. At the top of an elevator shaft, you are preparing to descend by rappel to reach a stuck elevator car. What equipment do you need?

3. In ascending over an edge where the rope takes a 90-degree bend, what additional equipment might be useful to you in getting over the transition point at the top?

CHAPTER 19

Technician

Pickoff and Litter Management

KNOWLEDGE OBJECTIVES

After studying this chapter, you should be able to:

- Determine a rescue approach based on the location and condition of a subject. (**NFPA 1006: 5.3.1**, **5.3.7**, pp. 417–422)
- Differentiate between the approaches to a secured subject versus an unsecured subject. (**NFPA 1006: 5.3.1**, **5.3.2**, pp. 417–422)
- Differentiate between the approaches to an ambulatory subject versus a nonambulatory subject. (**NFPA 1006: 5.3.1**, **5.3.2**, pp. 417–422)
- Describe when pickoff rescues are performed. (**NFPA 1006: 5.3.4**, pp. 422–430)
- Evaluate feasibility of attempting a pickoff rescue. (**NFPA 1006: 5.3.1**, **5.3.2**, **5.3.3**, pp. 422–430)
- Describe methods of accessing a subject for a pickoff rescue. (**NFPA 1006: 5.3.1**, **5.3.2**, **5.3.3**, pp. 422–430)
- Direct the use of a rope system to rescue a stranded individual. (**NFPA 1006: 5.3.1**, pp. 422–430)
- Direct the use of a rope system to rescue a suspended individual. (**NFPA 1006: 5.3.2**, pp. 422–430)
- Describe the consideration for subject management in litters. (**NFPA 1006: 5.3.4**, pp. 430–434)
- Describe passing knots with a loaded litter. (**NFPA 1006: 5.3.4**, pp. 434–439)
- Explain the process of tending a subject-loaded litter during a lower. (**NFPA 1006: 5.3.3**, **5.3.4**, pp. 434–439)
- Explain the process of managing a litter across a highline. (**NFPA 1006: 5.3.4**, p. 439)
- Identify considerations for managing a distraught or distressed individual. (**NFPA 1006: 5.3.1**, **5.3.2**, **5.3.4**, **5.3.8**, pp. 439–440)

SKILL OBJECTIVES

After studying this chapter, you should be able to:

- Apply harnesses to a clinging subject. (**NFPA 1006: 5.3.1**, pp. 419–421)
- Perform a pickoff rescue. (**NFPA 1006: 5.3.1**, **5.3.2**, pp. 424–429)
- Transfer a subject between two lines. (**NFPA 1006: 5.5.3**, p. 429)
- Manage a litter loaded with a patient during a lowering operation. (**NFPA 1006: 5.3.4**, pp. 431–436)
- Pass knots with a loaded litter. (**NFPA 1006: 5.3.4**, pp. 437–438)
- Move a litter loaded with a subject across a horizontal highline. (**NFPA 1006: 5.3.4**, p. 439)
- Manage a distraught or distressed individual. (**NFPA 1006: 5.3.8**, pp. 439–440)

You Are the Rescuer

Your organization responds to a suicide attempt where someone has leapt from a bridge. Their fall was interrupted by impact with a lower part of the structure, upon which they have now become hung up. They are alive but immobile, about 20 feet (6.1 m) beneath the deck. Command decides to lower a rescuer to prepare the subject for packaging, and then to lower a litter with an attendant. There is no access for control of the load with tag-lines, and there is no egress from under the bridge or from the sides. Once loaded, the litter will have to be raised back up to the deck and managed away from the structure by the attendant.

1. What skills will the rescuer who is lowered initially need to have?
2. How will this rescuer attach to the rope?
3. How should the rescuer(s) approach the subject?

 *Access Navigate for more practice activities.*

Introduction

One of the key differentiators between a rescuer at the operations level and one performing at the technician level is the ability to effectively manage a subject while suspended vertically from a rope. It is one thing to be able to ascend and descend rope or even protect oneself while climbing a structure, but to also be responsible for the rescue and care of another person while performing these maneuvers introduces a whole new set of challenges. It is often preferable for the rescuer to be lowered or raised from above, rather than controlling their own descent/ascent. This allows them to focus more directly on the needs of the subject, whether physical, medical, or psychological, without interrupting or compromising progress.

In this chapter, we will explore personal vertical skills, to include methods for securing and managing a subject for rescue while interacting with them while suspended or secured in position with vertical exposure.

The Suspended Rescuer

While good ascending and descending skills are considered foundational skills for technician-level rope rescuers, there are clear advantages to performing a rescue while being lowered by a rope rescue system that is controlled from above by other rescuers. It is a bit like the difference between riding an elevator and taking the stairs. Freed from the physical exertion and mental attention required for ascending and descending, the rescuer riding a "rope elevator" is at liberty to focus more on surroundings; hazards; and of course, the subject.

A technician-level rescuer may be called upon to manage a subject who is packaged in a litter, or to manage a subject who is connected to the rescue system by means of a harness or extrication seat of the suspended rescuer. In addition, the rescuer could also be assigned the role of the third man. This role is not always utilized, but when it is, the purpose is to make initial contact with the subject and prepare them for

> **TIP**
>
> **Prerequisites for Performing this Chapter's Skills**
> Before attempting the activities described in this chapter, you must have demonstrated that you can properly perform all of the skills and knowledge contained in the previous chapters of this book. This includes (but is not limited to) the following:
> 1. Proper selection, care, and use of life safety rope and equipment.
> 2. Tie correctly and without hesitation the knots necessary for safe, effective work in the vertical environment.
> 3. Select and rig an appropriate anchor point for a rope rescue system.
> 4. Correctly operate the equipment and systems that will be used in the operations.
> 5. Communicate appropriately with other rescuers/co-workers in a rescue operation.
> 6. Apply the principles of descending and ascending, including adequate control, tie-off, and changeovers.
> 7. Extricate oneself from a jammed rappel device or similar problem.
> 8. Select, construct, and use an appropriate high-angle rope lowering system, raising system, and belay systems; conduct a system safety check and direct the operation of such systems.
> 9. Select, construct, and use an appropriate rope mechanical advantage haul system; conduct a system safety check and direct the operation of such a system.

extrication. Another role is the litter attendant. The person with this role is responsible for guiding and maneuvering the suspended litter through terrain and obstacles. This role will be discussed in detail later in this chapter. The third man role is discussed in Chapter 16, *Working in Suspension*.

Selecting the Rescue Approach

Initially, there are several decisions to be made about how to effect the rescue. The best approach will depend, at least in part, on how urgently the subject is in need of intervention:

- Are they suspended by rope, or relatively secure on a ledge?
- Are they clinging desperately to a vertical surface, in danger of falling at any moment?
- Are they injured? If so, how badly? What is the nature of their injury?
- Are they conscious or unconscious? Dead or alive?
- Are there other hazards contributing to the scene—weather, contaminants, unstable surface?
- Are there multiple subjects, or just one?

The answers to these questions, and more, will be obtained in the initial and ongoing size-up, and will help determine whether a simple snatch-and-go rescue is appropriate, whether a litter will be required, and what evacuation method will follow the extrication effort.

The order of business for most rescue operations is defined by the mnemonic LAST: Locate, Access, Stabilize, Transport. During a high-angle rope rescue operation, the rescuer may be required to negotiate obstacles, provide emergency medical care, manipulate or reposition the subject, or deal with emotionally distraught or unstable persons. For this reason, a key factor in selecting the rescue method is understanding the ability of the rescuer(s) to first maintain their own security in that environment as they gain access to the subject. Methods of protection might include conventional fall protection, lead climbing with a bottom belay, use of single or twin fall arrest lanyards, or lower from above. The rescue method chosen must be consistent with the skills and abilities of the rescuers present.

The term used to describe the transfer of a subject to a rescue system in a vertical environment is often referred to as a **pickoff**. Pickoff may be from a surface to the rescuer's rope system or from a suspended condition to the rescuer's rope system.

Accessing the Subject

When possible, performing a rescue from above, by lowering down to the subject, is generally preferred for the security and hands-free operation it permits, and for the fact that the effort is aided by gravity. With a lower from above, rescuers must first be able to reach the area above the subject to set anchor systems and rescue systems. Then, the rescuer assigned with first contact simply connects to a rope system and is securely lowered and/or raised into place by other rescuers. See Chapters 8, *Rescue Equipment* and 11, *Anchorages* for more information.

For a rescue from above, it may also be possible for the rescuer to descend (and ascend) a fixed rope under their own power. This may seem an attractive option as it offers the rescuer direct control of their own system, but in fact this is the very disadvantage of the approach. Having to manage their own descent/ascent system places that much greater physical and mental demand on the rescuer, dividing their attention with the care of the subject. While a self-controlled descending and ascending system may seem more "heroic," a system controlled from above offers greater security for both rescuer and subject, and allows the rescuer to place their unencumbered focus on hazard awareness, care of the subject, and extrication planning.

Free-climbing up a surface or structure to reach a subject is sometimes required, but this method is least preferred as it is generally slower and more physically taxing than being lowered or raised by rope. The anxiety of the subject can increase as they perceive the rescuer moving slowly toward their position, and the stress on the rescuer can be likewise compounded. Still, in some situations—such as on a communications tower or water tank support structure—there may be no other alternative.

The term **free-climbing**, which describes the act of climbing directly on a surface or structure, should not be confused with the term **free-soloing**, which is defined as unprotected free-climbing. A rescuer can protect themselves while free-climbing by use of a conventional fall protection system, personal fall arrest lanyards, or lead climbing techniques such as those rock climbers use. Lead climbing is a specialized skill that involves placing anchor points into the climbing surface as one ascends a surface. To make progress, the climber must climb past each placed anchor point, as illustrated in **FIGURE 19-1**, placing additional protection along their climb path as they proceed.

Regardless of the access method used, as the rescuer approaches the subject, they should speak calmly and remain out of reach, as there can be a tendency for a panicked person to want to reach out and grab

FIGURE 19-1 To make progress, the climber must climb past each placed anchor point.

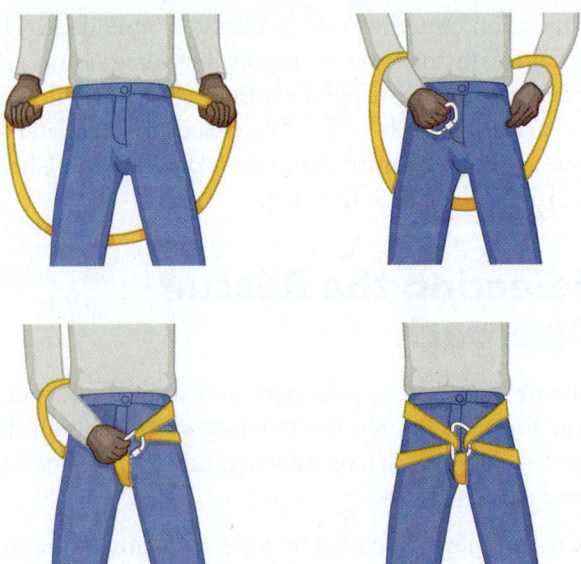

FIGURE 19-2 Diaper-style hasty harness.

hold of the rescuer for security. It is imperative that the subject not be permitted to encroach or encumber the rescuer, but that they wait for the rescuer to secure them properly to the rescue system.

Rescue of a subject who is clinging to a vertical or steep surface takes on a different sense of urgency than rescue of a subject who is relatively secure—for example, suspended by rope. Likewise, the rescue of a subject in medical distress also drives rescuers toward a more rapid intervention. The more urgent a rescue seems, the more important it is for rescuers to take a breath and clear their thinking before they act. Maintaining a sense of calm composure is essential to rescuer and subject safety.

Rescue of a Subject Clinging to a Surface

When a rescuer initially responds to a call, there is often no way for them to know exactly what they will face on arrival. In some situations, such as when responding to a popular rock climbing area or a local company known to use fall protection, rescuers might assume that the subject will be wearing a safety harness, but this assumption may or may not be accurate. In urban circumstances, rescuers may be even less certain what conditions they are likely to face on arrival. When responding to a potential rope rescue situation, whether urban or remote, it is best to assume that the subject is not wearing a safety harness, and to mount the rescue accordingly. In general, the means of holding and suspending the body of a subject for rescue consist of either a stretcher or litter, a manufactured extrication seat or harness, or an improvised system. These tools are described in detail in previous chapters.

A clinging subject who is ambulatory can be quickly secured using a hasty evacuation seat. Placing the evacuation seat on the subject can be a bit of a challenge, particularly if the rescuer is suspended from a rope at a difficult angle, dealing with a frightened or even injured subject, and/or in difficult environmental conditions. These skills are best practiced first on a manikin, initially on the ground and then at height. There is a good argument for also practicing pickoff rescue with a live subject before attempting a real rescue. For those organizations utilizing live subject practice, the live practice subject should be protected by a secondary system when appropriate.

Evacuation seats that are designed so that the subject does not have to step into them or through leg loops are preferred. These may be designed like a diaper, or more like a seat harness (**FIGURE 19-2**). When feasible, a manufactured evacuation seat or harness is preferable to an improvised one. The following points should be considered when selecting a manufactured evacuation seat or harness for use in high-angle rope rescue events:

- It should be designed so that it can be placed on the subject with as little disruption or movement as possible.
- It must be intuitive, and quick to put on. It should not have a multitude of straps, buckles, or adjustments.
- It should fit a wide range of wearers.
- The device should impart a sense of security to the subject.

Applying the Evacuation Seat to a Clinging Subject

Ideally, the rescuer should be able to capture and secure the subject quickly and easily in one motion, even if full adjustment may take a little more time. Manufactured versions are usually easiest to work with. A well-designed diaper-style hasty harness will cause the subject to ride very low in the seat to prevent inversion. Pickoff seats with a high center of gravity may require shoulder straps.

Approaching the subject from behind provides maximum control and minimizes the probability that they will try to grab the rescuer. Having the diaper-style hasty harness deployed and ready is helpful. Be sure that it is clipped into something—either its own rescue rope, or to a rope grab on the rescuer's rope—during the donning exercise to help prevent dropping it. A locking carabiner may be connected to one of the front connection loops and to a rescue line. To apply a diaper style harness, follow the steps in **SKILL DRILL 19-1**:

1. Grasp one of the waist tie-in loops in each hand, with the inside of the harness facing away from you. One of the tie-in loops should have a locking carabiner attached, and be connected to a safety line.

2. Wrap your arms around the subject, as if giving them a large bear hug, and clip the carabiner into the second waist tie-in loop. Now the subject at least has a horse-collar type connection to something.

3. Reach down between the subjects legs from the front and grasp the leg loops of the harness. Pull these between the legs toward the front, and up toward the carabiner that is already attached to the other two loops.

4. Clip these to the same carabiner.

It is also possible to use an improvised harness for a clinging subject. Although any number of improvised seat harness styles may be used as very viable methods of securing a subject, for a subject who is clinging precariously to safety it is best to use a hasty method. While the full-body arrangement shown in Chapter 14, *Lowering Systems*, is preferred, one quick way to apply a hasty improvised seat for rescue is to tie a long (10–15 ft [3–46. m]) length of webbing into a large loop using a ring bend. Then, approaching the subject from behind, apply it as shown in **SKILL DRILL 19-2**.

SKILL DRILL 19-2
Tying a Hasty Seat Harness

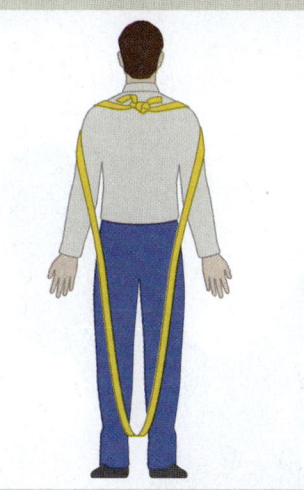

1. Place the loop across the subject's shoulder so that the sides of the loop hang down along their side, and the top of the loop runs across the back of the subject's neck.

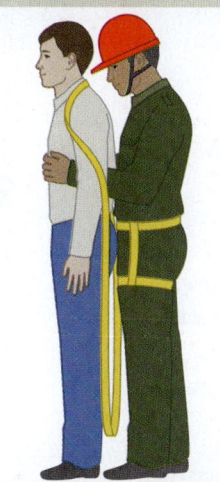

2. With both hands, reach around the sides and under the arms of the subject and the vertically hanging sides of the loop.

3. Reach down between the subject's legs from the front, and grasp the bottom of the loop behind.

(continues)

SKILL DRILL 19-2 (continued)
Tying a Hasty Seat Harness

4 Pull the loop back between the subject's legs and up toward the front of their waist.

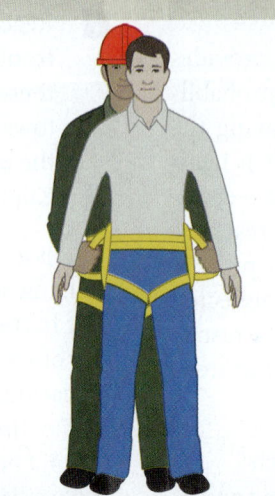

5 As you pull the slack out of the loop, the webbing will slide down to form the harness. Take a loop in each hand and pull each one to an opposite side so that the webbing is contoured around the subject's body.

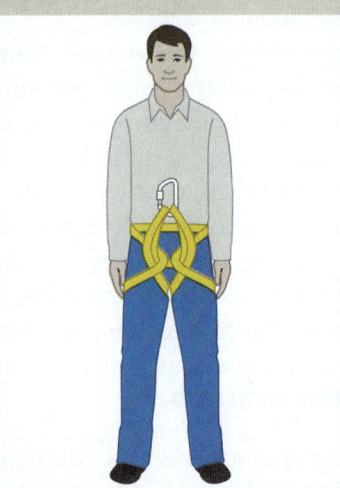

6 Bring the two loops back to the center and clip them together with a locking carabiner. Twist as needed to take up slack.

TIP

When performing pickoff rescue techniques, do not clip any attachments supporting the weight of the subject directly into your own seat harness. Doing so could exert painful and possibly damaging pressure on your body, particularly in the groin area. This also loads the harness in directions for which it was not designed, creating potential for failure, and impeding the movement.

Rescue Chest Harness

A rescue chest harness may be useful for subjects who have trouble remaining upright in a seat harness (**FIGURE 19-3**). This may include people with a relatively large upper body size or with large waists. A person who tends to be top heavy will be suspended in

FIGURE 19-3 Rescue chest harness.

a backward leaning position when sitting in a typical seat harness, whether manufactured or improvised. Placing a chest harness on a rescue subject can help them stay upright, which in turn will help reduce anxiety, facilitate maintaining an open airway, and make them easier to maneuver.

> **TIP**
> A chest harness must never be used alone, but only in conjunction with an appropriate seat harness.

To devise a simple rescue chest harness, follow the steps in **SKILL DRILL 19-3**.

This quick chest harness can be used together with a seat harness for a pickoff rescue. Connect the chest harness to the rescue system by clipping the carabiner from the chest harness into the end of the rescue sling (where you have previously clipped the carabiner for the seat harness.) In this way, both the seat harness and the chest harness may be adjusted independently of one another. Do not clip the two carabiners together.

A clinging subject typically does not need to be raised or lifted from the surface to which they are clinging. Rescue is usually a matter of simply and safely accessing their location (by lower, raise, ascent,

SKILL DRILL 19-3
Tying a Rescue Chest Harness

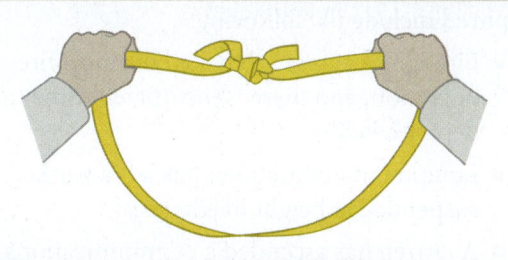

1 Tie a length of webbing into a continuous loop with a ring bend (water knot) backed up with an overhand knot.

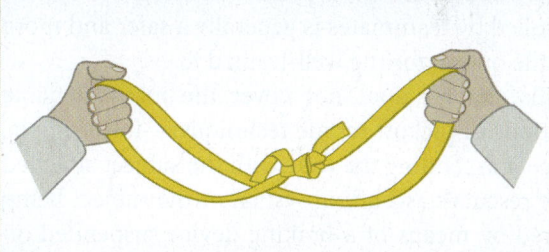

2 Twist the loop into a figure 8.

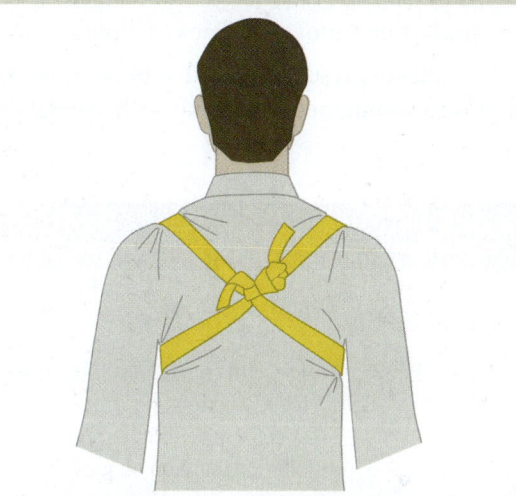

3 Lay the loop across the subject's back with the loop crossing on the back at armpit height and put the subject's arms through each loop.

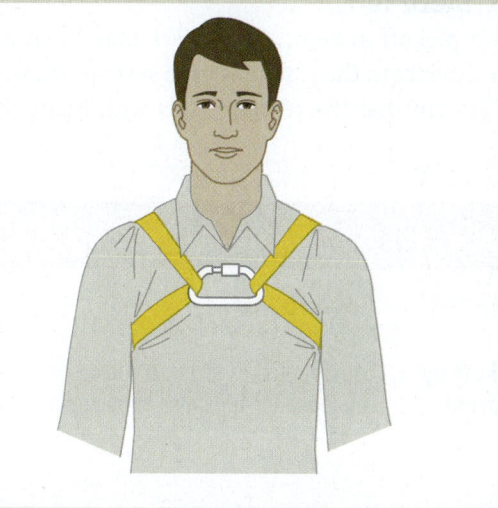

4 Bring each loop to the center of the chest and clip the loops together with a carabiner. Make sure the seat and chest system is equalized.

or descent), donning a harness (if necessary), attaching them to a rescue system, and prying them loose from whatever they are holding onto so that you may descend to safety.

Pickoff Rescue

The term *pickoff*, or *snatch rescue*, generally refers to the act of a single rescuer quickly accessing a subject in the high-angle environment and removing them from their hazardous situation. This is not to say that the individual rescuer is working alone—usually there is an entire team involved—but the simplicity of not using a litter expedites matters greatly.

The rescuer who is accessing the subject may be raised and lowered by other rescuers in a standard rescue system, or they may be controlling their own rappel (or ascent). In this section, we will discuss variations of rescue pickoff from above, both by lower and by descent, with preference toward rescue by lowering. There is little advantage to the rescuer controlling their own descent for a pickoff. Having the system controlled by teammates is generally a safer and more versatile option for the well-trained team.

This chapter does not cover the many possible variations of pickoff rescue techniques—for example, a rescuer ascending the rope with the subject attached to the rescuer's ascending system or the subject being lowered by means of a braking device suspended on rope or attached to midface anchor points while the rescuer remains in position on the rope. Follow the best practices and techniques of the authority having jurisdiction (AHJ) and train often to keep your skills sharp (**TABLE 19-1**).

In a pickoff system, the subject may be attached either directly to the rescuer's rope system or to a separate system that the rescuer trails with them. These techniques are most appropriate when the subject has only minimal (if any) injuries. The risk of further compromising an injured subject is simply not worth the risk, except perhaps in the most dire circumstances that pose an immediate threat to life, such as a hazardous atmosphere, explosion, fire, or violence.

Most pickoff techniques require that the rescuer begin the procedure above the subject and rappel or be lowered to the subject's area. When access is available only from below, the rescuer begins below the subject and ascends a rope to the area.

Pickoff rescue techniques may be appropriate in the following situations:

- It is appropriate for only one person to perform the rescue.
- There is a shortage of resources (personnel, equipment, time).
- The urgency of the situation dictates fast action.
- The benefits of using this method outweigh the risks involved.

Some examples of when a pickoff rescue may be required include the following:

- Firefighter egress is blocked during a fireground operation, and there is insufficient time to deploy a ladder
- Equipment malfunction has left a worker suspended at height in a harness
- A citizen has ascended a communications tower and cannot get down
- A rock climber has fallen and is only slightly injured but needs assistance
- A person has blundered into dangerous terrain and cannot move for fear of falling

In a pickoff rescue, several sets of rope skills are employed simultaneously, along with several different

TABLE 19-1 Advantages and Disadvantages of Rappel and Lowering Pickoffs

	Advantages	Disadvantages
Pickoff by Descent	Usually requires fewer rescuers Less need for communication among rescuers	Not easy to correct mistakes, such as the rescuer getting too far down to effectively perform rescue Greater pressure on one rescuer
Pickoff by Lower	Rescuer hands remain free Extra rescuers mean greater flexibility in choice of technique - No rope below rescuer for victim to grab and endanger the rescuer - Can easily be converted to a raise	Requires additional personnel and gear

kinds of equipment. Choosing appropriate methods for the skills, abilities, and equipment available is a crucial first step to success. The pickoff rescuer and all supporting rescuers must have appropriate knowledge of the equipment, the ability to perform the necessary skills, and adequate judgment to make decisions on the fly. This combination comes only with frequent practice and challenge.

Medical Considerations

Prior to attempting a pickoff, rescuers must evaluate and stabilize the subject's injuries. This requires their having at least some medical training, such as to the level of emergency medical responder (EMR). Determination of whether a subject is a good candidate for a pickoff-style rescue will depend at least on part on their medical considerations.

A subject suitable for pickoff rescue has minor injuries and faces no risk of disability or life-threatening injuries. The assessment of the subject should be thorough and as complete as possible considering the environment and precarious situation. Assessment of the subject follows a simple format of airway, breathing, circulation, and spine. If the subject has a problem in any of these subject assessment areas, consider whether it would be preferable to begin treatment and move the person with a litter evacuation rather than a pickoff rescue. Where a subject is in dire medical condition, or where level of consciousness is waning, a rescuer with more advanced medical skills may be deployed with the primary rescuer to focus primarily on medical care. Just keep in mind that as more rescuers, and more activity, are deployed onto a high-angle rescue site, the potential for dropped objects becomes an increasing hazard.

Even if the subject could potentially be rescued by pickoff, as long as no immediate threat to life or limb exists, it may be advisable to use an organized rescue team and follow proper spinal motion precautions with a litter. Erring on the side of subject care is always the best approach. For many groups, it takes just as long to perform a safe pickoff rescue as it does to perform a litter evacuation, because the latter procedure is practiced more often.

Pickoff rescues should be performed only by rescuers who have been trained in the technique and who have demonstrated the required skills.

To Belay or Not

As with any rescue scenario, the question of whether or not to incorporate a separate belay system will be determined by careful analysis of the potential risks and the potential outcomes. While belay may be desirable in some rappel pickoff rescues, in some cases it may not be feasible—and may even be contraindicated. If the use of a belay could impede the operation or even endanger the individuals involved, other options should be considered.

For example, in a rescue approach where both rescuer and subject are placed on their own rope, use of separate belays for both would require the operation of four separate rope systems. The potential for entanglement and confusion, additional rope management issues, requirement for more personnel and equipment, and overall complexity of the operation might suggest that separate belays would not be ideal. However, for additional protection, the operations leader may decide that a lanyard between rescuer and subject would suffice to offer dual protection to both, with the additional benefit of optimizing subject control. Situations in which a belay may be required include the following:

- Initial training and/or practice sessions
- Prolonged operations, in which fatigue is a factor
- Exposure to objective hazards, such as falling debris

The question of whether to use a belay in a rappel pickoff must be answered with decision making based on training, experience, department standard operating procedures, and the specific situation.

Teamwork and Communication

Pickoff rescue usually involves only one rescuer in direct contact with the person in distress. That said, it is quite likely that other rescuers will be involved. As with many rope rescue operations, the rescue process will proceed more safely and efficiently if other skilled and knowledgeable people are involved in essential tasks, such as belaying, lowering, or raising, spotting, and communications.

Also, if the rescue subject is to be rappelled or lowered to the ground and has injuries, personnel are needed at the arrival spot to attend to the individual's medical needs. Several people may be needed to perform a litter evacuation to an ambulance or to provide other medical care.

Protecting the Rescuer

Rescuers involved in pickoff rescues must be equipped with adequate personal protective equipment to the situation at hand. This includes the standard items of personal protective equipment (PPE), such as harness, helmet, gloves, and other equipment. However, PPE does not begin and end with rope-related gear.

Depending on the environment where the rescue will take place, rescuers may also require protection from the elements, from workplace hazards, or from contaminants. In a workplace rescue, would-be rescuers should be protected in a manner similar to, and at least to the extent of, workers who would normally be working in that environment. If possible, the subject should also be provided with adequate personal protective equipment, including a helmet, harness, gloves, and other PPE items. PPE is discussed in greater detail in Chapter 7, *Hazard Specific Personal Protective Equipment*.

Pickoff by Lowering

Lowering a rescuer from above to access the subject is perhaps the simplest and most expeditious approach to a pickoff rescue. Rescuers who practice litter lowers frequently will be adept in this method because it is very similar to a litter lower, just without the litter.

With this technique, a lowering team at the top lowers a rescuer to the subject. Most often, two rope systems are used: a mainline rope for the rescuer and subject and a separate line for safety. Several different alternatives for two-line systems will be discussed in this chapter.

A rescuer preparing to perform a pickoff by lower will connect to the lowering system, and to the belay if applicable, and then the lowering team lowers the rescuer to the subject. The rescuer attaches the subject to the mainline system and to the separate belay system. The team at the top then either lowers both individuals to the ground or converts to a haul system and hauls both to the top.

This method allows the rescuer versatility because the hand they would otherwise use to control a rappel is free to engage in care of the subject, and for maneuverability. As long as sufficient personnel and equipment are available, this approach should be considered.

To perform a pickoff by lowering, see **SKILL DRILL 19-4**.

SKILL DRILL 19-4
Pickoff by Lowering

1 Establish a lowering system (with ability to convert to raise) and a belay system that will position the rescuer next to the rescue subject.

2 With both systems locked off, connect the primary system and the backup to the rescuer.

SKILL DRILL 19-4 (continued)
Pickoff by Lowering

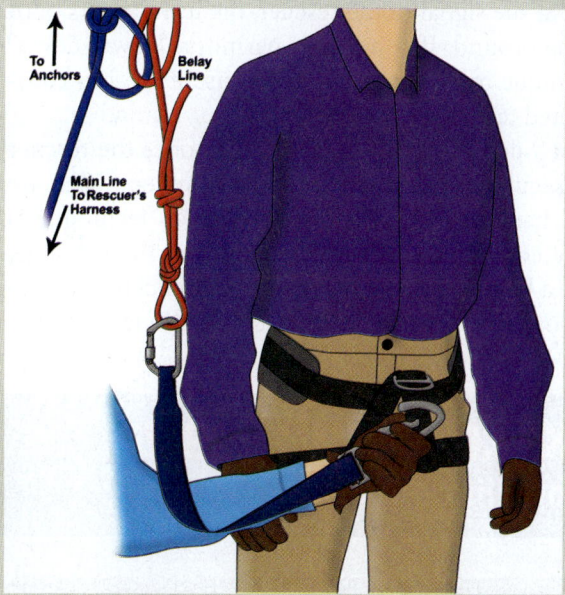

3 Station a spotter in a location where they can see the lowering rout and the subject, if possible. Perform a safety check. Commence lowering using standard communications and methods. When the rescuer is a few feet above the subject, they should call for "Stop." Once the rescuer is above the level of the subject, stop the lowering.

4 The rescuer attaches a pickoff sling from the main attachment point of the primary system to subject's seat harness and locks the carabiner.

5 The rescuer attaches a secondary sling from the main attachment point of the belay system to the subjects seat harness, and locks the carabiner. Upon command of the rescuer, the rescuer and subject are lowered (or raised) to a secure location. All rescuers remain at devices until told "Off main" and "Off belay."

In sizing up the rescue scene, the operations leader should determine whether they will use a "down, down" or a "down, up" technique. In a down, down pickoff technique, once the rescuer has reached the victim and secured the individual into the rescue system, at the signal of the rescuer, the team lowers both to the ground. If conditions permit, a "down, down" technique may be preferable because it is a less complicated technique.

In a down, up pickoff technique, once the rescuer has secured the individual into the system, both are then hauled to the top. The down, up technique is more complicated because the team at the top must have a hauling system that they then quickly attach to the lowering line and haul rescuer and victim to the top. Situations requiring a down, up technique may include the following:

- Distance to the bottom is too great to lower them with available line
- There is a hazardous area at the bottom
- Egress from the top is better

Pickoff by Rappel

Pickoff by rappel may be appropriate when time is of the essence, or when resources are slim. Although there are variations in technique for different pickoff rescue situations, the basic sequence is as demonstrated in **SKILL DRILL 19-5**.

SKILL DRILL 19-5
Rappelling Pickoff

1 The rescuer anchors a fixed rope for descent above the subject so that they will not rappel straight down onto the subject, but will be within reach. The rescuer anchors a second fixed rope for backup safety, very near the first.

Courtesy of Pigeon Mountain Industries.

2 The rescuer stops rappelling slightly above the subject and ties off the descent. The rescuer clips the pickoff strap to the subject's harness waist connection.

SKILL DRILL 19-5 (continued)
Rappelling Pickoff

Courtesy of Pigeon Mountain Industries.

3 The rescuer transfers the weight of the subject from whatever is holding them to the braking device. The rescuer and subject descend to ground and are removed from the system.

It is important that the rescuer rig their system off to the side of the subject so that it does not endanger them from the rope, the rescuer, or falling objects, but is close enough so that they can easily pendulum over to the subject.

Note that using two carabiners between the rescuer's harness and the braking device offers multiple advantages (**FIGURE 19-4**). First, extending the distance from the descender will make things more controllable when the subject is attached to the same descender, and second, the extra carabiner allows either the rescuer or the subject to be unclipped from the descender without compromising one another while under load.

Either an autolocking descender or an adjustable friction descender may be used. Be sure that the device used will withstand the weight of both rescuer and subject, and that it will be controllable under that load. For the purposes of this text, we will assume that the subject is wearing a seat harness; if they are not, first employ the methods for "Donning an Emergency Harness" found earlier in this chapter. We will also focus on pickoff of a suspended person; once this skill is attained, picking off a person

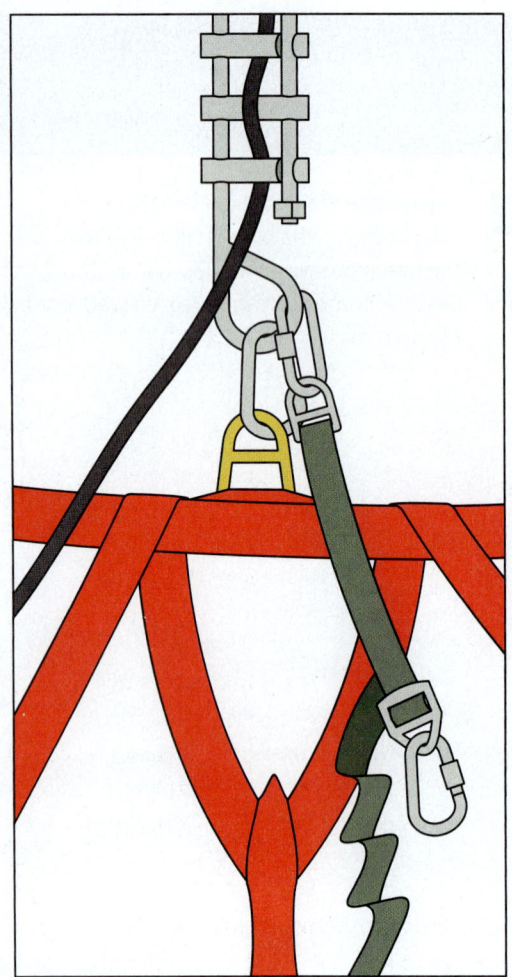

FIGURE 19-4 An extra carabiner between the harness and descender adds versatility.

who is more or less comfortably seated on a ledge, or otherwise not suspended, is relatively easy. To perform a pickoff rescue, follow the steps in **SKILL DRILL 19-6**:

1. Station a practice rescue subject or manikin wearing a full-body harness on rope. Rig fixed lines for primary and secondary systems. Attach an appropriate descender on the descent line (with two carabiners between the rescuer's harness and the descender, and an appropriate backup device on the safety line.
2. Clip a pickoff strap to the carabiner on the descender with another large locking carabiner.
3. Descend to the subject, stopping slightly above them but where you can just barely reach down and clip the pickoff sling to their harness. Clip the pickoff sling to their seat harness attachment point and pull out the slack. Perform a medical assessment as needed. Remove the subject from their system.
4. Brace with your legs spread wide, feet against the vertical surface, and subject between your legs. Awkward though it may seem, having the subject facing your crotch affords greatest protection and control. Descend to safety, protecting the subject as you do so by pushing with your feet against the surface and slowly "walking" backward down the face. As you approach the ground, instruct the subject to not stand, but to remain seated as the ground rises to meet them. When at the ground, disconnect yourself from the system while leaving the subject attached. The system will continue to hold them in a slightly upright position until you can effect further evacuation.

See Chapter 8, *Rescue Equipment* for a discussion of the advantages and disadvantages of types of descenders.

Pickoff by Traveling Brake

It is also possible to use dual mainline systems, known also as dual-tension rescue systems (DTRS) or twin-tensioned rescue systems (TTRS), for pickoff, but the traditional form of having two mainlines tended from the top can be more difficult to operate with the lower loads associated with pickoff rescue. One type of DTRS that works well for pickoff is the "traveling brake." Either autolocking or free-running braking devices may be used for traveling brake systems.

> **TIP**
>
> During pickoff rescue, anchorages, rope, hardware, and personnel are subjected to suddenly increased loads, shock loading, and loads that may come from directions other than those anticipated. For these reasons, the following rules apply:
> 1. Anchor points must be rigged for increased and multidirectional loading.
> 2. Carabiners must be locked, aligned in manner of function, and monitored so that they remain in manner of function.
> 3. The rescue system must have an adequate safety factor.
> 4. Rescuers must be prepared to handle sudden increases in weight and to provide extra friction on rappel devices.

A traveling brake system is essentially a cross between rescue by rappel and rescue by lower, combining the advantages and safety features of both into one system. To set up for a traveling brake system, the

FIGURE 19-5 Pickoff by traveling brake.

rescuer prepares for rappel and, at the same time, a brakeman prepares for rescue by lower. Both systems are attached to the rescuer, with the rescuer calling all commands and initiating all actions. Although the goal is to maintain consistent loading on both ropes, the system works more smoothly when the rescuers agree on which brake will be "primary" and the second brakeman takes their cues from the primary (**FIGURE 19-5**).

Initially, the rescuer who is on rope should serve as primary brake, but primary control will usually be transferred back and forth between the two throughout the operation. Serving as primary gives the rescuer on rope most control of their movement, while transferring primary to the top brake can be especially useful when negotiating obstacles or tending to the subject. Control is transferred back and forth by the brakeman stating, "Your load," which is then acknowledged by the new brakeman responding with "My load." Regardless of who is serving as primary brake at any given time, the rescuer who is on rope still always remains in command.

To practice a traveling brake pickoff, follow the steps in **SKILL DRILL 19-7**:

1. Station a practice rescue subject or manikin wearing a full body harness on a ledge or rope. Rig one fixed line for descent, keeping the bulk of the rope in a rope bag. The rescuer will carry this rope and deploy it as they descend. Rig another line for lower, with a fixed brake at the top.
2. Attach an appropriate descender on the descent line; connect it to the rescuer's harness with two carabiners between the rescuer's harness and the descender. Attach a pickoff strap to the carabiner on this descender with another large locking carabiner. Attach an appropriate descender to the top brake; connect the end of this rope to the rescuer's harness.
3. Rescuer positions for descent, then calls the command "Down Slow." The fixed brake opens their braking device and allows the traveling brake to take the majority of load, then matches their pace.
4. Rescuer descends; a few feet above the subject the traveling brake gives the command "Your Load." Fixed brake responds, "My Load" and takes control of the lower; traveling brake allows just enough rope to slip through their descender to ensure that the fixed brake has primary control. When just slightly above the subject, the traveling brake calls "Stop." Rescuer reaches down and clips the pickoff sling to the subject's harness, pulling out as much slack as possible. NOTE: If subject is not wearing a harness, the rescuer may need to descend to the same level, or even a bit lower than the subject, to put on an evacuation seat.
5. Perform a medical assessment as needed. Remove the subject from their system, and position them just below and between rescuers legs for optimum load management. Rescuer/traveling brake gives the command "My Load" and begins descent; fixed brake matches their rate of lower.
6. As ground approaches, instruct the subject to land on the ground in a seated position; rescuer should remain standing. When fully secure on the ground, rescuer/traveling brake calls "Stop" and "Your Load." Fixed brake maintains a secure hold on the rope until the rescuer and subject are ready to go off belay.

Releasing a Suspended Subject

If the subject is suspended from a rope by their harness, transferring them onto your system could be difficult. If they are suspended by their own autolocking descender, activating its opening mechanism should be enough to lower them onto your system. Otherwise, you may need to raise them slightly to get them onto your system.

The easiest way to do this is usually to place a haul system of some sort above them on their own rope, connecting it with a rope grab or Prusik. While small, prepackaged 4:1 or greater haul systems are commonly available for this, even an inefficient 2:1 using a piece of webbing can sometimes provide just enough force to raise the subject high enough to get them onto your line (**FIGURE 19-6**). Cutting their line should be avoided if at all possible, as wielding sharp instruments near fiber ropes can be a recipe for disaster.

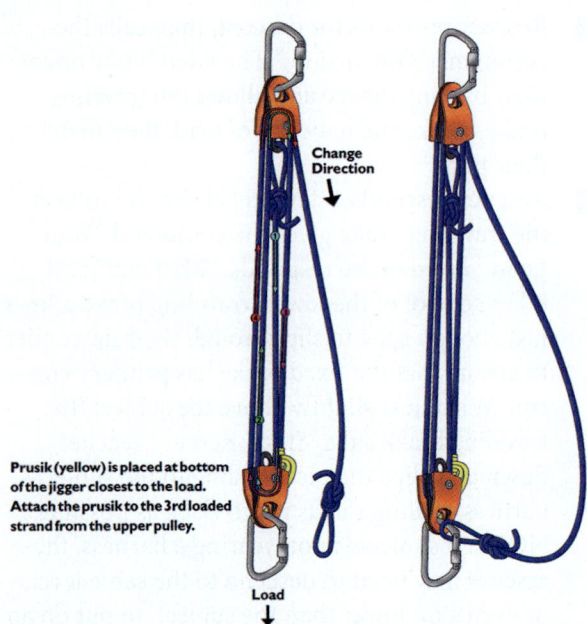

FIGURE 19-6 Prerigged haul system.

TIP

1. Before you go over the edge in a rappel, particularly in a rescue, check for loose personal items or vertical gear. All items and gear must be secured so that they do not fall out of pockets, packs, or gear slings. In addition to possibly injuring the rescue subject or other rescuers, you may lose an essential piece of equipment just when you need it the most.
2. Always perform a last-minute safety check:
   - Make sure harness buckles are correctly secured and all carabiners are locked and aligned in the correct manner of function.
   - Make sure knots are tied correctly and anchor systems are secure.
   - Check for any loose clothing or hair that may be drawn into the descender.
   - Make sure your helmet is secure.

Managing a Loaded Litter

When the condition of a subject warrants, a litter may need to be used for their extrication and evacuation. An operations-level rescuer might move a loaded litter from one stable location to another in preparation for rescue. With technician-level rescuers, it may become necessary for the rescuer to manage the litter with a subject loaded into it. See Chapter 14, *Lowering Systems* for more on operations-level litter management.

Litter lowering systems bear higher loads than systems where only one person is suspended. In a litter lowering system, the load includes the combined weight of the litter, hardware, and other rescue gear; the rescue subject; and one or two litter tenders. This increased load means greater stress on the entire vertical system, including ropes, carabiners, knots, anchor systems, braking systems, and the belay system.

When possible, a litter is lowered in a horizontal position. This usually is the more comfortable and reassuring position for the rescue subject. It also is less likely to complicate most medical conditions, and it makes it easier for litter attendants to tend to the subject's medical needs. However, in some cases, the litter may have to be lowered in a vertical position, usually in a confined space or when obstructions on the face require a small cross-section for the litter.

Methods for rigging and lowering litters in both horizontal and vertical orientations are addressed in Chapters 14, *Lowering Systems* and 15, *Mechanical Advantage Systems*. The methods in those chapters are discussed in context of an empty litter, but even when the litter is loaded the general principles remain the same. The primary difference is the physical exertion and awkwardness of maneuvering the litter.

A Note on Litter Bridles (aka Spiders)

Adjustable litter bridles are ideal, as they permit them to be optimized for conditions and circumstances. In many cases, very short bridle legs are desirable for the control they offer. In some situations, such as a mid-wall litter load, longer legs make loading the subject easier. The lengths of legs are a compromise between the following two factors:

1. The litter bridle must be long enough to allow loading and unloading of a subject on the midface if necessary.
2. It must be short enough to allow litter tenders to maneuver the litter in difficult terrain, to enable the tenders to reach the main attachment point (MAP), and to prevent the litter from flopping about while the tenders are trying to maneuver it around obstacles.

Tending a Litter Lower

Before practicing this technique, you should have mastered the skills throughout this book, especially those found in Chapters 14, *Lowering Systems* and 15, *Mechanical Advantage Systems*. Methods for executing a single point litter lower without a subject were covered in Chapter 14, *Lowering Systems*.

CHAPTER 19 Pickoff and Litter Management 431

For this exercise, the primary difference will be inclusion of a subject in the litter. During training exercises, efforts should be undertaken with something other than a real person—perhaps a bag of rocks or a manikin. If your agency permits, allowing experienced personnel to practice with live loads provides more realistic experience that is applicable to actual rescue.

To begin, rig rope systems for a single point litter lower and belay. Then, follow the steps in **SKILL DRILL 19-8**.

SKILL DRILL 19-8
Preparing the Litter for Lower

1. Rig a litter with a four-point spider for lowering. (See Chapter 14, *Lowering Systems*.)

2. Attach the belay line to the master attachment point with a figure 8 on a bight. Note: It is left to the discretion of the AHJ as to whether or not to leave a tail for attachment to the head end of the litter.

3. Attach the main lowering line to the spider MAP.

4. Connect a rescuer to the spider using an attendant rig.

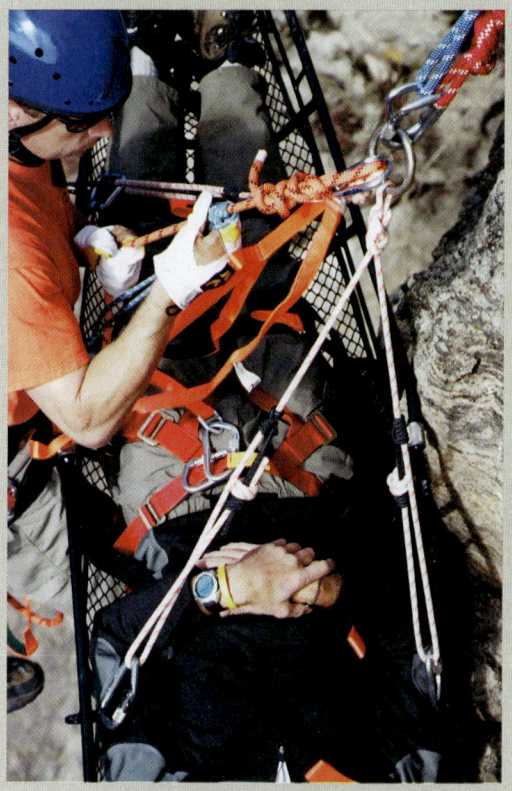

5. Lower the litter/load and attendant, on the same system.

Personnel requirements and positions should be similar to what is described in Chapter 14, *Lowering Systems*, except that there is but one rescuer attached to the litter. Review Chapter 14, *Lowering Systems* and Chapter 15, *Mechanical Advantage Systems* for the standard terminology used to communicate during litter lowering operations.

The operation begins with the team acknowledging their individual readiness. After the readiness roll call, communications are primarily between the rescuer and brakeman, however, any team member may call "Stop!" if a hazard is perceived.

If there is a subject waiting to get into the litter, once the belay is confirmed, the litter should be loaded. While the litter remains on belay the subject may be transferred into the litter. If the subject will be loaded mid-face, omit this step. Then, when the litter attendant is ready, they should initiate further action to the brakeman. Once the litter and rescuers are on the ground, the rescuer announces that they are "Off belay," and the brakeman acknowledges this statement. Communication commands and order may vary according to local standard operating protocol and the situation.

Prepare to practice the DTRS litter lower by setting up the system as for Skill Drill 19-8, then follow the steps in **SKILL DRILL 19-9**.

Traveling Brake

When resources are particularly slim, the number of personnel required can be reduced by reconfiguring

SKILL DRILL 19-9
Performing a Dual Tension Litter Lower

1. Package the subject into the litter that is properly secured by two lowering systems.

2. NOTE: If the devices are conducive to doing so, both brake devices may be attached to the same anchor point and run by one brakeman; the downside of this is that the shared anchor point and personnel translate to reduced redundancy. If two separate brakemen are used, for clarity in communications, designate one of these as primary and the other as safety.

3. The operations leader initiates the roll call.

4. The litter attendant initiates the lowering sequence.

5. The brakeman (and secondary, if required) holds the system in place as the litter attendant pretensions the system.

6. A check is made by the safety officer to ensure that all parts of the system have been rigged correctly, no lines will tangle, and that slack is removed from the system.

7. Once readiness is confirmed, the litter attendant calls for "Down slow."

8. The primary brakeman lets rope through the braking device to assert speed control; meanwhile, if a second brakeman is used, they should feed rope into their braking device, striving to maintain as equal tension as possible between the two systems.

9. At the top, before the load goes over the edge, the weight on the brake system may be insufficient to pull the rope through. The litter attendant may need to lean back, pulling the litter with them, while the brakeman reduces the friction. However, once the litter and attendant go over the edge, there will be greater weight on the system, and greater friction will be needed.

the dual-tension lowering system into a traveling brake litter lowering system. In a traveling brake system, the litter attendant also serves as a brakeman, in harmony with a standard braking system operated by a brakeman at the top. This method is most effective on vertical to overhanging drops where minimal litter attendant intervention and maneuvering is required, and is particularly conducive to mid-wall loading.

For this method, the legs of the litter spider should be extended so that the litter attendant can position themselves above the litter, between it and the MAP. One fixed line and one lowering line are anchored at the top of the drop. The end of the lowering line is terminated with a figure 8 on a bight and attached to the MAP. A braking device is attached to the MAP, into which the fixed line is reeved.

The litter attendant positions themselves inside the litter spider, between the MAP and the litter, attached with a very short attendant rig. An adjustable pickoff strap works well here. As with any other lower, the litter attendant commands the movement of the litter, and as with any other dual-tensioned system both brakemen (one of whom is also the litter attendant) strive to maintain approximately equal tension in their respective systems (**FIGURE 19-7**). This is easiest to achieve if one serves as primary brake and the other matches their control to that one.

Initially, the litter attendant/traveling brake should serve as primary brake; however, anytime they need to give extra attention to some function (such as getting over an edge, lowering a litter, etc.), they can transfer control to the other party by saying "Your load," which is acknowledged by reflective listening as the other party replies "My load." Regardless of who is serving as primary control at any given time, the litter attendant still always remains in command.

FIGURE 19-7 A traveling brake system.

To practice a traveling brake system, rig as described at the beginning of Skill Drill 19-8, then commence lowering as described in **SKILL DRILL 19-10**:

1. The litter attendant/traveling brake initiates the lowering sequence by saying "Down Slow."
2. The fixed brake opens their braking device and allows the traveling brake to take the majority of load, then matches their pace accordingly.
3. When the traveling brake gets to a point at which they wish for the fixed brake to take control, they say "Your Load."
4. Fixed brake replies "My Load."
5. Traveling brake requests "Down Slow"/"Down Faster"/"Stop" as applicable.
6. Fixed brake replies "Down Slow"/"Down Faster"/"Stop" as applicable.
7. Traveling brake requests primary control back by stating "My Load."
8. Fixed brake replies "Your Load."
9. Traveling brake calls out "Down Slow"/"Down Faster"/"Stop" as applicable.
10. Fixed brake replies "Down Slow"/"Down Faster"/"Stop" as applicable.
11. The exercise continues to the bottom.

Free-running, adjustable friction devices like brake racks work well for traveling brake litter systems, as they offer optimum control with one handed operation. Traveling brake operations are still feasible with autolocking devices that require two handed operation, but smooth operation requires more practice.

Scaffold-Type Litter Lowers

Another method is the use of scaffold-type litter lower systems, in which two litter attendants work together to manage the litter. This method incorporates a six-point litter spider (instead of four-point) and two MAPs (instead of one). It is most useful when particularly rugged vertical terrain poses additional difficulties, such as overhangs, breakdown blocks, and broken surfaces. A two-person vertical litter team is also useful when medical management of the subject is particularly difficult.

The downsides of this method is that having three people (two rescuers plus one subject) on the system imparts greater forces and stress on the equipment and rigging, rope management can be more complex, and more resources are required.

For two-attendant lowers, typically one serves as medical attendant while the other serves as litter captain. The former is typically positioned at the head, with the latter at the foot. That said, there are numerous regional variations and no right or wrong assignments here.

Litter Bridles for Scaffold-Type Lowers

Positioning two attendants on the litter works best when there is a certain amount of separation between them. The best way to achieve this is with a six-legged spider with two MAPs (**FIGURE 19-8**). For this type of bridle, there are three legs at the head end coming up to meet one lowering rope and three at the foot end coming up to meet the other lowering rope. This, in effect, creates a TTRS lower—but with a bit of space between the MAPs, and a bit more fall distance in the event that one fails. A tether between the two MAPs helps to prevent extreme fall distances in such an event.

With this type of litter rigging, the two litter attendants can work together to provide more maneuverability and strength in handling the litter. If necessary, one end or the other can be lowered at a different rate of speed, or to a different level, than the other to maneuver around obstacles. The litter can even be taken temporarily into a vertical configuration, although in this case, the tether between the two MAPs must be managed so that it does not impede the subject.

It is easiest to maintain a level litter if both lowering lines are run through a single lowering device, such as a brake rack or a brake tube. If autolocking devices are preferred, these can be used in tandem and operated by either one or two brakemen. To even out the loading in a two-line system, the lighter litter tender can be placed at the head end and the heavier tender at the foot end. Using different color ropes at head and foot helps with communications.

Practicing this type of lower is quite similar to practicing any of the others, but the interesting part to practice is that of lowering the two ends unevenly to negotiate obstacles. The steps for practicing this skill are in **SKILL DRILL 19-11**. For this exercise, we will assume that the litter has already gone over the edge and is being lowered with a yellow rope at the head and a blue rope at the foot.

With practice, teamwork, and good communication, the "Stops" between commands can be eliminated, making the operation continuous and smooth.

Passing Knots for Litters

If the distance of the litter lowering is more than one rope length, it may be necessary to go through a procedure known as a knot pass. In such a situation, a second length of rope (or more) must be tied to the first rope for the load to reach the bottom. However, some brake systems jam if knots enter them. Therefore, a bypass procedure must be used.

Use **SKILL DRILL 19-12** for passing knots with a litter. The system shown assumes a single-line lower with a belay (which, for clarity, is not shown) and a brake bar rack, but the skill can be used with other devices, such as an autolocking descender.

In addition to a regular lowering system, and the new rope(s) to be connected, an auxiliary braking system should be prepared using the following equipment:

- A separate, anchored auxiliary braking system
- A short length of auxiliary lowering rope (about 25 feet [7.6 m])
- A locking carabiner
- A rope grab for each rope in the main lowering system

For optimum efficiency, the separate brake system should be rigged and ready before the knot pass is imminent.

1. Anchor the auxiliary braking system near the main braking system so that it can be cammed onto the main rope just below the main brake, with at least 20 feet (6.1 m) of running room.
2. Rig the short length of auxiliary lowering rope through the brake device and tie a figure 8 on a bight at the lower end, with a locking carabiner attached.

With the auxiliary system at the ready, the knot pass is executed as follows in Skill Drill 19-11.

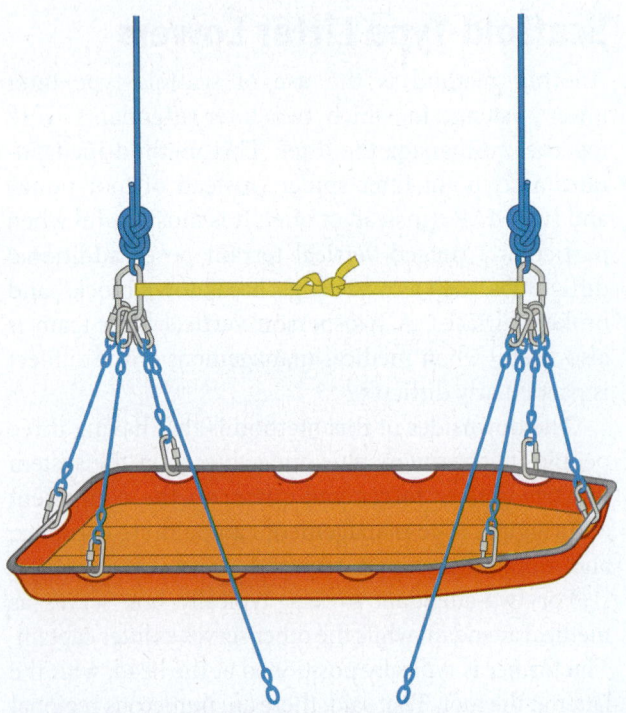

FIGURE 19-8 Six-point/two-MAP spider.

SKILL DRILL 19-11
Performing a Scaffold-Type Litter Lower

1 Litter captain to brakeman: "Stop!"

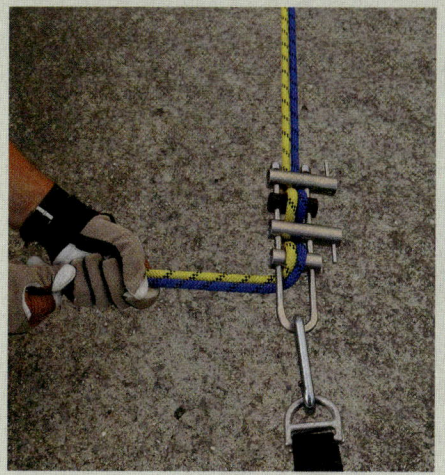 

2 Brakeman repeats "Stop."

3 Litter captain: "Down on Blue only" (or, "Down on Foot only")

4 Brakeman repeats: "Down on Blue only" (or, "Down on Foot only"); holding the yellow head rope still, they lower the blue foot rope.

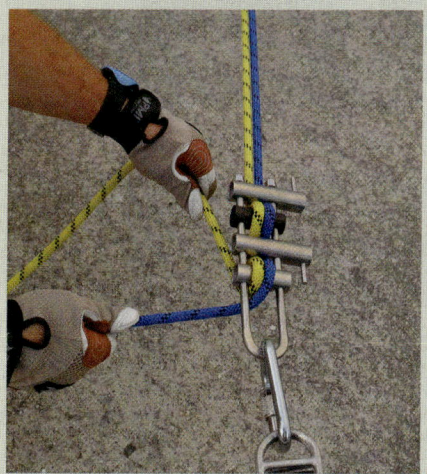 

5 Litter captain, when litter has reached the desired angle: "Stop."

6 Brakeman repeats, "Stop" and holds both ropes where they now are.

(continues)

SKILL DRILL 19-11 (continued)
Performing a Scaffold-Type Litter Lower

7 Litter captain: "Down slow."

8 Brakeman repeats "Down slow," and begins to lower both ropes at the same speed.

9 Litter captain, when obstacle is past: "Stop".

10 Brakeman repeats "Stop."

11 Litter captain: "Down on Yellow only."

12 Brakeman repeats: "Down on Yellow only."

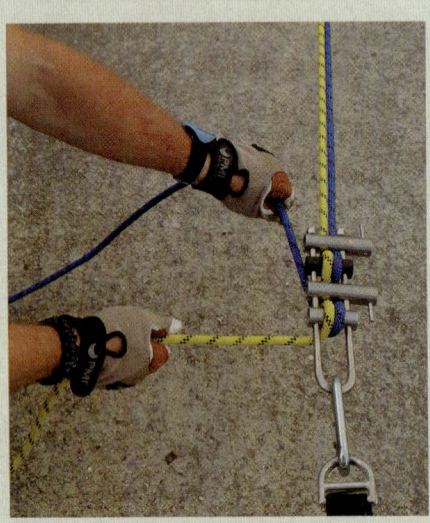

13 Litter captain, when litter has leveled out: "Stop"

14 Brakeman repeats: "Stop."

15 Litter captain: "Down slow."

16 Brakeman repeats "Down slow" until operation is complete or litter captain instructs otherwise.

SKILL DRILL 19-12
Performing a Knot Pass with a Litter

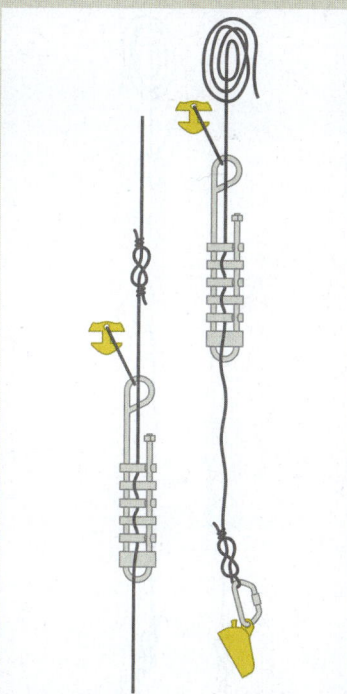

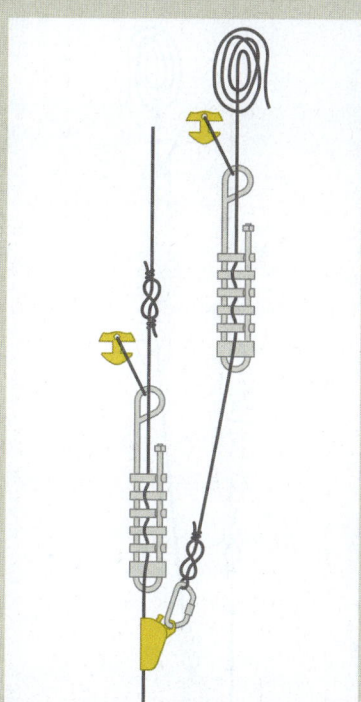

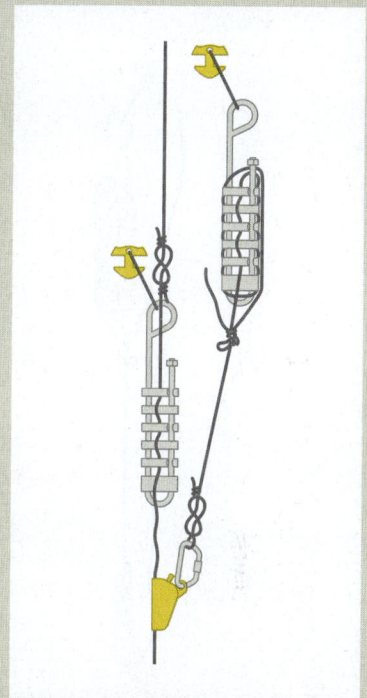

1 With at least 20 feet remaining on the main rope, attach the second rope to it with a neat and well-dressed knot.

2 When this knot is about 3 feet (0.9 m) from the main brake, stop the lowering operation.

NOTE: Ensure that the knot does not get too close to the main brake. If the system slips and the knot enters the brakes, it will jam and will require a difficult hauling procedure to get it unjammed. Place the rope grab on the mainline lowering rope just below the main brake so that it grips the rope when pulled back up slope.

3 The main brakeman calls: "Ready for knot pass." Auxiliary brakeman calls: "Ready!" Main brakeman calls: "Your load." Auxiliary brakeman calls, "My load," and maintains tension, without releasing or lowering the load. Main brakeman lets out rope until there is slack in the main system and the load is held solely by the auxiliary brake.

(continues)

SKILL DRILL 19-12 (continued)
Performing a Knot Pass with a Litter

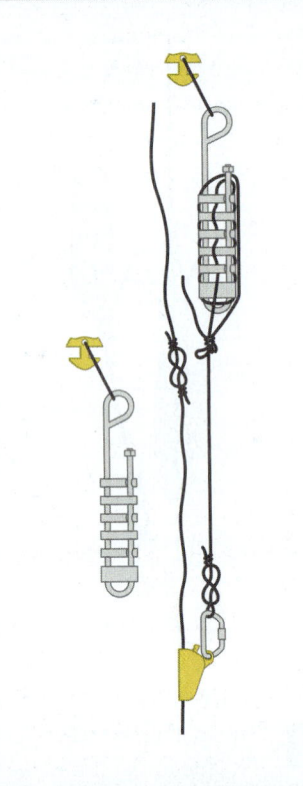

4 When the main rope is slack and the load is fully taken by the auxiliary brake, the main brakeman removes it from the brake system.

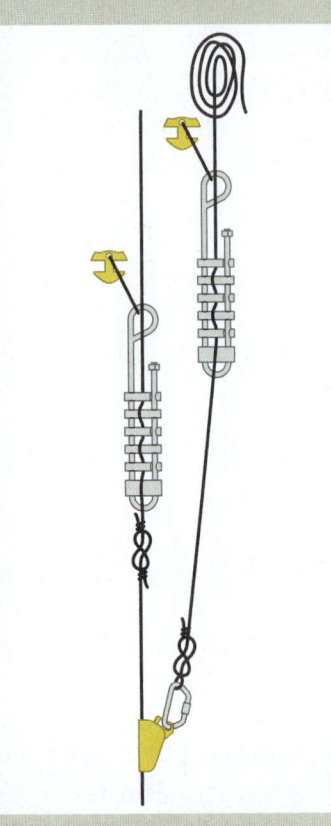

5 The knotted part of the main rope is carried downhill so it is well past the brake, and the mainline is re-reeved through the main brake, above the knot.

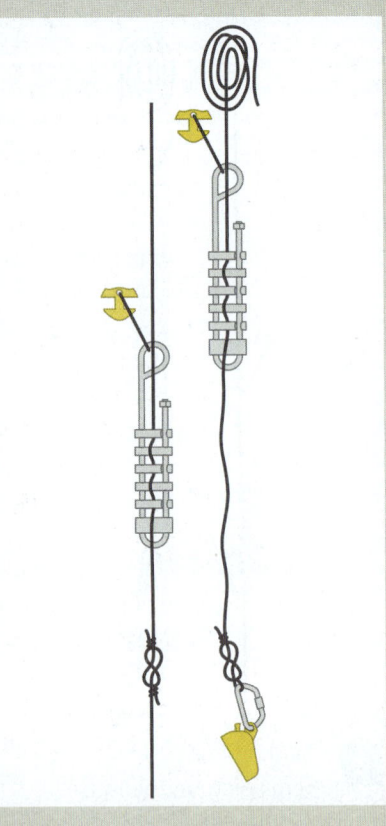

6 When ready, the main brakeman calls, "Ready for lower," and maintains tension on the mainline brake. Auxiliary brakeman calls, "Lowering," and begins to lower the auxiliary brake rope through the auxiliary brake until the load is once again taken by the mainline brake. When the load is fully on the mainline again, the main brakeman calls, "My load," and continues to hold steady. The auxiliary brakeman continues to feed rope through the auxiliary brake until it is fully slack, at which time the auxiliary rope is removed from the mainline. The lower continues. If there are two lines, as with a dual-tensioned system, the knot passes should be staggered so that they happen consecutively, not simultaneously

Subject Care

The procedures for loading the subject into the litter depend both on the individual's medical condition and on where the subject is to be loaded. Loading the subject into the litter is easier if it takes place on flat ground, such as at the top of the drop or on a ledge. More personnel may be available to assist, and all members of the rescue team will have solid footing. However, the litter loaded with a subject may be more difficult for the litter attendant to manhandle over an edge. In this situation, edge attendants can be of great assistance in helping to facilitate a smooth transition.

If the subject was injured partway down the steep or vertical face, the person will need to be loaded into the litter at that point. However, this type of litter loading can be very difficult, especially if it is not possible to place additional rescuers on the wall to assist with loading. It is difficult for a free-hanging rescuer to get any leverage with which to lift or shift the subject. In some cases, performing a pickoff-type rescue and lowering the subject into the litter may be the best solution.

When initiating a midface load, stop the litter a bit higher than you think you will need to. It is better to start the loading attempt with the litter too high, because the brakeman can always let a bit more rope out to lower. However, if you start out much too low, you may not get another chance. In addition, even low-stretch rope stretches some when the subject is loaded into the litter. Because it is difficult to lift the subject up to clear the litter rails, the optimum level for the litter is equal to the level of the subject.

Before going over the edge, be sure the litter is rigged so that it is in line for the subject's position (litter head and foot pointing in same direction as subject's head and feet). Consider placing an Oregon Spine Splint on the subject before movement, as this offers additional handholds for moving the subject, as well as a bit of immobilization.

For security, it is always a good idea to clip a safety line onto the subject with a seat harness before they are moved for loading into the litter. Once the subject is in the litter, a safety sling can be run from their seat harness to the carabiners at the top of the litter harness. When possible, it is nice to have a third man to assist in loading a litter partway down a drop.

Managing a Litter Across a Highline

At the technician level, it may be necessary to manage a loaded litter across a highline. The rigging methods and techniques are quite similar, the main difference being the awkwardness of the heavier, loaded litter. In Chapter 15, *Mechanical Advantage Systems*, we discussed skills for an operations-level rescuer to use in managing an unloaded litter across a highline.

That said, managing a litter across a highline is actually a good bit easier than managing a litter up or down a vertical face, simply because there is not much you can really do when you are hanging free midspan. Of course, this is the very reason that litters rarely need to be tended across a highline. On the rare occasion it is deemed necessary, the main reason for doing so is cited as being to comfort and reassure the subject.

Tying into the system is substantially the same as for the unattended litter. The litter attendant should wear gloves, and should have an etrier on hand in case they need to climb up onto the litter for any reason.

The litter should be rigged onto the highline in a manner that minimizes head down positioning for the subject. Of course, every highline has a "downhill" and an "uphill," so rescuers should take into consideration how long the litter is likely to be sloped in which direction. In a steep-angle highline, the litter will stay at a more or less constant angle until near the end, so rigging head up from the get-go is preferred. In a horizontal system that will require a lot of hauling on the far side, it may be predicted that the litter will spend more time sloped in that direction, and the subject positioned accordingly.

Managing a Distraught Subject

Many rope rescuers consider the rigging part of rescue to be the fun part. What can be more challenging is the mental and emotional aspects of dealing with a subject who is not happy about their predicament, or perhaps even afraid. Depending on the situation, the subject could be hanging upside down, exhausted from trying to right themselves, or even unconscious. Alternatively, they may be injured, in a drug-induced state, euphoric, angry, sad, belligerent, or any combination of these. Sizing up a pickoff rescue scenario should include a quick analysis of the both the physical and the emotional state of the subject.

Consider first whether the subject is in any immediate danger of further harm. Might they fall (further), are they exposed to any hazards (water, weather, contaminants, electricity, etc.), or is there loose material or debris above them that could be dislodged by the rescue?

What about their medical condition? Have they suffered a medical or traumatic injury? What are their physical complaints, if any? If the subject is not breathing, that is clearly an urgent problem—likewise if there is uncontrolled bleeding. Analyze the obvious injuries to determine what level of medical intervention they need, and how quickly. Whether or not they are conscious and alert may help drive your decision-making processes toward rescue methods.

Finally, consider their emotional state. While this may be the loudest or most visual part of their condition, it may or may not be the most critical. Consider whether they are calm or afraid, or even angry. A subject who is potentially hostile toward you can place you and other rescuers at risk, and should be treated accordingly. Is the subject able to follow simple directions, or are they panicked and likely to reach out as soon as a rescuer approaches? Again, these will help drive your decision-making processes.

Communicating with the Subject

Initially, most of your communications should be focused on assessing the situation and the condition of the subject. Once the process has begun, helping the subject to remain (or become) calm will contribute greatly to a successful outcome. Be sure to tell the subject who you are. Provide your name, and let them know you are a professional. Ask them questions—starting with their name—but avoid asking things that will rile them up or stress them out.

As the rescue progresses, the subject can be reassured by your communicating the intentions of the rescuers. Let them know what, exactly, is happening, and what steps are being taken toward their rescue. Be sure that someone stays in communication with them as often as possible. Make sure the subject knows how they can help—specifically, by not moving or grabbing anything unless told to do so! Good communications will help to ensure a successful outcome for the rescue.

After-Action REVIEW

IN SUMMARY

- A technician-level rescuer may be called upon to manage a subject who is packaged in a litter, manage a subject who is connected to the rescue system by means of a harness or extrication seat of the suspended rescuer, or assigned the role of the third man.
- When responding to a potential rope rescue situation, whether urban or remote, it is best to assume that the subject is not wearing a safety harness, and to mount the rescue accordingly. In general, the means of holding and suspending the body of a subject for rescue consist of either a stretcher or litter, a manufactured extrication seat or harness, or an improvised system.
- Pickoff, or snatch rescue, generally refers to the act of a single rescuer quickly accessing a subject in the high-angle environment and removing them from their hazardous situation.
- The rescuer who is accessing the subject may be raised and lowered by other rescuers in a standard rescue system, or they may be controlling their own descent or ascent.
- There is little advantage to the rescuer controlling their own descent for a pickoff. Having the rope rescue system controlled by teammates is generally a safer and more versatile option for the well-trained team.
- In a pickoff system, the subject may be attached either directly to the rescuer's rope system or to a separate system that the rescuer trails with them.
- Most pickoff techniques require that the rescuer begin the procedure above the subject and rappel or be lowered to the subject's area. When access is available only from below, the rescuer begins below the subject and ascends a rope to the area.
- Lowering a rescuer from above to access the subject is perhaps the simplest and most expeditious approach to a pickoff rescue.
- Pickoff by rappel may be appropriate when time is of the essence, or when resources are limited.
- Either autolocking or free-running braking devices may be used for traveling brake systems. A traveling brake system is essentially a cross between rescue by rappel and rescue by lower, combining the advantages and safety features of both into one system.

- In a litter lowering system, the load includes the combined weight of the litter, hardware, and other rescue gear; the rescue subject; and one or two litter tenders. This increased load means greater stress on the entire vertical system, including ropes, carabiners, knots, anchor systems, braking systems, and the belay system.
- When possible, a litter is lowered in a horizontal position. However, in some cases the litter may have to be lowered in a vertical position, usually in a confined space or when obstructions on the face require a small cross-section for the litter.
- If the distance of the litter lowering is more than one rope length, it may be necessary to go through a procedure known as a knot pass.
- The procedures for loading the subject into the litter depend both on the individual's medical condition and on where the subject is to be loaded.
- At the technician level, it may be necessary to manage a loaded litter across a highline.
- Keeping the subject calm is part of subject care.

KEY TERMS

Free-climbing The act of climbing a structure while using a rope or safety system for backup, but not as an aid to climbing.

Free-soloing The act of climbing a structure without using a rope or safety system for backup safety.

Pickoff The act of rescuing a subject from a precarious location while suspended by rope.

On Scene

1. Under what conditions might you choose to use each of the following?
   A. Lowering system with belay
   B. Dual-tensioned lowering system
   C. Scaffold-type lower

2. When is a pickoff contraindicated?

3. When might you choose to use a free running, adjustable friction braking device rather than an autolocking device for a rescue lower?

Chapter Opener: © Jones & Bartlett Learning. Courtesy of Loui McCurley; On Scene siren: © Bildgigant/Shutterstock.

CHAPTER 20

Technician

Special Rescue Disciplines for Additional Training

KNOWLEDGE OBJECTIVES

After studying this chapter, you should be able to:
- Identify other rescue disciplines with rope rescue skills recommendations or requirements. (pp. 444–448)

SKILL OBJECTIVES

There are no skills objective for this chapter.

You Are the Rescuer

Following a period of intense rain, your organization is called to a local park, where there is a known cave. There is a gate across the entrance and no one ever goes into the cave, except now the gate is missing and two teenagers are reported to have entered several hours ago. From the cave entrance, you can hear faint yelling, but you have no idea how far into the cave the teens have gone, nor do you know whether they are stuck, injured, or merely lost.

1. Will you send anyone in from your team?
2. Aside from darkness, what hazards might you be concerned with?
3. If rope rescue is required inside the cave, how will you anchor your rope rescue system?

 *Access Navigate for more practice activities.*

Introduction

The general principles outlined in this text apply to rope rescue in a very broad sense. A rescuer should take care to not overestimate their abilities or knowledge, because rope rescue methods as applied in specific environments warrant consideration of additional, environment-specific criteria. NFPA 1006, *Standard for Technical Rescue*, and NFPA 2500 (1670), *Standard for Operations and Training for Technical Search and Rescue Incidents*, reference rope rescue skills for various other disciplines, including the following:

- Structural collapse
- Confined space
- Swiftwater
- Trench and excavation
- Mine and tunnel
- Wilderness rescue
- Cave rescue
- Tower rescue
- Animal rescue

Rope Rescue for Structural Collapse

According to NFPA 2500 (1670), organizations operating at the operations level for structural collapse incidents must also meet the requirements of operations level for rope rescue, and organizations operating at the technician level for structural collapse incidents must also meet the requirements of technician level for rope rescue. The professional qualifications standard mirrors this requirement, stating that individual responders operating at the operations level for collapse must also meet the operations requirements for rope rescue, and individual responders operating at the technician level for collapse must also meet the technician requirements for rope rescue.

The intent of the requirement is to ensure that responders to collapse incidents have not only the technical knowledge of how to extricate victims from the rubble, but also the ability to use rope systems to transfer subjects away from the scene (**FIGURE 20-1**). A collapse zone may well be the height of the remaining structure or elements plus one-third that height again. At the very least, low-angle rope rescue methods are often required to lower extricated persons from the collapse zone. For responders capable of rigging them, a sloping highline can be an ankle-saver and help to expedite transfer of subjects, particularly if there are several. This also helps to reduce hazards to rescuers from subsequent collapse or slides.

Rope Rescue for Confined Spaces

Because the need for rope rescue is so prevalent in confined-space incidents, NFPA 2500 (1670) calls for organizations at the awareness level in confined

FIGURE 20-1 Rope systems may be utilized in structural collapses.
© AlenaPaulus/E+/Getty Images.

space to also meet awareness-level requirements for rope rescue. Organizations operating at operations or technician level for confined space will need to meet technician-level requirements for rope rescue.

A **confined space** is defined as one that is of sufficient breadth and configuration for a person to enter, but is not designed for continuous human occupancy. Another defining feature is that a confined space has limited means for entry/exit. Tanks, underground vaults, silos, sewage pits, marine engine rooms, and numerous other areas qualify as confined spaces, and rescues are not uncommon in these environments.

Rescuers operating in a confined space must be particularly aware of potential atmospheric hazards, among other things. Such hazards can result from dead-air space, built-up scale encapsulation of hazardous vapors, and other means. Air should be monitored prior to and during entry, and appropriate protection used to maintain the safety of the entrants. Note that respirators or self-contained breathing apparatus provide protection only against contaminants being ingested or inhaled; for complete protection, specialized protective clothing may need to be used to prevent absorption or injection hazards. Where use of a hazardous materials suit is warranted, care should be taken to ensure that it is rated to provide sufficient protection against whatever hazardous substances are likely to be present.

Personal protective equipment (PPE) for rope rescue and other equipment selected for confined space use may need to be tailored for the intended use. A well-fitting, secure helmet that does not shift around on the head is imperative. Fire helmets are not appropriate for use in confined-space rope rescue operations. A typical NFPA 1951 utility helmet is generally an appropriate choice unless other specific hazards dictate otherwise. While lighting is really not considered to be PPE in and of itself, care should be taken prior to entry to ensure that the helmet functions appropriately with and supports the intended light source. Some rescue helmets even offer a sleeve for a light source to slide into. Heavy, exterior mounted light sources can create additional hazards.

It is not always feasible to use standard harnesses, litters, or other patient packaging in confined spaces. Rescuers should practice in confined spaces with whatever **transfer devices** they may be likely to use to move the subject, whether it be short immobilization devices, pickoff seats, flexible litters, confined space litters (short or long), full-body harnesses with shoulder lift points, or other methods (**FIGURE 20-2**). One method that is unique to confined spaces is the use of wristlets (or anklets) for when it is not feasible to reach the torso of a subject to apply a body holding device.

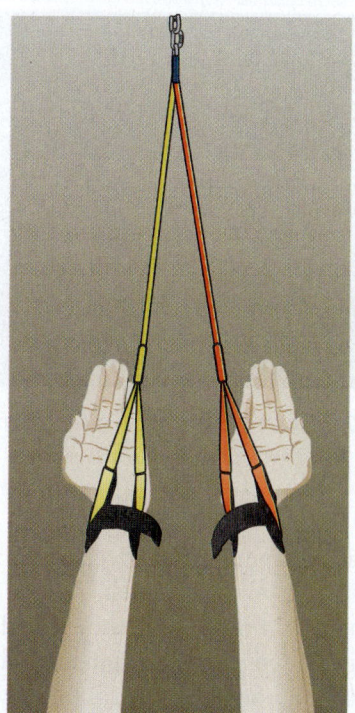

FIGURE 20-2 Wrislet transfer device.

Wristlets and anklets are painful for the subject to use, and are best practiced with manikins.

Rope systems may be used to insert rescuers into and remove rescuers and victims from confined spaces. Because utilization of lowering and raising systems in a confined space may be hampered by lack of space as well as twists and turns within the space, rigging considerations may need to be modified accordingly. Rope rescuers intending to operate in confined space environments should train and practice rope rescue techniques in every size, type, and configuration of confined space to which they are likely to respond.

Rope Rescue for Swiftwater Environments

Organizations operating at the operations level at **swiftwater** search and rescue incidents are well-served to meet rope rescue requirements at the operations level, and for technician-level swiftwater responders, meeting technician-level requirements for rope rescue is also advised. The use of throw ropes, tag-lines, and hauling systems are widespread in swiftwater rescue operations, so every member of the swiftwater response team would do well to understand and be capable of assisting rope rescuers with the construction and operation of rope rescue systems (**FIGURE 20-3**). Rope rescue training and knowledge are not specified by the NFPA standards for surf, flatwater, or other

FIGURE 20-3 Swiftwater rescue methods may incorporate rope techniques.

water rescue disciplines, but special training may be useful to these responders as well.

At the bare minimum, this should include an understanding of rope rescue equipment and its use, the ability to tie knots, and how to utilize the types of rope systems frequently used within their jurisdiction.

Even when NFPA standards do not identify a requirement for rope rescue knowledge and training for a given discipline, the authority having jurisdiction (AHJ) should analyze their response types and rescue standard operating procedures to determine whether additional training over and above NFPA recommendations is warranted. For example, many water rescue organizations frequently implement highline or sloping highline rescue methods—which, according to NFPA 1006, is a technician-level skill. Within these response agencies, a strong contingent trained to a rope rescue technician level would seem imperative.

Rope Rescue for the Wilderness Rescue Organization

Because of the terrain hazards commonly associated with **wilderness** rescue incidents, rope rescue training at the operations level for the operations-level wilderness responders and at the technician level for the technician-level wilderness responders is advised. Hazards including steep embankments, cliffs, swiftwaters, caves, and gullies are commonly encountered during wilderness search and rescue operations, requiring rescuers to be capable of negotiating these hazards both prior to and during extrication and transport of the subject(s).

Wilderness incidents can be very time consuming and resource intensive, so all members of the responding agency should be trained at the appropriate level in whatever disciplines they might be likely to use, including rope rescue. Having to call in specialized personnel to perform just one function is impractical and less likely to result in a positive outcome. Wilderness rescue is such a specialized discipline that areas where wilderness rescue occurs frequently are likely to have specialized resources, such as a team accredited by the Mountain Rescue Association (MRA), in place. These teams are likely to be operating under standards that are more specifically relevant to their work, such as those promulgated by the MRA or ASTM International (ASTM), rather than NFPA standards.

Personal equipment and rescue gear commonly used for urban response is generally not well suited to wilderness operations, and in some cases can even introduce or increase potential for injury. Typically, rope rescue equipment carried and used in wilderness operations is lighter weight that what is often found in urban rope rescue situations, with rope diameters as small as 7.5 mm (0.3 inch) and rarely larger than 11 mm (0.43 inch) common. Responding agencies who will interface with highly technical or specialized wilderness teams should take the time and make the effort to familiarize themselves with equipment commonly used, even training together where possible.

Urban-based response organizations on the edge of wilderness areas should establish a preplan for reaching out to specialized wilderness rescue resources through the MRA in the event that they are called to a wilderness rescue operation that extends beyond their level of comfort in wilderness–urban interface areas.

Rope Rescue for Trench and Excavation

NFPA standards recommend that responders at operations and technician levels for **trench** and excavation also be trained to at least the operations level for rope rescue. This is because rope systems are frequently used to haul a subject from a lower level in these types of incidents. Many trench and excavation incidents occur on construction sites, where trips and falls are high-risk hazards and confined spaces are likely present.

In a trench or excavation response, rope rescuers should block off a wide area to help mitigate potential for ground shock to initiate further collapse, to give trench rescue responders room to work, and to protect others in the area from additional hazards. The general area around an incident site is considered to be anything within 300 feet (91.4 m) (or even more, as determined by the incident commander). Rope rescuers will also be instrumental in these types of responses for setting anchors, accessing the subject, and extricating the subject to ground elevation (**FIGURE 20-4**).

CHAPTER 20 Special Rescue Disciplines for Additional Training

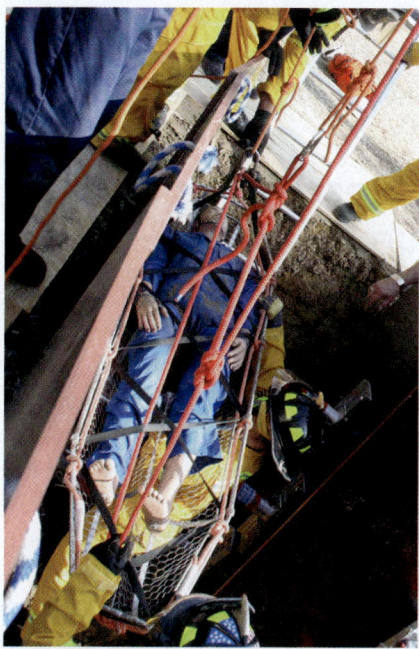

FIGURE 20-4 Rope rescue skills are used during trench rescue operations.

The hazards associated with trench and excavation collapse are difficult to enumerate because circumstances can vary so widely. It is incumbent on the AHJ to familiarize themselves with potential incident sites in their area, along with applicable laws and appropriate response protocols. All responders should be trained at least at the awareness level for any hazards likely to be present, and at least at the operations level for any associated rescue methods that they might be called upon to use.

Rope Rescue for Cave Rescue

Caves are another environment where rescue operations frequently demand a higher level of skill in rope rescue operations. While NFPA 2500 (1670) calls for organizations at the operations level in cave rescue to also maintain rope rescue abilities at the operations level, and for organizations at the technician level in cave rescue to also maintain rope rescue abilities at the technician level, it is wise for a cave rescue response organization to obtain even deeper training in rope rescue specific to cave rescue. Caves are especially challenging environments in which to rig, as anchorage points are difficult to analyze and seldom in the location(s) you want them, and the required path of travel for the subject is dictated by terrain and often convoluted. Response agencies operating in an area where caves exist and where cave rescue is likely to occur should establish a plan for contacting appropriate specialized resources for cave rescue incidents.

Rope Rescue for Mine and Tunnel Incidents

Mine and tunnel incidents are unique and distinctly different from cave incidents. Training and preparation to perform rescue in one does not constitute sufficient training or preparation for rescue in the other. Although a mine or tunnel may at first seem to fit the definition of a confined space, those requirements also do not apply here.

By definition, the term **tunnel** is used to describe a covered excavation (including shafts and trenches) through which people might travel. Underground mines, on the other hand, typically consist of a series of underground tunnels and shafts where material is excavated from the soil. The distinguishing characteristic between a mine and a tunnel is that in a mine the tunnel is the by-product of the soil removal process, whereas with a tunnel, the final product is the hole itself.

Guidance for active underground mines and excavations are well covered by existing standards and regulations, such as those of the Mine Safety and Health Administration (MSHA) in the United States. Because active mines are required by MSHA to have response capabilities on site, NFPA 1006 and NFPA 2500 (1670) guidance is targeted primarily toward abandoned mines and excavations. Organizations whose response area includes inactive or abandoned underground mines, excavations, and related structures should preplan for potential search and rescue to those areas. Surface mines such as quarries and open pits are outside the scope of these NFPA standards.

Rope rescue capabilities may be required in a mine or excavation rescue to lower rescuers into, and to remove rescuers and/or subjects from a lower space. NFPA-compliant rescue organizations that purport to address mine and excavation rescue will ensure that all awareness-level mine and excavation rescuers are cross trained at the awareness level in rope rescue, and that all operations- and technician-level mine rescuers will be trained in rope rescue at the operations level. Consider, too, that operating underground requires a certain level of comfort with underground spaces that not all rope rescuers will possess.

Rope Rescue for Tower Incidents

Municipal responders are regularly called upon to execute rescue operations on towers in the community. One of the difficulties with tower rescue is that there are so many different types of towers. Rescuers might encounter the need for rescue from a communications tower, power transmission or distribution tower, water tower, and more. In such environments, it is very important to consult with local or industry safety experts for maximum safety. If the incident involves a workplace, soliciting input from workplace safety managers can be invaluable.

In addition to the workers who regularly access tower structures, for whatever reason, they also seem to attract civilians who are feeling adventurous, depressed, or are otherwise disposed to climb them. Rope rescue is linked quite closely with tower operations in that most such rescues do require the use of rope for personal safety and for removal of the subject.

While a rope rescuer may consider themselves quite capable of figuring out how to execute a rescue from a tower structure, understanding the risks associated with doing so requires specific training. Hazards associated with tower rescue can be related to the rescue, or to design or structural integrity of the tower itself, trappings on the tower that are associated with the purpose and function of the tower, or even environmental hazards that tend to crop up in the area of towers.

Nearly any municipal responder has at least some likelihood of becoming involved in a tower rescue response at some point in their career. Safety is predicated on their having appropriate training in both rope rescue and tower-specific operations. NFPA recommends a close link between training levels, with rope rescue awareness being recommended for awareness-level tower rescuers, rope rescue operations training recommended for operations-level tower rescuers, and rope rescue technician training recommended for technician-level tower rescuers.

Rope Rescue for Animal Rescue Incidents

The emerging discipline of animal rescue is one born of need. While the cliché cat-in-a-tree scenario might be what leaps immediately to mind, the reality is much more widespread than this, with incidents ranging from farm animals falling into wells or vertical cave shafts, to dogs stranded on cliffs while hiking with their owners, to all manner of animals swept away by floodwaters. As responders find themselves being called upon more and more to respond to animal rescues, NFPA 1006 and NFPA 2500 (1670) committee members have opted to incorporate this specialty discipline into those standards.

Ropes are almost always a foundational element of animal rescue, whether for restraint of the creature, for fashioning a body-holding device, or for lifting. The weights of animals and associated forces on systems vary widely, from a dog on a cliff edge to a draft horse in a canal and almost everything between. For this reason, a close association between levels of response are called for in the NFPA standards. For operations-level animal rescuers, rope rescue capability at the operations level is recommended, and for technician-level animal rescuers, rope rescue capability at the technician level is recommended.

After-Action REVIEW

IN SUMMARY

- Special training is required to use rope rescue skills in specific environments.
- Structural collapse responders sometimes utilize rope rescue methods to extricate victims.
- Rope techniques are an important part of confined space rescue.
- Safe practices for using ropes differ greatly in swiftwater rescue.
- While rope is not always used in trench and excavation, operations skills are useful.
- Roped techniques may be, but are not always, a part of mine and tunnel rescue; these environments pose unique hazards.
- Wilderness environments should be approached with caution, as anchors and hazards can be deceiving.
- Roped methods are frequently required in cave rescue, and training for this differs greatly from training for mine and tunnel rescue.
- When responding to rope rescue at industrial sites, such as towers, co-workers of the affected person may already be involved in rescue attempts.
- Animal rescue is a specialized discipline that requires at least some level of knowledge about special packaging and handling methods for the animals.

KEY TERMS

Confined space A space that is large enough and so configured that a person can enter and perform assigned work, that has limited or restricted means for entry or exit (e.g., tanks, vessels, silos, storage bins, hoppers, vaults, and pits), and that is not designed for continuous human occupancy. (NFPA 1006)

Swiftwater Water moving at a rate greater than 1 knot. (NFPA 1006)

Trench (or trench excavation) A narrow (in relation to its length) excavation made below the surface of the earth. (NFPA 1006)

Transfer devices Components of equipment used by rescuers to package and transport a subject from a rescue environment. May include litters, harnesses, backboards, etc. (NFPA 1006)

Tunnel A covered excavation used for the conveyance of people or materials, typically no smaller than 0.91 m (36 in.) in diameter and within 20 degrees of horizontal. (NFPA 1006)

Wilderness A setting in which the delivery of services including search, rescue, and patient care by response personnel is adversely affected by logistical complications, such as an environment that is physically stressful or hazardous to the patient, response personnel, or both; remoteness of the patient's location, such that it causes a delay in the delivery of care to the patient; anywhere the local infrastructure has been compromised enough to experience wilderness-type conditions, such as lack of adequate medical supplies, equipment, or transportation; remoteness from public infrastructure support services; poor to no medical services or potable water; compromised public safety buildings, public utilities or communications systems; city, county, state or national recreational areas or parks with mountains, trails; areas they define as wilderness. (NFPA 1006)

On Scene

1. Under what circumstances might it be acceptable to allow a person not trained in a specific discipline to operate as a rope rescuer in that discipline?

2. What variables should a rope rescuer consider as potentially different, or difficult to analyze, from one environment to another?

3. When called to a tower rescue incident, what community resources might you consider reaching out to for collaboration during the operation?

Appendix A
Correlation Grid
NFPA 1006, 2021 Edition

Objectives	Corresponding Chapter(s)	Corresponding Page(s)
5.1	1, 2, 3, 4, 5	3–78
5.1.1	5	73–78
5.1.1(A)	5	73–78
5.1.1(B)	5	74–76
5.1.2	1, 2	6–7, 22–30
5.1.2(A)	1, 2	6–7, 22–30
5.1.2(B)	2	22–28, 30
5.1.3	3	34–46
5.1.3(A)	3, 4	34–36, 54–55, 56–57
5.1.3(B)	4	54–55
5.1.4	4	50–61
5.1.4(A)	4	50–61
5.1.4(B)	4	50–60
5.1.5	1, 5	4, 67–78
5.1.5(A)	1, 5	4, 67–78
5.1.5(B)	5	66–78
5.2	1, 2, 3, 4, 5, 6, 7, 8, 9, 10, 11, 12, 13, 14, 15, 16	3–336
5.2.1	6	84–90
5.2.1(A)	6	84–90

Objectives	Corresponding Chapter(s)	Corresponding Page(s)
5.2.1(B)	6	90–92
5.2.2	7	108–109
5.2.2(A)	7	98–103, 108–109
5.2.2(B)	7	108–109
5.2.3	8, 9	114–115, 142–145, 172–180
5.2.3(A)	8, 9	117–145, 172–180
5.2.3(B)	8, 9	114–115, 142–145, 175, 178–180
5.2.4	9	148–177
5.2.4(A)	9	148–177
5.2.4(B)	9	159–173
5.2.5	10, 11	185–192, 199–205, 212–216
5.2.5(A)	10, 11	185–192, 199–205, 212–216
5.2.5(B)	11	202, 216
5.2.6	10, 11	192–194, 199–201, 206–216
5.2.6(A)	10, 11	186–187, 192–194, 199–201, 206–216
5.2.6(B)	10	208, 210, 213, 216
5.2.7	10	185–191
5.2.7(A)	10	185–191
5.2.7(B)	10	216
5.2.8	10	188–191
5.2.8(A)	10	188–191
5.2.8(B)	10	188–191
5.2.9	10, 12	192–193, 220–237
5.2.9(A)	10, 12	192–193, 220–237
5.2.9(B)	10, 12	191–192, 194, 228–229, 232–234, 237
5.2.10	12	220–239

Appendix A

Objectives	Corresponding Chapter(s)	Corresponding Page(s)
5.2.10(A)	12	220–239
5.2.10(B)	12	226–229, 235–237
5.2.11	12	220–239
5.2.11(A)	12	220–239
5.2.11(B)	12	228–231
5.2.12	14, 16	270, 285–288, 327–330
5.2.12(A)	14, 16	270, 285–288
5.2.12(B)	14, 16	274–279, 285–288, 327–330, 332–334
5.2.13	14, 16	270, 285–288, 327–330
5.2.13(A)	14, 16	270, 285–288, 327–330
5.2.13(B)	14, 16	273–279, 285–288, 327–330, 332–334
5.2.14	14, 16	285–295, 324, 326–327, 331, 334, 335
5.2.14(A)	14, 16	285–295, 324, 326–327, 331, 334, 335
5.2.14(B)	14	274–275, 293
5.2.15	15	300–307, 311–317
5.2.15(A)	15	300–307, 311–317
5.2.15(B)	15	306–310, 316
5.2.16	15	300–301, 310–317
5.2.16(A)	15	300–301, 310–317
5.2.16(B)	15	310–312, 316
5.2.17	10, 15	194, 300–301, 311–317
5.2.17(A)	10, 15	194, 300–301, 311–317
5.2.17(B)	15	309–310, 313–316
5.2.18	15	300–301, 311–317
5.2.18(A)	15	300–301, 311–317
5.2.18(B)	15	310–312, 316

Objectives	Corresponding Chapter(s)	Corresponding Page(s)
5.2.19	16	324, 326–327, 331, 334, 335
5.2.19(A)	16	324, 326–327, 331, 334, 335
5.2.19(B)	16	326–327
5.2.20	13	244–259
5.2.20(A)	13	244–259
5.2.20(B)	13	251–258
5.2.21	14, 16	270–271, 275–277, 280, 283–285, 330–335
5.2.21(A)	14, 16	270–271, 275–277, 280, 283–285, 227–335
5.2.21(B)	14, 16	276–277, 227–283, 331–332
5.2.22	13, 16	259–264, 227–335
5.2.22(A)	13, 16	259–264, 227–335
5.2.22(B)	14, 16	278–279, 331–332
5.2.23	14, 16	285–288, 291–295, 330–335
5.2.23(A)	14, 16	285–288, 291–295, 330–335
5.2.23(B)	16	331–332, 335
5.2.24	6	92–94
5.2.24(A)	4, 5, 6	52–54, 76–78, 84–90,
5.2.24(B)	6	92–94
5.3	6, 7, 8, 9, 10, 11, 12, 13, 14, 15, 16, 17, 18, 19	84–412
5.3.1	19	417–430, 439–440
5.3.1(A)	19	417–430, 439–440
5.3.1(B)	19	419–421, 424–429
5.3.2	19	417–430, 439–440
5.3.2(A)	19	417–430, 439–440
5.3.2(B)	19	424–429
5.3.3	19	422–430, 434–439

Objectives	Corresponding Chapter(s)	Corresponding Page(s)
5.3.3(A)	19	422–430, 434–439
5.3.3(B)	19	429
5.3.4	19	422–440
5.3.4(A)	19	422–440
5.3.4(B)	19	437–438
5.3.5	17	343–352, 354–360, 362–364
5.3.5(A)	17	343–352, 354–360, 362–364
5.3.5(B)	17	352, 355, 357–360
5.3.6	17	343–345, 350–351, 354–357
5.3.6(A)	17	343–345, 350–351, 354–357
5.3.6(B)	17	352, 355, 360–361
5.3.7	12, 18, 19	220–239, 384–412, 417–418
5.3.7(A)	12, 18, 19	220–239, 384–412, 417–418
5.3.7(B)	18	384–412
5.3.8	19	439–440
5.3.8(A)	19	439–440
5.3.8(B)	19	439–440a
5.3.9	19	378–379, 382–406
5.3.9(A)	19	378–379, 382–406
5.3.9(B)	19	386–388, 391–392, 395–406, 414
5.3.10	19	372–384, 391–406
5.3.10(A)	19	372–384, 391–406
5.310(B)	19	377–384, 395–396, 398–406
5.3.11	19	398–404, 406–407
5.3.11(A)	19	398–404, 406–407
5.3.11(B)	19	406–412

Glossary

3-sigma A statistical method of calculating minimum breaking strength (MBS); this method results in a figure that is 99.87 percent reliable.

30-30 Tubular Restraint A method of securing a subject into a litter using two 30 foot (9.1 m) lengths of tubular webbing, one from shoulder to waist, and the other from waist to feet.

A

Active belay A means of protecting another person from a fall by means of a system that requires active participation on the part of a belayer, as in by exerting tension on a trailing rope.

Active fall protection A form of protecting a person from a fall through the use of harness-based personal protective equipment, and involving active participation of the user.

Adjustable low-angle litter tie-in A low-angle litter tie-in that can be adjusted.

A-frame Another name for a bi-pod (also called a sideways A-frame).

Ambulatory subject lower method The process of helping an ambulatory subject help themselves through a roped evacuation.

Anchorage connector How rope rescue components are secured to an anchor.

Anchorage (or anchor point) A single, structural component used either alone or in combination with other components to create an anchor system capable of sustaining the actual and potential load on the rope rescue system.

Anchor system One or more anchor points rigged to provide a structurally significant connection point for rope rescue system components.

Angle The intersection of two vectors.

Arm rappel (guide's rappel) A type of rappel in which the rope wraps around both outstretched arms and across the person's back. The technique is sometimes used in sloping terrain. It does not give enough control for vertical situations.

Ascender slings Attachments, usually webbing or rope, that connect a climber to the ascenders.

Ascenders Rope grab devices used to ascend a fixed rope or, with specific types of ascenders, to create a hauling system. Ascenders are also known as *ascending devices*.

Ascending A means of traveling up a fixed line using either mechanical devices or friction hitches attached with slings to the climber's body.

Ascending system An arrangement of two or more ascenders used for traveling up a rope.

Authority having jurisdiction (AHJ) An organization, office, or individual responsible for enforcing the requirements of a code or standard, or for approving equipment, materials, an installation, or a procedure.

Average arrest force The average amount of force that a person using a fall arrest system will experience over the duration of the catch.

Awareness level A functional level of technical rescue capability that represents the minimum capability of an organization or an individual to provide response to technical search and rescue incidents.

B

Back-tie A connector from a primary anchor to a second, backup anchor.

Backup system Secondary rope system, or "belay" used to provide protection against potential failure of a primary system.

Belay device A braking mechanism through which a secondary line, also called the belay line, is rigged. The device must allow free run of the belay rope through it when the system is operating correctly; it must exert a slowing or stopping action on the belay line if an uncontrolled descent or fall occurs on the mainline.

Belay To protect against falling by managing an unloaded rope (the belay rope) in a way that secures one or more individuals in case the mainline rope or support fails.

Belayer Person who operates the belay system.

Belaying To protect against falling by managing an unloaded rope (the belay rope) in a way that secures one or more individuals in case the mainline rope or support fails.

Belay line The line attached to one or more individuals that provides protection against a fall or system failure.

Belay rope The line attached to one or more individuals that provides protection from a fall.

Bend A class of knot that joins two ropes or webbing pieces together.

Bight The open loop in a rope or piece of webbing formed when the rope is doubled back on itself.

Body rappel (Dulfersitz rappel) A type of rappel that uses the body as friction by running the rope through the legs, across one hip, over the opposite shoulder, and to a braking hand. Because of the discomfort involved and the potential injury to body parts, this rappel has largely been supplanted by other techniques.

Body substance isolation The use of protective barriers to reduce the risk of transmission of infectious agents to healthcare personnel.

Bombproof Jargon for an anchor or anchor system believed to be very secure.

Brake bar rack A descending device (also known as a rappel rack) that consists of a J- or U-shaped metal bar to which are attached several metal bars, which create friction on the rope. Some racks are limited to use in personal rappelling, whereas others may also be used to lower rescue loads.

Brake hand The hand that grasps the rope to help control the speed of descent during a rappel. The dominant hand (e.g., the right hand in a right-handed individual) usually is the brake hand.

Brakeman Rescuer responsible for operating a braking device in a rescue system.

Braking/belay device A piece of hardware used to help manage the rate at which a load is lowered in a system or at which a rescuer descends.

C

Carabiners Load-bearing metal connectors with a self-closing gate used to link the elements of the high-angle system. Also sometimes called *biners*, *snap links*, or *krabs*.

Carabiner wrap A rappel technique that uses several rope wraps around a seat harness carabiner to create friction and control the descent. It generally is not considered a safe or secure technique for rappelling.

Carriage system A traveling pulley used on a track-line to enable a load to travel with minimal friction.

Carrier The piece of equipment that carries the bundles of sheath yarn around and around the core to form the outer braid during the manufacturing process.

Changeover To transfer from an ascending mode to a rappelling mode or from a rappelling mode to an ascending mode while on rope.

Chest harness A type of harness worn around the chest for upper body support. In the high-angle environment, it should never be used as the only source of support; it should always be used in conjunction with a seat harness.

Chicken loop A safety loop that fits around the ankle to secure the ascender sling and to prevent the foot from slipping out of the sling should an upper connection fail and the climber fall backward.

Class II harnesses Assemblies composed of webbing and worn primarily around the waist and thighs.

Class III harnesses Assemblies composed of webbing and worn around the torso, upper body, and shoulders.

Closed-loop communication A direct communication style in which a person requesting an action makes their request directly and explicitly, and then the person on the receiving end acknowledges the request and, if applicable, states when it has been completed.

Command and Coordination A NIMS term describing the structural interaction among leadership roles, processes, and recommended organizational parameters for incident management.

Command emphasis A section within the Federal Emergency Management Agency (FEMA) form ICS 202 that describes expected outcomes or milestones for the operational period, a list of priorities, or the key message(s) that underpin the effort for that operational period.

Command staff Positions that provide support and report directly to the incident commander, including public information officer, safety officer, and liaison officer.

Commodity ropes Generic, nonspecialty ropes used in household and commercial applications.

Communications and Information Management Procedures used to establish and maintain a common operating picture and systems interoperability during an incident.

Complex mechanical advantage systems Mechanical advantage systems in which traveling pulleys move in the opposite direction from the load or even from one another.

Component safety factor The ratio between the minimum breaking strength and an equipment component.

Compound mechanical advantage system A system in which one haul system is devised to piggyback onto another haul system, thereby multiplying force as one pulls on the other; see also *stacked mechanical advantage*.

Compression A physical force that presses on an object.

Confined space A space that is large enough and so configured that a person can enter and perform assigned work, that has limited or restricted means for entry or exit (e.g., tanks, vessels, silos, storage bins, hoppers, vaults, and pits), and that is not designed for continuous human occupancy. (NFPA 1006)

Counterbalance haul system A procedure for hauling that uses a 1:1 ratio mechanical advantage, employed by placing a pulley at an anchor a haul team at one end of the rope that moves in a direction opposite to the load using the haul team's own weight to offset the load.

Critical incident stress management (CISM) A program designed to reduce acute and chronic effects of stress related to job function (NFPA 450).

Critical point test An assessment that determines if there is a point in the system that, if it fails, will result in failure of the entire system.

Cross haul Multiple haul systems that act in opposition to one another to achieve horizontal movement.

D

Descender A rappel device that creates friction by means of a rope running through it; it is attached to a rapeller to control descent on a rope. Most descenders can also be used as a fixed brake lowering device. Also called a descent control device.

Descending The act of using friction to control the speed at which rope rescuers slide down a rope.

Design load The weight for which a product is designed to manage.

Directional A technique for bringing a rope into a more favorable position or angle.

Direction change A method of deviating the rope path by creating an angle, usually via a pulley placed into the system.

Direction of pull The specific direction in which a vector is applied.

Dual rope system A rope system in which access equipment is used on one rope and backup/safety equipment is attached to another.

Dual tension rope system (DTRS) A two-rope system in which both the primary and the backup are both loaded.

Dynamic directional system A type of track-line that involves two or more systems applying force in opposition to one another.

E

Emergency operations center (EOC) Location from which staff provide support to incident management with information and resourcing.

Emergency seat harness A temporary tied harness that is used when a manufactured, sewn seat harness is not available; also known as a *hasty harness*.

English reeve One of several rigging systems that can be added to a highline to control the load from either side.

Equipment standards NFPA equipment standards are test methods developed by consensus agreement among a balanced committee of persons representing user, regulatory, manufacturer, and other interests.

Escape devices Rope rescue equipment designed to be more compact, and suited to a smaller diameter of rope (usually 7–9 mm) so that they may be easily carried by a rescuer in a pocket or small pouch, ready to be deployed in a life-threatening emergency.

Etrier A short ladder made of webbing which is attached to a harness or on the rope above the attachment point to provide additional stability.

F

Fail-safe The failure of a device or system will not endanger life or property.

Fall arrest The form of fall protection that stops a load (or person) that is falling.

Fall line The path a ball would following rolling down an incline.

Fall protection Any equipment, device, or system that prevents an accidental fall from elevation or that mitigates the effect of such a fall.

Federal Emergency Management Agency (FEMA) An agency of the United States Department of Homeland Security intended to build, sustain, and improve our capability to prepare for, protect against, respond to, recover from, and mitigate all hazards.

Festoons Short cordage hangars used along a highline to suspend and control the droop of control lines beneath the track-line.

Figure 8 descender A commonly used descender made roughly in the shape of a figure 8.

Five Why's A technique used to explore fundamental cause-and-effect relationships of an occurrence.

Fixed and focused A multi-point anchor system in which the anchor points are rigged such that the load-connection point does not move even as the direction of pull might change.

Fixed pulley A pulley that is rigged into a system in a position where it will remain stationary at a specific location.

Flexible litter A litter that does not have an inherent rigid structure, but rather is made of a pliable material (usually plastic) that can be wrapped closely around the subject.

Flying W tensioning system Device used to pre-tension two track-lines as one.

Focal point The point where rigging comes together for maximum strength.

Footprint The area between the imaginary lines between the legs of a tripod that form a triangle.

Force The push or pull an object experiences as a result of its interaction with another object.

Free-climbing The act of climbing a structure while using a rope or safety system for backup, but not as an aid to climbing.

Free-soloing The act of climbing a structure without using a rope or safety system for backup safety.

Full-body harness A type of harness that offers pelvic and upper body support as one unit.

G

General staff Chiefs of each of the sections of ICS (Operations, Planning, Logistics, and Finance/Administration) who report directly to the incident commander.

Gin pole A monopod.

Go-Fever The response to take action without thoroughly considering potential risks or errors, to the extent that the drive to act overshadows clear thinking.

Guide hand The hand that cradles the rope to help balance the rescuer. The nondominant hand (e.g., the right hand in a left-handed individual) is usually the guide hand.

Guiding-line A low- to medium-tensioned line used to clear obstacles such as trees, rocks, or debris. Often has variable tension; the tension is tightened just enough to clear the obstacle and then released when the obstacle is cleared. Also known as a sloping highlines.

H

Hand How a rope feels in a person's palm; influenced by the number of yarn bundles used to create the sheath, tightness of weave, and material.

Handline A fixed rope anchored at the top of a slope and grasped by a subject(s) and/or rescuer(s) for stability.

Hauling systems Rope systems generally constructed from life safety rope, pulleys, and other rope-rescue system components capable of lifting or moving a load across a given area.

Haul cam A camming device used to grip a rope for the purpose of hauling.

Haul line A type of tag-line that the haul team uses for pulling on a load; with highlines, it is used to transport the load across the highline.

Haul system An arrangement of ropes and pulleys and connectors that is arranged in a manner to provide mechanical advantage for the purpose of raising a load under control.

Haul team The group of individuals who provide the power to raise the load.

Hazard identification and risk assessment Intentional exercise to identify all potential hazards within a given context, and then to assess the likelihood and possible severity of injury or harm.

Hazard identification Often referred to in context of risk assessment, this is an evaluation and analysis of environmental and physical factors influencing the scope, frequency, and magnitude of technical rescue incidents and the impact and influence they can have on the ability of the authority having jurisdiction to respond to and safely operate at these incidents. An examination of *what* things might cause danger.

Hazard-specific PPE Personal protective equipment particularly suited for protection against hazards common to rope rescue, and those additionally inherent in rescue disciplines addressed by NFPA 1006 and NFPA 2500, where rope rescue is also likely to be a factor.

Helmet A head covering that protects against head injury both from falling objects and from head impact.

High-angle rope rescue Use of a rope to make a controlled lowering of a rescue subject down a slope that is steep enough that the load is dependent on the rope for support.

Highline A system of using a rope suspended between two points to move people or equipment over an area that is a barrier to a rescue operation. Also known as a *telpher* or *tyrolean*.

Highline pulley A special large pulley used for the travelling pullet on a track-line, also called *Kootenay carriage* or *knot-passing pulley*.

Hitch A knot that attaches to or wraps around an object or rope in such a way that when the object or rope is removed, the knot falls apart.

Horizontal highline A highline in which the two suspension points are nearly on the same level.

Hot debrief An open, collective constructive analysis of an incident that takes place with all responders as a group immediately following completion of the operation.

I

Incident action plan (IAP) A written preplan that outlines key aspects of preparing for, responding to, and managing a given type of incident.

Incident commander (IC) The person responsible for managing an emergency response, including (but not limited to) developing incident objectives, assigning other roles and duties, and applying available resources to resolve the incident.

Incident command system (ICS) A standardized on-scene emergency management construct promulgated by the Federal Emergency Management Agency and specifically designed to provide for the adoption of an integrated organizational structure that reflects the complexity and demands of single or multiple incidents, without being hindered by jurisdictional boundaries.

Incident communications plan The Federal Emergency Management Agency (FEMA) form ICS 205 that enables rescuers to collect contact information for all personnel assigned to the incident including phone numbers, pager numbers, radio frequencies, call signs, etc., and functions as an incident directory.

Incident map A graphical representation, sketch, photograph, or actual map that shows the total area of operations, the incident site/area, impacted and threatened areas, or other graphics depicting situational status and resource assignment.

Incident objectives form The Federal Emergency Management Agency (FEMA) form ICS 202 enables rescuers to describe the basic incident strategy, incident objectives, command emphasis/priorities, and safety considerations for use during an operational period.

Incident safety officer (ISO) An individual assigned by the incident commander (IC) to help identify hazards and manage risk during an operation.

In-line haul system A manner of arranging ropes, pulleys, and connectors directly on the primary support line so that mechanical advantage is provided.

Installed anchorages Anchor points that need to be created and that do not exist in nature or engineered structures.

Interagency agreement (IAA) A legal instrument used between two response agencies to formalize the terms under which resources might be shared or exchanged.

J

Job performance requirements (JPRs) Written statements that describe a specific job task, list the items necessary to complete the task, and define measurable or observable outcomes and evaluation areas for the specific task.

K

Kernmantle A rope design consisting of two elements: an interior core (kern), which usually supports the major portion of the load on the rope, and an outer sheath (mantle), which serves primarily to protect the core but also may support a minor portion of the load.

Knots Fastenings made by tying rope or webbing together in a prescribed way. Knots include bights, bends, and hitches.

Kootenay carriage A modified knot-passing pulley that is used on highlines as the track-line pulley for single and double track-line systems or as a high-strength tie-off.

L

Ladder belts Devices that fasten around the waist and are intended for use as a positioning device for a person on a ladder. They should never be used as the sole means of suspension.

Laid rope made by twisting three or more strands together with the twist direction opposite that of the strands. Plain, or hawser, laid ropes have three strands, whereas shroud laid ropes have four strands.

Land search The search of terrain by Earth-bound personnel.

Lanyard with force absorber A length of rope or webbing with connectors at each end that is designed with an integrated means of mitigating arresting force on the body in the event of a fall.

Last known point (LKP) This is the point where the person was last known to have been, with a positive identification. It might be a trailhead, hunting camp, boat dock, parking lot, etc. Some agencies use the term *point last seen* instead.

Leapfrogging The use of two sets of rigging teams and brake teams to alternate the rigging of anchors and the operation of the brake for an extended evacuation.

Line A rope that is being utilized in a rope rescue system.

Litter A transfer device designed to support and protect a subject during movement.

Litter attendants Persons assigned to participate in the movement of a litter.

Litter bridle The mechanism by which a rope system is attached to a litter for the purpose of supporting and stabilizing the load.

Litter captain A litter attendant appointed to be responsible for the actions taken by a litter team during the time they are on the litter.

Litter spider An arrangement of straps and connectors used to connect to a litter so that it is appropriately balanced and stabilized and to provide a means by which a litter can be connected to a rescue rope; also called a litter bridle.

Load-distributing anchor system See *self-adjusting anchor system*.

Load ratio Term used to express the difference between the strength of something and the anticipated load that is expected to be placed on it.

Load-sharing anchor system See *fixed and focused anchor system*.

Locking off The technique of jamming a rope into a descender or tying off securely so that the rescuer can stop the descent and operate hands free of the rope.

Loop A portion of rope formed into a circle with the ends crossing one another.

Low-angle evacuation A rescue over any inclined or rugged area where a litter must be carried and where it is difficult or dangerous to do so without the assistance of a rope.

Low-angle litter tie-ins Method used by rescuers to attach themselves to a low-angle litter lowering system.

Lowering/belay line The line attached to a load on a highline that is used to control the load from the near-side point.

Low stretch A quality of a type of rope designed to be used in applications such as rescue, rappelling, and ascending in which high stretch would be a disadvantage and no falls, or only very short falls, are expected before the climber is caught by the rope; refers to ropes with slightly more elongation than the traditional static ropes.

M

Manner of function The method in which a particular piece of equipment was designed to be used.

Manufactured eye termination A permanent formed loop in the end of a rope created by the rope manufacturer, as an alternative to a knot.

Mass-casualty incident (MCI) An emergency situation where the number of patients exceeds the capacity of available resources. Generally defined as at least three patients.

Mass The amount of material in an object.

Master attachment point The main point at which a load (litter, spider, rescuer) is attached to a rescue system.

Maximum arrest force The largest amount of force that a person using a fall arrest system will experience as a result of being caught by the safety system.

Mechanical advantage A force created through mechanical means including, but not limited to, a system of levers, gearing, or ropes and pulleys usually creating an output force greater than the input force and expressed in terms of a ratio of output force to input force.

Medical attendant A member of the rescue team who provides emergency medical care to the subject during rescue and evacuation.

Medical plan The Federal Emergency Management Agency (FEMA) form ICS 206 enables rescuers to collect information on incident medical aid stations, transportation services, hospitals, and medical emergency procedures.

Memorandum of understanding (MOU) A formal agreement between two or more parties that outlines conditions or terms of a mutual agreement. Typically not legally binding.

Minimum breaking strength (MBS) The force at which tested samples of equipment actually fail during testing; calculated by a testing laboratory as three standard deviations below the mean breaking strength of a minimum number of samples tested in manner of use.

Multiagency coordination (MAC) groups Groups of representatives from different organizations who have authorization to commit resources from their respective agencies to a collaborative agreement.

Multi-point anchor system An anchor system in which multiple anchor points are connected together to form a single connection point.

Münter hitch A simple method of configuring a rope around a carabiner in a way that applies friction for the purpose of controlling a load, as for belaying.

N

National Incident Management System (NIMS) A comprehensive, national approach to incident management promulgated by FEMA for application at all jurisdictional levels and across functional disciplines.

Norwegian reeve One of several rigging systems that can be added to a highline to control the load; in this case, control occurs from just one side of the highline, allowing rescuers to raise or lower a load from the carrier at any position along the full length of the highline.

Nylon 6,6 A type of nylon used in rope manufacturing. With its resistance to wear and reduced elongation under load, most static or low-stretch ropes are constructed of type 6,6. In North America it is manufactured by DuPont and the Monsanto Corporation.

Nylon 6 A type of nylon used in rope manufacturing. Because of its shock-absorbing qualities, nylon type 6 is found in most climbing ropes. One trade name for this type of nylon is Perlon.

O

Operational period A period of time scheduled for execution of a given set of operational actions that are specifically specified in the incident action plan (IAP). Operational periods can vary in lengths, although do not normally exceed 24 hours.

Operational period briefing A meeting held at the beginning of each operational period with Command and general staff, along with other stakeholders, to share the incident action plan for the upcoming period

Operational Risk Management A systematic process used to assess and manage risks continuously during an operation.

Operation O The repetitive cycle of planning and operations outlined by the oval part of the Planning P, which continues and is repeated each operational period.

Operations briefing A meeting conducted at the beginning of each operational period to inform supervisory personnel within the operations section of the incident action plan (IAP) for the upcoming period.

Operations leader The person in charge of the rescue operations.

Operations level A functional level of technical rescue capability that represents the capability of an organization or an individual to respond to technical search and rescue incidents and to identify hazards, use equipment, and apply limited techniques specified in this standard to support and participate in technical search and rescue incidents.

Oregon Spine Splint A vest-like torso splint that wraps around and can be secured to a patient's upper body to help remove them from injury sites without doing further damage to the spine.

P

Packaging Process and materials used to protect and keep a subject from shifting in a litter during evacuation and transport.

Passive fall protection A form of protecting a person from a fall through the use of methods or equipment that does not require the wearing of personal protective harness or gear, and that does not require interaction on the part of the user.

PCD tender The person assigned to tend the haul cam, ensuring that it sets when appropriate and that it does not set when it should not.

Personal escape An escape harness typically worn during work-at-height operations explicitly for the purpose of enabling the rescuer to escape if needed to a lower position of safety; typically used with an emergency descent system.

Personal protective equipment (PPE) Those items of equipment that can be worn or used directly by rescuers to protect themselves against a recognized hazard.

Personnel accountability reporting (PAR) A radio-based roll-call system initiated by command at predetermined intervals to ensure that all personnel are safe and accounted for.

Pickoff The act of rescuing a subject from a precarious location while suspended by rope.

Pickoff seat A harness constructed of a combination of webbing straps and escape belt; also called a *rescue triangle*.

Piggyback rope system Any rope system that is used to exert force on another.

Pike and pivot bridle A method of rigging into a litter using straps and connectors to facilitate the pike and pivot technique.

Point last seen (PLS) This is the point on the map where the person was last spotted by a witness with a positive identification. It might be a trailhead, hunting camp, boat dock, parking lot, etc. Some agencies use the term *last known point* instead.

Polyester A type of fiber used in some rope manufacturing. Also known by the trade name Dacron. Dacron is often found in the core of very low-stretch static ropes due to polyester's much lower elongation compared to nylon fiber.

Polyolefin A group of fiber types (e.g., polypropylene, polyethylene) used in the manufacture of ropes often used in water applications.

Prerigging During a rescue operation that requires transfer of the load from one system to another, the act of setting up a subsequent system while waiting for the load to arrive.

Probability of area (POA) The likelihood that a subject is in a given area or search segment. POA changes for each segment after any portion of the total search area is searched

Probability of detection (POD) The likelihood of a subject (or a clue) being found if it is in a given search area.

Progress-capture device (PCD) A rope-grab device, general-use ascender, or hitch placed on the rope in a hauling system to prevent the rope (and the load) from slipping back down as the haul system is reset. Also commonly referred to as the ratchet.

Progress capture The act of preventing a rope from traveling in a direction so that a load is maintained in a static location.

Protected self-adjusting anchor system A self-adjusting anchor with built-in redundancy.

Prusik hitch A type of friction hitch used in ascending and belaying. The term Prusik also is used by some individuals as a verb, meaning to ascend, even when mechanical devices are used (e.g., to Prusik a slope).

Prusik loops Continuous loops of rope in which a Prusik hitch is tied.

Public information officer (PIO) Individual assigned with communications relative to both press and members of the public to provide a consistent, clear, and accurate message about the incident.

Pulley A device with a free-turning, grooved metal wheel (sheave) used to reduce rope friction; it also has side plates to which a carabiner may be attached.

Q

QAS (Quick Attachment Safety) A short sling attached on one end to the climber's seat harness and on the other end to an ascender, which can be easily attached with one hand to a secure point. Also known as a *quick attachment point*.

R

Ratchet cam A device that permits rope to travel through in one direction but not the other, commonly used as a progress capture device.

Reeve A broad term used to describe the act of threading rope through pulleys to create a haul system.

Reporting party (RP) The person(s) who report an incident.

Rescue equipment Equipment or gear used to perform a rescue task.

Rescue incident safety control (RISC) chart A visual tool to help incident managers and responders to identify specific hazards at a rescue site, and to visualize strategies for contending with them.

Rescue load Generally accepted amount of weight expected to be borne by a rescue system.

Rescue preplan An emergency response plan that is prepared prior to an incident occurring, based on a concept and generalized parameters.

Rescue solutions A complete, prepackaged system that can be used to execute a rescue.

Resource Management A term used in NIMS to describe coordination and oversight of personnel, tools, processes, and systems used during an incident.

Restraint A system that prevents a person at height from reaching an edge where a fall might occur.

Resultant The net force exerted on an object by the combination of all force vectors acting upon it.

Rigger Rescuer responsible for building, analyzing, and adapting rescue systems in the field.

Rigid basket litter A patient transport device with a protective shell and depth of several inches used to carry and protect a subject during technical rescue operations.

Risk assessment A determination of *how great a danger* is posed by a given hazard.

Rope access A set of techniques where hardware is used as the primary means of providing access and support to workers. Generally, a two-rope system is employed: the working (main) rope supports the worker, and the safety (belay) rope provides backup fall protection. The two ropes may be switched back and forth between being the main and belay, but the worker must always be attached to two ropes. Rope access workers must be trained in self- and partner rescue; outside rescue sources may also be needed.

Rope access backup A vertical lifeline-like type of safety system used in rope access, wherein the backup system is completely interchangeable with the progress system.

Rope grabs Device that grips the rope.

Rope rescuer Rescuer primarily tasked with extrication and removal of the subject from hazard.

Rope rescue system Rope system used to effect a rescue.

Rope system Rope and equipment that is rigged together to perform a specified function.

Running end The portion of rope used for lowering or hauling, or that will touch the bottom when rappelling.

S

Safety knot A second knot used to secure the tail of a primary knot; also known as a backup or keeper knot.

Safety message The Federal Emergency Management Agency (FEMA) form ICS 208 is an optional form that may be included and completed as part of the incident action plan (IAP) to expound or emphasize key safety information regarding safety hazards and specific precautions to be observed during a given operational perod.

Safety officer (SO) The person responsible for monitoring current and projected hazards associated with dangerous conditions affecting rescuers and others.

Scope As pertains to search and rescue, the extent of the area or subject matter that is relevant to the incident.

Scree evac A type of low-angle evacuation over a loose, rocky slope, typically in wilderness terrain.

Seat harness A system of nylon or polyester webbing that wraps and supports the pelvic region to attach the wearer to the rope or other protection in the high-angle environment. There are two classes of NFPA harnesses: Class II (meant for heavy-duty work by one person or in rescue situations in which another person's weight may be added in the course of the rescue) and Class III (a full-body harness meant for fall protection and rescue where inversion may occur).

Self-adjusting anchor A multi-point anchor system that is rigged so that the load-connection point floats as the direction of pull changes.

Self-belay A form of fall protection that a person manages themselves, such as a rope grab.

Self-retracting lifeline A type of fall arrest device that contains a drum-wound line that pays out from and automatically retracts back onto an enclosed drum during use, and that automatically locks at the onset of a fall to arrest the user.

Single rope technique (SRT) An approach to vertical ropework in which one rope is used for ascending and descending, without a second rope for backup safety or belay.

Situational awareness The ability to know what is going on around us.

Size-up The ongoing observation and evaluation of factors that are used to develop strategic goals and tactical objectives.

Skate block A rescue system in which one continuous rope is used to form both the track-line and the lowering system for a load.

S-M-A-R-T acronym A tool used in goal-setting to help guide planners toward setting objectives that are Specific, Measurable, Achievable, Realistic, and Timely.

Span of control The ratio of subordinate personnel to one supervisor during an emergency incident. The optimum span of control is considered to be 5:1; this may be increased to a maximum of 7:1.

Stacked mechanical advantage A system in which one haul system is devised to piggyback onto another haul system, thereby multiplying force as one pulls on the other; see also *compound mechanical advantage system*

Standard operating procedures Written organizational directives that establish or prescribe specific operational or administrative methods to be followed routinely for the performance of designated operations or actions (NFPA 1521).

Standing part The portion of rope between the running end and the working end.

Static rope A type of rope designed to be used in applications such as rescue, rappelling, and ascending in which high or moderate stretch would be a disadvantage and in which no falls, or only very short falls, are expected before the climber is caught by the rope. Static ropes have slightly less elongation than low-stretch ropes built to the same standard.

Static system safety factor (SSSF) The ratio of the overall rope system strength divided by the anticipated load when it is not moving. It is based on the weakest link on the system, not the strongest.

Steep-angle highline A highline in which one of the suspension points is considerably higher than the other.

Strategy An approach to achieving the objective.

String line A search area containment method that utilizes lines of string to bound an area.

Structural anchorages Anchor points that already exist in nature or in engineered structures.

Structured debrief An in-depth review of a response that considers a specific area of concern; usually takes place a short time after the incident, so that participants can have time to research.

Subject The individual who is being rescued; the preferred term for a victim in rope rescue.

SUDOT A standardized whistle-command protocol used in rope rescue where different numbers/types of whistle blasts each has a specific meaning.

Suspension-induced injury Physiologic shock response resulting from the effects of hanging motionless in a harness for a period of time.

Suspension trauma A life-threatening condition where venous blood flow from the extremities to the right side of the heart is reduced due to lack of movement and harness strap compression.

Swiftwater Water moving at a rate greater than 1 knot. (NFPA 1006)

System safety factor A load ratio expressing the difference between the breaking strength of the weakest point of a rope rescue system as compared with the maximum anticipated load at that point.

System safety factor (SSF) The ratio between the least strong point in the rope rescue system and the potential load at that point.

T

Tactics The steps used to execute the strategy.

Tag-line A non–load-bearing line that connects to the litter to help move it side to side to avoid obstacles, prevent swing-fall, or attain a more advantageous position.

Tail The unused portion of rope remaining behind the working end of a rope.

Tandem pulleys Two in-line pulleys on the same rope. Used in highlines to stabilize a load and to distribute its weight along the rope.

Tap litter attendant rotation method A method used by litter transport teams to switch out litter attendants without stopping the litter.

Technical rescue The application of special knowledge, skills, and equipment to resolve unique and/or complex rescue situations.

Technical search and rescue The application of special knowledge, skills, and equipment to resolve unique and/or complex search and rescue situations.

Technician level A functional level of technical rescue capability that represents the capability of an organization or an individual to respond to technical search and rescue incidents and to identify hazards, use equipment, and apply advanced techniques specified in this standard necessary to coordinate, perform, and supervise technical search and rescue incidents.

Tension member Component parts of a system used to support force in a rope-based system (rope, webbing, and other similar equipment).

Tension Term used in rope rescue to describe the stress in a system or system component due to transmitted forces that pull in equal and opposite directions.

Third man A rescuer who is assigned to access the subject quickly to provide assessment and assistance with patient care and rescue tasks.

Threat and Error Management System (TEMS) An approach to safety management developed by the aviation industry that is grounded in the assumption that risk happens, and can be managed in a way so as to not impair safety.

Time Critical Risk Management (TCRM) An Operational Risk Management system that takes into account the understanding that risk management is especially challenging when time and resources are limited.

Track-line The line (or lines) that provide the support of the load

Transfer devices Components of equipment used by rescuers to package and transport a subject from a rescue environment. May include litters, harnesses, backboards, etc. (NFPA 1006)

Transfer line A system used to temporarily control a load during a knot pass.

Traveling pulley A pulley that travels with the load.

Trench (or trench excavation) A narrow (in relation to its length) excavation made below the surface of the earth. (NFPA 1006)

Tunnel A covered excavation used for the conveyance of people or materials, typically no smaller than 0.91 m (36 in.) in diameter and within 20 degrees of horizontal. (NFPA 1006)

Turn A single pass of a rope behind an object.

Two-tensioned rescue system (TTRS) A particular form of a dual rope system in which the primary system and the belay line are more-or-less identical.

Tying off short A safety technique that creates an extra point of attachment during ascending by tying the person directly into the mainline rope.

U

Unified command Management of an emergency incident through the incident command system in which multiple agencies or jurisdictions are involved, each one having an incident commander assigned to the operation.

V

Vector Direction and magnitude of a given force.

Vertical evacuation Use of a rope to make a controlled lowering of a rescue subject down a slope that is steep enough that the load is dependent on the rope for support.

Vertical lifeline A rope, a cable, or a track to which a worker is attached via a full body harness (usually dorsal attachment), a lanyard, and a traveling rope grab.

W

Weight The effect of gravity attracting two masses.

Wilderness A setting in which the delivery of services including search, rescue, and patient care by response personnel is adversely affected by logistical complications, such as an environment that is physically stressful or hazardous to the patient, response personnel, or both; remoteness of the patient's location, such that it causes a delay in the delivery of care to the patient; anywhere the local infrastructure has been compromised enough to experience wilderness-type conditions, such as lack of adequate medical supplies, equipment, or transportation; remoteness from public infrastructure support services; poor to no medical services or potable water; compromised public safety buildings, public utilities or communications systems; city, county, state or national recreational areas or parks with mountains, trails; areas they define as wilderness. (NFPA 1006)

Working end The end of the rope that is used to tie and form the knot.

Working system Primary rope system used to move a load; may be a raising system, lowering system, horizontal system, etc.

Y

Yosemite Litter Packaging A method of securing a subject into a litter using two 18-foot (5.5-m) runners and several additional circumferential cross straps (approximately 12 feet (3.7 m) each in length).

Index

Note: Page numbers followed by *f* and *t* indicate figures and tables respectively.

A

abrasion, 176
   resistance, 156–157
accountability, 76, 76*f*
   briefings, 76, 76*f*
action review, 60
active belay, 221, 223, 223*f*
active fall protection, 221
adjustable friction devices, 275
adjustable low-angle litter tie-in, 278
agency liaison, 54
American Society of Safety Professionals (ASSP) Fall Protection Code, 11
anchorage connector, 139, 139*f*, 198
   anchor slings, 139
   beam clamps, 139–140
   bolt anchors, 140, 140*f*
   pickets, 140
   portable anchors, 140
   pulleys, 140–141, 141*f*
   specialized pulleys, 141
anchorages, 198, 345–347, 346*f*
   adapting to lack of anchors
      extending anchor points, 215–216
      using vehicles for anchorages, 216
   backing up, 200, 201*f*
   back-tie, 213, 213*f*
      pretensioned, 213–214, 214*f*
   carriage system, 348–349, 349*f*
   connecting to, 201
   directionals, 212–213, 212*f*
   evaluating anchor system, 216
   far-side track-line, 347, 347*f*
   knots, 203
   less-obvious anchorages, 214
      elevator and machine housings, 214, 215*f*
      installed anchorages, 215
      scuppers (roof drain holes), 214–215
      stairwell beams, 215
      wall sections between windows/doors, 215
   multi-point anchor systems, 206–211, 207*f*, 209*f*
   near-side track-line anchor/tensioning system, 347–348, 348*f*
   placement of anchor systems, 216
   portable, 205–206, 206*f*
   positioning of, 200
   principles of, 199
   redundancy, 200–201, 201*f*
   rigging plates, 206, 206*f*
   single point anchorages, 202–203
   soft slings as, 203
      anchor straps, 203–205, 204*f*, 205*f*
      presewn slings, 203, 203*f*
   strength, 199
   structural anchorages, 214
   tag-line anchor systems, 349
   terminology, 198–199, 199*f*
anchor point, 417, 418*f*
anchor straps, 203–205, 204*f*, 205*f*
anchor systems, 200
   evaluating, 216
   placement of, 216
anchor terminology, 198–199, 199*f*
angles in rescue systems, 186–187, 187*f*
animal rescue incidents, 448
aramids, 150
arm rappel (guide's rappel), 372
ascenders, 385
   Prusik ascending system, 385–387, 385*f*
ascending, 391
   basics, 384–385
   changeovers to descending, 398–403
   edge with gradual rollover, 391–392
   with Prusik, 387–388
   system, 389–390, 391
   undercut edge, 393
assisted-braking belay devices, 225–226
attendant rig, 330
attention to detail, 16–17
attraction, 29, 29*f*
authority having jurisdiction (AHJ), 4–5, 98
autolocking
   descender, 427
   devices, 274
   Petzl EXO, 135
   Sterling F4, 135
average arrest force, 38
awareness level, 6, 6*f*

B

back-tie, 213, 213*f*
   pretensioned, 213–214, 214*f*
backup system, 200
bagged rope, descending with, 375, 375*f*
bagging, 174–175, 175*f*
beanbag-style subject immobilization system, 254, 255*f*

belay/belaying, 136, 220
   active belay, 223, 223f
   arranging direction, 238
   assisted-braking belay devices, 225–226
   communications, 228–229
   decisions on when to, 220–221
   devices, 136–137, 136f, 223
   dual-tension rope systems (DTRS), 239
   failure, 239
   free-running belay device, 225
   light load
      belay system, 226–228
      vs. rescue load, 224–229
   line, 223, 345
   load-releasing hitches, 231–232
   maintaining proper slack, 238
      belay failure, 239
      body belays, 239
      bottom belay (for rappelling only), 238–239, 238f
      securing belayer, 238
   manually operated assisted, 229–230, 230f
   Münter hitch, 225
   Prusik minding pulley (PMP), 236–237
   question, 284
   releasing radium-release hitch, 236
   rescue load, 229–231, 230f
   rope, 136
   as safety backups, 221
   self-belay, 221–223, 222f, 223f
   situations requiring, 220
   system, 223–224, 224f
   tandem Prusik belay system, 231, 231f, 232, 234–235
   two-tensioned rescue system (TTRS), 239–240, 240
belayer, 223
bending, effects of, 177, 177f
bight, 158
bi-pod, 352
body belays, 239
body rappel (Dulfersitz rappel), 372, 410–411
body substance isolation (BSI), 56
bombproof, 199
bottom belay (for rappelling only), 238–239, 238f
braided construction, 150–151, 150f
brake bar rack, 131–133, 131f, 132f, 376, 378–381
   Alpine Brake Tube, 134
   Figure-8 descender, 134–135, 134f, 135f
   lock off, 383–384
   scarab, 133
brake hand, 372, 372f
brakeman, 271
braking devices and descenders, 127–128, 127f
   additional considerations, 128
   autolocking, 128, 129–130, 129f
   climbing technology sparrow, 130
   CMC MPD, 130
   friction and heat, 129
   guidelines for selecting, 129
   Harken clutch, 130
   ISC D4/D5 Work Rescue Descenders, 130
   manual devices, 128
   Petzl ID, 130
   Petzl Maestro, 130–131
   Skylotec Lory, 131
   skylotec sirius, 131
   SMC Spider, 131
   types of, 129

breaking strength, 154–155, 154t
bridge engineer, using rope access, 222, 223f
BSI. *See* body substance isolation (BSI)
Burrito wrap, 256
butterfly knot, 167

C

carabiners, 119, 119f, 427, 428f
   components, 119–120, 120f
   gate openings, 122
   inspection and maintenance, 125, 125f
   locking mechanisms, 122–124, 123f, 124t
   material, 119
   shapes, 120, 120f
   strength, 120–122, 120f, 121f, 122t
   use considerations, 124–125, 124f, 125f
   wrap, 412
care of equipment, 16
carriage system, 348–349, 349f
cascade trail technician litter wheel, 261, 262f
cave rescue, rope rescue for, 447
caving, 9, 10f
changeovers, 385, 397–398
   from ascending to descending, 398–403
   getting off rope, 406
   intermediate anchorage point, 405–406
   passing deviation, 404–405
   passing knot, 404
   rope-to-rope transfers, 403–404
chest harness, 103–104
chicken loop, 389, 389f
class II harnesses, 69
class III harnesses, 69
clinging subject, evacuation seat to, 418–419
clinging surface, 418, 418f
closed-loop communication, 42
clove hitch, 170, 172
coiling, 175, 176f
command and coordination, 52
command emphasis, 85–86
command staff information, 88
command structure, 52
   additional command staff, 54, 54f
   incident command system, 52–54, 53f
   multiagency operations, 54
commodity ropes, 149
communication, 41, 42f, 90–92, 92f, 228–229, 324–325, 324f
   explicit, 42–43, 42f
   IC, 92
   incident name, 92
   and information management, 52
   in raising operations, 317
   for risk management, 40
   verbal, 41, 41f
complex mechanical advantage systems, 315
component safety factor, 118
composite basket litter, 245, 245f
compound mechanical advantage system, 313
conduct postincident analysis, 94
confined space, 445
   rescue, 11–12, 11f
containment, 29
counterbalance haul system, 301–302, 302f
critical incident stress management (CISM), 45
critical point test, 229
cross-haul, 365

D

data collection, 94
demobilization, 92–93, 93f
    recordkeeping and documentation, 94
    rescuer risk and site safety, 93
    scene security and custody transfer, 93
descenders, 372
    types of, 373
descending techniques, 373, 373f
    brake bar rack, 376, 380–381, 382–384
    descending with bagged rope, 375, 375f
    initiating descent, 375–376, 376f
    locking off midline, 376, 381
    lock off descent, 382
    practicing, 376
    setting up system, 373–374, 374f
    throwing rope, 374–375
design load, 118
detailed search methods, 29
diaper emergency harness, 105–106, 106f
diaper-style hasty harness, 418, 418f
directionals, 212–213, 212f
    anchorage, forces on, 200, 201f
direction change, 301, 301f
    counterbalance haul system, 301–302, 302f
    progress capture, 302, 302f
direction of pull, 199
distraught subject, managing, 439–440
double braid construction, 151, 151f
dressing rope ends, 180, 180f
dual-tensioned rope systems (DTRS), 192–193, 193f
dual tension litter lower, 432
dual-tension rescue systems (DTRS), 428
dual-tension rope systems (DTRS), 239
dynamic directional system, 342
dynamic systems, 363
    English reeve, 363, 364f
    opposing systems, 365
    skate block, 363–365, 364f

E

edge protection, 188–191, 189f, 190f, 190t, 191f
edge transitions, 394, 395–396
emergency descent systems, 407
    arm rappel, 407, 407f
    body rappel (Dulfersitz), 410–411
    carabiner wrap, 412
    Münter hitch rappel, 412, 412f
emergency harnesses, 105
    diaper emergency harness, 105–106, 106f
    Swiss-seat emergency harness, 106, 106f
emergency medical responder (EMR), 423
emergency response system, 51–52
emergency seat harness, 105
EMR. *See* emergency medical responder (EMR)
enclosed basket litters, advantages and disadvantages of, 245, 246t
English reeve, 363, 364f
equipment standards, 4
escape belts, 106–107, 106f
escape devices, 135
    autolocking
        Petzl EXO, 135
        Sterling F4, 135
    Escape 8, 135–136
    nonautolocking, 135
    PMW Hook, 136

etrier, 107
evacuation
    seats, 418
    signals, 55

F

fabric softeners, 180
fail-safe design, 304
fall arrest, 69, 137–138, 221, 222, 222f
    self-retracting lifeline for, 222, 222f
falling objects, damage from, 175–176, 176t
falls
    line, 191
    protection, 10
    protection against, 38–39, 38t
far-side track-line anchorage, 347, 347f
fire service rescue, 8
Five Why's approach, 61
fixed and focused multi-point anchor system, 207–208
fixed pulley, 301
"flash" rappels, 177
flexible litters, 253
    advantages and disadvantages of, 246, 246t
Flying W tensioning system, 348
focal point, 207
footloops, 389
footprint, 353
footwear, 101–102, 102f
force, 185
    optimizing, 193
forces in highline system
    analyzing system for force, 354–355
    anchorages, 354
    load, 354
    loaded line guidelines, 357
    pre-tension guidelines, 355, 356f
    sag, 354
    10 percent loaded line sag rule, 355–357, 356f
    10 percent sag rule (pre-loading), 355
free-climbing, 417
free-running belay device, 225
free-soloing, 417
friction, 187–188, 187f
    wrap, 275
full-body harness, 104, 104f
full-body vacuum splint, 253, 253f

G

gaining extra friction, rappel, 396
    control with body, 397
    control with rope angle, 397
    control with spare carabiner, 397, 397f
general staff, 53
gloves, 102, 102f
Go-Fever, 34
gradual rollover, 391–392
guide hand, 372
guiding-lines, 343, 343f, 362

H

handline, 278
hand pass, 263–264
hand signals, 41
hard goods, 143–144, 143f
harnesses for rescue, 103–105, 104f
hasty search, 29
hasty seat harness, 419–420

Index

hauling systems
    communications for, 74–75t
    haul team, 74
    overview, 73
    working of, 73–74
haul line, 345
haul team, 301
hazards
    identification, 26, 34–35, 35f
    preparing for, 35–37, 36f
    specific PPE, 98
hazards associated with rope rescue, 34
    clear communication, 41, 42f
    communicating for risk management, 40
    explicit communication, 42–43, 42f
    hand signals, 41
    hazard identification and risk assessment, 34–35, 35f
    incident safety control chart, 37, 37f
    incident safety officer (ISO), 43–44, 43f
    preparing for hazards, 35–37, 36f
    protection against falls and other hazards, 38–39, 38t
    risk management, 39–40, 40t
    shared mental model, 41–42
    situational awareness, 39
    specific hazards, 37–38
    stress management, 44–46
    team briefings, 42
    verbal communications, 41, 41f
head lamp, 108, 108f
helmet, 102–103, 102f
high-angle litter lower, 290
    attaching litter to main rescue line, 291–292, 291f
    litter orientation, 290–291, 290f, 291f
    practicing high-angle litter lower techniques, 293–295, 293f
    rigging without litter attendant, 292–293
    securing subject in litter, 291
high directionals, 349–350
    feet, 351
    head, 350, 350f
    hobbles, 351, 351f
    legs, 350
highline, 342
    pulley, 349
hitches, 158, 169
    clove hitch, 170, 172
    Münter hitch, 172, 173f
    Prusik hitch, 169–170, 171
    Truckers hitch, 172, 173f
horizontal highline, 343
    elements of, 343f
horizontal litter, 328
horizontal rescue systems, 343
horizontal span, 352
horizontal systems, 342–343
    anchorages, 345–347, 346f
        carriage system, 348–349, 349f
        far-side track-line, 347, 347f
        near-side track-line anchor/tensioning system, 347–348, 348f
        tag-line anchor systems, 349
    attaching load to highline, 357–358, 357f
    configurations
        bi-pod, 352
        horizontal span, 352
        monopod, 352
        quad-pod, 352
        tripod, 351, 352f
    dynamic systems, 363
        English reeve, 363, 364f
        opposing systems, 365
        skate block, 363–365, 364f
    forces in highline system
        analyzing system for force, 354–355
        anchorages, 354
        load, 354
        loaded line guidelines, 357
        10 percent loaded line sag rule, 355–357, 356f
        10 percent sag rule (pre-loading), 355
        pre-tension guidelines, 355, 356f
        sag, 354
    guiding-lines, 362–363, 362f
    high directionals, 349–350
        additional rigging points, 351
        feet, 351
        head, 350, 350f
        hobbles, 351, 351f
        legs, 350
        parts of, 350, 350f
    rigging litter to highline, 358–361
    track-line systems, 343–344, 343f
        rigging, 344
        ropes for highlines, 344–345
        spanning gap, 344
    using multi-pods, 352
        assembly, 352
        managing forces, 353–354, 353f
        raising and securing portable high anchorage, 352–353
hot debrief, 60–61

I

IAA. *See* interagency agreement (IAA)
IAP. *See* incident action plan (IAP)
IC. *See* incident commander (IC)
ice climbing, 9–10
ICS. *See* incident command system (ICS)
implementation
    chain of command, 90
    multijurisdictional/multiagency operations, 90, 91f
    span of control, 90, 90f
incident action plan (IAP), 27, 57–59, 58f, 84–85, 84f
    communications, 90–92, 92f
        IC, 92
        incident name, 92
    conduct postincident analysis, 94
    crafting incident objectives in, 98
    data collection and management systems, 94
    demobilization, 92–93, 93f
        recordkeeping and documentation, 94
        rescuer risk and site safety, 93
        scene security and custody transfer, 93
    implementing, 89
        chain of command, 90
        multijurisdictional/multiagency operations, 90, 91f
        span of control, 90, 90f
    incident objectives form, 85
    Planning P concept, 85–89, 86f, 87f
incident commander (IC), 43, 52
incident command system (ICS), 27, 52
incident communications plan, 88
incident location concerns, 66, 67f
incident map, 88
incident objectives, 85, 88
incident planning forms, 58t, 59
incident safety control chart, 37, 37f
incident safety officer (ISO), 43–44, 43f

industrial sites, 11
in-line haul system, 305
installed anchorages, 199, 215
interagency agreement (IAA), 35
ISO. *See* incident safety officer (ISO)

J

job performance requirements (JPRs), 7–8
    critical components, 7
JPRs. *See* job performance requirements (JPRs)

K

kernmantle, 151, 151f
    static and low-stretch, 151–152
kneeling transition method, 396
knives, 108
knotability, 157
knots, 159, 203
    anatomy of, 157–159, 158f
    bowline knots, 163–164, 164f, 166, 166f
    butterfly knot, 167
    completing, 159–160
    double figure 8 loop, 162–163
    double sheet bend, 168, 169f
    figure 8 bend knot, 168
    figure 8 family of, 161, 161f
    figure 8 on bight, 161–162
    figure 8 stopper knot, 161, 161f
    grapevine bend, 168
    hitches, 169
        clove hitch, 170, 172
        Münter hitch, 172, 173f
        Prusik hitch, 169–170, 171
        Truckers hitch, 172, 173f
    inline figure 8, 166
    overhand, 160–161
    passing pulley, 349
    pass, performing, 282–283
    pass with litter, 437–438
    qualities of, 159, 159f
    removing, 159
    ring bend (water knot), 168–169
    strength loss through, 177
Kootenay carriage, 349

L

ladder belts, 107
laid rope construction, 150
land search, 28
Lanyard with force absorber, 221
last known point (LKP), 24
LAST mnemonic, 27, 59–60, 417
leapfrogging, 283
less-obvious anchorages
    elevator and machine housings, 214, 215f
    installed anchorages, 215
    scuppers (roof drain holes), 214–215
    stairwell beams, 215
    wall sections between windows/doors, 215
lift-assist sling carry, 261–262, 262f
light load
    belay practice system, 226–228
    belay system, 226–228
    vs. rescue load, 224–229
litter, 251
    across a highline, 439
    attendants, 259

basket litter strength and performance, 246
bridles, 276, 430
captain, 260
carrying, 259
choosing litter for rescue operations, 247
control movement of, 294
definition of, 244
enclosed basket litters, 245–246, 245f, 246t
flexible litters, 246, 246t
function, 244
handles, 263
lifting and carrying, 260
lower, tending, 430–432
metal basket litters, 245
movement, body management with, 331–332, 332f
negotiating obstacles with
    hand pass, 263–264
    traveling control lines, 263, 263f
passing knots for, 434–438
protecting subject in, 254–255, 256
restraint, 256–258, 259
rigid, 224f, 244–245, 245f
securing subject in, 255–256
spider, 291, 328–329
team roles, 259–260, 259f
team transitions, 261
wheels, 262–263, 263f
LKP. *See* last known point (LKP)
load distributing anchor system, 209
loaded line guidelines, 357
load ratios, 117–119
load release, 303
load-releasing hitches, 231–232
load-sharing anchor system, 207
locking off, 376
lock off descent, 382
loop, 158
lost person behavior, 28, 28f
low-angle evacuation, 268
low-angle litter bridle, 263
low-angle litter tie-ins, 278
low-angle rope rescue systems, 269–270, 274
low-angle system braking devices, 274
    adjustable friction devices, 275
    autolocking devices, 274
    friction wrap, 275
    non-autolocking, 274–275
lowering systems, 268–269
    aerial ladder slide lowers, 295
    ambulatory subjects, 272–273
        litters and nonambulatory subjects, 274
        lower method, 273–274
    belays for high-angle lowering, 286–287
    braking systems for high-angle lowering, 286
    communications, 280
    elements of low-angle evacuation system, 270, 270f
    high-angle rope rescue system, 285–286, 285f
    line, 345
    litter, 275
        head of, 275–277, 276f, 277f
    low-angle evacuation with a litter wheel, 279
    low-angle litter management, 277
        no tie-in, 277–278
        tie in to litter, 278–279, 278f
        tie in to rope, 278
    low-angle rope rescue overview, 269–270
    low-angle rope rescue systems, 274

low-angle system braking devices, 274
   adjustable friction devices, 275
   autolocking devices, 274
   friction wrap, 275
   non-autolocking, 274–275
pickoff by, 424–426
practicing high-angle litter lower, 290
   attaching litter to main rescue line, 291–292, 291f
   litter orientation, 290–291, 290f, 291f
   rigging without litter attendant, 292–293
   securing subject in litter, 291
   techniques, 293–295, 293f
practicing low-angle litter evacuation, 279–280
primary anchor system, 286
rigging system for, 288–290, 289f
roles and responsibilities in, 270
   brakeman, 271
   operations leader, 270–271
   optional additional roles, 271–272
   rigger, 271
   safety officer, 271
ropes for, 275, 286
safety considerations in low-angle rope rescues, 283
   Belay question, 284
   slope safety lines, 283–284, 284f
summarizing low-angle evacuations, 284
two-tensioned rescue system (TTRS), 287–288, 287f
typical low-angle lowering operation
   evacuation of more than one rope length, 281–283
   preparation, 280–281
low stretch, 152

M
MAF. *See* maximum arrest force (MAF)
main attachment point (MAP), 328
management systems, 94
manner of function, 124
manufactured eye termination, 158
MAP. *See* main attachment point (MAP)
mass, 185
mass-casualty incident (MCI), 23
master attachment point, 206, 291
maximum arrest force (MAF), 38
MBS. *See* minimum breaking strength (MBS)
MCI. *See* mass-casualty incident (MCI)
mechanical advantage systems, 300, 301, 304–305
   communications in raising operations, 317
   converting, 307
   with direction change, 301, 301f, 306–307
      counterbalance haul system, 301–302, 302f
      progress capture, 302, 302f
   in-line haul system with braking device PCD, 309–310
   load release, 303
   operating haul system, 317
   operating mechanical advantage system, 310–311, 310–311f
   piggyback rope systems, 311–312, 312f
      compound systems, 312–315, 315f
   positioning the PCD, 302–303, 302f, 303f
   principles of, 300–301, 301f
   safety check, 316–317
   simple in-line systems and beyond, 307, 308f
   simple systems and beyond, 305–306, 305f, 306f
   two-tensioned rescue systems (TTRS), 315–316
   winch for, 304, 304f
medical attendant, 249

medical care, patient evacuation, 247
   assessment and interventions
      airway and breathing, 250
      head and spine considerations, 250–251
      severe bleeding, 250
      triage, 249–250
   benefit to subjects, 248–249
   medical technologies, 247–248
   technical rescue medical kit (TMK), 249
medical plan, 89
medical technologies, 247–248
memorandum of understanding (MOU), 35
minimum breaking strength (MBS), 118
monopod, 352
motionless suspension, 322, 323f
MOU. *See* memorandum of understanding (MOU)
multiagency coordination (MAC) groups, 52
multi-pods, 352
   assembly, 352
   managing forces, 353–354, 353f
   raising and securing portable high anchorage, 352–353
multi-point anchor systems, 206–211, 207f, 209f
Münter hitch rappel, 172, 173f, 225, 412, 412f

N
National Fire Protection Association (NFPA), 4, 67, 268, 370, 371, 444
   levels of rescue, 6–7, 6–7f
   resources, 5–6
   Standards Addressing Personal Protective Equipment, 100t
National Incident Management System (NIMS), 52
natural fibers, 148, 148f
near-side track-line anchor/tensioning system, 347–348, 348f
NFPA. *See* National Fire Protection Association (NFPA)
NIMS. *See* National Incident Management System (NIMS)
non-autolocking, 131, 274–275
Norwegian reeve, 363
nylon, 149

O
Occupational Safety and Health Administration (OSHA), 11, 36
operating haul system, 317
operational period, 27, 59
operational risk management (ORM), 39
operational tactics, 88
Operation O, 85
operations briefing, 89
operations leader, 270–271
operations level, 6, 6f
opposing systems, 365
Oregon spine splint, 253, 253f
ORM. *See* operational risk management (ORM)

P
PAR. *See* personnel accountability reporting (PAR)
passive fall protection, 222
patient evacuation, 244
   carrying litter, 259
   lift-assist sling carry, 261–262, 262f
   lifting and carrying litter, 260
   litter function, 244
   litter handles, 263
   litter team roles, 259–260, 259f
   litter team transitions, 261
   litter wheels, 262–263, 263f
   medical care, 247
      assessment and interventions. *See* medical care, patient evacuation
      benefit to subjects, 248–249

patient evacuation (*Continued*)
    medical technologies, 247–248
    technical rescue medical kit (TMK), 249
  negotiating obstacles with litter
    hand pass, 263–264
    traveling control lines, 263, 263*f*
  packaging subject in litter, 251
    improvised litter restraint, 256–258
    manufactured litter restraint, 259
    protecting subject in litter, 254–255, 256
    securing subject in litter, 255–256
    spinal motion restriction and medical considerations, 251–254, 252*f*, 253*f*
  rest breaks, 260
  transferring subject to EMS, 264–265, 264*f*
  types of litters
    basket litter strength and performance, 246
    choosing litter for rescue operations, 247
    enclosed basket litters, 245–246, 245*f*, 246*t*
    flexible litters, 246, 246*t*
    metal basket litters, 245
    rigid litters, 224*f*, 244–245, 245*f*
    Sked litter, 246–247, 247*f*
patient transfer report, 264
  verbal communication, 264, 264*f*
  written communication, 264–265
PCD. *See* progress-capture device (PCD)
personal escape, 68
personal handled ascender, 138–139
personal protective equipment (PPE), 56, 56*f*, 98, 98*f*
  emergency harnesses, 105
    diaper emergency harness, 105–106, 106*f*
    Swiss-seat emergency harness, 106, 106*f*
  escape belts, 106–107, 106*f*
  footwear, 101–102, 102*f*
  gloves, 70–72, 71*f*, 102, 102*f*
  harnesses for rescue, 103–105, 104*f*
  helmet, 72–73, 72*f*, 102–103, 102*f*
  inspection, maintenance, and recordkeeping, 108–109, 108*f*
  knives, 108
  maintenance, 109
  National Fire Protection Association Standards, 67
  pickoff seat (rescue triangle), 106–107*f*
  preventing dropped objects, 107–108, 108*f*
  protective garments, 99–101, 100*t*, 101*f*
  rope rescues, 68–70, 68*f*, 69*f*, 98–99, 99*f*
  signs of damage, 109
  suspension trauma, 107
  victim extrication device, 105
personal red flags, 14
personal vertical skills, 370–371, 370*f*
  ascending basics, 384–385
  ascending over edge, 391
    edge with gradual rollover, 391–392
    undercut edge, 393
  ascending system, 389–390, 391
  changeovers, 397–398
    from ascending to descending, 398–403
    getting off rope, 406
    intermediate anchorage point, 405–406
    passing deviation, 404–405
    passing knot, 404
    rope-to-rope transfers, 403–404
  choosing ascender, 385–388
  customizing slings, 390–391, 391*f*
  descending techniques, 373, 373*f*
    brake bar rack, 376, 380–381, 382–384
    descending with bagged rope, 375, 375*f*
    initiating descent, 375–376, 376*f*
    locking off midline, 376, 381
    lock off descent, 382
    practicing, 376
    setting up system, 373–374, 374*f*
    throwing rope, 374–375
  developing vertical skills, 393
  edge transitions, 394, 395–396
  effect of rope angle, 393–394, 393*f*
  equipment management, 394
  gaining extra friction when on rappel, 396
    control with body, 397
    control with rope angle, 397
    control with spare carabiner, 397, 397*f*
  importance of control, 372
  mechanical ascenders, 388, 388*f*
    ascender slings, 388–389, 389*f*
    footloops, 389
    parts of, 388
  passing obstacles, 400, 403
  personal vertical skills, 371–372, 372, 372*f*
  single rope technique (SRT), 371
  stance, 394
  troubleshooting
    emergency descent systems. *See* emergency descent systems
    secondary backup systems, 406–407
    self-belay techniques, 406
    self-rescue, 407
    using Prusik safety, 406
  tying off short, 393
  undercut edges, 394, 396
personnel accountability reporting (PAR), 90
personnel rehabilitation, 77
Petzl ID, 130
Petzl Maestro, 130–131
PHD footprint/resultant, 353, 353*f*
physical safety concepts, 14, 14*f*
physics, principles of, 184, 184*f*
pickoff and litter management, 416, 417
  communicating with subject, 440
  managing distraught subject, 439–440
  managing litter across a highline, 439
  passing knots for litters, 434–438
  pickoff rescue. *See* pickoff rescue
  selecting rescue approach, rescue approach, selecting
  subject care, 439
  suspended rescuer, 416–417
pickoff rescue, 422–423, 422*t*
  to belay/not, 423
  litter bridles for scaffold-type lowers, 434, 434*f*
  by lowering, 424–426
  medical considerations, 423
  note on litter bridles, 430
  protecting rescuer, 423–424
  by rappel, 426–428, 428*f*
  releasing suspended subject, 429–430, 430*f*
  scaffold-type litter lowers, 433
  teamwork and communication, 423
  techniques, 422
  tending litter lower, 430–432
  by traveling brake, 428–429, 429*f*, 432–433, 433*f*
pickoff seat (rescue triangle), 106–107*f*
pickoff technique, 426
piggyback rope systems, 311–312, 312*f*
  compound systems, 312–315, 315*f*
pike and pivot bridle, 332–333
pike and pivot lower, 333–334
pike and pivot raise, 334–335

Index

plaited construction, 150, 150f
Planning P concept, 85–89, 86f, 87f
PLS. *See* point last seen (PLS)
PMP. *See* Prusik minding pulley (PMP)
POD. *See* probability of detection (POD)
point last seen (PLS), 24
polyolefins (polypropylene and polyethylene), 149
portable anchorages, 205–206, 206f
PPE. *See* personal protective equipment (PPE)
prerigged haul system, 429, 430f
prerigging, 281
presewn slings, 203, 203f
pre-tension guidelines, 355, 356f
primary anchor system, 286
principles of rigging, 184
probability of detection (POD), 29
progress-capture device (PCD), 301–302, 301–303, 302f, 303f
protective garments, 99–101, 100t, 101f
Prusik ascending system, 386–388
Prusik hitch, 169–170, 171, 385, 385f
Prusik loops, 386–387
Prusik minding pulley (PMP), 236–237, 237
Prusik safety, 406
public information officer (PIO), 54
pull, direction of, 199
pulleys, 140–141, 141f

Q

QAS (Quick Attachment Safety), 391
quad-pod, 352

R

radium-release hitch, 232–233, 236
rappelling, 9
rappel, pickoff by, 426–428, 428f
ratchet cam, 302
recordkeeping, 142–144, 143f
recreational incidents
    caving, 9, 10f
    ice climbing, 9–10
    rappelling, 9
    rock climbing, 9, 9f
reeve, 363
reference materials, 27–28
relay information, 30, 30f
reporting party (RP), 24, 24f
rescue approach, selecting, 417
    accessing subject, 417–418, 418f
    evacuation seat to clinging subject, 418–420, 418f
    rescue chest harness, 420–422, 420f
    subject clinging to surface, 418, 418f
rescue chest harness, 420–422, 420f
rescue equipment, 114, 114f
    anchorage connectors, 139, 139f
        anchor slings, 139
        beam clamps, 139–140
        bolt anchors, 140, 140f
        pickets, 140
        portable anchors, 140
        pulleys, 140–141, 141f
        specialized pulleys, 141
    belay devices, 136–137, 136f
    brake bar racks, 131–133, 131f, 132f
        Alpine Brake Tube, 134
        Figure-8 descender, 134–135, 134f, 135f
        scarab, 133
    braking devices and descenders, 127–128, 127f
        additional considerations, 128
        autolocking, 128–130, 129f
        climbing technology sparrow, 130
        CMC MPD, 130
        friction and heat, 129
        guidelines for selecting, 129
        Harken clutch, 130
        ISC D4/D5 Work Rescue Descenders, 130
        manual devices, 128
        Petzl ID, 130
        Petzl Maestro, 130–131
        Skylotec Lory, 131
        skylotec sirius, 131
        SMC Spider, 131
        types of, 129
    carabiners, 119, 119f
        components, 119–120, 120f
        gate openings, 122
        inspection and maintenance, 125, 125f
        locking mechanisms, 122–124, 123f, 124t
        material, 119
        shapes, 120, 120f
        strength, 120–122, 120f, 121f, 122t
        use considerations, 124–125, 124f, 125f
    equipment selection considerations, 115–117, 117f
    equipment strength and safety factors, 117–119
    escape devices, 135
        autolocking. *See* autolocking
        Escape 8, 135–136
        nonautolocking, 135
        PMW Hook, 136
    maintenance and inspection
        care, 142
        and recordkeeping, 142–144, 143f
    nonautolocking, 131
    rigging plates, 141–142, 142f
    rope grabs and ascenders, 137
        fall arrest, 137–138
        personal handled ascender, 138–139
        rigging, 138, 138f
    rope rescue equipment program, 115, 115f
    screw links, 126–127, 126f
    swivels, 142, 142f
rescue harness, 68, 69f
rescue incident safety control (RISC) chart, 37, 37f
rescue line to litter, 329
rescue load, 229–231, 230f
    belay practice, 230–231
rescue operations, scope of, 23
rescue preplan, 35
rescuer, protecting, 423–424
resource management, 52
response, initiating, 50
    activate emergency response system, 51–52
    after action review, 60
    command structure, 52
        additional command staff, 54, 54f
        incident command system, 52–54, 53f
        multiagency operations, 54
    Five Why's approach, 61
    hot debrief, 60–61
    incident action plan, 57–59, 58f
    incident planning forms, 58t, 59
    initial response, 51
    LAST mnemonic, 59–60
    personal protective equipment, 56, 56f
    preplanning, 50–51, 50f

response, initiating (*Continued*)
   recognize and mitigate hazards, 55
   recognizing need, 51, 51*f*
   rope rescue resources, 56
      awareness level, 56
      operations level, 56–57, 57*f*
      technician level, 57, 57*f*
   scene management, 55
      evacuation signals, 55
   site control, 54–55, 55*f*
   structured debrief, 61
   threat and error management system (TEMS), 61
rest breaks, 260
restraint, 68
resultant, 186, 186*f*
retiring, 178
rigger, 271
rigging, 138, 138*f*
   dual-tensioned rope systems (DTRS), 192–193, 193*f*
   fall line, 191
   optimizing force, 193
   plates, 141–142, 142*f*, 206, 206*f*
   principles, 185
      angles in rescue systems, 186–187, 187*f*
      edge protection, 188–191, 189*f*, 190*f*, 190*t*, 191*f*
      force, 185
      friction, 187–188, 187*f*
      resultant, 186, 186*f*
      tension, 187
      vector, 186
   rope systems, 192, 192*f*
   safety factors, 191–192
rigid backboard, 252, 252*f*
rigid basket litter, 244, 244*f*
risk assessment, 34–35, 35*f*
risk management, 39–40, 40*t*
rock climbing, 9, 9*f*
rope, 148, 275
   abrasion, 156–157, 176
   anatomy of knot, 157–159, 158*f*
   bagging, 174–175, 175*f*
   braided construction, 150–151, 150*f*
   breaking strength, 154–155, 154*t*
   care of, 172–173
   cleaning, 179, 179*f*
   coiling, 175, 176*f*
   color, 157
   damage from falling objects, 175–176, 176*t*
   damage through "flash" rappels, 177
   diameter, 152–154, 153*f*
   double braid construction, 151, 151*f*
   dressing rope ends, 180, 180*f*
   effects of bending, 177, 177*f*
   elongation, 155–156, 156*t*
   fabric softeners, 180
   fibers used to make rope
      aramids, 150
      natural fibers, 148, 148*f*
      nylon, 149
      polyolefins (polypropylene and polyethylene), 149
      synthetic fiber ropes, 149
      UHMPE (extended chain, high-modulus polyethylene), 149–150
   hand, 157
   harmful substances, 175
   history log, 173, 174*f*
   identification and marking of, 173, 174*f*
   inspecting, 177–178, 178*f*

kernmantle, 151, 151*f*
   static and low-stretch, 151–152
laid construction, 150
large-diameter ropes, 154
length, 157
life safety devices, 178
for lowering, 286
overloading, 175
performance characteristics, 152
plaited construction, 150, 150*f*
retiring, 178
rotation of, 177
selection, 152
special cleaning problems, 180
storing, 174
systems, 184, 192, 192*f*, 444, 444*f*
thermal damage, 176–177
washing, 178
   devices, 178–179, 179*f*
rope access, 10
   backup, 221
rope rescue, 56, 98–99, 99*f*, 184, 325–326, 325*f*
   for animal rescue incidents, 448
   awareness level, 56
   for cave rescue, 447
   common protective equipment at, 68–70, 68*f*, 69*f*
   for confined spaces, 444–445, 444*f*
   environments, 66, 67*f*
   for mine and tunnel incidents, 447
   operations level, 56–57, 57*f*
   personal protective equipment (PPE) for, 445
   skills, 446, 447*f*
   for structural collapse, 444, 444*f*
   for swiftwater environments, 445–446, 446*f*
   technician level, 57, 57*f*
   for tower incidents, 447–448
   for trench and excavation, 446–447, 447*f*
   for wilderness rescue organization, 446
rope rescuer, 4, 4*f*
   aptitude for high-angle environment, 13, 13*f*
   attention to detail, 16–17
   authority having jurisdiction (AHJ), 4–5
   backup of other rescuers, 16
   becoming, 12–13
   care of equipment, 16
   characteristics of effective rescuer, 13
   definition of, 8, 8*f*
   fire service rescue, 8
   job performance requirements (JPRs), 7–8
   levels of rescue, 6–7, 6–7*f*
   NFPA resources, 5–6
   preparation for self-rescue, 16
   recreational incidents
      caving, 9, 10*f*
      ice climbing, 9–10
      rappelling, 9
      rock climbing, 9, 9*f*
   safety, 13–14
      personal red flags, 14
      physical safety concepts, 14, 14*f*
   systems thinking, 14–15
   team concepts, 17
   training expectations, 17
   use of low-risk methods first, 15–16, 15*f*
   warning call, 17
   workplace incidents
      confined space rescue, 11–12, 11*f*
      industrial sites, 11

tower rescue, 12
water rescue, 12, 12f
wind turbine rescue, 11
work at height and rope access, 10–11, 10f
RP. *See* reporting party (RP)
running end, 158

S

safety, 13–14
  backups, 221
  check, 316–317
  factors, 191–192
  message, 89
  personal red flags, 14
  physical safety concepts, 14, 14f
safety officer (SO), 54, 271
scaffold-type litter lowers, 433, 435–436
scaffold-type lowers, litter bridles for, 434
scene management, 55
  evacuation signals, 55
scope of operation, 22
scope of rescue, 22–23, 23f
scree evac, 268
screw links, 126–127, 126f
scuppers (roof drain holes), 214–215
search parameters, 28
seat harness, 104, 104f
secondary backup systems, 406–407
securing belayer, 238
self-adjusting anchor system, 209–211
self-belay, 221–223, 222f, 223f
  techniques, 406
self-rescue, 407
  preparation for, 16
self-retracting lifeline, 221
shared mental model, 41–42
signs of damage, 109
simple in-line systems and beyond, 307, 308f
simple systems and beyond, 305–306, 306f
single attachment point, 328
single rope technique (SRT), 371
single track-line tensioning system, 348, 348f
situational awareness, 39
six-point/two-MAP spider, 434, 434f
size-up, 22
  assessing resource needs, 25–26, 26f
  detailed search methods, 29
  determine scope of rescue, 22–23, 23f
  establish last known location of subjects, 23–24
  identify and interview witnesses and reporting parties, 24–25, 24f
  identifying appropriate resources, 26–27
  identify search parameters, 28
  incident action plan (IAP), 27
  initial goals
    attraction, 29, 29f
    containment, 29
    hasty search, 29
  LAST mnemonic, 27
  lost person behavior, 28, 28f
  relay information, 30, 30f
  time of day, 28–29
  using reference materials, 27–28
  Who-What-When-Where-Why-How group of questions, 30
skate block system, 363–365, 364f
Sked litter, 246–247, 247f
Skylotec Lory, 131
slings, customizing, 390–391

slope safety lines, 283–284, 284f
sloping highlines, 362
S-M-A-R-T acronym model, 87
SMC Spider, 131
SO. *See* safety officer (SO)
Society of Professional Rope Access Technicians (SPRAT), 10–11
soft goods, 143, 143f
SOPs. *See* standard operating procedures (SOPs)
spanning gap, 344
span of control, 53
specialized pulleys, 141
specific hazards, 37–38
spinal motion restriction and medical considerations
  airway management, 251–252
  immobilization, 252–253, 252f, 253f
  splinting, 253–254
splinting, 253–254
SRT. *See* single rope technique (SRT)
SSF. *See* system safety factor (SSF)
stacked mechanical advantage, 313
stairwell beams, 215
standard operating procedures (SOPs), 66
standing part, 158
stationary rope technique, 371
strategy, 88
stress management, 44–46
structural anchorages, 198, 214
structured debrief, 60, 61
subject, 8
subject care, 439
SUDOT, 92
surface, subject clinging to, 418, 418f
suspended rescuer, 416–417
suspended subject, 429–430, 430f
suspension-induced injury, 322–323, 323f
suspension trauma, 107
swiftwater environments, 445–446, 446f
Swiss-seat emergency harness, 106, 106f
swivels, 142, 142f
synthetic fiber ropes, 149
system safety factor (SSF), 118
systems thinking, 14–15

T

tactics, 88
tag-line(s), 292, 345
tag-line anchor systems, 349
tail, 157
tandem Prusik belay system, 231, 231f, 232, 234–235
tap litter attendant rotation method, 261
TCRM, ABCD Mnemonic for, 40f
team briefings, 42
team concepts, 17
technical rescue, 4
technical rope rescue, 370
technical search and rescue hazards, 6
technician level, 6, 7f
technician-level rescue incident, 66
  accountability, 76, 76f
  guidance on providing support, 66–67
  hauling systems
    communications for, 74–75t
    haul team, 74
    overview, 73
    working of, 73–74
  incident facilities and stations, 76
  incident location concerns, 66, 67f

technician-level rescue incident (*Continued*)
   personal protective equipment selection
      common protective equipment at rope rescues, 68–70, 68f, 69f
         gloves, 70–72, 71f
         helmet, 72–73, 72f
         National Fire Protection Association Standards, 67
   personnel rehabilitation, 77
   reporting progress, 77
   supporting incident action plan, 74–76
technician-level rescuer, 370, 371f, 416
TEMS. *See* threat and error management system (TEMS)
tension, 187
tensionless hitch, 202
thermal damage, 176–177
third man operations, 323–324, 324f
30-30 tubular restraint method, 256, 257–258
threat and error management system (TEMS), 61
tied webbing harness, 272–273
time critical risk management (TCRM), 39
time of day, 28–29
tower incidents, rope rescue for, 447–448
tower rescue, 12
track-line systems, 342, 343–345, 343f
   rigging, 344
   ropes for highlines, 344–345
   spanning gap, 344
training expectations, 17
transfer devices, 445
traveling brake, pickoff by, 428–429, 429f
traveling control lines, 263, 263f
traveling pulley, 305
trench and excavation, 446
triage, 249–250
tripod, 351, 352f
troubleshooting, emergency descent systems, 407
   arm rappel, 407, 407f
   body rappel (Dulfersitz), 410–411
   carabiner wrap, 412
   Münter hitch rappel, 412, 412f
Truckers hitch, 172, 173f
TTRS. *See* two-tensioned rescue systems (TTRS)
tubular-type belay device, 223, 223f
tunnel incidents, 447
turn, 158
two-piece litter, 245, 245f
two-tensioned rescue systems (TTRS), 239–240, 287–288, 287f, 315–316, 428
   operating, 240
tying off short, 393
typical low-angle lowering operation
   evacuation of more than one rope length, 281–283
   preparation, 280–281

U

UHMPE (extended chain, high-modulus polyethylene), 149–150
undercut edges, 394, 396
unified command, 54, 90

V

vector, 186
verbal communications, 41, 41f, 264, 264f
vertical evacuation, 268
vertical lifeline, 221
vertical litter configuration, 332
   pike and pivot bridle, 332–333
   pike and pivot lower, 333–334
   pike and pivot raise, 334–335
victim extrication device, 105

W

washing, 178
   devices, 178–179, 179f
water rescue, 12, 12f
weight, 185
Who-What-When-Where-Why-How group of questions, 30
wilderness rescue organization, 446
wind turbine rescue, 11
work at height and rope access, 10–11, 10f
working end, 158
working in suspension, 322
   attendant rig, 330
   body management with litter movement, 331–332, 332f
   commencing lower, 330–331
   communications, 324–325, 324f
   connecting to rope rescue system, 325–326, 325f
   horizontal litter, single attachment point, 328
   moving litter while suspended, 327, 327f
   negotiating edges
      landing, 327
      lowering over, 326
      raising over, 327
   rescue line to litter, 329
   suspended work on highline, 335
      riding with litter, 336, 336f
      rigging to horizontal system, 335–336
   suspension-induced injury, 322–323, 323f
   third man operations, 323–324, 324f
   using litter spider, 328–329
   vertical litter configuration, 332
      pike and pivot bridle, 332–333
      pike and pivot lower, 333–334
      pike and pivot raise, 334–335
working system, 200
wrislet transfer device, 445, 445f
written communication, 264–265

Y

Yosemite litter packaging method, 257–258